Edited by
Evgeny Katz

**Molecular and Supramolecular
Information Processing**

Related Titles

Katz, Evgeny (Ed.)

Biomolecular Information Processing

From Logic Systems to Smart Sensors and Actuators

2012
ISBN: 978-3-527-33228-1

Katz, Evgeny (Ed.)

Information Processing Set

(comprising "Biomolecular Information Processing" and "Molecular and Supramolecular Information Processing")

2 Volumes
2012
ISBN: 978-3-527-33245-8

Samori, P., Cacialli, F. (Eds.)

Functional Supramolecular Architectures

for Organic Electronics and Nanotechnology

2011
ISBN: 978-3-527-32611-2

Feringa, B. L., Browne, W. R. (Eds.)

Molecular Switches

Second, Completely Revised and Enlarged Edition

2011
ISBN: 978-3-527-31365-5

Cosnier, S., Karyakin, A. (Eds.)

Electropolymerization

Concepts, Materials and Applications

2010
ISBN: 978-3-527-32414-9

Matta, C. F. (Ed.)

Quantum Biochemistry

2010
ISBN: 978-3-527-32322-7

Wolf, E. L.

Quantum Nanoelectronics

An Introduction to Electronic Nanotechnology and Quantum Computing

2009
ISBN: 978-3-527-40749-1

Stolze, J., Suter, D.

Quantum Computing

A Short Course from Theory to Experiment

2008
ISBN: 978-3-527-40787-3

Helms, V.

Principles of Computational Cell Biology

From Protein Complexes to Cellular Networks

2008
ISBN: 978-3-527-31555-0

Edited by Evgeny Katz

Molecular and Supramolecular Information Processing

From Molecular Switches to Logic Systems

WILEY-VCH Verlag GmbH & Co. KGaA

The Editor

Prof. Dr. Evgeny Katz
Clarkson University
Department of Chemistry
and Biomolecular Science
8, Clarkson Avenue
Potsdam, NY 13699-5810
USA

Cover
The cover page picture was designed by Dr. Vera Bocharova (Clarkson University) and represents artistic vision of the chapter "From Sensors to Molecular Logic: A Journey" by A. Prasanna de Silva.

All books published by **Wiley-VCH** are carefully produced. Nevertheless, authors, editors, and publisher do not warrant the information contained in these books, including this book, to be free of errors. Readers are advised to keep in mind that statements, data, illustrations, procedural details or other items may inadvertently be inaccurate.

Library of Congress Card No.: applied for

British Library Cataloguing-in-Publication Data
A catalogue record for this book is available from the British Library.

Bibliographic information published by the Deutsche Nationalbibliothek
The Deutsche Nationalbibliothek lists this publication in the Deutsche Nationalbibliografie; detailed bibliographic data are available on the Internet at <http://dnb.d-nb.de>.

© 2012 Wiley-VCH Verlag & Co. KGaA, Boschstr. 12, 69469 Weinheim, Germany

All rights reserved (including those of translation into other languages). No part of this book may be reproduced in any form – by photoprinting, microfilm, or any other means – nor transmitted or translated into a machine language without written permission from the publishers. Registered names, trademarks, etc. used in this book, even when not specifically marked as such, are not to be considered unprotected by law.

Composition Laserwords Private Limited, Chennai, India

Printing and Binding Markono Print Media Pte Ltd, Singapore

Cover Design   Adam Design, Weinheim

Print ISBN: 978-3-527-33195-6
ePDF ISBN: 978-3-527-64546-6
ePub ISBN: 978-3-527-64545-9
mobi ISBN: 978-3-527-64547-3
oBook ISBN: 978-3-527-64544-2

Printed in Singapore
Printed on acid-free paper

Contents

Preface *XIII*
List of Contributors *XV*

1 **Molecular Information Processing: from Single Molecules to Supramolecular Systems and Interfaces – from Algorithms to Devices – Editorial Introduction** *1*
Evgeny Katz and Vera Bocharova
References *7*

2 **From Sensors to Molecular Logic: A Journey** *11*
A. Prasanna de Silva
2.1 Introduction *11*
2.2 Designing Luminescent Switching Systems *11*
2.3 Converting Sensing/Switching into Logic *13*
2.4 Generalizing Logic *15*
2.5 Expanding Logic *16*
2.6 Utilizing Logic *17*
2.7 Bringing in Physical Inputs *20*
2.8 Summary and Outlook *21*
Acknowledgments *21*
References *21*

3 **Binary Logic with Synthetic Molecular and Supramolecular Species** *25*
Monica Semeraro, Massimo Baroncini, and Alberto Credi
3.1 Introduction *25*
3.1.1 Information Processing: Semiconductor Devices versus Biological Structures *25*
3.1.2 Toward Chemical Computers? *26*
3.2 Combinational Logic Gates and Circuits *27*
3.2.1 Basic Concepts *27*
3.2.2 Bidirectional Half Subtractor and Reversible Logic Device *28*
3.2.3 A Simple Unimolecular Multiplexer–Demultiplexer *32*
3.2.4 An Encoder/Decoder Based on Ruthenium Tris(bipyridine) *36*

3.2.5	All-Optical Integrated Logic Operations Based on Communicating Molecular Switches 38	
3.3	Sequential Logic Circuits 41	
3.3.1	Basic Concepts 41	
3.3.2	Memory Effect in Communicating Molecular Switches 42	
3.3.3	A Molecular Keypad Lock 43	
3.3.4	A Set–Reset Memory Device Based on a Copper Rotaxane 46	
3.4	Summary and Outlook 48	
	Acknowledgments 49	
	References 49	

4	**Photonically Switched Molecular Logic Devices** 53	
	Joakim Andréasson and Devens Gust	
4.1	Introduction 53	
4.2	Photochromic Molecules 54	
4.3	Photonic Control of Energy and Electron Transfer Reactions 55	
4.3.1	Energy Transfer 55	
4.3.2	Electron Transfer 59	
4.4	Boolean Logic Gates 61	
4.5	Advanced Logic Functions 64	
4.5.1	Half-Adders and Half-Subtractors 65	
4.5.2	Multiplexers and Demultiplexers 68	
4.5.3	Encoders and Decoders 69	
4.5.4	Sequential Logic Devices 71	
4.5.5	An All-Photonic Multifunctional Molecular Logic Device 75	
4.6	Conclusion 75	
	References 76	

5	**Engineering Luminescent Molecules with Sensing and Logic Capabilities** 79	
	David C. Magri	
5.1	Introduction 79	
5.2	Engineering Luminescent Molecules 80	
5.3	Logic Gates with the Same Modules in Different Arrangements 83	
5.4	Consolidating AND Logic 84	
5.5	"Lab-on-a-Molecule" Systems 87	
5.6	Redox-Fluorescent Logic Gates 90	
5.7	Summary and Perspectives 95	
	References 96	

6	**Supramolecular Assemblies for Information Processing** 99	
	Cátia Parente Carvalho and Uwe Pischel	
6.1	Introduction 99	
6.2	Recognition of Metal Ion Inputs by Crown Ethers 100	
6.3	Hydrogen-Bonded Supramolecular Assemblies as Logic Devices 102	

6.4	Molecular Logic Gates with [2]Pseudorotaxane- and [2]Rotaxane-Based Switches *103*	
6.5	Supramolecular Host-Guest Complexes with Cyclodextrins and Cucurbiturils *110*	
6.6	Summary *116*	
	Acknowledgments *117*	
	References *117*	
7	**Hybrid Semiconducting Materials: New Perspectives for Molecular-Scale Information Processing** *121*	
	Sylwia Gawęda, Remigiusz Kowalik, Przemysław Kwolek, Wojciech Macyk, Justyna Mech, Marek Oszajca, Agnieszka Podborska, and Konrad Szaciłowski	
7.1	Introduction *121*	
7.2	Synthesis of Semiconducting Thin Layers and Nanoparticles *122*	
7.2.1	Microwave Synthesis of Nanoparticles *123*	
7.2.2	Chemical Bath Deposition *124*	
7.2.2.1	Sulfide Ion Precursors *124*	
7.2.2.2	Commonly Used Ligand *124*	
7.3	Electrochemical Deposition *125*	
7.3.1	Nanoheterostructure Preparation *133*	
7.3.2	Nanoparticles Directed Self-Assembly *135*	
7.4	Organic Semiconductors–toward Hybrid Organic/Inorganic Materials *136*	
7.4.1	Self-Organization Motifs Exhibited by Acenes and Acene-Like Structures *137*	
7.4.2	Applications of Acenes in Organic Electronic Devices *141*	
7.5	Mechanisms of Photocurrent Switching Phenomena *142*	
7.5.1	Neat Semiconductor *143*	
7.5.2	Composite Semiconductor Materials *144*	
7.5.3	Semiconductor–Adsorbate Interactions *148*	
7.5.4	Surface-Modified Semiconductor *152*	
7.5.5	Optoelectronic Devices Based on Organic Molecules/Semiconductors *160*	
7.6	Digital Devices Based on PEPS Effect *161*	
7.7	Concluding Remarks *167*	
	Acknowledgments *168*	
	References *168*	
8	**Toward Arithmetic Circuits in Subexcitable Chemical Media** *175*	
	Andrew Adamatzky, Ben De Lacy Costello, and Julian Holley	
8.1	Awakening Gates in Chemical Media *175*	
8.2	Collision-Based Computing *176*	
8.3	Localizations in Subexcitable BZ Medium *176*	
8.4	BZ Vesicles *180*	

8.5	Interaction Between Wave Fragments 181
8.6	Universality and Polymorphism 183
8.7	Binary Adder 186
8.7.1	Sum 188
8.7.2	Carry Out 191
8.8	Regular and Irregular BZ Disc Networks 193
8.8.1	Elementary Logic Gates 194
8.8.2	Half Adder 198
8.9	Memory Cells with BZ Discs 201
8.10	Conclusion 204
	Acknowledgments 204
	References 205
9	**High-Concentration Chemical Computing Techniques for Solving Hard-To-Solve Problems, and their Relation to Numerical Optimization, Neural Computing, Reasoning under Uncertainty, and Freedom of Choice** 209
	Vladik Kreinovich and Olac Fuentes
9.1	What are Hard-To-Solve Problems and Why Solving Even One of Them is Important 209
9.1.1	What is so Good About Being Able to Solve Hard-To-Solve Problems from Some Exotic Class? 209
9.1.2	In Many Applications Areas –In Particular in Chemistry –There are Many Well-Defined Complex Problems 210
9.1.3	In Principle, There Exist Algorithms for Solving These Problems 210
9.1.4	These Algorithms may Take Too Much Time to be Practical 210
9.1.5	Feasible and Unfeasible Algorithms: General Idea 210
9.1.6	Solving Equations of Chemical Kinetics: An Example of a Feasible Algorithm 211
9.1.7	Straightforward Solution of Schrödinger Equation: An Example of an Unfeasible Algorithm 212
9.1.8	Straightforward Approach to Protein Folding: Another Example of an Unfeasible Algorithm 213
9.1.9	Feasible and Unfeasible Algorithms: Toward a Formal Description 213
9.1.10	Maybe the Problem Itself is Hard to Solve? 213
9.1.11	What Is a Problem in the First Place? 213
9.1.12	What is a Problem: Mathematics 214
9.1.13	A Description of a General Problem 214
9.1.14	What About Other Activity Areas? 214
9.1.15	What is a Problem: Theoretical Physics 215
9.1.16	What is a Problem: Engineering 215
9.1.17	Class NP 215
9.1.18	Class P and the $P\stackrel{?}{=}NP$ Problem 215
9.1.19	Exhaustive Search: Why it is Possible and Why it is Not Feasible 216

9.1.20	Notion of NP-Complete Problems *216*	
9.1.21	Why Solving Even One NP-Complete (Hard-To-Solve) Problem is Very Important *216*	
9.1.22	Propositional Satisfiability: Historically the First NP-Complete Problem *217*	
9.1.23	What We Do *217*	
9.2	How Chemical Computing Can Solve a Hard-To-Solve Problem of Propositional Satisfiability *218*	
9.2.1	Chemical Computing: Main Idea *218*	
9.2.2	Why Propositional Satisfiability was Historically the First Problem for Which a Chemical Computing Scheme was Proposed *218*	
9.2.3	How to Apply Chemical Computing to Propositional Satisfiability: Matiyasevich's Original Idea *219*	
9.2.4	A Precise Description of Matiyasevich's Chemical Computer: First Example *219*	
9.2.5	A Precise Description of Matiyasevich's Chemical Computer: Second Example *221*	
9.2.6	A Precise Description of Matiyasevich's Chemical Computer: General Formula *221*	
9.2.7	A Simplified Version (Corresponding to Catalysis) *222*	
9.2.8	Simplified Equations: Example *223*	
9.2.9	Chemical Computations Implementing Matiyasevich's Idea Are Too Slow *223*	
9.2.10	Natural Idea: Let us Use High-Concentration Chemical Reactions Instead *223*	
9.2.11	Resulting Equations *224*	
9.2.12	Discrete-Time Version of These Equations Have Already Been Shown to be Successful in Solving the Propositional Satisfiability Problem *225*	
9.2.13	Conclusion *225*	
9.2.14	Auxiliary Result: How to Select the Parameter Δt *226*	
9.3	The Resulting Method for Solving Hard Problems is Related to Numerical Optimization, Neural Computing, Reasoning under Uncertainty, and Freedom of Choice *228*	
9.3.1	Relation to Optimization: Why it is Important *228*	
9.3.2	Relation to Optimization: Main Idea *229*	
9.3.3	Relation to Numerical Optimization: Conclusion *231*	
9.3.4	Relation to Numerical Optimization: What Do We Gain from It? *231*	
9.3.5	Relation to Neural Computing *231*	
9.3.6	Relation to Reasoning Under Uncertainty *232*	
9.3.7	Relation to Freedom of Choice *233*	
	Acknowledgments *234*	
	References *234*	

10	**All Kinds of Behavior are Possible in Chemical Kinetics: A Theorem and its Potential Applications to Chemical Computing** *237*	
	Vladik Kreinovich	
10.1	Introduction *237*	
10.1.1	Chemical Computing: A Brief Reminder *237*	
10.1.2	Chemical Computing: Remaining Theoretical Challenge *238*	
10.1.3	What We Do *238*	
10.2	Main Result *239*	
10.2.1	Chemical Kinetics Equations: A Brief Reminder *239*	
10.2.2	Chemical Kinetics Until Late 1950s *240*	
10.2.3	Belousov – Zhabotinsky Reaction and Further Discoveries *240*	
10.2.4	A Natural Hypothesis *240*	
10.2.5	Dynamical Systems *241*	
10.2.6	W.l.o.g., We Start at Time $t = 0$ *241*	
10.2.7	Limited Time *241*	
10.2.8	Limited Values of x_i *242*	
10.2.9	Limited Accuracy *242*	
10.2.10	Need to Consider Auxiliary Chemical Substances *242*	
10.2.11	Discussion *244*	
10.2.12	Effect of External Noise *245*	
10.3	Proof *246*	
	Acknowledgments *256*	
	References *257*	
11	**Kabbalistic–Leibnizian Automata for Simulating the Universe** *259*	
	Andrew Schumann	
11.1	Introduction *259*	
11.2	Historical Background of Kabbalistic–Leibnizian Automata *259*	
11.3	Proof-Theoretic Cellular Automata *264*	
11.4	The Proof-Theoretic Cellular Automaton for Belousov–Zhabotinsky Reaction *268*	
11.5	The Proof-Theoretic Cellular Automaton for Dynamics of *Plasmodium* of *Physarum polycephalum* *271*	
11.6	Unconventional Computing as a Novel Paradigm in Natural Sciences *276*	
11.7	Conclusion *278*	
	Acknowledgments *278*	
	References *278*	
12	**Approaches to Control of Noise in Chemical and Biochemical Information and Signal Processing** *281*	
	Vladimir Privman	
12.1	Introduction *281*	
12.2	From Chemical Information-Processing Gates to Networks *283*	
12.3	Noise Handling at the Gate Level and Beyond *286*	

12.4	Optimization of AND Gates	290
12.5	Networking of Gates	294
12.6	Conclusions and Challenges	296
	Acknowledgments	297
	References	297

13 Electrochemistry, Emergent Patterns, and Inorganic Intelligent Response *305*

Saman Sadeghi and Michael Thompson

13.1	Introduction	305
13.2	Patten Formation in Complex Systems	306
13.3	Intelligent Response and Pattern Formation	308
13.3.1	Self-Organization in Systems Removed from the Equilibrium State	309
13.3.2	Patterns in Nature	310
13.3.3	Functional Self-Organizing Systems	310
13.3.4	Emergent Patterns and Associative Memory	312
13.4	Artificial Cognitive Materials	314
13.5	An Intelligent Electrochemical Platform	315
13.6	From Chemistry to Brain Dynamics	321
13.6.1	Understanding the Brain	321
13.6.2	Brain Dynamics	323
13.6.3	Electrochemical Dynamics	324
13.6.4	Experimental Paradigm for Information Processing in Complex Systems	325
13.7	Final Remarks	327
	References	328

14 Electrode Interfaces Switchable by Physical and Chemical Signals Operating as a Platform for Information Processing *333*

Evgeny Katz

14.1	Introduction	333
14.2	Light-Switchable Modified Electrodes Based on Photoisomerizable Materials	334
14.3	Magnetoswitchable Electrodes Utilizing Functionalized Magnetic Nanoparticles or Nanowires	336
14.4	Potential-Switchable Modified Electrodes Based on Electrochemical Transformations of Functional Interfaces	339
14.5	Chemically/Biochemically Switchable Electrodes and Their Coupling with Biomolecular Computing Systems	343
14.6	Summary and Outlook	350
	Acknowledgments	351
	References	352

15 Conclusions and Perspectives *355*
Evgeny Katz
References *357*

Index *359*

Preface

The use of molecular systems for processing information, performing logic operations, and finally making computation attracts substantial recent research efforts. The entire field was named with the general buzzwords, "molecular computing" or "chemical computing." Exciting advances in the area include the development of molecular, supramolecular, and nanostructured systems operating as "hardware" for unconventional computing, the use of reaction-diffusion media for computational operations, as well as the creation of novel algorithms and computational theories for the new "hardware" based on molecules rather than electronics. Another general scientific and engineering effort is directed to the integration of unconventional chemical computing systems with electronic or optical devices for transduction of the computational results obtained in the form of chemical concentration changes to electronically readable signals. The various topics covered highlight key aspects and future perspectives of molecular computing. The book discusses experimental work done by chemists and theoretical approaches developed by physicists and computer scientists. The different topics addressed in this book will be of interest to the interdisciplinary community active in the area of unconventional computing. It is hoped that the collection of the different chapters will be important and beneficial for researchers and students working in various areas related to chemical computing, including chemistry, materials science, computer science, and so on. Furthermore, the book is aimed to attract young scientists and introduce them to the field while providing newcomers with an enormous collection of literature references. I, indeed, hope that the book will spark the imagination of scientists to further develop the topic.

Finally, the Editor (E. Katz) and the Publisher (Wiley-VCH) express their thanks to all authors of the chapters, whose dedication and hard work made this book possible, hoping that the book will be interesting and beneficial for researchers and students working in various areas related to unconventional chemical computing. It should be noted that the field of chemical unconventional computing extends to the fascinating area of biomolecular systems, consideration of which was outside the scope of the present book. This complementary area of biomolecular computing is covered in another new book of Wiley-VCH: *Biomolecular Information Processing: from Logic Systems to Smart*

Sensors and Actuators – E. Katz, Editor. Both books are a must for the shelves of specialists interested in various aspects of molecular and biomolecular information processing.

Potsdam, NY, USA *Evgeny Katz*
October 2011

List of Contributors

Andrew Adamatzky
University of the West of England
Unconventional Computing Centre
Bristol BS16 1QY
UK

Joakim Andréasson
Chalmers University of Technology
Department of Chemical and Biological Engineering
Physical Chemistry
412 96 Göteborg
Sweden

Massimo Baroncini
Università di Bologna
Dipartimento di Chimica "G. Ciamician"
Via Selmi 2
40126 Bologna
Italy

Vera Bocharova
Clarkson University
Department of Chemistry and Biomolecular Science
8 Clarkson Avenue
Potsdam, NY 13699-5810
USA

Cátia Parente Carvalho
Universidad de Huelva
Departamento de Ingeniería Química
Química Física y Química Orgánica
Campus de El Carmen
21071 Huelva
Spain

and

Universidad de Huelva
Centro de Investigación en Química Sostenible (CIQSO)
Campus de El Carmen
21071 Huelva
Spain

Alberto Credi
Università di Bologna
Dipartimento di Chimica
"G. Ciamician"
Via Selmi 2
40126 Bologna
Italy

and

SOLARCHEM – Interuniversity
Center for the Chemical
Conversion of Solar Energy
Bologna Unit
Via Selmi 2
40126 Bologna
Italy

Ben De Lacy Costello
University of the West of England
Unconventional Computing
Centre
Bristol BS16 1QY
UK

A. Prasanna de Silva
Queen's University
School of Chemistry and
Chemical Engineering
Stranmillis Road
Belfast BT9 5AG
Northern Ireland

Olac Fuentes
University of Texas at El Paso
Department of Computer Science
500 W. University
El Paso, TX 79968
USA

Sylwia Gawęda
Uniwersytet Jagielloński
Wydział Chemii
ul. Ingardena 3
30-060 Kraków
Poland

Devens Gust
Arizona State University
Department of Chemistry and
Biochemistry
Tempe, AZ 85287-1604
USA

Julian Holley
University of the West of England
Unconventional Computing
Centre
Bristol BS16 1QY
UK

Evgeny Katz
Clarkson University
Department of Chemistry and
Biomolecular Science
8 Clarkson Avenue
Potsdam, NY 13699-5810
USA

Remigiusz Kowalik
Akademia Górniczo-Hutnicza
Wydział Metali Nieżelaznych
al. Mickiewicza 30
30-059 Kraków
Poland

Vladik Kreinovich
University of Texas at El Paso
Department of Computer Science
500 W. University
El Paso, TX 79968
USA

Przemysław Kwolek
Akademia Górniczo-Hutnicza
Wydział Metali Nieżelaznych
al. Mickiewicza 30
30-059 Kraków
Poland

Wojciech Macyk
Uniwersytet Jagielloński
Wydział Chemii
ul. Ingardena 3
30-060 Kraków
Poland

David C. Magri
University of Malta
Department of Chemistry
Msida MSD 2080
Malta

Justyna Mech
Akademia Górniczo-Hutnicza
Wydział Metali Nieżelaznych
al. Mickiewicza 30
30-059 Kraków
Poland

Marek Oszajca
Uniwersytet Jagielloński
Wydział Chemii
ul. Ingardena 3
30-060 Kraków
Poland

Uwe Pischel
Universidad de Huelva
Departamento de Ingeniería Química
Química Física y
Química Orgánica
Campus de El Carmen
21071 Huelva
Spain

and

Universidad de Huelva
Centro de Investigación en
Química Sostenible (CIQSO)
Campus de El Carmen
21071 Huelva
Spain

Agnieszka Podborska
Uniwersytet Jagielloński
Wydział Chemii
ul. Ingardena 3
30-060 Kraków
Poland

Vladimir Privman
Clarkson University
Department of Physics
Potsdam, NY 13699
USA

Saman Sadeghi
David Geffen School of Medicine
at UCLA
Department of Molecular and
Medical Pharmacology
Crump Institute for
Molecular Imaging
570 Westwood Plaza
Los Angeles, CA 90095
USA

Andrew Schumann
University of Information
Technology and
Management in Rzeszow
ul. Sucharskiego 2
35-225 Rzeszów
Poland

Monica Semeraro
Università di Bologna
Dipartimento di Chimica
"G. Ciamician"
Via Selmi 2
40126 Bologna
Italy

Konrad Szaciłowski
Uniwersytet Jagielloński
Wydział Chemii
ul. Ingardena 3
30-060 Kraków
Poland

and

Akademia Górniczo-Hutnicza
Wydział Metali Nieżelaznych
al. Mickiewicza 30
30-059 Kraków
Poland

Michael Thompson
University of Toronto
Department of Chemistry and
Institute for Biomaterials and
Biomedical Engineering
80 St. George Street
Toronto
Ontario M5S 3H6
Canada

1
Molecular Information Processing: from Single Molecules to Supramolecular Systems and Interfaces – from Algorithms to Devices – Editorial Introduction

Evgeny Katz and Vera Bocharova

Fast development of electronic computers with continuous progress [1] resulting in doubling their complexity every two years was formulated as the Moore's law in 1965 [2] (Figure 1.1). However, because of reaching physical limits for miniaturization of computing elements [3], the end of this exponential growth is expected soon. Economic [4] and fundamental physical problems (including limits placed on the miniaturization by quantum tunneling [5] and by the universal light speed [6]), which cannot be overcome in the frame of the existing paradigm, abolish all forms of future sophistication of computing systems. The inevitably expected limit to the development of the computer technology based on silicon electronics motivates various directions in unconventional computing [7] ranging from quantum computing [8], which aspires to achieve significant speedup over the conventional electronic computers for some problems, to biomolecular computing with a "soup" of biochemical reactions inspired by biology and usually represented by DNA-based computing [9].

Chemical computing, as a research subarea of unconventional computing, aims at using molecular or supramolecular systems to perform various computing operations that mimic processes typical of electronic computing devices [7]. Chemical reactions observed as changes of bulk material properties or structural reorganizations at the level of single molecules can be described in terms of information processing language, thus allowing for formulation of chemical processes as computing operations rather than traditional chemical transformations.

The idea of using chemical transformations for information processing originated from the concept of "artificial life" as early as in 1970s, when *artificial molecular machines* were inspired by chemistry and brought to computer science [10]. Theoretical background for implementing logic gates and finite-state machines based on chemical flow systems and bistable reactions was developed in 1980s and 1990s [11], including application of chemical computing to solving a hard-to-solve problem of propositional satisfiability [12]. However, the practical realization of the theoretical concepts came later when already known Belousov–Zhabotinsky chemical oscillating systems [13] (Figure 1.2) were applied for experimental design of logic gates [14]. Extensive research in the area of reaction-diffusion computing systems [15] resulted in the formulation of conceptually novel circuits performing

Molecular and Supramolecular Information Processing: From Molecular Switches to Logic Systems, First Edition. Edited by Evgeny Katz.
© 2012 Wiley-VCH Verlag GmbH & Co. KGaA. Published 2012 by Wiley-VCH Verlag GmbH & Co. KGaA.

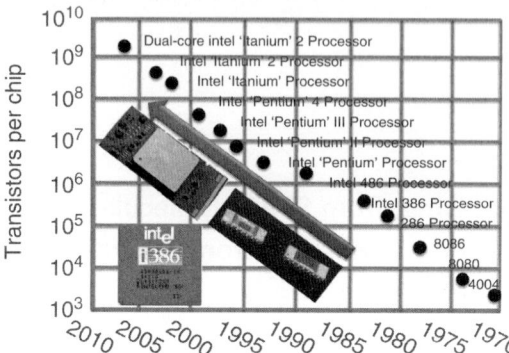

Figure 1.1 Moore's law demonstrating exponential growth of sophistication of electronic computing systems resulting in doubling of their complexity every two years.

Figure 1.2 Spiral waves in the Belousov–Zhabotinsky reaction – background for many chemical computing systems. (Adapted with permission from [13b]; Copyright (2006) National Academy of Sciences, USA.)

information processing with the use of subexcitable chemical media [16]. Developments in this area resulted not only in new chemical "hardware" for information processing but also in novel approaches to computing algorithms absolutely different from those presently used in silicon-based electronics. The major difference and advantage comparing with presently used electronic computers is the possibility of using 10^{23} molecules performing computations in parallel, thus resulting in great parallelization and acceleration of computing, which is not achievable in the present electronic paradigm. It should be noted that upon appropriate design, chemical systems can realize any kind of nonlinear behavior, which leads to the possibility to emulate any computational device needed for assembling information processing systems.

Information processing can be performed at the level of a single molecule or in a supramolecular complex. On the basis of early ideas to use molecules as logic gates [17], one of the first examples, a molecular AND gate, was reported in 1993 by de Silva *et al.* [18]. This novel research direction has been developed rapidly from the formulation of single logic gates mimicking Boolean operations, including AND, OR, XOR, NOR, NAND, INHIB, XNOR, and so on, to small logic networks [19]. Combination of chemical logic gates in groups or networks resulted in simple computing devices performing basic arithmetic operations [20] such as half-adder/half-subtractor or full-adder/full-subtractor [21]. Sophisticated molecular design has resulted in reversible [22], reconfigurable [23], and resettable [24] logic gates for processing chemical information. Other chemical systems mimicking various components of digital electronic devices were designed, including molecular comparator [25], digital multiplexer/demultiplexer [26], encoder/decoder [27], keypad lock [28], as well as flip-flop and write/read/erase memory units [29].

Many of the chemical systems used for information processing were based on molecules or supramolecular ensembles that exist in different states. These states can switch reversibly from one to another upon application of various external physical or chemical inputs. Ingenious supramolecular ensembles operating as molecular machines with translocation of their parts upon external signals were designed and used to operate as chemical switchable elements performing logic operations. Rotaxane supramolecular complexes (Figure 1.3) designed by the group of Prof. Stoddart as early as 1990s [30] can be mentioned as examples of such signal-switchable systems – research that later received numerous extensions and applications [31].

Chemical systems can solve computing problems at the level of a single molecule resulting in nanoscaling of the computing units and allowing parallel computations performed by numerous molecules involved in various reactions. Chemical transformations in switchable molecular systems used for mimicking computing

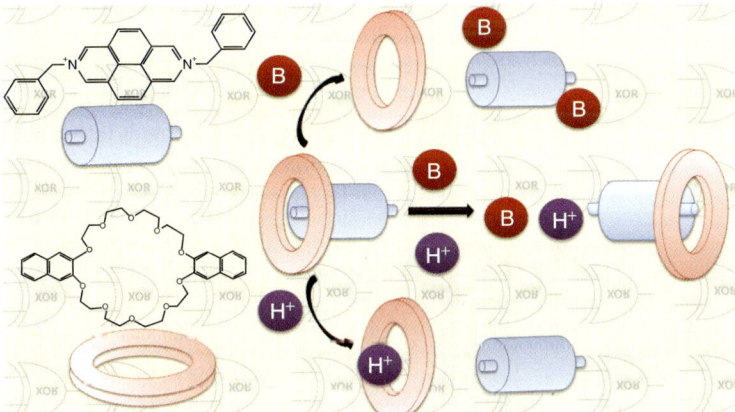

Figure 1.3 An example of a pseudorotaxane supramolecular complex activated by light, producing fluorescence only when the molecular hoop is released from the axle. The complex dissociates, thus resulting in fluorescence on application of an acid (H^+) or a base (B), while in the absence or presence of both chemical inputs the hoop remains in place and quenches the fluorescence. The system mimics the Boolean logic XOR (eXclusive OR), which is activated only in the presence of 0,1 or 1,0 input combinations, while being mute in the cases of 0,0 and 1,1 inputs.

operations can be based on redox changes, acid–base or chelating reactions, and isomerization processes [32]. The most representative research performed by the group of Prof. de Silva yielded various Boolean logic gates based on reconfiguration of switchable supramolecular complexes [33]. Chemical reactions in switchable systems have been induced by external physical signals, for example, light, magnetic field, or electrochemical potential, and by chemical signals, for example, pH changes or metal cation additions. Some of the studied switchable systems can respond to two kinds of physical or physical/chemical signals, for example, potential applied to an electrode and illumination, pH change and illumination, or ion addition and applied potential. The output signals generated by the chemical switchable systems are usually read by optical methods: UV–vis or fluorescence spectroscopies or by electrochemical means: current or potential generated at electrodes or in field-effect transistors. Association of chemical logic systems with electrode interfaces [34] or nanowires [35] resulted in electronic devices with implemented molecular logic performing on-chip data processing functions for lab-on-a-chip devices [36].

To some extent, the design of chemical "hardware" for information processing and future built-up of molecular computers depends on the success of molecular electronics [37] – the subarea of nanotechnology aiming at coupling of single molecules and nanoobjects for performing various electronic functions [38]. This approach resulted in numerous hybrid molecular-nanoobject systems performing various electronic functions potentially applicable to molecular computers. For example, nanowires made of conducting molecules [39] or templated by polymeric molecules [40] were designed for nanoelectronic applications (Figure 1.4) [41]. Single-molecule transistors and other functional molecular devices integrated with nanoscale electronic circuitries became possible (Figure 1.5). However, this approach copying the present conceptual design of electronic systems to the novel molecular architecture might be counterproductive for creating computers of next generation. A more effective way of assembling and operating molecular computing systems, aiming at massively parallel nonlinear computers mimicking human brain operation, requires absolutely novel approaches to the "hardware" and computing algorithms – their design is presently at a very preliminary stage. Novel

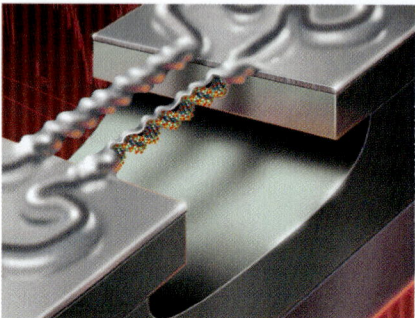

Figure 1.4 The artistic view of molecular-templated nanowires connecting microelectrodes. (Adapted from [40b] with permission; courtesy of Prof. Alexey Bezryadin.)

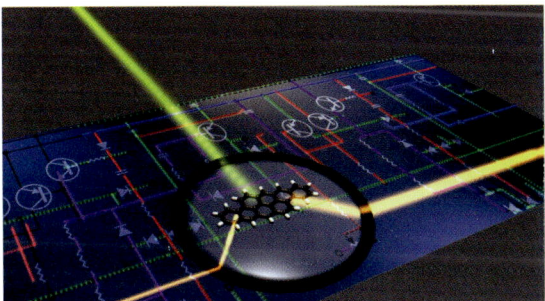

Figure 1.5 An artistic representation of a single-molecule optical transistor integrated into a nanocircuitry. (Adapted from *http://www.opfocus.org/index.php?topic=story&v=7&s=2*; courtesy of Prof. Vahid Sandoghdar.)

algorithms in information processing, particularly for hard-to-solve computational problems, are emerging in this study. Application of these algorithms with the use of chemical massively parallel computing might be much more efficient for solving hard-to-solve problems rather than the use of presently available supercomputers.

Before chemical computing becomes practically possible, many issues addressing the architecture and operation of molecular systems have to be addressed. Particularly important are the scaling up of the systems complexity and management of noise in chemical systems [42]. As an information processing network becomes larger and information is processed in greater quantities and at higher levels of complexity, noise inevitably builds up and can ultimately degrade the useful "signal," which is the intended result of the logic processing or computation. One then has to develop approaches to achieve what is known as *"fault-tolerant"* information processing that involves noise control and suppression. Chemical systems are much more prone to noise than electronic computer components. Their applications are in environments where the inputs (reactant chemicals' concentrations) and the "gate machinery" (other chemicals' concentrations) are all expected to fluctuate within at least a couple of percent of the range of values between the "digital" 0 and 1. Therefore, consideration of control of noise is required already in concatenating as few as two to three logic gates. While noise analysis in large chemical networks might be very complicated and requires heavy computational facilities [43], in single chemical gates and small networks noise analysis resulting in the systems optimization and noise suppression can be achieved using relatively simple computational and chemical approaches [42] (Figure 1.6).

Several comprehensive review articles have already covered to various degrees the molecular computing systems (mostly addressing chemical aspects of switchable signal-responsive systems operating as logic gates and small circuits) [19]. A good collection of review articles covering chemical and biochemical computing systems has been recently published in the special issue of *Israel Journal of Chemistry* (Wiley-VCH) – "Molecular and Biomolecular Information Processing Systems" (February 2011, Vol. 51, Issue 1, Guest Editor – E. Katz). However, taking into account rapid developments in this area, the editor and the publisher believe that

Figure 1.6 Theoretical and experimental response functions of a chemical AND logic gate representing noise-suppressing (top-left schematic) and noise-amplifying (top-right schematic) operation. Two experimental response surfaces recently realized in the parameter regime of no noise amplification are also shown (lower schematics).

another comprehensive summary of the results would be beneficial for the multidisciplinary research community, which includes fields of chemistry, materials science, and computer science, thus bringing to your attention the present book.

This book represents a unique collection of review accounts written by major contributors to this newly emerged research area overviewing the state of the art in unconventional chemical computing and its potential applications. In this book, following introductory Chapter 1, Chapters 2–6 authored by A. Credi *et al.*, J. Andréasson and D. Gust, A.P. de Silva, D.C. Magri, C.P. Carvalho and U. Pischel represent an overview of a broad research area utilizing molecular and supramolecular species for chemical computing. Chapter 7 written by K. Szaciłowski *et al.* extends the chemical computing research area to semiconductive species (thin films and nanoparticles). Theoretical and experimental approach to reaction-diffusion computing systems is described in Chapter 8, written by A. Adamatzky *et al.* Various aspects of theoretical approaches to novel computing algorithms based on chemical computing can be found in Chapters 9 and 10, written by V. Kreinovich *et al.* A very unusual approach to unconventional chemical computing is offered by A. Schumann in Chapter 11. In this chapter, he connected Kabbalah, the esoteric teaching of Judaism, with massively parallel chemical computing, coming to the conclusion that any physical, chemical, or biological phenomena could be simulated with the help of chemical computing. Chapter 12, authored by V. Privman, offers theoretical consideration and practical realization of noise management in chemical computing systems. Electrochemical systems employed for information processing are outlined in Chapter 13 of S. Sadeghi and M. Thompson, while Chapter 14 prepared by E. Katz describes electrode interfaces switchable by external signals considering them as a platform for information processing systems. Chapter 15 offers the Editorial (E. Katz) conclusions and speculates about future perspectives of chemical computing systems.

The Editor (E. Katz) and Publisher (Wiley-VCH) express their gratitude to all authors of the chapters, whose dedication and hard work made this book possible, hoping that the book will be interesting and beneficial for researchers and students working in various areas related to unconventional chemical computing, including chemistry, materials science, computer science, and so on. It should be noted that the field of chemical unconventional computing extends to the fascinating area of biomolecular systems, consideration of which is outside the scope of the present book. This complementary area of biomolecular computing is covered in another new book of Wiley-VCH: *Biomolecular Information Processing: From Logic Systems to Smart Sensors and Actuators* – E. Katz, Editor. Both books are a must for the shelves of specialists interested in various aspects of molecular and biomolecular information processing.

References

1. Mack, C.A. (2011) *IEEE Trans. Semicond. Manuf.*, **24**, 202–207.
2. Moore, G.E. (1965) *Electronics*, **38**, 114–117.
3. Freebody, M. (2011) *Photon. Spectra.*, **45**, 45–47.
4. (a) Rupp, K. and Selberherr, S. (2010) *Proc. IEEE*, **98**, 351–353; (b) Rupp, K. and Selberherr, S. (2011) *IEEE Trans. Semicond. Manuf.*, **24**, 1–4.
5. Powell, J.R. (2008) *Proc. IEEE*, **96**, 1247–1248.
6. Choi, C. (2004) *New Sci.*, **182**, 12–12.
7. (a) Calude, C.S., Costa, J.F., Dershowitz, N., Freire, E., and Rozenberg, G. (eds) (2009) *Unconventional Computation*, Lecture Notes in Computer Science, Vol. 5715, Springer, Berlin; (b) Adamatzky, A., De.Lacy.Costello, B., Bull, L., Stepney, S., and Teuscher, C. (eds) (2007) *Unconventional Computing 2007*, Luniver Press.
8. (a) The U.S. Department of Energy Quantum Information Science and Technology Roadmapping Project, maintained online at *http://qist.lanl.gov*; (b) Ezziane, Z. (2010) *Int. J. Quantum Chem.*, **110**, 981–992.
9. (a) Xu, J. and Tan, G.J. (2007) *J. Comput. Theor. Nanosci.*, **4**, 1219–1230; (b) Soreni, M., Yogev, S., Kossoy, E., Shoham, Y., and Keinan, E. (2005) *J. Am. Chem. Soc.*, **127**, 3935–3943; (c) Stojanovic, M.N. and Stefanovic, D. (2003) *Nat. Biotechnol.*, **21**, 1069–1074; (d) Stojanovic, M.N., Stefanovic, D., LaBean, T., and Yan, H. (2005) in *Bioelectronics: from Theory to Applications*, Chapter 14 (eds I. Willner and E. Katz), Wiley-VCH Verlag GmbH, Weinheim, pp. 427–455.
10. Laing, R. (1972) *J. Cybern.*, **2**, 38–49.
11. (a) Hjelmfelt, A., Weinberger, E.D., and Ross, J. (1991) *Proc. Natl. Acad. Sci. U.S.A.*, **88**, 10983–10987; (b) Hjelmfelt, A., Weinberger, E.D., and Ross, J. (1992) *Proc. Natl. Acad. Sci. U.S.A.*, **89**, 383–387; (c) Hjelmfelt, A. and Ross, J. (1993) *J. Phys. Chem.*, **97**, 7988–7992; (d) Hjelmfelt, A. and Ross, J. (1995) *Physica D*, **84**, 180–193; (e) Hjelmfelt, A., Schneider, F.W., and Ross, J. (1993) *Science*, **260**, 335–337.
12. Matiyasevich, Y. (1987) *Problems of Cybernetics*, vol. 131, Moscow (English translation in: Kreinovich, V. and Mints, G. (eds) (1996) *Problems of Reducing the Exhaustive Search*, American Mathematical Society, Providence, RI, pp. 75–77).
13. (a) Belousov, B.P. (1959) *Collection of Abstracts on Radiation Medicine in 1958*, Medicine Publisher, Moscow, pp. 145–147 (in Russian). (translated in: Field, R.J. and Burger, M. (eds) (1985) *Oscillations and Traveling Waves in Chemical Systems*, John Wiley & Sons, Inc., New York); (b) Epstein, I.R.

(2006) *Proc. Natl. Acad. U.S.A.*, **103**, 15727–15728.
14. (a) Lebender, D. and Schneider, F.W. (1994) *J. Phys. Chem.*, **98**, 7533–7537; (b) Tóth, A., Gáspár, V., and Showalter, K. (1994) *J. Phys. Chem.*, **98**, 522–531; (c) Tóth, A. and Showalter, K. (1995) *J. Chem. Phys.*, **103**, 2058–2066.
15. Adamatzky, A. (2011) *J. Comput. Theor. Nanosci.*, **8**, 295–303.
16. (a) Adamatzky, A., Costello, B., Bull, L., and Holley, J. (2011) *Isr. J. Chem.*, **51**, 56–66; (b) Costello, B.D. and Adamatzky, A. (2005) *Chaos Solitons Fractals*, **25**, 535–544.
17. Aviram, A. (1988) *J. Am. Chem. Soc.*, **110**, 5687–5692.
18. de Silva, A.P., Gunaratne, H.Q.N., and McCoy, C.P. (1993) *Nature*, **364**, 42–44.
19. (a) de Silva, A.P., Uchiyama, S., Vance, T.P., and Wannalerse, B. (2007) *Coord. Chem. Rev.*, **251**, 1623–1632; (b) de Silva, A.P. and Uchiyama, S. (2007) *Nat. Nanotechnol.*, **2**, 399–410; (c) Szacilowski, K. (2008) *Chem. Rev.*, **108**, 3481–3548; (d) Credi, A. (2007) *Angew. Chem. Int. Ed.*, **46**, 5472–5475; (e) Pischel, U. (2007) *Angew. Chem. Int. Ed.*, **46**, 4026–4040; (f) Pischel, U. (2010) *Aust. J. Chem.*, **63**, 148–164; (g) Andreasson, J. and Pischel, U. (2010) *Chem. Soc. Rev.*, **39**, 174–188.
20. Brown, G.J., de Silva, A.P., and Pagliari, S. (2002) *Chem. Commun.*, 2461–2463.
21. (a) Qu, D.H., Wang, Q.C., and Tian, H. (2005) *Angew. Chem. Int. Ed.*, **44**, 5296–5299; (b) Andréasson, J., Straight, S.D., Kodis, G., Park, C.D., Hambourger, M., Gervaldo, M., Albinsson, B., Moore, T.A., Moore, A.L., and Gust, D. (2006) *J. Am. Chem. Soc.*, **128**, 16259–16265; (c) Andréasson, J., Kodis, G., Terazono, Y., Liddell, P.A., Bandyopadhyay, S., Mitchell, R.H., Moore, T.A., Moore, A.L., and Gust, D. (2004) *J. Am. Chem. Soc.*, **126**, 15926–15927; (d) Lopez, M.V., Vazquez, M.E., Gomez-Reino, C., Pedrido, R., and Bermejo, M.R. (2008) *New J. Chem.*, **32**, 1473–1477; (e) Margulies, D., Melman, G., and Shanzer, A. (2006) *J. Am. Chem. Soc.*, **128**, 4865–4871; (f) Kuznetz, O., Salman, H., Shakkour, N., Eichen, Y., and Speiser, S. (2008) *Chem. Phys. Lett.*, **451**, 63–67.
22. (a) Pérez-Inestrosa, E., Montenegro, J.M., Collado, D., Suau, R., and Casado, J. (2007) *J. Phys. Chem. C*, **111**, 6904–6909; (b) Remón, P., Ferreira, R., Montenegro, J.M., Suau, R., Pérez-Inestrosa, E., and Pischel, U. (2009) *ChemPhysChem*, **10**, 2004–2007.
23. (a) Sun, W., Xu, C.H., Zhu, Z., Fang, C.J., and Yan, C.H. (2008) *J. Phys. Chem. C*, **112**, 16973–16983; (b) Li, Z.X., Liao, L.Y., Sun, W., Xu, C.H., Zhang, C., Fang, C.J., and Yan, C.H. (2008) *J. Phys. Chem. C*, **112**, 5190–5196; (c) Coskun, A., Deniz, E., and Akkaya, E.U. (2005) *Org. Lett.*, **7**, 5187–5189; (d) Jiménez, D., Martínez-Máñez, R., Sancenón, F., Ros-Lis, J.V., Soto, J., Benito, A., and García-Breijo, E. (2005) *Eur. J. Inorg. Chem.*, **2005**, 2393–2403.
24. (a) Sun, W., Zheng, Y.R., Xu, C.H., Fang, C.J., and Yan, C.H. (2007) *J. Phys. Chem. C*, **111**, 11706–11711; (b) Zhou, Y., Wu, H., Qu, L., Zhang, D., and Zhu, D. (2006) *J. Phys. Chem. B*, **110**, 15676–15679.
25. Pischel, U. and Heller, B. (2008) *New J. Chem.*, **32**, 395–400.
26. (a) Andreasson, J., Straight, S.D., Bandyopadhyay, S., Mitchell, R.H., Moore, T.A., Moore, A.L., and Gust, D. (2007) *J. Phys. Chem. C*, **111**, 14274–14278; (b) Amelia, M., Baroncini, M., and Credi, A. (2008) *Angew. Chem. Int. Ed.*, **47**, 6240–6243; (c) Perez-Inestrosa, E., Montenegro, J.M., Collado, D., and Suau, R. (2008) *Chem. Commun.*, 1085–1087.
27. Andreasson, J., Straight, S.D., Moore, T.A., Moore, A.L., and Gust, D. (2008) *J. Am. Chem. Soc.*, **130**, 11122–11128.
28. (a) Margulies, D., Felder, C.E., Melman, G., and Shanzer, A. (2007) *J. Am. Chem. Soc.*, **129**, 347–354; (b) Suresh, M., Ghosh, A., and Das, A. (2008) *Chem. Commun.*, 3906–3908.
29. (a) Chatterjee, M.N., Kay, E.R., and Leigh, D.A. (2006) *J. Am. Chem. Soc.*, **128**, 4058–4073; (b) Baron, R., Onopriyenko, A., Katz, E., Lioubashevski, O., Willner, I., Wang, S.,

and Tian, H. (2006) *Chem. Commun.*, 2147–2149; (c) Galindo, F., Lima, J.C., Luis, S.V., Parola, A.J., and Pina, F. (2005) *Adv. Funct. Mater.*, **15**, 541–545; (d) Bandyopadhyay, A. and Pal, A.J. (2005) *J. Phys. Chem. B*, **109**, 6084–6088; (e) Pina, F., Lima, J.C., Parola, A.J., and Afonso, C.A.M. (2004) *Angew. Chem. Int. Ed.*, **43**, 1525–1527.

30. Bissel, R.A., Córdova, E., Kaifer, A.E., and Stoddart, J.F. (1994) *Nature*, **369**, 133–137.

31. Flood, A.H., Ramirez, R.J.A., Deng, W.Q., Muller, R.P., Goddard, W.A., and Stoddart, J.F. (2004) *Aust. J. Chem.*, **57**, 301–322.

32. Shipway, A.N., Katz, E., and Willner, I. (2001) in *Molecular Machines and Motors*, Structure and Bonding, Vol. 99 (ed. J.P. Sauvage), Springer, Berlin, pp. 237–281.

33. (a) de Silva, A.P., McClenaghan, N.D., and McCoy, C.P. (2001) in *Handbook of Electron Transfer in Chemistry*, vol. 5 (ed. V. Balzani), Wiley-VCH Verlag GmbH, Weinheim, pp. 156–185; (b) de Silva, A.P., McClenaghan, N.D., and McCoy, C.P. (2001) in *Molecular Switches* (ed. B.L. Feringa) Wiley-VCH Verlag GmbH, New York, pp. 339–361; (c) de Silva, A.P. (2011) *Chem. Asian J.*, **6**, 750–766.

34. Gupta, T. and van der Boom, M.E. (2008) *Angew. Chem. Int. Ed.*, **47**, 5322–5326.

35. Mu, L.X., Shi, W.S., She, G.W., Chang, J.C., and Lee, S.T. (2009) *Angew. Chem. Int. Ed.*, **48**, 3469–3472.

36. Chang, B.Y., Crooks, J.A., Chow, K.F., Mavre, F., and Crooks, R.M. (2010) *J. Am. Chem. Soc.*, **132**, 15404–15409.

37. (a) Tyagi, P. (2011) *J. Mater. Chem.*, **21**, 4733–4742; (b) Del Nero, J., de Souza, F.M., and Capaz, R.B. (2010) *J. Comp. Theor. Nanosci.*, **7**, 503–516; (c) Heath, J.R. (2009) *Ann. Rev. Mater. Res.*, **39**, 1–23; (d) James, D.K. and Tour, J.M. (2004) *Chem. Mater.*, **16**, 4423–4435.

38. Ball, P. (2000) *Nature*, **406**, 118–120.

39. Kemp, N.T., Newbury, R., Cochrane, J.W., and Dujardin, E. (2011) *Nanotechnology*, **22**, article #105202.

40. (a) Hopkins, D., Pekker, D., Goldbart, P., and Bezryadin, A. (2005) *Science*, **308**, 1762–1765; (b) Bezryadin, A. and Goldbart, P.M. (2010) *Adv. Mater.*, **22**, 1111–1121.

41. Li, C., Lei, B., Fan, W., Zhang, D.H., Meyyappan, M., and Zhou, C.W. (2007) *J. Nanosci. Nanotechnol.*, **7**, 138–150.

42. (a) Privman, V. (2010) *Isr. J. Chem.*, **51**, 118–131; (b) Privman, V. (2011) *J. Comput. Theor. Nanosci.*, **8**, 490–502.

43. Intosalmi, J., Manninen, T., Ruohonen, K., and Linne, M.L. (2011) *BMC Bioinf.*, **12**, article # 252.

2
From Sensors to Molecular Logic: A Journey
A. Prasanna de Silva

2.1
Introduction

We can interpret the term *"nanotechnology"* as the useful science of nanometric objects, which only emerges in that size scale – neither smaller nor larger [1]. Chemistry allows us to reach the nanoscale by building up from smaller molecules and atomic species, whereas engineers can sculpt bulk materials down to the required size. Supramolecular chemistry offers a very natural way of building nanometric structures with emergent properties that are absent in smaller molecules.

A scientist can design a supermolecule to go to a small space that is humanly inaccessible and measure the levels of chosen atoms or molecules. Then a luminescence signal from the supermolecule informs the scientist about these levels. Supermolecules can indeed be designed so that they can perform chosen functions in chosen small spaces. Some of these uses take us across biology to medical diagnostics. Other potential uses take us in the opposite direction toward computer engineering.

2.2
Designing Luminescent Switching Systems

What is the design basis of these supermolecules [2, 3]? Photochemistry, which combines light and molecules, is a good starting point. Photoinduced electron transfer (PET) is the heart of photosynthesis and is a major channel of de-exciting excited states of molecules [4]. Luminescence is another such channel that is easily observed even by the naked eye. The controlled competition of luminescence with PET can switch the luminescence "on" or "off" by chemical means in an easy, predictable manner [5]. The modular nature of "lumophore-spacer-receptor" supramolecular systems [6] is vital not only for the occurrence of PET but also for the prediction of sensor/switch characteristics such as colors of the optical signals and the concentration range of the analyte (C^+ in Figure 2.1).

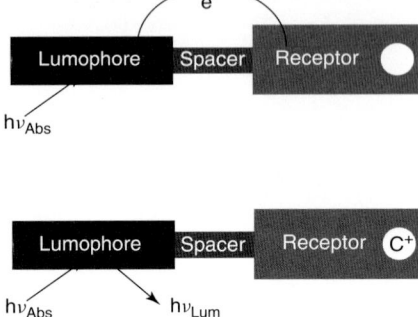

Figure 2.1 The basic design of luminescent PET (photoinduced electron transfer) sensors. No luminescence arises from the "lumophore-spacer-receptor" system until it captures the cation C^+.

Several laboratories published one or two examples, each starting in the mid-1970s [7–12], and then we had the chance to establish the generality of this design principle. The lumophores came from three categories; fluorophores, phosphors, and f–f lumophores. The fluorophores themselves emitted violet (7-methoxycoumarin [13]), blue (anthracene [14–22]), blue-green (1,3,5-triaryl-Δ^2-pyrazoline [23–27]), green-yellow (4-aminonaphthalimide [28, 29]), and orange (perylenedicarboximide [30]) light. 1-Bromonaphthalene [31] phosphors emitted green light, whereas f–f lumophores based on complexes of Tb^{3+} [32] and Eu^{3+} [33] gave green and red emissions, respectively. All of these had emission lifetimes in the millisecond range. Systems emitting in the orange and red regions of the spectrum were produced by employing tris(bipyridine)Ru(II) [34] and porphyrinSn(IV) [35] lumophores, respectively. The early receptors (amines [13, 14, 21, 28, 30–32], carboxylates [23], pyridines [13, 29]) targeted H^+, whereas the remaining receptor for H^+ (phenolate) was included in PET systems soon afterwards in Leeds [36] and in Regensburg [37]. Aza-18-crown-6 ether for K^+ [15, 33], benzo-15-crown-5 ether for Na^+ [16, 20], N,N'-diaryldiaza-18-crown-6 ether for Na^+ [27], dibenzo[2,2,2] cryptand for K^+ [18], Tsien's BAPTA ligand [38] for Ca^{2+} [17, 25], and London's APTRA ligand [39] for Mg^{2+} [24] were some of the other successful receptors during the early phase. Most spacer modules still tend to be methine, methylene, or dimethylene groups so that electron transfer rates remain high when thermodynamically allowed [4]. Since then, many laboratories around the world have joined in the effort by bringing many other lumophores and receptors on board [40–46]. For some recent examples, see Refs. [47–49].

Some of these sensor molecules can measure chemical species levels in small spaces [50, 51]. Some others [14] are currently used to monitor acidic compartments in living cells as foreign matter is ingested and trafficked [52]. Even acidic compartments in cells arising after radiotherapy have been imaged by these sensors [53]. An example is **1** [14]. Other sensor molecules based on the principles mentioned above are working in hospitals and in ambulances performing blood diagnostics within millimetric spaces [54]. They measure Na^+, K^+, and Ca^{2+}, besides pH, CO_2, and O_2. These arose from research carried out by scientists at Roche Diagnostics in consultation with us [55–58].

[Structure 1: 9,10-bis(diethylaminomethyl)anthracene, with NEt$_2$ groups]

1

Besides straightforward sensing of ions, the PET approach has other possibilities up its sleeve. First, PET separates charge in neutral molecules, and these charges would be stabilized in polar media. Hence the luminescence of some of these molecules is sharply controlled by solvent polarity [59–61]. Second, organic chemical reactions can be monitored if the functional group undergoing transformation also undergoes a change of PET activity. Say, a nucleophile attacks an enone and produces a saturated carbonyl compound according to the Michael addition. Since the enone is considerably more electron deficient than the product, the emission from an attached lumophore would be switched "on" [62, 63]. Older cases [64] also became understandable then. Third, a bit of the mystique can be cleared from perhaps the grandest example of PET: photosynthesis [65]. The path selectivity of PET within the bacterial photosynthetic reaction center [65] can be emulated in a small molecule by exploiting internal photogenerated electric fields to direct electron traffic [66]. Fourth, several matched sensors can be operated in parallel to create extended quasi-linear dynamic ranges for sensing [67]. Thus, a new functionality emerges from the system of components [68]. Such versatility of the PET approach is bound to encourage other molecular device designers to look for more ways in which the basic design can be manipulated so that desirable functions emerge.

2.3
Converting Sensing/Switching into Logic

The luminescent PET sensing/switching principle is so general that it can be adapted to build molecular-scale information processors that employ chemical species as inputs and light as output. Wireless interfacing of molecular-scale devices to human operators thus becomes possible. These processors are far smaller than the smallest silicon-based electronic devices. However, it is helpful to examine the chemical phenomena from a computer science viewpoint. The switching molecule is the processor device, the analyte is the input, the emission is the output, and the excitation is the power supply.

Table 2.1 Truth table for **2**.

Input₁	Input₂	Output
H⁺	Na⁺	Luminescence[a]
Low (0 M)	Low (0 M)	Low (0.012)
Low (0 M)	High (10^{-2} M)	Low (0.013)
High (10^{-3} M)	Low (0 M)	Low (0.020)
High (10^{-3} M)	High (10^{-2} M)	High (0.068)

[a]Quantum yields in methanol: 2-propanol (1 : 1 v/v), λ_{exc} 387 nm, λ_{em} 446 nm.

For instance, we can modify the "lumophore-spacer-receptor" system to a "lumophore-spacer₁-receptor₁-spacer₂-receptor₂" counterpart so that both receptors need to be blocked with their corresponding analytes before a light emission takes place. Thus two "high" input signals are required before the output becomes "high." In a Boolean sense [69], this is AND logic [70, 71] but at a molecular scale. The first example of this type that launched molecular logic and computation as an experimental field was **2** [72], which operated with H⁺ and Na⁺ as inputs while employing ultraviolet light as the power supply and blue light as the output (Table 2.1). Notably, molecule **2** could marshal all four of these agents without human assistance and operate correctly.

Suddenly, chemists could participate in the information technology revolution that was sweeping society. The substantial diversity of molecules could be harnessed to deliver cases with input–output patterns that fitted various situations arising from Boole's discovery [69]. Importantly, the concepts and devices of data processing could be brought into chemistry and vice versa. Such cross-fertilization would be welcome, especially to a rapidly growing web2.0 generation. The welcome would be even warmer when the molecular nature of the biological information processing was appreciated. Chemistry could then connect biology with engineering at the informational level.

2.4 Generalizing Logic

The discovery made with **2** was no flash in the pan because modification of its format to a "receptor$_1$-spacer$_1$-lumophore-spacer$_2$-receptor$_2$" version **3** led to a strongly improved AND logic performance [73]. Acceleration of PET from each receptor to the lumophore due to the smaller separation distance was held responsible. As alluded to earlier, Boole's work [69] led to a set of logic operations [74] numbering 4 for the single-input type, 16 for the double-input variety, and so on. AND was just one among the 16. Each of the other logic gates could also be emulated with suitable designed molecules. For instance, deliberately designed OR logic [24] arrived with **4**, which was a simple "lumophore-spacer-receptor," but the receptor was sufficiently nonselective so as to receive either of the two analytes, Ca^{2+} and Mg^{2+}. Importantly, the luminescence enhancement factor is nearly the same for each analyte (Table 2.2). The binding constants of **4** for each of the analytes are quite different, but the cation concentrations can be chosen appropriately for correct OR logic action.

Further, such logic ideas can be applied to improving the selectivity of sensing. A good example, **5**, is from Cooper and James [75], who employed a lumophore outfitted with two PET-active receptors so that emission switching "on" occurs only when the correct bifunctional guest (glucosammonium) is captured by

Table 2.2 Truth table for **4**.

Input$_1$	Input$_2$	Output
Ca^{2+}	Mg^{2+}	Luminescence[a]
Low (0 M)	Low (0 M)	Low (0.0042)
Low (0 M)	High (0.5 M)	High (0.24)
High (10^{-3} M)	Low (0 M)	High (0.28)
High (10^{-3} M)	High (0.5 M)	High (0.28)

[a] Quantum yields in water at pH 7.3, λ_{exc} 389 nm, λ_{em} 490 nm.

both receptors at once. Our approach to selectively sensing γ-aminobutyric acid (GABA) [76] is similar but less elegant. Targeting of multifunctional biomolecules should be an important application area for molecular logic systems in the near term.

5

2.5
Expanding Logic

Molecular devices always had the potential to operate in spaces far smaller than those accessible to semiconductor logic gates, but the demonstration did not occur until 2005 when a molecular computational device was successfully operated in a sphere 3 nm in diameter [77]. Remarkably, this sphere can be found in common soap solutions, where molecular interrogations are easily conducted [78, 79] and where logic gates can even be self-assembled from simpler components such as lumophores and receptors [80]. It is worth pointing out that semiconductor quantum dots can also enter nanometric spaces [81], but we are not aware of any work where their Boolean properties have been examined under such conditions. Miniaturized autonomous semiconductor logic systems such as "smart dust" operate in millimetric dimensions only [82]. In spite of this size advantage possessed by molecular logic devices, many difficulties can hinder progress if we only follow the path of semiconductor device mimicry. While some of these difficulties are slowly being tackled, it is important that several other worthwhile paths are available for exploration. All these explorations have yielded molecular logic devices of increasing complexity [83–101], including the demonstration of molecular-scale arithmetic operations [102–104]. For instance, the first synthetic molecules that could count, although only up the short number series 0, 1, and 2, were based on a molecular AND gate **6** and an XOR gate **7** integrated in parallel in a test tube [102]. This was important since natural molecule-based brains have been performing arithmetic operations on record for at least five millennia.

6

7

In an interesting development, chemical diversity has endowed molecular logic with reconfigurability options that are unimaginable for devices of semiconductor electronics [105, 106]. This has meant that a given molecular structure can display multiple logic faces depending on how it is interrogated. What excitation color? What input species? What interrogation technique? These are some of the many questions that can be productively asked. Important quantum computing issues are raised when a molecular device shows multiple logic configurations all at the same time [106]. This situation arises because it is easy for different people to watch a chemical phenomenon such as a pH indicator each through his/her own color filter to result in different results. The classic story of "the blind men and the elephant" is not dissimilar in terms of its ideas. It is clear that the productive expansion of molecular logic has only begun.

2.6
Utilizing Logic

There has been a growing effort to use computational ideas expressed in molecules for new and useful purposes. For instance, we have shown that some of these devices have potential uses as improved detectors, such as in the direct detection of two or even three analytes at once [107, 108]. The latter can serve as "lab-on-a-molecule" systems [108] that can enable do-it-yourself diagnosis of medical problems [109, 110], especially in a pandemic situation when health care professionals are overloaded. In these devices, the reciprocal of the binding constant of each receptor for its target analyte defines a threshold above which the analyte concentration will be read as "high." When each of the analytes is found to be "high," the lumophore releases a light signal, which conveys the overall outcome. If the light signal remains "low," this AND logic device **8** suggests that the condition of all the analyte concentrations (H^+, Na^+, and Zn^{2+}) being "high" has not been met (Table 2.3). In a similar way, cases conveying other combinations of analyte conditions can be built. For instance, an INHIBIT logic device **9** will produce a "high" emission intensity only if two specified analytes are "high" and another is "low" [111].

Another important use of molecular logic would be in the tracking of large populations of objects that are too small to be tagged by semiconductor-based radiofrequency labels [112]. Being far smaller than current radio frequency

Table 2.3 Truth table for **8**.

Input₁	Input₂	Input₃	Output
Na⁺	H⁺	Zn²⁺	Luminescence[a]
Low (0 M)	Low ($10^{-9.5}$ M)	Low (0 M)	Low (0.001)
Low (0 M)	High (10^{-6} M)	Low (0 M)	Low (0.001)
Low (0 M)	Low ($10^{-9.5}$ M)	High (10^{-3} M)	Low (0.002)
Low (0 M)	High (10^{-6} M)	High (10^{-3} M)	Low (0.003)
High (5 M)	Low ($10^{-9.5}$ M)	Low (0 M)	Low (0.006)
High (5 M)	High (10^{-6} M)	Low (0 M)	Low (0.007)
High (5 M)	Low ($10^{-9.5}$ M)	High (10^{-3} M)	Low (0.006)
High (5 M)	High (10^{-6} M)	High (10^{-3} M)	High (0.020)

[a]Quantum yields in water, λ_{exc} 379 nm, λ_{em} 435 nm.

identification (RFID) tags, molecular logic gates are also rich enough in information to handle a large diversity. The Boolean logic type, the nature of the inputs, and the level of input signal strength needed are all flexible, as are the colors of optimal excitation and emission. Further diversity is presented when pairs of logic gates are run in parallel on a given object. In each case, luminescence intensity is the output that is monitored. This is the molecular computational identification (MCID) method [113, 114], which has been demonstrated by covalently attaching luminescent logic gates to small beads [115]. For instance, the emission intensity–pH profile would serve as a distinct signature in the case of labeled objects **10** (YES logic) and **11** (PASS 1 logic) [113]. Developing these uses of molecular logic and finding other examples must remain an important priority for practitioners in this field.

10

11

2.7
Bringing in Physical Inputs

Physical properties such as temperature can also be brought into the fold of luminescent switches and gates by employing thermoresponsive polymers and polarity-sensitive lumophores [116, 117]. When dissolved in water, polymers such as **12** possess the property of curling up tightly as the temperature increases beyond a critical value. Under this condition, water molecules move around faster and lose their ability to solvate the amide groups of **12**. This situation forces the amide groups to bind to each other. This isolates the lumophore from the surrounding aqueous environment, which prevents water-centered quenching. Thus emission flares up beyond the critical temperature. Although simple, this scenario is sufficient for our purpose of building some of the most sensitive molecular thermometers [118, 119] known. They can also be tuned to suit a chosen temperature [120]. Others respond to temperature as well as to chemical species such as protons [121]. Logic functions driven by temperature as one input are created in this way.

Even small molecules can be successful as molecular thermometers [122], but without the sharp sensitivity described earlier. Nevertheless, the ease of their synthesis and incorporation into interesting structures is a favorable point. For instance, the excited state of **13** undergoes dissociation of the C5–N1 bond but only if a small amount of energy is supplied as heat. Thus, the luminescence of **13** is controlled by temperature in an Arrhenius fashion.

2.8
Summary and Outlook

Luminescent sensing and switching supermolecules can be designed according to basic photochemical principles to deliver robust experimental behavior, which is easily appreciated even by uninitiated observers. These molecules can also be inserted into tiny spaces of critical human importance, while still maintaining their functionality. Gradual increase of their complexity permits the addressing of mathematical and computational issues, especially within small spaces. It seems probable that these molecules will continue to serve fields as diverse as medical diagnostics, nanoscience, and information processing in the coming years [123–125].

Acknowledgments

The author thanks the Engineering and Physical Sciences Research Council, UK; Department of Employment and Learning, Northern Ireland; McClay Trust, InvestNI, European Union; Japan Society for the Promotion of Science; Procter and Gamble Co; and Avecia Ltd. for their support.

References

1. Schwarz, J.A., Contescu, C., and Putyera, K. (eds) (2004) *Encyclopedia of Nanoscience and Nanotechnology*, Marcel Dekker, New York.
2. Bryan, A.J., De Silva, A.P., De Silva, S.A., Rupasinghe, R.A.D.D., and Sandanayake, K.R.A.S. (1989) *Biosensors*, **4**, 169.
3. De Silva, A.P., Gunaratne, H.Q.N., Gunnlaugsson, T., Mccoy, C.P., Maxwell, P.R.S., Rademacher, J.T., and Rice, T.E. (1996) *Pure Appl. Chem.*, **68**, 1443.
4. Balzani, V. (ed.) (2001) *Electron Transfer in Chemistry*, Wiley-VCH Verlag GmbH, Weinheim.
5. Bissell, R.A., De Silva, A.P., Gunaratne, H.Q.N., Lynch, P.L.M., Maguire, G.E.M., Mccoy, C.P., and Sandanayake, K.R.A.S. (1993) *Top. Curr. Chem.*, **168**, 223.
6. Bissell, R.A., De Silva, A.P., Gunaratne, H.Q.N., Lynch, P.L.M., Maguire, G.E.M., and Sandanayake, K.R.A.S. (1992) *Chem. Soc. Rev.*, **21**, 187.
7. Wang, Y.C. and Morawetz, H. (1976) *J. Am. Chem. Soc.*, **98**, 3611.
8. Selinger, B.K. (1977) *Aust. J. Chem.*, **30**, 2087.
9. Beddard, G.S., Davidson, R.S., and Whelan, T.D. (1978) *Chem. Phys. Lett.*, **56**, 54.
10. Shizuka, H., Nakamura, M., and Morita, T. (1979) *J. Phys. Chem.*, **83**, 2019.
11. Shizuka, H., Ogiwara, T., and Kimura, E. (1985) *J. Phys. Chem.*, **89**, 4302.
12. Konopelski, J.P., Kotzyba-Hibert, F., Lehn, J.-M., Desvergne, J.-P., Fages, F., Castellan, A., and Bouas-Laurent, H. (1985) *J. Chem. Soc., Chem. Commun.*, 433.
13. De Silva, A.P., Gunaratne, H.Q.N., Lynch, P.L.M., Patty, A.L., and Spence, G.L. (1993) *J. Chem. Soc. Perkin Trans. 2*, 1611.
14. De Silva, A.P. and Rupasinghe, R.A.D.D. (1985) *J. Chem. Soc., Chem. Commun.*, 1669.
15. De Silva, A.P. and De Silva, S.A. (1986) *J. Chem. Soc., Chem. Commun.*, 1709.

16. De Silva, A.P. and Sandanayake, K.R.A.S. (1989) *J. Chem. Soc., Chem. Commun.*, 1183.
17. De Silva, A.P. and Gunaratne, H.Q.N. (1990) *J. Chem. Soc., Chem. Commun.*, 186.
18. De Silva, A.P., Gunaratne, H.Q.N., and Sandanayake, K.R.A.S. (1990) *Tetrahedron Lett.*, **31**, 5193.
19. De Silva, A.P. and Sandanayake, K.R.A.S. (1990) *Angew. Chem. Int. Ed. Engl.*, **29**, 1173.
20. De Silva, A.P. and Sandanayake, K.R.A.S. (1991) *Tetrahedron Lett.*, **32**, 421.
21. Bissell, R.A., Calle, E., De Silva, A.P., De Silva, S.A., Gunaratne, H.Q.N., Habib-Jiwan, J.-L., Peiris, S.L.A., Rupasinghe, R.A.D.D., Samarasinghe, T.K.S.D., Sandanayake, K.R.A.S., and Soumillion, J.-P. (1992) *J. Chem. Soc., Perkin Trans. 2*, 1559.
22. De Silva, A.P., Gunaratne, H.Q.N., Gunnlaugsson, T., and Lynch, P.L.M. (1996) *New J. Chem.*, **20**, 871.
23. De Silva, A.P., De Silva, S.A., Dissanayake, A.S., and Sandanayake, K.R.A.S. (1989) *J. Chem. Soc., Chem. Commun.*, 1054.
24. De Silva, A.P., Gunaratne, H.Q.N., and Maguire, G.E.M. (1994) *J. Chem. Soc., Chem. Commun.*, 1213.
25. De Silva, A.P., Gunaratne, H.Q.N., Kane, A.T.M., and Maguire, G.E.M. (1995) *Chem. Lett.*, 125.
26. De Silva, A.P., Gunaratne, H.Q.N., and Lynch, P.L.M. (1995) *J. Chem. Soc., Perkin Trans. 2*, 685.
27. De Silva, A.P., Gunaratne, H.Q.N., Gunnlaugsson, T., and Nieuwenhuyzen, M. (1996) *Chem. Commun.*, 1967.
28. De Silva, A.P., Gunaratne, H.Q.N., Habib-Jiwan, J.-L., Mccoy, C.P., Rice, T.E., and Soumillion, J.-P. (1995) *Angew. Chem. Int. Ed. Engl.*, **34**, 1728.
29. De Silva, A.P., Goligher, A., Gunaratne, H.Q.N., and Rice, T.E. (2003) *ARKIVOC*, 229, part vii.
30. Daffy, L.M., De Silva, A.P., Gunaratne, H.Q.N., Huber, C., Lynch, P.L.M., Werner, T., and Wolfbeis, O.S. (1998) *Chem. Eur. J.*, **4**, 1810.
31. Bissell, R.A. and De Silva, A.P. (1991) *J. Chem. Soc., Chem. Commun.*, 1148.
32. De Silva, A.P., Gunaratne, H.Q.N., and Rice, T.E. (1996) *Angew. Chem. Int. Ed. Engl.*, **35**, 2116.
33. De Silva, A.P., Gunaratne, H.Q.N., Rice, T.E., and Stewart, S. (1997) *Chem. Commun.*, 1891.
34. Grigg, R. and Norbert, W.D.J.A. (1992) *J. Chem. Soc., Chem. Commun.*, 1300.
35. Grigg, R. and Norbert, W.D.J.A. (1992) *J. Chem. Soc., Chem. Commun.*, 1298.
36. Grigg, R., Holmes, J.M., Jones, S.K., and Norbert, W.D.J.A. (1994) *J. Chem. Soc., Chem. Commun.*, 185.
37. Gareis, T., Huber, C., Wolfbeis, O.S., and Daub, J. (1997) *Chem. Commun.*, 1717.
38. Tsien, R.Y. (1980) *Biochemistry*, **19**, 2396.
39. Raju, B., Murphy, E., Levy, L.A., Hall, R.D., and London, R.E. (1989) *Am. J. Physiol.*, **256**, C540.
40. Czarnik, A.W. (ed.) (1993) *Fluorescent Chemosensors for Ion and Molecule Recognition*, ACS Publications, Washington, DC.
41. Desvergne, J.-P. and Czarnik, A.W. (eds) (1997) *Chemosensors for Ion and Molecule Recognition*, Kluwer Academic Publishers, Dordrecht.
42. De Silva, A.P., Gunaratne, H.Q.N., Gunnlaugsson, T., Huxley, A.J.M., Mccoy, C.P., Rademacher, J.T., and Rice, T.E. (1997) *Chem. Rev.*, **97**, 1515.
43. De Silva, A.P., Mccaughan, B., Mckinney, B.O.F., and Querol, M. (2003) *Dalton Trans.*, 1902.
44. De Silva, A.P., Fox, D.B., Huxley, A.J.M., Mcclenaghan, N.D., and Moody, T.S. (2000) *Coord. Chem. Rev.*, **205**, 41.
45. Domaille, D.W., Que, E.L., and Chang, C.J. (2008) *Nat. Chem. Biol.*, **4**, 168.
46. Que, E.L., Domaille, D.W., and Chang, C.J. (2008) *Chem. Rev.*, **108**, 1517.
47. Li, P., Duan, X., Chen, Z.Z., Liu, Y., Xie, T., Fang, L.B., Li, X.R., Yin, M., and Tang, B. (2011) *Chem. Commun.*, **47**, 7755.
48. Xu, Z.C., Yoon, J., and Spring, D.R. (2011) *Chem. Soc. Rev.*, **40**, 2593.
49. Koide, Y., Urano, Y., Hanaoka, K., Terai, T., and Nagano, T. (2011) *ACS Chem. Biol.*, **6**, 600.

50. De Silva, A.P., Fox, D.B., Moody, T.S., and Weir, S.M. (2001) *Pure Appl. Chem.*, **73**, 503.
51. De Silva, A.P., Eilers, J., and Zlokarnik, G. (1999) *Proc. Natl. Acad. Sci. U.S.A.*, **96**, 8336.
52. Haugland, R.P. (2012) *Molecular Probes. The Handbook*, Invitrogen, http://www.invitrogen.com/site/us/en/home/References/Molecular-Probes-The-Handbook.html.
53. Paglin, S., Hollister, T., Delohery, T., Hackett, N., Mcmahill, M., Sphicas, E., Domingo, D., and Yahalom, J. (2001) *Cancer Res.*, **61**, 439.
54. (2012) *OPTI Medical Systems*, Roswell, GA, USA.
55. He, H.R., Mortellaro, M., Leiner, M.J.P., Young, S.T., Fraatz, R.J., and Tusa, J. (2003) *Anal. Chem.*, **75**, 549.
56. He, H.R., Mortellaro, M., Leiner, M.J.P., Fraatz, R.J., and Tusa, J. (2003) *J. Am. Chem. Soc.*, **125**, 1468.
57. Tusa, J. and He, H.R. (2005) *J. Mater. Chem.*, **15**, 2640.
58. He, H.R., Jenkins, K., and Lin, C. (2008) *Anal. Chim. Acta*, **611**, 197.
59. Bissell, R.A., De Silva, A.P., Fernando, W.T.M.L., Patuwathavithana, S.T., and Samarasinghe, T.K.S.D. (1991) *Tetrahedron Lett.*, **32**, 425.
60. Callan, J.F., De Silva, A.P., Fox, D.B., Mcclenaghan, N.D., and Sandanayake, K.R.A.S. (2005) *J. Fluoresc.*, **15**, 769.
61. Hall, M.J., Allen, L.T., and O'shea, D.F. (2006) *Org. Biomol. Chem.*, **4**, 776.
62. De Silva, A.P., Gunaratne, H.Q.N., and Gunnlaugsson, T. (1998) *Tetrahedron Lett.*, **39**, 5077.
63. Chen, X., Zhou, Y., Peng, X.J., and Yoon, J. (2010) *Chem. Soc. Rev.*, **39**, 2120.
64. Weltman, J.K., Szaro, R.P., Frackelt, A.R., Dowben, R.M., Bunting, J.R., and Cathou, R.E. (1973) *J. Biol. Chem.*, **248**, 3173.
65. Deisenhofer, J. and Norris, J.R. (eds) (1993) *The Photosynthetic Reaction Center*, vol. 1 & 2, Academic Press, San Diego, CA.
66. De Silva, A.P. and Rice, T.E. (1999) *Chem. Commun.*, 163.
67. De Silva, A.P., De Silva, S.S.K., Goonesekera, N.C.W., Gunaratne, H.Q.N., Lynch, P.L.M., Nesbitt, K.R., Patuwathavithana, S.T., and Ramyalal, N.L.D.S. (2007) *J. Am. Chem. Soc.*, **129**, 3050.
68. Reek, J.H.R. and Otto, S. (eds) (2010) *Dynamic Combinatorial Chemistry*, Wiley-VCH Verlag GmbH, Weinheim.
69. Boole, G. (1958) *An Investigation of the Laws of Thought*, Dover Publications, New York.
70. Malvino, A.P. and Brown, J.A. (1993) *Digital Computer Electronics*, 3rd edn, Glencoe, New York.
71. Ben-Ari, M. (1993) *Mathematical Logic for Computer Science*, Prentice-Hall, Hemel Hempstead.
72. De Silva, A.P., Gunaratne, H.Q.N., and Mccoy, C.P. (1993) *Nature*, **364**, 42.
73. De Silva, A.P., Gunaratne, H.Q.N., and Mccoy, C.P. (1997) *J. Am. Chem. Soc.*, **119**, 7891.
74. Gregg, J.R. (1998) *Ones and Zeros*, IEEE Press, New York.
75. Cooper, C.R. and James, T.D. (1997) *Chem. Commun.*, 1419.
76. De Silva, A.P., Gunaratne, H.Q.N., Mcveigh, C., Maguire, G.E.M., Maxwell, P.R.S., and O'hanlon, E. (1996) *Chem. Commun.*, 2191.
77. Uchiyama, S., Mcclean, G.D., Iwai, K., and De Silva, A.P. (2005) *J. Am. Chem. Soc.*, **127**, 8920.
78. Bissell, R.A., Bryan, A.J., De Silva, A.P., and Mccoy, C.P. (1994) *J. Chem. Soc., Chem. Commun.*, 405.
79. Uchiyama, S., Iwai, K., and De Silva, A.P. (2008) *Angew. Chem. Int. Ed. Engl.*, **47**, 4667.
80. De Silva, A.P., Dobbin, C.M., Vance, T.P., and Wannalerse, B. (2009) *Chem. Commun.*, 1386.
81. Silva, G.A. (2006) *Nat. Rev. Neurosci.*, **7**, 65.
82. Warneke, B., Last, M., Liebowitz, B., and Pister, K.S.J. (2001) *Computer*, **34**, 44.
83. De Silva, A.P., Mcclenaghan, N.D., and Mccoy, C.P. (2001) *Handbook of Electron Transfer in Chemistry*, vol. 5, Wiley-VCH Verlag GmbH, Weinheim, p. 156.
84. De Silva, A.P., Mcclenaghan, N.D., and Mccoy, C.P. (2001) in *Molecular*

Switches (ed. B.L. Feringa), Wiley-VCH Verlag GmbH, Weinheim, p. 339.
85. Raymo, F.M. (2002) *Adv. Mater.*, **14**, 401.
86. De Silva, A.P. and Mcclenaghan, N.D. (2004) *Chem. Eur. J.*, **10**, 574.
87. De Silva, A.P. (2005) *Nat. Mater.*, **4**, 15.
88. De Silva, A.P., Leydet, Y., Lincheneau, C., and Mcclenaghan, N.D. (2006) *J. Phys. Condens. Matter.*, **18**, S1847.
89. De Silva, A.P. and Uchiyama, S. (2007) *Nat. Nanotechnol.*, **2**, 399.
90. Credi, A. (2007) *Angew. Chem. Int. Ed.*, **46**, 5472.
91. Magri, D.C., Vance, T.P., and De Silva, A.P. (2007) *Inorg. Chim. Acta*, **360**, 751.
92. De Silva, A.P., Uchiyama, S., Vance, T.P., and Wannalerse, B. (2007) *Coord. Chem. Rev.*, **251**, 1623.
93. Szacilowski, K. (2008) *Chem. Rev.*, **108**, 3481.
94. Balzani, V., Venturi, M., and Credi, A. (2008) *Molecular Devices and Machines*, 2nd edn, Wiley-VCH Verlag GmbH, Weinheim.
95. Benenson, Y. (2009) *Mol. Biosyst.*, **5**, 675.
96. Andreasson, J. and Pischel, U. (2010) *Chem. Soc. Rev.*, **39**, 174.
97. De Silva, A.P. (2010) *Aust. J. Chem.*, **63**, 146.
98. Katz, E. and Privman, V. (2010) *Chem. Soc. Rev.*, **39**, 1835.
99. Amelia, M., Zou, L., and Credi, A. (2010) *Coord. Chem. Rev.*, **254**, 2267.
100. Pischel, U. (2010) *Aust. J. Chem.*, **63**, 148.
101. Tian, H. (2010) *Angew. Chem. Int. Ed.*, **49**, 4710.
102. De Silva, A.P. and Mcclenaghan, N.D. (2000) *J. Am. Chem. Soc.*, **122**, 3965.
103. Pischel, U. (2007) *Angew. Chem. Int. Ed.*, **46**, 4026.
104. Brown, G.J., De Silva, A.P., and Pagliari, S. (2002) *Chem. Commun.*, 2461.
105. De Silva, A.P. and Mcclenaghan, N.D. (2002) *Eur. J. Chem.*, **8**, 4935.
106. Callan, J.F., De Silva, A.P., and Mcclenaghan, N.D. (2004) *Chem. Commun.*, 2048.
107. De Silva, A.P., Mcclean, G.D., and Pagliari, S. (2003) *Chem. Commun.*, 2010.
108. Magri, D.C., Brown, G.J., Mcclean, G.D., and De Silva, A.P. (2006) *J. Am. Chem. Soc.*, **128**, 4950.
109. Konry, T. and Walt, D.R. (2009) *J. Am. Chem. Soc.*, **131**, 13232.
110. Wang, J. and Katz, E. (2010) *Anal. Bioanal. Chem.*, **398**, 1591.
111. De Silva, A.P., Dixon, I.M., Gunaratne, H.Q.N., Gunnlaugsson, T., Maxwell, P.R.S., and Rice, T.E. (1999) *J. Am. Chem. Soc.*, **121**, 1393.
112. Shepard, S. (2005) *RFID: Radio Frequency Identification*, McGraw-Hill, New York.
113. De Silva, A.P., James, M.R., Mckinney, B.O.F., Pears, D.A., and Weir, S.M. (2006) *Nat. Mater.*, **5**, 787.
114. Brown, G.J., De Silva, A.P., James, M.R., Mckinney, B.O.F., Pears, D.A., and Weir, S.M. (2008) *Tetrahedron*, **64**, 8301.
115. Ayadim, M., Habib-Jiwan, J.-L., De Silva, A.P., and Soumillion, J.-P. (1996) *Tetrahedron Lett.*, **37**, 7039.
116. Uchiyama, S., Matsumura, Y., De Silva, A.P., and Iwai, K. (2003) *Anal. Chem.*, **75**, 5926.
117. Iwai, K., Matsumura, Y., Uchiyama, S., and De Silva, A.P. (2005) *J. Mater. Chem.*, **15**, 2796.
118. Chandrasekharan, N. and Kelly, L.A. (2004) *Rev. Fluoresc.*, **1**, 21.
119. Uchiyama, S., De Silva, A.P., and Iwai, K. (2006) *J. Chem. Educ.*, **83**, 720.
120. Uchiyama, S., Matsumura, Y., De Silva, A.P., and Iwai, K. (2004) *Anal. Chem.*, **76**, 1793.
121. Uchiyama, S., Kawai, N., De Silva, A.P., and Iwai, K. (2004) *J. Am. Chem. Soc.*, **126**, 3032.
122. De Silva, A.P., Gunaratne, H.Q.N., Jayasekera, K.R., O'callaghan, S., and Sandanayake, K.R.A.S. (1995) *Chem. Lett.*, 123.
123. De Silva, A.P. (2011) *Top. Curr. Chem.*, **300**, 1.
124. De Silva, A.P. (2011) *Chem. Asian J.*, **6**, 750.
125. De Silva, A.P. (2011) *J. Comput. Theor. Nanosci.*, **8**, 409.

3
Binary Logic with Synthetic Molecular and Supramolecular Species

Monica Semeraro, Massimo Baroncini, and Alberto Credi

3.1
Introduction

3.1.1
Information Processing: Semiconductor Devices versus Biological Structures

When a molecule is subjected to a stimulus, such as a pulse of light, an electric potential, or an encounter with another molecule, it has to make a choice. It can either do nothing or undergo a reaction that leads to a new stable or metastable state with physicochemical properties differing from those of the initial state. In this sense, any chemical reaction, even if carried out on a statistically significant ensemble of molecules rather than on a single entity, can be viewed as a transformation of "inputs" into "outputs," and can thus be used to process information [1]. This concept is best illustrated by living organisms, in which information is elaborated, transported, and stored using "soft" molecular or ionic substrates [2, 3]. Conventional computers, on the other hand, operate in a radically different manner: they are made of "hard" semiconductor devices, signals are carried by electric charges [4], and two states are used to store and process data because such binary or Boolean logic [5] is robust with respect to fluctuations of the physical signals.

Within cells, and also between cells, information is carried from place to place by the use of messenger molecules and relies on Brownian motion. At the scale of the cell, diffusive motion is remarkably fast: any two molecules within a micrometer-size cell meet each other every second [2]. Over large distances, however, this simple kind of chemical signaling would be too slow, so evolution has created a longer range communication and information-processing system, the brain and nervous system. Its signaling mechanism is again quite different compared to that of electronic computers. In fact, although a nerve pulse manifests itself as a pulse of electrical potential, it is not the movement of electrons that is carrying the information, but ultimately the movement of ions through the membrane [3].

Another difference between silicon-based and biological information processing should be emphasized. In an electronic computer, there is a clear distinction

Molecular and Supramolecular Information Processing: From Molecular Switches to Logic Systems,
First Edition. Edited by Evgeny Katz.
© 2012 Wiley-VCH Verlag GmbH & Co. KGaA. Published 2012 by Wiley VCH Verlag GmbH & Co. KGaA.

between hardware and software. Hardware is a physical entity, useless without the direction of software. By contrast, software is a transient and noncorporeal entity, but capable of animating the otherwise inert hardware to perform the desired task. In biological information-processing systems, the basic elements of logics are well defined chemical entities, and a clear distinction between hardware and software is impractical. For example, the genetic code, that is, the "software" of an organism, is embodied in DNA molecules, which are physical objects with very specific properties, and the "hardware" consists again of molecules, proteins that assemble to form the machines that carry out the functions of the organism. There are also molecules such as RNA that can properly be considered as both the software and the hardware, because they can read and carry information from the DNA to the protein machines that create new molecules according to the specification of the genetic code, and also act as machines themselves.

3.1.2
Toward Chemical Computers?

The primary motivation behind the search for computing strategies based on molecules is to develop novel paradigms for information processing that, by moving beyond silicon-based technology, could lead to computers of extremely small size, low power consumption, and unprecedented performance. The term *molecular computer*, however, still sounds weird to most chemists in spite of the fact that more than 25 years ago the Pimentel report was explicit and optimistic in this regard [6].

Clearly, the development of a computing machine based on molecules that can make current solid-state devices obsolete is a very ambitious objective even for basic research. While it is true that conventional scaling methods of the semiconductor industry face increasing technical and fundamental challenges as device features are pushed toward the deep sub-100-nm regime [7–9], recent progresses in transistor technology will enable continued trends in downscaling and improvements in performance of logic complementary metal-oxide semiconductor (CMOS) transistors until at least the middle of this decade [9, 10].

Nevertheless, the advances in (bio)chemical synthesis and characterization have allowed the design and construction of molecule-based systems capable of performing quite complex functions. Several research laboratories have undertaken the design, synthesis, and characterization of chemical systems that mimic the operation of semiconductor circuit components (wires, switches, memories, logic gates, and logic circuits), and books [11, 12], journal issues [13], and review articles [14] focusing on molecular logic have been published. It should be noted, however, that the components of a "molecular processor" do not necessarily need to follow the blueprint of microelectronic circuits; other choices such as multivalued [15] and fuzzy [16] logic or variable threshold (neural) mechanisms [17] could also be made.

The chemical approach to information processing described above is typically based on the modular design and construction of multicomponent (i.e., supramolecular) species made of a discrete number of structurally and/or functionally integrated components, each playing a specific role. In this chapter we illustrate the concept of information processing with artificial supramolecular systems by presenting a few examples, mostly taken from our recent work. These systems are based on binary logic, operate in solution environments, and are investigated at the ensemble level. Perspectives and limitations of such kinds of devices are critically discussed. For space reasons, and because the topic is thoroughly illustrated elsewhere in the book, we will not deal with information processing based on biomolecules (nucleic acids, enzymes, etc.).

3.2
Combinational Logic Gates and Circuits

3.2.1
Basic Concepts

Logic gates are devices that perform binary (Boolean) operations on one or more inputs to produce an output [4, 5]. Logic circuits can be viewed as a network of basic logic gates that implement a more complex Boolean function. Each type of logic gate or circuit possesses a specific input–output signal correlation pattern described in the so-called truth table, which lists all the possible combinations of input–output states.

The first proposal to execute logic operations at the molecular level was made in 1988 [18], but the field developed only five years later when the analogy between molecular switches and logic gates was experimentally demonstrated [19]. In fact, as molecular switches convert input stimulations into output signals [11], the principles of binary logic can be applied to the signal transduction operated by molecules under appropriate conditions [1]. Following this approach, the most common Boolean functions (PASS, YES, NOT, AND, NAND, OR, NOR, XOR, XNOR, INH) have been implemented with chemical systems [11–14].

A critical issue of molecular logic gates is the interconnection of basic elements to create complex circuits, which is instead a key feature of electronic logic gates owing to full input/output homogeneity. The construction of molecular logic networks, however, can take advantage of functional integration that can be achieved within one molecule by rational chemical design, rather than by relying on intensive physical connection of elementary gates [12–14, 20]. The fact that unsophisticated dye molecules in solution can perform complex logic functions [21–23], which, in silicon-based systems, require circuits made of several interconnected gates, is a demonstration of this idea.

Such a high functional integration is possible thanks to an important feature of molecular logic gates and circuits, namely, reconfigurability [12, 14, 24]. This property refers to the possibility of defining different logic operations for the same gate. In silicon-based circuits, reconfiguration of the logic operation is carried out by interrupting selected connections in gate arrays, usually in an irreversible manner [25]. Molecular logic systems can be reconfigured by changing the type of input/output signals or by using a separate switching stimulus. For instance, the logic expression for molecular logic gates with optical output signals and significant input-induced spectral shifts can be conveniently reconfigured by monitoring the output at a different wavelength.

In wavelength-reconfigurable logic gates, different logic types are observed depending on the wavelength at which the output signal is monitored. Because multiple wavelengths can be observed at the same time, such gates perform simultaneously different logic functions on a given set of inputs, a property that has been referred to as *superposition* or *multiplicity* of logic types [12, 14, 24]. Because logic superposition is a consequence of the multichannel nature of light, it does not occur with electronic systems. These concepts are better illustrated with the help of the examples described in the following sections. A molecular logic device characterized by a high degree of reconfiguration and superposition of logic functions, based on a multicomponent photochromic compound, has been reported recently (see also Chapter 10) [26].

The logic functions performed by the systems that will be described in this section are called *combinational* because the state of the output(s) is determined only by the input combination, irrespective of the order in which the inputs are applied. Conversely, in *sequential* logic functions, the output signal turns on only when the correct inputs are applied in the correct order. As described in Section 3.3, this behavior implies a memory function of the device.

3.2.2
Bidirectional Half Subtractor and Reversible Logic Device

In 1997, some of us reported [27] the first example of a molecular XOR (eXclusive OR) gate based on the controlled assembly and disassembly of a host–guest pseudorotaxane-type species in solution. The article received considerable attention [28] because the XOR gate is of particular importance as it compares the digital state of two signals. At the time, however, the field of molecular logic was still in its infancy and the potentialities of the system were not fully explored. Very recently, we described an improved version of such a system and investigated in detail its UV–visible spectroscopic properties in response to acid/base stimulation in solution, showing that it can operate as a bidirectional half subtractor and that it exhibits logic reversibility [29].

In CH_2Cl_2-CH_3CN (9 : 1 v/v) the electron-rich macrocycle **1** can be threaded by an electron-deficient wire-type molecule such as the 2,7-diazapyrenium derivative 2^{2+} (Figure 3.1). In organic solution both individual molecular components exhibit intense and structured absorption spectra and are strongly fluorescent (Figure 3.2). The resulting pseudorotaxane complex $[1 \supset 2]^{2+}$ is held together by a charge-transfer (CT) interaction. Complexation is signaled by as many as three different optical channels: (i) appearance of a yellow color because of the presence of a CT absorption tail in the visible region; (ii) disappearance of the blue-green fluorescence of 2^{2+} (λ_{max} = 428 nm); and (iii) disappearance of the UV fluorescence of **1** with

Figure 3.1 The self-assembly of compounds **1** and **2**$^{2+}$ to yield the pseudorotaxane [1⊃2]$^{2+}$.

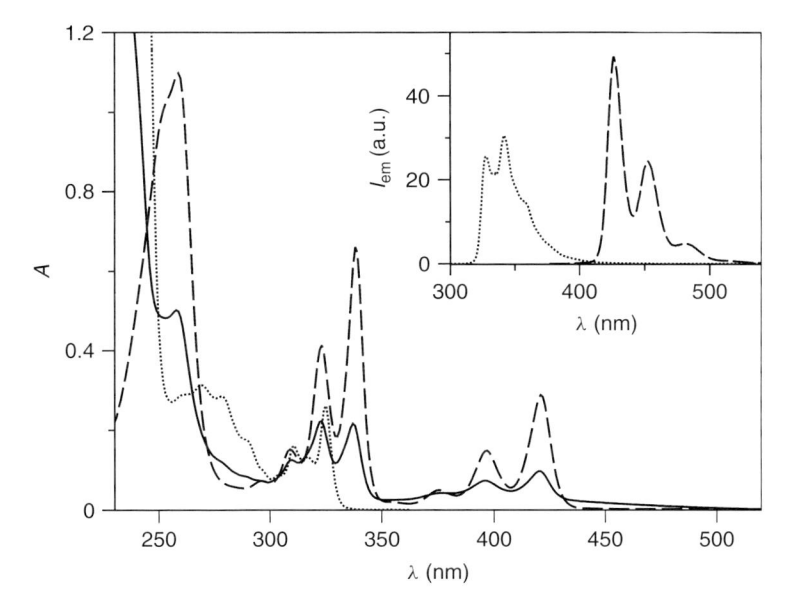

Figure 3.2 Absorption and luminescence (inset; λ_{ex} = 264 nm) spectra of **1** (dotted line) and **2**$^{2+}$ (dashed line) at a concentration of 18 μM in CH$_2$Cl$_2$–CH$_3$CN (9:1 v/v) at room temperature. The solid line is the absorption spectrum of a 1:1 mixture of **1** and **2**$^{2+}$ under the same conditions.

λ_{max} = 340 nm (the disappearance of the two fluorescent signals is due to the presence of the lower lying CT state in the complex).

For the XOR function, acid (CF$_3$SO$_3$H) and base (n-Bu$_3$N) are the inputs and the fluorescence of **1** at 340 nm is the output. The working mechanism of this system is illustrated schematically in Figure 3.3 and Table 3.1. As mentioned above, in the absence of the two inputs, the fluorescence of **1** is quenched in the pseudorotaxane (output {0}). When the n-Bu$_3$N input (B in Figure 3.3) alone is applied

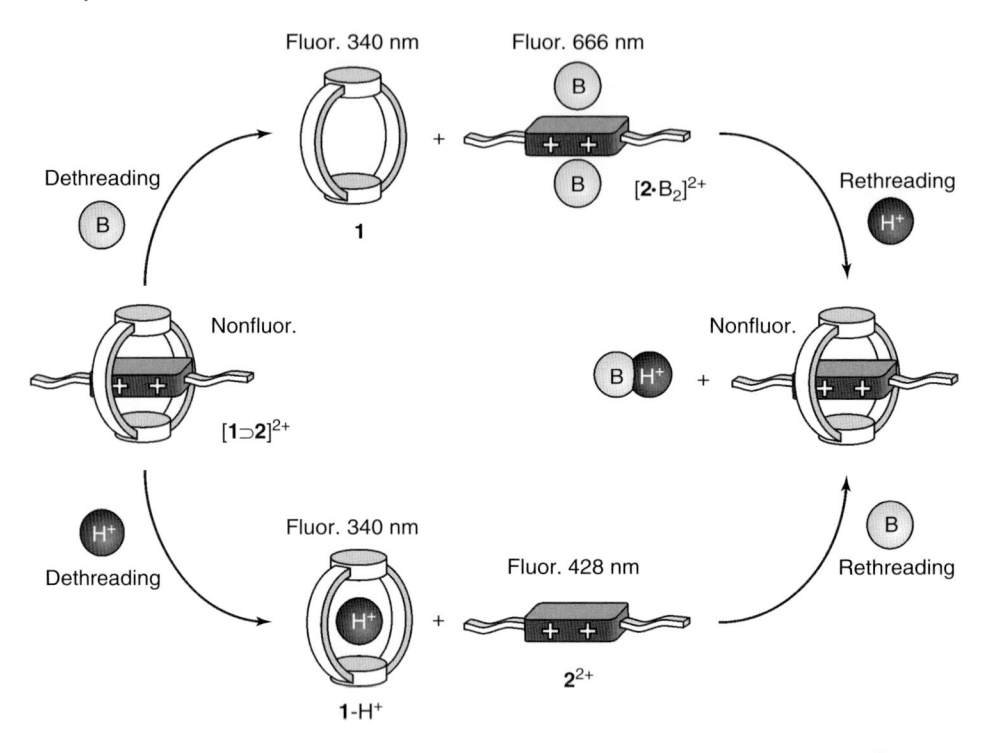

Figure 3.3 Working mechanism of the logic system based on compounds **1** and **2**$^{2+}$ whose formulas are shown in Figure 3.1.

Table 3.1 Truth table of the logic operations performed by the chemical ensemble shown in Figure 3.3.

In$_1$ (amine)^a	In$_2$ (acid)^b	Out$_1$ (340 nm)^{c,d}	Out$_2$ (428 nm)^{c,d}	Out$_3$ (666 nm)^{c,d}	Out$_1$' (340 nm)^{c,e}
0	0	0 (23)	0 (10)	0 (1.0)	0 (23)
1	0	1 (35)	0 (3.0)	1 (94)	1 (35)
0	1	1 (36)	1 (62)	0 (2.0)	1 (36)
1	1	0 (27)	0 (18)	0 (2.0)	1 (27)
		↓	↓	↓	↓
		XOR	INH(In$_1$)	INH(In$_2$)	OR

Experimental conditions are those described in Figure 3.2.
^{a}Binary state {1} corresponds to addition of 30 equiv. of n-Bu$_3$N.
^{b}Binary state {1} corresponds to addition of 30 equiv. of CF$_3$SO$_3$H.
^{c}The values in parentheses indicate the experimental fluorescence intensity values at the wavelength indicated, in arbitrary units, upon excitation at 264 nm.
^{d}Binary states determined by applying a threshold value I_{em} = 30 a.u.
^{e}Binary states determined by applying a threshold value I_{em} = 25 a.u.

(30 equiv. with respect to the pseudorotaxane), the complex dethreads because of the formation of a stronger CT interaction between the amine and 2^{2+}. Under such conditions, **1** is free and its fluorescence is not quenched (amine input {1}, output {1}). The fluorescence typical of free 2^{2+} ($\lambda_{max} = 428$ nm) is still quenched, and a luminescence band with $\lambda_{max} = 666$ nm, arising from the complex between 2^{2+} and the amine, is observed. Application of the H$^+$ input (30 equiv. with respect to the pseudorotaxane) causes protonation of **1** and, again, dethreading of the pseudorotaxane. As a consequence, the fluorescence of free 2^{2+} ($\lambda_{max} = 428$ nm) is restored. Since protonation of **1** (presumably at the aliphatic ether oxygens) does not perturb its emission compared to the neutral form, activation of the H$^+$ input switches on the output at 340 nm. Therefore, the output achieves logic state {1} in the two situations in which *exclusively* one of the two inputs is present. However, when both inputs are applied, acid–base neutralization results, the complex remains intact, and the 340 nm emission is quenched (output {0}).

Interestingly, wavelength reconfiguration of the luminescence readout affords double Inhibit (INH) behavior (Table 3.1). If the output is monitored at 428 nm (fluorescence of free 2^{2+}), the system behaves as an INH gate with tris-*n*-butylamine as the disabling input. By monitoring the emission of the CT complex between 2^{2+} and *n*-Bu$_3$N ($\lambda_{max} = 666$ nm), an INH gate disabled by a proton input is obtained. The superposition of the XOR and two complementary INH functions enables Boolean subtraction of two 1-bit digits, x and y, in either order ($x - y$ and $y - x$), thereby mimicking the behavior of a bidirectional half subtractor. This logic feature corresponds to the physical inversion of the input channel that represents the minuend and the subtrahend data in a conventional half subtractor [30]. It should be noted that this molecular ensemble can exhibit such advanced Boolean functionalities because of the high degree of logic integration, which in turn is made possible because of the peculiar – and for some aspects unusual – physicochemical properties of the system.

The truth table displayed in Table 3.1 shows that two different input strings, namely, {0,0} and {1,1}, produce the same combination of Out$_1$, Out$_2$, and Out$_3$, namely, {0,0,0}. Such a behavior, which occurs for most molecular logic gates and circuits reported so far [12–14], determines a loss of information upon performing the operation, as distinct input strings can no longer be distinguished on the basis of the output state [30c]. Logic gates that erase information in their operation are said to be irreversible. Landauer showed that, because of the increase in the entropic content of the system, irreversible logic computations dissipate $kT\ln2$ joules of heat energy for each bit of information lost [31]. In contrast, reversible logic operations do not erase information, and therefore, they do not generate heat as a result of entropy increase. As recently pointed out [32], the study of molecular systems exhibiting logically reversible behavior is interesting for basic science reasons and, in a perspective, also because heat dissipation is a main issue in the construction of ultraminiaturized information processing devices [9].

The present system would become logically reversible if a signal that enables distinguishing between the {0,0} (no inputs added) and {1,1} (amine and acid added together in stoichiometric amounts) states could be found. A careful analysis of the luminescence spectra obtained on sequential addition of acid and base reveals that the signals corresponding to Out$_1$ (340 nm) and Out$_2$ (428 nm) do not go back to the initial values upon reset, because the triflate anions generated by input annihilation compete with **1** for the **2**$^{2+}$ guest, thus diminishing the apparent stability of the pseudorotaxane [33]. Such an interference hampers the reversibility of the threading–dethreading process but allows distinguishing the {0,0} from the {1,1} state. For example, if a new threshold is applied to Out$_1$ (see Out$_1'$ in Table 3.1), an OR response can be obtained, complementing the set of the already discussed XOR and INH functions and allowing reversible logic. Therefore, in this case, *chemical irreversibility* of the processes caused by the application of the inputs translates into *logic reversibility* of the Boolean operations performed [29].

3.2.3
A Simple Unimolecular Multiplexer–Demultiplexer

An important function in information technology is signal multiplexing/demultiplexing. A 2:1 multiplexer (MUX) is a circuit with two data inputs, one address input, and one output. The MUX selects the binary state from one of the data inputs and directs it to the output; the selected input depends on the binary state of the address input (Figure 3.4a). Conversely, a 1:2 demultiplexer (DEMUX) is a circuit that possesses one data input, one address input, and two outputs. The DEMUX routes the data input to one of the output lines; the selected output is determined by the binary state of the address input (Figure 3.4b). Hence, a MUX allows the encoding of multiple data streams into a single data line for transmission, and a DEMUX can decode such entangled data streams from the received single signal. The logic circuits corresponding to a 2:1 MUX and 1:2 DEMUX are shown, respectively, in Figure 3.4c,d.

Molecules that can mimic the function of a 2:1 MUX [34] or a 1:2 DEMUX [35, 36] have been reported in the past few years. However, these systems either rely on carefully designed multicomponent species [34, 35] and coupling to an external optical device [35] or imply a dependence of the data input on the binary state of the address input [36]. We have shown that the reversible acid–base switching of the absorption and photoluminescence properties of a fluorophore as simple as 8-methoxyquinoline (**3** in Figure 3.5) in solution can form the basis for molecular 2:1 multiplexing and 1:2 demultiplexing with a clear-cut digital response [22].

In CH$_3$CN, **3** shows an absorption band with $\lambda_{max} = 301$ nm and an intense fluorescence band with $\lambda_{max} = 388$ nm (Figure 3.6, ab). The addition of 1 equiv. of triflic acid (CF$_3$SO$_3$H) to **3** affords the protonated form **3**-H$^+$ (Figure 3.5), whose absorption and fluorescence spectra are markedly different from those of **3**. Specifically, the absorbance in the region of the absorption band of **3** (270–320 nm) decreases substantially, and a new band ($\lambda_{max} = 358$ nm) is observed in a spectral region where **3** does not absorb light. The fluorescence band of **3** ($\lambda_{max} = 388$ nm) is replaced by a weaker emission band with $\lambda_{max} = 500$ nm (Figure 3.6, cd). The nonprotonated form can be regenerated on the addition of 1 equiv. of tris-*n*-butylamine (*n*-Bu$_3$N) to **3**-H$^+$. The acid-base-controlled switching between **3**

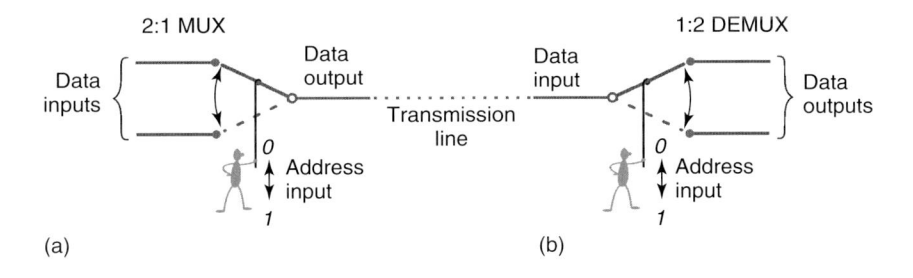

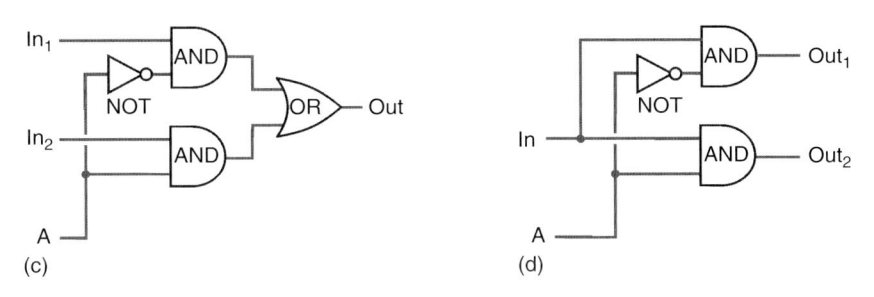

Figure 3.4 Operation scheme (top) and equivalent logic circuit (bottom) corresponding to a 2:1 multiplexer (a and c) and a 1:2 demultiplexer (b and d).

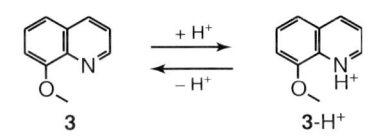

Figure 3.5 The acid-base-controlled switching between 8-methoxyquinoline **3** and its protonated form **3-H$^+$**.

and **3-H$^+$** is fully reversible and can be repeated many times on the same solution without appreciable losses in the absorption and fluorescence spectra. This peculiar spectroscopic behavior and the chemical reversibility of the acid–base switching can be used to obtain both 2:1 MUX and 1:2 DEMUX functions, using proton concentration as the address input (A), and excitation and emission optical signals as the data inputs and outputs, respectively.

In the case of the 2:1 MUX, data inputs are represented by the incident light intensity at 285 nm (In$_1$) and 350 nm (In$_2$); these wavelengths are chosen in order to afford selective excitation of **3** and **3-H$^+$**, respectively (Figure 3.6, ac). The output (Out) is coded for by the fluorescence intensity at 474 nm, a wavelength at which both **3** and **3-H$^+$** fluoresce (Figure 3.6, bd). The fluorescence output levels measured for a CH$_3$CN solution of **3** under the conditions corresponding to the eight combinations of the binary data and address inputs are listed in Table 3.2. A threshold value can be easily identified in order to assign binary output states such that the truth table corresponds to that of a 2:1 MUX. If the address input is {0} (no H$^+$ added, **3** is present), the output reports the state of data input In$_1$,

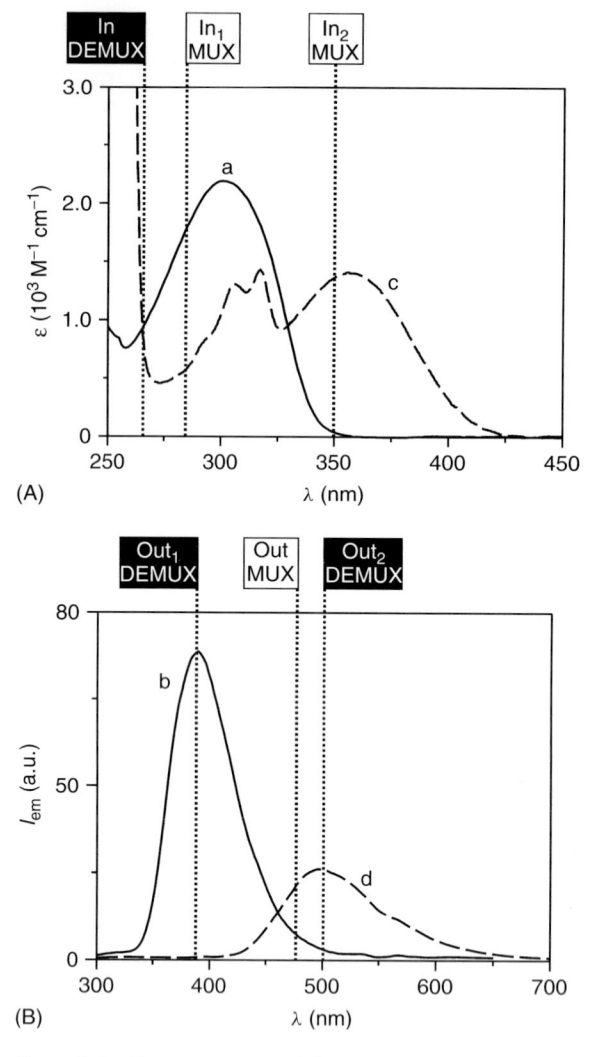

Figure 3.6 Absorption (A) and fluorescence ($\lambda_{ex} = 262$ nm, B) spectra of **3** (a and b) and **3**-H$^+$ (c and d). The wavelengths of the inputs and outputs signals relevant for the MUX/DEMUX binary functions are indicated. Conditions: CH$_3$CN, 15 µM, room temperature.

whereas if the address input is {1} (1 equiv. of H$^+$ added, **3**-H$^+$ is present), the output mirrors the state of data input In$_2$.

The system can be straightforwardly reconfigured to behave as a 1:2 DEMUX by changing the optical input and output channels. In this case, the data input (In) is coded for by the incident light intensity at 262 nm, which is an isosbestic point for **3** and **3**-H$^+$ (Figure 3.6, ac). The two output signals are represented by the fluorescence intensity at 388 nm (λ_{max} for **3**, Out$_1$) and 500 nm (λ_{max} for **3**-H$^+$,

3.2 Combinational Logic Gates and Circuits

Table 3.2 Truth table of the 2:1 multiplexer function based on compound **3**.

Data inputs		Address input	Data output
In$_1$ (285 nm)^a	In$_2$ (350 nm)^a	A (acid)^b	Out (474 nm)^c
0	0	0	0 (0)
1	0	0	1 (21)
0	1	0	0 (5.1)
1	1	0	1 (26)
0	0	1	0 (0)
1	0	1	0 (9.2)
0	1	1	1 (21)
1	1	1	1 (30)
0	0	0	0 (0)

Experimental conditions are those described in Figure 3.6.
^{a}Binary state {1} corresponds to irradiation with the excitation lamp of the spectrofluorimeter; state {0} corresponds to no excitation (lamp off).
^{b}Binary state {1} is obtained on addition of 1 equiv. of CF$_3$SO$_3$H.
^{c}The values in parentheses indicate the experimental fluorescence intensity values in arbitrary units; the corresponding binary states are determined by applying a threshold value $I_{em} = 15$ a.u.

Table 3.3 Truth table of the 1:2 demultiplexer function based on compound **3**.

Data input	Address input	Data outputs	
In (262 nm)^a	A (acid)^b	Out$_1$ (388 nm)^c	Out$_2$ (500 nm)^c
0	0	0 (0)	0 (0)
1	0	1 (71)	0 (0.5)
0	1	0 (0)	0 (0)
1	1	0 (2.1)	1 (21)

Experimental conditions are those described in Figure 3.6.
^{a}Binary state {1} corresponds to irradiation with the excitation lamp of the spectrofluorimeter; state {0} corresponds to no excitation (lamp off).
^{b}Binary state {1} is obtained on addition of 1 equiv. of CF$_3$SO$_3$H.
^{c}The values in parentheses indicate the experimental fluorescence intensity values in arbitrary units; the corresponding binary states are determined by applying a threshold value $I_{em} = 10$ a.u.

Out$_2$) (Figure 3.6, bd). Table 3.3 shows the fluorescence output levels measured for a CH$_3$CN solution of **3** in the conditions corresponding to the four combinations of the binary data and address inputs. On fixing an appropriate threshold for the fluorescence output, the truth table of the 1:2 DEMUX is obtained. If the address input is {0} (**3** is present), the binary state of the data input is transmitted to Out$_1$;

conversely, if the address input is {1} (**3**-H$^+$is present), the binary data input is transmitted to Out$_2$.

Interestingly, the acid–base switching of the absorption and fluorescence bands observed for **3** in solution can also be performed if the molecule is embedded in polystyrene thin films and exposed to solutions of CF_3COOH and n-Bu$_3$N [22]. More importantly, this study demonstrates that functions achieved by circuits whose logic design requires the interconnection of several basic elements can be implemented with simple molecules [21], taking advantage of logic reconfiguration.

3.2.4
An Encoder/Decoder Based on Ruthenium Tris(bipyridine)

The function performed by a digital encoder is that of converting data into a code, an operation that is useful to compress information that have to be transmitted or stored. A 4-to-2 encoder, for example, compresses four inputs into two outputs channels (Figure 3.7a). Conversely, a 2-to-4 decoder converts two coded inputs into four readable outputs (Figure 3.7b). Molecular systems that emulate the function of an encoder or a decoder have been reported [37, 38].

It has been shown recently that simple $[Ru(bpy)_3]^{2+}$ hexafluorophosphate in acetonitrile can mimic the operation of both 4-to-2 encoder and 2-to-4 decoder operations [23]. These functions are obtained by exploiting the unique physico-chemical properties of this complex [39], and in particular the possibility of (i) absorbing visible light; (ii) forming a long-lived and luminescent triplet excited state, $^*[Ru(bpy)_3]^{2+}$, by light irradiation; (iii) undergoing reversible monoelectronic

In$_1$ (+1.4 V)^a	In$_2$ (450 nm)^b	In$_3$ (−1.4 V)^c	In$_4$ (+1.4/−1.4 V)^d	Out$_1$ (530 nm)^e	Out$_2$ (620 nm)^f
1	0	0	0	0	0
0	1	0	0	0	1
0	0	1	0	1	0
0	0	0	1	1	1

^{a} Oxidation ^{b} Excitation ^{c} Reduction ^{d} Alternating oxidation/reduction ^{e} Absorption ^{f} Luminescence

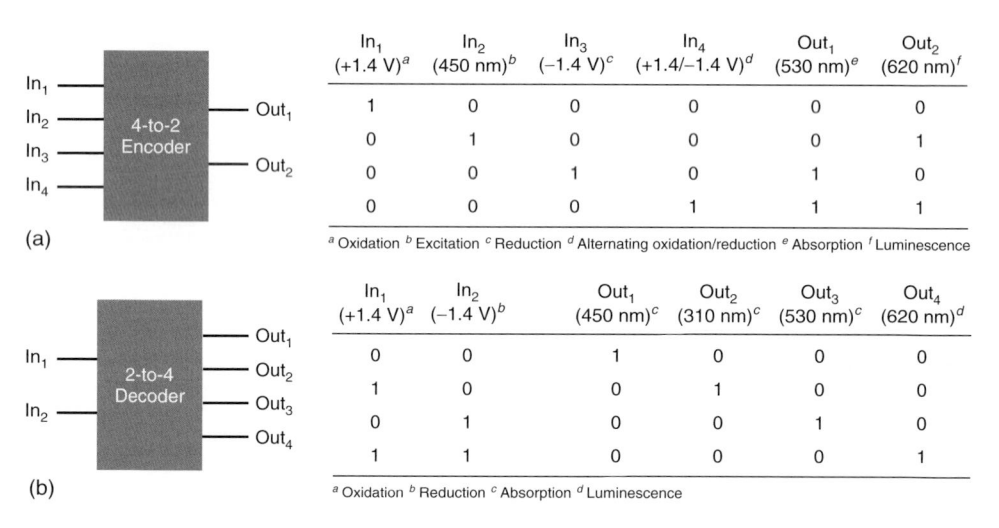

In$_1$ (+1.4 V)^a	In$_2$ (−1.4 V)^b	Out$_1$ (450 nm)^c	Out$_2$ (310 nm)^c	Out$_3$ (530 nm)^c	Out$_4$ (620 nm)^d
0	0	1	0	0	0
1	0	0	1	0	0
0	1	0	0	1	0
1	1	0	0	0	1

^{a} Oxidation ^{b} Reduction ^{c} Absorption ^{d} Luminescence

Figure 3.7 Schematic representation (left) and truth table (right) of a 4-to-2 encoder (a) and a 2-to-4 decoder (b) based on $[Ru(bpy)_3]^{2+}$. See text for details. (Reprinted from Ref. [14o] with permission from Elsevier.)

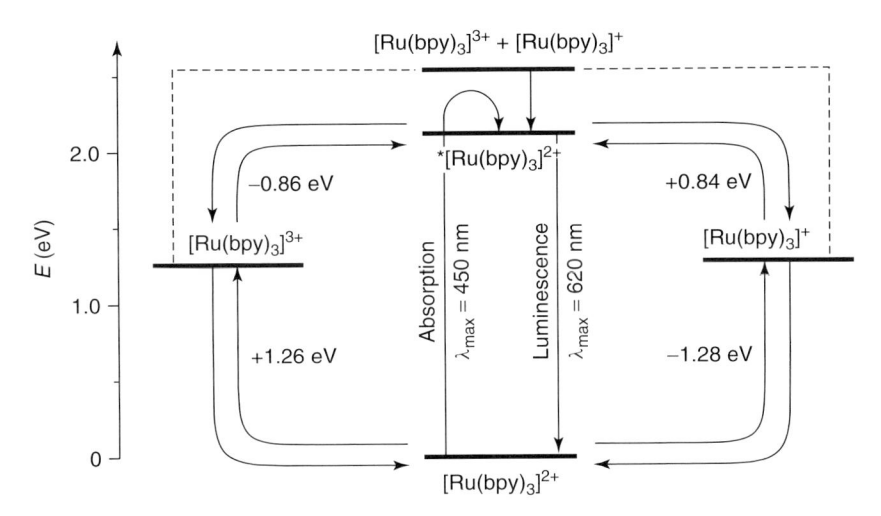

Figure 3.8 Simplified energy diagram illustrating the main spectroscopic and redox processes for [Ru(bpy)$_3$]$^{2+}$. The energy levels of the oxidized and reduced species are obtained from the redox potential values referred to the SCE electrode. The diagram also shows that the reaction between the one-electron oxidized and reduced forms of this species is sufficiently exoergonic to generate luminescence. (Reprinted from Ref. [14o], with permission from Elsevier.)

oxidation and reduction processes to yield [Ru(bpy)$_3$]$^{3+}$ and [Ru(bpy)$_3$]$^+$, respectively, characterized by distinct absorption spectra; and (iv) generating the triplet excited state via the comproportionation reaction between the oxidized [Ru(bpy)$_3$]$^{3+}$ and the reduced [Ru(bpy)$_3$]$^{3+}$ species [40], thereby enabling electrochemiluminescence generation (Figure 3.8). The starting state is [Ru(bpy)$_3$]$^{2+}$; the system is reset to this state (if required, by application of a potential value of 0 V) before each input operation.

For the encoder function (Figure 3.7a), the four inputs are oxidation at +1.4 V versus the SCE (In$_1$), low-power excitation at 450 nm (In$_2$), reduction at −1.4 V (In$_3$), and alternating oxidation/reduction by a square wave potential oscillating between +1.4 and −1.4 V (In$_4$). The two outputs are coded for by absorbance at 530 nm (Out$_1$) and luminescence intensity at 620 nm (Out$_2$). Application of In$_1$ (+1.4 V) converts [Ru(bpy)$_3$]$^{2+}$ into [Ru(bpy)$_3$]$^{3+}$, which exhibits neither absorption at 530 nm (Out$_1$ = {0}) nor emission at 620 nm (Out$_2$ = {0}). In$_2$, which involves low-power excitation of [Ru(bpy)$_3$]$^{2+}$ at 450 nm, causes no change in absorbance (Out$_1$ = {0}) and produces 620 nm luminescence (Out$_2$ = {1}). Application of In$_3$ (−1.4 V) converts [Ru(bpy)$_3$]$^{2+}$ into [Ru(bpy)$_3$]$^+$, which absorbs at 530 nm (Out$_1$ = {1}) and does not emit (Out$_2$ = {0}). Finally, application of In$_3$ (alternating +1.4 and −1.4 V) leads to an electrostationary state containing [Ru(bpy)$_3$]$^{2+}$, [Ru(bpy)$_3$]$^{3+}$, [Ru(bpy)$_3$]$^+$, and *[Ru(bpy)$_3$]$^{2+}$. On optimization of the experimental conditions, an absorption at 530 nm higher than the fixed threshold (Out$_1$ = {1}) and emission at 620 nm (Out$_1$ = {1}) can be obtained.

For the 2-to-4 decoder function (Figure 3.7b), the two inputs are oxidation at +1.4 V (In$_1$) and reduction at −1.4 V (In$_2$); the outputs are coded for by the absorbance at 450 nm (Out$_1$), 310 nm (Out$_2$), and 530 nm (Out$_3$) and the emission at 620 nm (Out$_4$). If no input is applied (In$_1$ and In$_2$ = {0}), the solution exhibits the absorption band at 450 nm (Out$_1$ = {1}), while all other output channels are in state {0}. Activation of In$_1$ (+1.4 V) converts [Ru(bpy)$_3$]$^{2+}$ into [Ru(bpy)$_3$]$^{3+}$, which shows substantial absorption at 310 nm (Out$_2$ = {1}) and neither absorption at 450 and 530 nm nor emission at 620 nm (Out$_1$, Out$_3$, and Out$_4$ = {0}). Application of In$_2$ (−1.4 V) generates [Ru(bpy)$_3$]$^+$, which has an absorbance higher than the threshold at 530 nm (Out$_3$ = {1}) but lower than the threshold at 450 and 310 nm, and no emission at 620 nm (Out$_1$, Out$_2$, and Out$_4$ = {0}). Finally, concomitant oxidation at +1.4 V and reduction at −1.4 V of [Ru(bpy)$_3$]$^{2+}$ with a bipotentiostat (In$_1$ and In$_2$ = {1}) produces an electrostationary state that contains [Ru(bpy)$_3$]$^{2+}$, [Ru(bpy)$_3$]$^{3+}$, [Ru(bpy)$_3$]$^+$, and *[Ru(bpy)$_3$]$^{2+}$. Again, by adjusting the experimental conditions, absorbance values at 450, 310, and 530 nm below the respective threshold can be obtained (Out$_1$, Out$_2$, and Out$_3$ = {0}), together with an emission at 620 nm, which is continuously produced by annihilation of the electrogenerated [Ru(bpy)$_3$]$^{3+}$, [Ru(bpy)$_3$]$^+$ species (Out$_4$ = {1}).

As for the previous example, it is worthwhile to note that this logic system is based on an unsophisticated molecule, it is reconfigurable in situ without addition of chemical reagents, and it can be cycled without formation of waste products.

3.2.5
All-Optical Integrated Logic Operations Based on Communicating Molecular Switches

In most molecular logic gates, the input and output signals have a different physical nature; typically, they operate on ionic or molecular inputs and provide a photonic output, as described in the two previous sections. This input/output inhomogeneity is a severe problem if complex circuits are to be created because it prevents the interconnection of basic elements [21b, 41]. In contrast, serial connection of logic gates, also termed *cascading* – namely, the output of an upstream gate sent as the input for a downstream gate – is extensively used in semiconductor circuits [4]. Molecular logic devices that use optical input and output signals [14a,e,h, 42] are particularly interesting in this regard, and also because (i) access for chemicals or wires is not required, (ii) no waste products are formed on repeated cycling of the device, (iii) operation in rigid or semirigid media is possible, and (iv) the multichannel nature of light can be exploited to configure the device for different logic functions. Prototypical examples of this type of molecular logic gates based on the control of porphyrin fluorescence by covalently linked photochromic units have been reported [14e, 42d].

We reported a different approach to devise all-optical molecular logic systems [43]. This strategy is based on the coupled operation of two molecular switches, namely, (i) a photoacid that transduces optical inputs into ionic outputs (Sw1) and (ii) a switch that responds to ionic inputs and provides optical outputs (Sw2). The two switching units communicate with one another by chemical signals [44] and

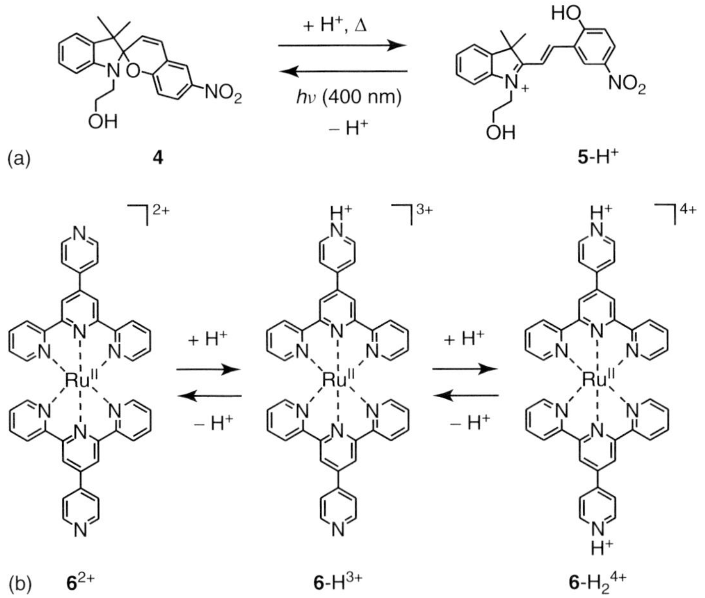

Figure 3.9 (a) The acid- and light-controlled equilibrium between the spiropyran **4** and the protonated merocyanine **5-H**$^+$ (Sw1). (b) The acid-base-driven interconversion between the three protonation states of metal complex **6**$^{2+}$ (Sw2).

are connected serially, that is, the photogenerated ionic output of Sw1 is the input of Sw2. The concept was realized with the system shown in Figure 3.9. Sw1 is based on a colorless spiropyran derivative, **4**, which is converted into the yellow protonated merocyanine form **5-H**$^+$ in acid solution [45]. On irradiation with light of 400 nm, **5-H**$^+$ isomerizes back to **4** and releases a proton into the solution. Sw2 is **6**$^{2+}$, a [Ru(tpy)$_2$]$^{2+}$ (tpy = 2,2′ : 6′,2″-terpyridine) complex having 4-pyridyl substituents in the 4″-position of each tpy ligand. In solution, this complex can exist in three different protonation states, namely, **6**$^{2+}$, **6-H**$^{3+}$, and **6-H**$_2^{4+}$. Spectroscopic titrations showed that these three forms (i) exhibit distinct and characteristic absorption spectra, luminescence spectra, and lifetimes, (ii) can be populated selectively because the two pendant pyridyl units of **6**$^{2+}$ can be sequentially protonated in two consecutive steps, and (iii) are reversibly interconverted by stoichiometric additions of acid and base [46]. Therefore, **6**$^{2+}$ behaves as an acid-base-controlled three-state luminescent switch. Because **4** and **5-H**$^+$ exhibit smaller and larger pK_a values than that of the pyridinium ion, respectively [47], the protonation state of the ruthenium complex could be controlled by light using **5-H**$^+$ as a photoacid.

As shown in Figure 3.10 (thick lines), irradiation of **5-H**$^+$ at 400 nm in acetonitrile such that 1 equiv. of protons is transferred to the metal complex generates the **6-H**$^{3+}$ state. Further irradiation of **5-H**$^+$ can cause the transfer of another equivalent of protons to the metal complex, thereby generating the **6-H**$_2^{4+}$ state (Figure 3.10, thin lines). Subsequent thermal equilibration in the dark regenerates the initial state

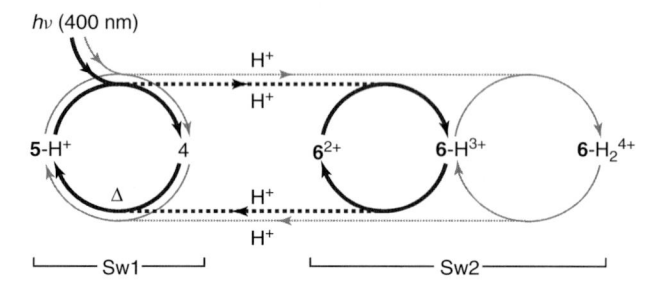

3 Binary Logic with Synthetic Molecular and Supramolecular Species

Figure 3.10 Coupled operation of the two-state switch Sw1 and three-state switch Sw2 shown in Figure 3.9 by means of light-induced proton exchange. Irradiation of 5-H$^+$ such that 1 equiv. of protons is transferred to the metal complex generates the 6-H^{3+} state (thick lines). Further irradiation of 5-H$^+$ can cause the transfer of another equivalent of protons to the metal complex, thereby generating the 6-H$_2^{4+}$ state (thin lines). Subsequent thermal equilibration in the dark regenerates the initial state by reverse proton exchange.

by reverse proton exchange. Several irradiation/thermal equilibration cycles can be repeated on the same solution without appreciable degradation of the components.

The analysis of the behavior of this system in terms of binary logic shows that all-optical two-input AND, OR, and XNOR functions are obtained (Figure 3.11 and Table 3.4). Both inputs are coded for by irradiation with 400 nm light. The input string {0,0} corresponds to dark conditions, whereas the input strings {1,0} and {0,1} correspond to irradiation with the amount of 400 nm light needed to transfer

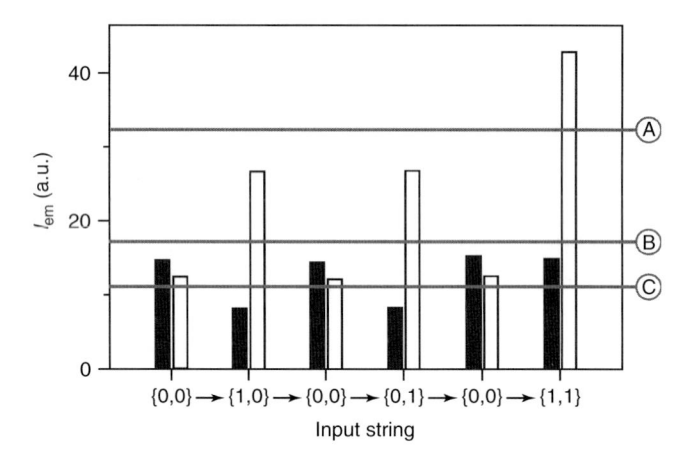

Figure 3.11 Response of the Sw1/Sw2 device to different input strings and cycling. Empty and filled bars represent the experimental luminescence intensities at 732 nm (Out$_1$) and 626 nm (Out$_2$), respectively. A, B, and C represent the threshold values for the AND, OR, and XNOR operations, respectively (Table 3.4). The experimental conditions are 48 µM 5-H$^+$ and 24 µM 6^{2+}, CH$_3$CN, room temperature; $\lambda_{ex} = 493$ nm. Input conditions: {0,0}, 10 h at 318 K; {1,0} and {0,1}, irradiation at 400 nm for 15 min; {1,1}, exhaustive irradiation at 400 nm.

Table 3.4 Truth table of the logic operations performed by the system Sw1/Sw2 shown in Figures 3.9 and 3.10.

In$_1$ (400 nm)^a	In$_2$ (400 nm)^a	Out$_1$ (732 nm)^{b,c}	Out$_1'$ (732 nm)^{b,d}	Out$_2$ (626 nm)^{b,e}
0	0	0	0	1
0	1	0	1	0
1	0	0	1	0
1	1	1	1	1
		↓	↓	↓
		AND	OR	XNOR

Experimental conditions are those described in Figure 3.11.
^{a}Binary state {1} corresponds to irradiation with the amount of 400 nm light needed to generate 1 equiv. of protons with respect to the metal complex; state {0} corresponds to no irradiation.
^{b}The actual values of the luminescence intensity and thresholds are indicated in Figure 3.11.
^{c}Binary states determined by applying threshold A.
^{d}Binary states determined by applying threshold B.
^{e}Binary states determined by applying threshold C.

1 equiv. of protons from **5**-H$^+$ to **6**$^{2+}$ so that the complex **6**-H^{3+} is formed. It should be noted that the molecular system cannot distinguish between these input strings but the operator does, because the two light inputs can be supplied by two physically independent channels. The input string {1,1} corresponds to irradiation with a dose of 400 nm photons such that 2 equiv. of protons are transferred from **5**-H$^+$ and the complex **6**-H$_2^{4+}$ is obtained. The optical output is represented by the luminescence intensity of the metal complex upon excitation at 493 nm. If the emission is monitored at 732 nm (Out$_1$), application of an appropriate threshold enables emulation of the AND operation; the OR function can also be obtained by applying a different threshold (Out$_1'$, Table 3.4). The XNOR function can be performed by reconfiguring the monitored emission wavelength to 626 nm (Out$_2$). The same approach was employed to control the luminescence of the Os(II) analog of **6**$^{2+}$ and the photoinduced generation of singlet oxygen [48].

3.3
Sequential Logic Circuits

3.3.1
Basic Concepts

In contrast with combinational circuits (Section 3.2), whose outputs are determined only by the current state of the inputs, sequential circuits are logic networks whose output depends on both the *current* and *past* values of the inputs [4]. In other words, combinational circuits such as adder, subtractor, MUX/DEMUX, and encoder/decoder combine the current input states in some way to produce the output and hence have no memory, whereas sequential circuits use the sequence of the input states over time to determine the output. In fact, memory circuits are inherently sequential. In logic design, sequentiality is achieved by introducing feedback loops, that is, the output of a certain logic gate is sent back as one of its inputs or as the input of another upstream gate. This feedback mechanism disrupts the unidirectional information flow that is a fundamental characteristic of combinational circuits. Under a chemical point of view, the time variable is put into the game by exploiting the different rates of the processes involved in the gate operation. In chemical terms, one could say that combinational logic can be achieved with molecules by relying on equilibrium states, whereas sequential logic requires also the analysis (and possibly optimization) of the kinetic behavior of the system. Examples of synthetic chemical systems whose properties are interpreted in terms of sequential logic operation have started to appear only in the past few years [26, 49].

3.3.2
Memory Effect in Communicating Molecular Switches

In the chemical system $5\text{-H}^+/6^{2+}$ (Figures 3.9 and 3.10, Section 3.2.5), under the conditions employed, the switching by light irradiation is complete in about 1 h, whereas several days are required for the dark equilibration process [43]. These features can be used to implement memory effects typical of sequential logic circuits. Figure 3.12 shows the luminescence intensity at 732 nm (Out) as

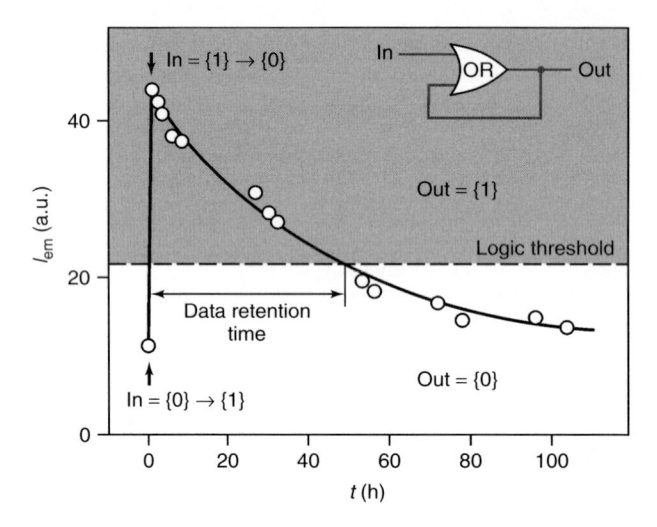

Figure 3.12 Time-dependent changes of the luminescence intensity at 732 nm ($\lambda_{ex} = 493$ nm) for the Sw1/Sw2 device (Figures 3.9 and 3.10) upon switching the light irradiation on (In = {1}) and off (In = {0}). The solid line is a guide to the eye, not a fit. The dashed line shows the threshold value chosen to determine the binary state of Out. The logic behavior of this system corresponds to the sequential circuit displayed in the top right and is summarized by the truth table shown in Table 3.5.

Table 3.5 Truth table of the sequential logic function performed by the system Sw1/Sw2 (Figures 3.9 and 3.10) in the experiment described in Figure 3.12.

Current input	Previous input[b]	Out[c]
In_{curr}[a]	In_{prev}[a]	
0	0	0
0	1	1
1	0	1
1	1	1

[a] Binary state {1} corresponds to irradiation with the amount of 400 nm light needed to generate 2 equiv. of protons with respect to the metal complex; state {0} corresponds to no irradiation.
[b] Input applied within a previous period of time corresponding to the data retention time.
[c] The actual values of the luminescence intensity and threshold are indicated in Figure 3.12.

a function of time in an irradiation/equilibration experiment. The output can be rapidly switched from {0} to {1} by activating the input (light irradiation); however, when the input is changed back to {0}, the output remains above the chosen logic threshold for about 50 h (Figure 3.12). Hence, the written information is retained for a rather long time after removal of the writing input, thus determining a memory effect.

The equivalent logic circuit and truth table corresponding to this behavior are, respectively, shown in Figure 3.12 and Table 3.5. If the input state is {1} (system under light irradiation), the output can only be {1} because the emissive Ru complex **6**-H_2^{4+} is formed. However, if the input state is {0} (no light irradiation), the state of the output within the data retention time cannot be solely determined from the current input state. In fact, as shown in Figure 3.12 and Table 3.5, the output is {0} if the input in the previous data retention time period was {0}, that is, the system has been given enough time to equilibrate in the dark, yielding 6^{2+} and **5**-H^+. Conversely, if in the previous data retention period the input was switched to {1}, the current output is {1} because **6**-H_2^{4+} and **4** have been formed, and their disappearance to give 6^{2+} and **5**-H^+ is very slow. Therefore, the system not only responds to the current input state but also "remembers" what happened in a previous period of time corresponding to the data retention time.

3.3.3
A Molecular Keypad Lock

As discussed above, sequential logic differs fundamentally from combinational logic, in which either input may be applied first. The keypad lock is a typical example of a sequential logic device, as the lock opens (output switches on) only

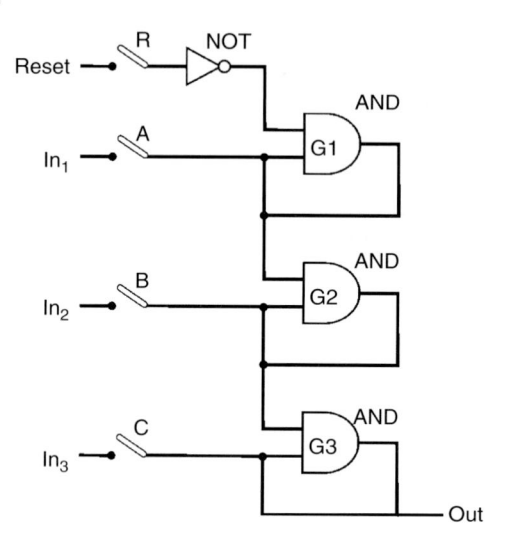

Figure 3.13 Sequential logic circuit for a keypad lock, whose output is switched on by three input signals supplied in a defined sequence ($In_1 \rightarrow In_2 \rightarrow In_3$). The activation of In_1 by pushing key A locks the output of gate G1 on and enables gate G2. When key B is pressed, In_2 is activated: the output of gate G2 is locked on, and gate G3 is enabled. Finally, on activation of In_3 by pushing key C, the output of gate G3 goes on and Out switches to 1. Reset is accomplished by pushing the R key.

on receiving the correct sequence of inputs; a simplified representation of the corresponding logic circuit is shown in Figure 3.13. The first important step toward the construction of a molecular version of the keypad lock is represented by the investigation on the Fe(III) complex 7^{2+} (Figure 3.14a). This compound comprises a fluorescein (FL) and a pyrene (PY) fluorophore connected by a linker that is also a ligand (siderophore) capable of binding ferric ions strongly and selectively [50, 51]. In ethanol, 7^{2+} contains the FL moiety in its monoanionic form and most likely exhibits a folded structure in which the FL and PY moieties can approach each other. The inputs are coded for by a chelating agent for Fe(III) (EDTA, In_1), a basic reactant (sodium acetate, In_2), and UV irradiation (In_3), whereas the output channel is identified as the FL emission at 525 nm.

In the starting state, the emission of the FL anion moiety is quenched by the metal ion (Figure 3.14a). Addition of EDTA not only extracts the iron ion from the siderophore but also protonates the FL unit, transforming it to the nonemissive neutral state (Figure 3.14b). Subsequent addition of acetate ions leads to the formation of the FL dianion, which is strongly fluorescent (Figure 3.14c). Alternatively, addition of acetate ions to 7^{2+} causes the formation of the FL dianion, whose fluorescence is quenched by the still bound metal ion (Figure 3.14d). It is clear that inverting the order of the inputs (first base, then EDTA) leads ultimately to the same state, resulting in strong emission at 525 nm (Figure 3.14c). However, as the extraction of a ferric ion from a siderophore by EDTA is inhibited in a basic environment, a substantial difference in the reaction rate between

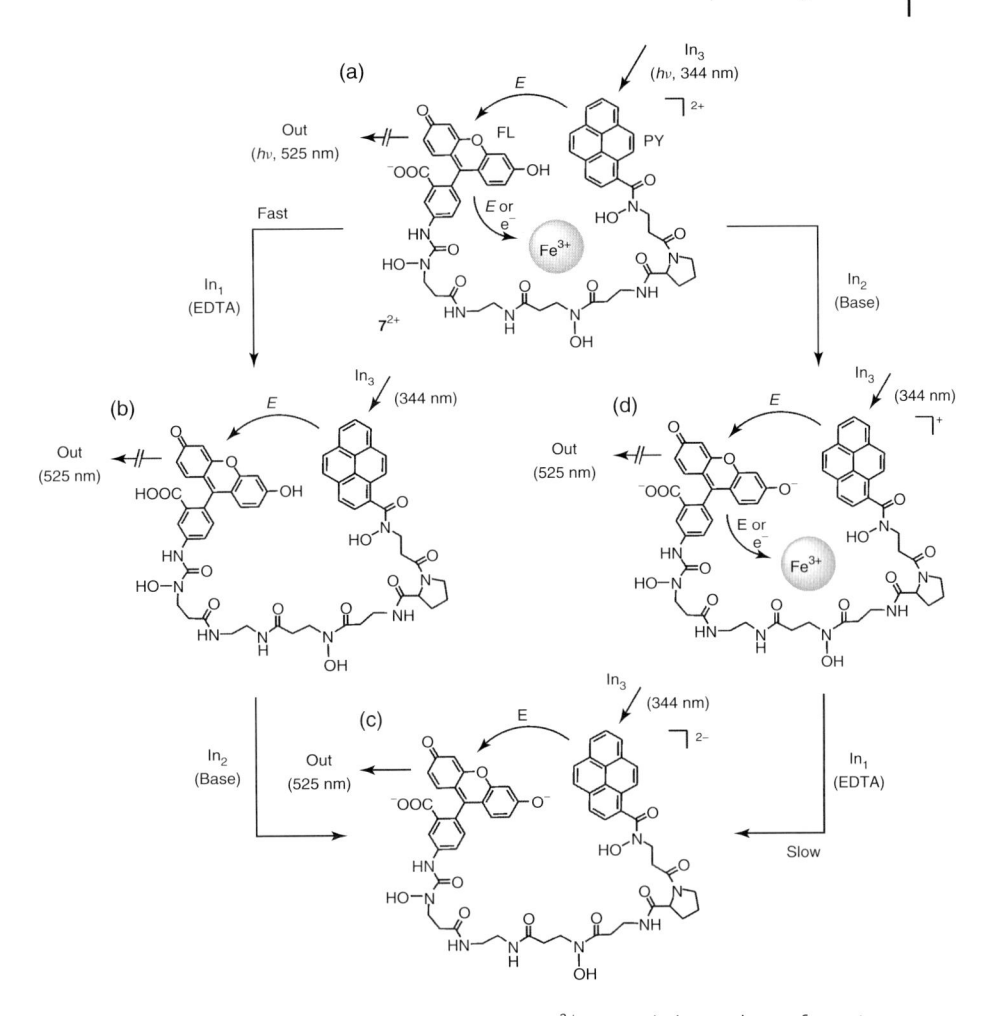

Figure 3.14 Structural formula of the Fe(III) complex 7^{2+} (a) and chemical transformations induced by addition of EDTA (b), base (d), and both (c). This molecular device can recognize the correct sequence of three input signals (EDTA, base, and 344 nm excitation).

the two paths (a → b → c and a → d → c) is observed. On the other hand, no photoluminescence can be monitored without light excitation. this provides the basis for the third input signal, namely, irradiation at 344 nm. Light of this wavelength is mainly absorbed by the PY unit, which then sensitizes the FL luminescence by energy transfer. Therefore, by reading the response of the device not later than a few minutes after activation of the inputs, fluorescence at 525 nm is observed only if the correct sequence In_1 (EDTA), In_2 (base), and In_3 (UV excitation) is supplied. Any other input combination leads to weak or no fluorescence. In other words, activation of In_3 generates an output only if the current state of In_1 and In_2 is {1,1} and the previous state was {1,0}, showing the sequential behavior of the system.

3.3.4
A Set–Reset Memory Device Based on a Copper Rotaxane

Catenanes and rotaxanes based on Cu(I)-polypyridine complexes represent an important class of interlocked compounds [52] that have revealed interesting as molecular switches and machines [12, 53]. An example of such compounds is rotaxane 8^+ (Figure 3.15), composed of an axle containing a bidentate ligand and two bulky stopper groups, and a wheel containing a bidentate (1,10-phenanthroline) and a tridentate (2,2′ : 6′,2″-terpyridine) ligand to which Cu^+ ions can bind [54]. The electrochemical oxidation of Cu(I) to Cu(II) results in the reorganization of the tetracoordinated complex 8^{2+} (4) into a pentacoordinated one, 8^{2+} (5), via the pirouetting of the wheel around the axle. Cyclic voltammetric experiments showed that the process is reversible and were employed to determine the rotation rates (Figure 3.15). The ability of rotaxane 8^+ to undergo reversible geometrical reorganization between the two different metal oxidation states makes it an attractive candidate for implementing a set–reset memory device [55].

A set–reset device is a finite-state machine that starts in an initial state and processes inputs one at a time; given the current state and input, the new state and output are determined by a transition scheme as depicted in Figure 3.16 [4, 5]. The two stable states, 8^+ (4) and 8^{2+} (5), are shown in squares and correspond to logic {0} and {1}, respectively.

An essential feature of the operation of this system is the hysteresis due to the nature of the reorganization of the coordination about the Cu ion following redox stimulation [56]. Such a phenomenon leads to a large difference between the potential values for oxidation and reduction of the copper ion (Figure 3.15) and enables distinguishing the set and reset transitions. The set–reset implementation requires three potential values to be defined: set, reset, and do nothing. The set potential was taken as +1.0 V (vs an Ag quasi-reference electrode), a value at which 8^{2+} (5) is the stable species. Conversely, the reset potential was fixed at −0.4 V, a value at which the stable species is 8^+ (4). The do-nothing potential has to be taken in the region of bistability; for example, at +0.2 V the device is in state {0} [8^+ (4)] for the oxidation process and in state {1} [8^{2+} (5)] for the reduction process. As shown in Figure 3.16, application of a set input (+1.0 V) switches the device from state {0} to {1}; however, if the device is already in state {1} the set input has no effect. Application of a reset input (−0.40 V) switches the device from state {1} to {0} but has no effect if the device is already in state {0}. The do-nothing potential (+0.2 V) does not change the state irrespective of whether it is {0} or {1}. The state of the device can be read by monitoring the voltammetric current in response to the applied potential input or by spectroscopic measurements. Clearly, this system operates as a memory because an information bit can be written by applying a set potential and remains memorized even when the writing potential is removed. The written data can be erased by applying a reset potential pulse. Owing to its chemical reversibility, the device can be cycled a large number of times.

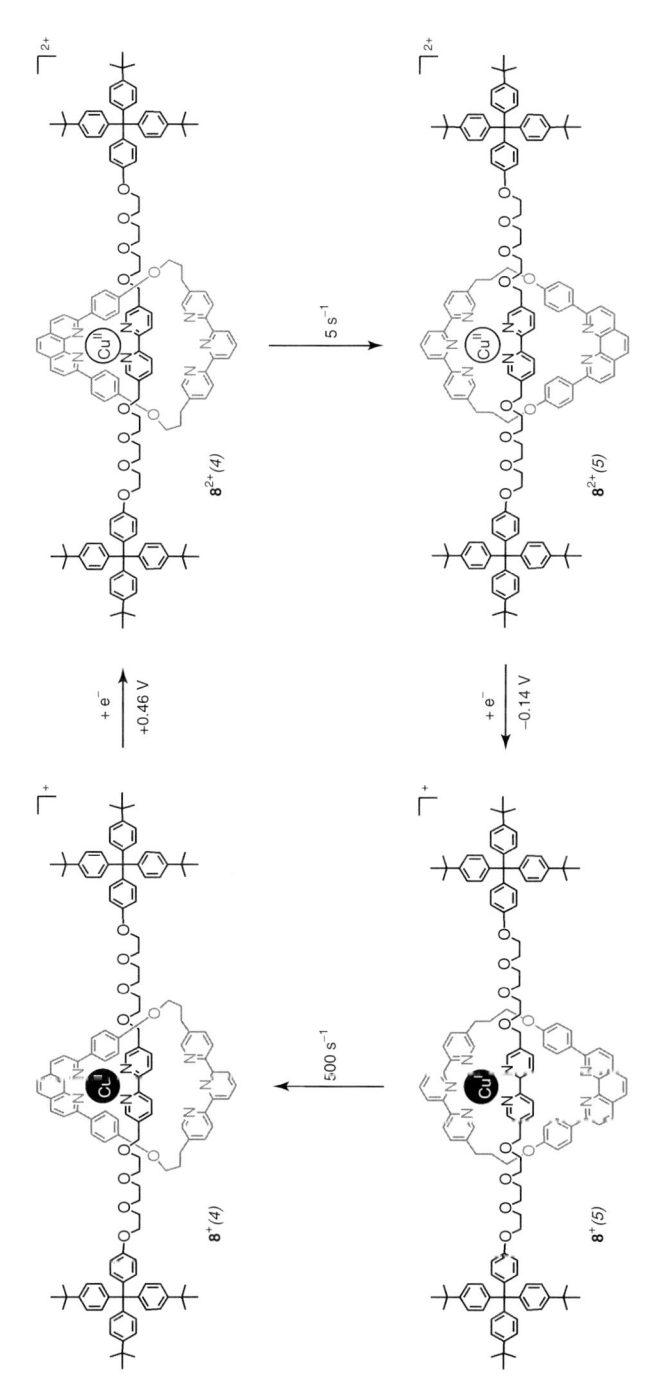

Figure 3.15 The electrochemical switching of the redox states in the Cu rotaxane 8^+ (horizontal reactions) is followed by ligand reorganization (vertical reactions). Peak potential values for the redox processes (in V vs an Ag quasi-reference electrode) and rate constants for the intramolecular rearrangements in acetonitrile at room temperature are reported. (Reprinted from Ref. [14o], with permission from Elsevier.)

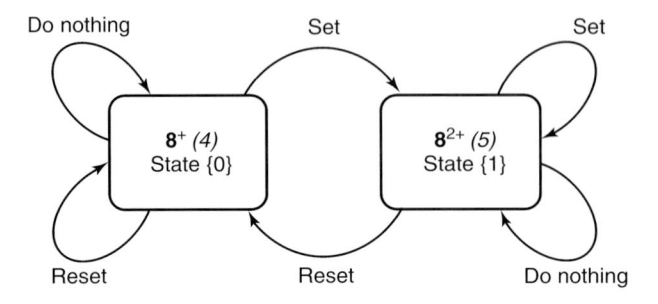

Figure 3.16 Operation of a set–reset device based on the electrochemical switching and ligand reorganization processes represented in Figure 3.15 for rotaxane 8^+. See text for details. (Reprinted from Ref. [14o], with permission from Elsevier.)

3.4
Summary and Outlook

We have shown that multicomponent molecular systems operating in solution and using chemical or photonic input/output signals are capable of executing advanced switching and logic functions, if appropriately designed and constructed. Obviously, the development of chemical computers 28d] – that is, relying on molecule-based information processing – that can rival with current semiconductor devices in terms of computing power is a very ambitious objective even for basic research. It is therefore more realistic to point out that such kind of molecular devices could be interesting for specific applications in fields such as diagnostics, medicine, and materials science. In fact, in these areas there are problems that can be solved by simple computations that are at hand for current molecular processors, and need to be addressed in places where a silicon-based computer cannot go (e.g., inside a cell or in a membrane) [57]. Examples include the encoding [58] and networking [59] of microscopic objects, parallel chemical analyses in microfluidic systems [60], control of the function of biomolecules [61], screening of receptors [62] and diagnostic markers [63], and intelligent drug delivery *in vitro* [64] and *in vivo* [65].

While molecule-based logic systems in solution are useful for proof-of-principle studies and for some applications (e.g., sensing, drug delivery), they are unpractical for integration into real devices that can be addressed through an external interface. Moreover, compatibility with microelectronic circuits would be required in order to develop a hybrid technology that would constitute a viable alternative to continue pushing the top-down fabrication techniques toward their physical limits [8, 9]. In such a context, the development of molecule-based logic gates that function in heterogeneous environments, for example, on surfaces or at interfaces, is an important task [66]. A first step in this direction is represented by the construction of suitably functionalized electrodes and the analysis of their behavior in terms of Boolean logic. For example, electrodes derivatized with chemical compounds are extensively used in solar cells, display devices, and switching applications, but their potential as logic systems has been recognized only recently [49e,f, 67, 68].

Finally, we would like to emphasize that many logic devices described in this article are based on simple chemical compounds and well-documented physicochemical processes taking place in solution, and the methods used to obtain the illustrated results rely on the simultaneous stimulation of large numbers of molecules. From this viewpoint, one could argue that we have simply illustrated some types of chemical reactions. We have tried to stress, however, the novel conceptual interpretation of the observed processes. Regardless of the possibility of short-term applications as discussed above, we believe that research in this field is interesting for several reasons: (i) at least in principle, some of the described effects can be scaled down to single molecules; (ii) synthetic multistate–multifunctional systems may play the role of models to begin understanding the chemical basis of complex biological processes; (iii) integration of molecular-level devices may be successfully achieved by intermolecular communication based on chemical and light signals, thereby overcoming the difficulty in establishing electrical communication between different molecules; and (iv) last, but not least, these studies introduce new concepts in the "old" field of chemistry and stimulate the ingenuity of research workers engaged in the bottom-up approach to nanotechnology.

Acknowledgments

Financial support from MIUR (PRIN 2008), Fondazione CARISBO, and the University of Bologna is gratefully acknowledged.

References

1. Balzani, V., Credi, A., and Venturi, M. (2008) *Chem. Eur. J.*, **14**, 26.
2. Goodsell, D.S. (2004) *Bionanotechnology – Lessons from Nature*, John Wiley & Sons, Inc., Hoboken, NJ.
3. Jones, R.A.L. (2004) *Soft Machines: Nanotechnology and Life*, Oxford, University Press, New York.
4. Mitchell, R.J. (1995) *Microprocessor Systems: An Introduction*, Macmillan, London.
5. Gregg, J.R. (1998) *Ones and Zeros: Understanding Boolean Algebra, Digital Circuits, and the Logic of Sets*, Wiley-IEEE Press, New York.
6. Pimentel, G.C. and Coonrod, J.A. (1985) *Opportunities in Chemistry*, National Academy of Sciences, National Academy Press, Washington, DC.
7. Lu, W. and Lieber, C.M. (2007) *Nat. Mater.*, **6**, 841.
8. Cerofolini, G.F. (2009) *Nanoscale Devices – Fabrication, Functionalization and Accessibility from the Macroscopic World*, Springer, Berlin.
9. The International Technology Roadmap for Semiconductors (ITRS) (2009) edition and 2010 update, www.itrs.net/reports.html (accessed August 2011).
10. Chau, R., Doyle, B., Datta, S., Kavalieros, J., and Zhang, K. (2007) *Nat. Mater.*, **6**, 810.
11. Feringa, B.L. (ed.) (2011) *Molecular Switches*, 2nd edn, Wiley-VCH Verlag GmbH, Weinheim.
12. Balzani, V., Credi, A., and Venturi, M. (2008) *Molecular Devices and Machines – Concepts and Perspectives for the Nanoworld*, 2nd edn, Wiley-VCH Verlag GmbH, Weinheim.
13. (a) Credi A. (Guest ed.) (2010) *Aust. J. Chem.*, **63** (2), 145–342, research

front on molecular logic; (b) Katz, E. (Guest ed.) (2011) *Isr. J. Chem.*, **51** (1), 1–164, special issue on molecular and biomolecular information-processing systems.

14. (a) Raymo, F.M. (2002) *Adv. Mater.*, **14**, 401; (b) Steinitz, D., Remacle, F., and Levine, R.D. (2002) *ChemPhysChem*, **3**, 43; (c) Balzani, V., Credi, A., and Venturi, M. (2003) *ChemPhysChem*, **3**, 49; (d) de Silva, A.P. and McClenaghan, N.D. (2004) *Chem. Eur. J.*, **10**, 574; (e) Gust, D., Moore, T.A., and Moore, A.L. (2006) *Chem. Commun.*, **8**, 1169; (f) de Silva, A.P. and Uchiyama, S. (2007) *Nat. Nanotechnol.*, **2**, 399; (g) Pischel, U. (2007) *Angew. Chem. Int. Ed.*, **46**, 4026; (h) Szaciłowski, K. (2008) *Chem. Rev.*, **108**, 3481; (i) Willner, I., Shlyahovsky, B., Zayats, M., and Willner, B. (2008) *Chem. Soc. Rev.*, **37**, 1153; (j) Wagner, N. and Ashkenasy, G. (2009) *Chem. Eur. J.*, **15**, 1765; (k) Benenson, Y. (2009) *Mol. Biosyst.*, **5**, 675; (l) Wenger, O. (2009) *Coord. Chem. Rev.*, **253**, 1439; (m) Andréasson, J. and Pischel, U. (2010) *Chem. Soc. Rev.*, **39**, 174; (n) Katz, E. and Privman, V. (2010) *Chem. Soc. Rev.*, **39**, 1835; (o) Amelia, M., Zou, L., and Credi, A. (2010) *Coord. Chem. Rev.*, **254**, 2267.

15. (a) Klein, M., Rogge, S., Remacle, F., and Levine, R.D. (2007) *Nano Lett.*, **7**, 2795; (b) Ferreira, R., Remon, P., and Pischel, U. (2009) *J. Phys. Chem. C*, **113**, 5805; (c) de Ruiter, G., Motiei, L., Choudhury, J., Oded, N., and van der Boom, M. (2010) *Angew. Chem. Int. Ed.*, **122**, 4890.

16. Gentili, P.L. (2008) *J. Phys. Chem. A*, **112**, 11992.

17. Pina, F., Melo, M.J., Maestri, M., Passaniti, P., and Balzani, V. (2000) *J. Am. Chem. Soc.*, **122**, 4496.

18. Aviram, A. (1988) *J. Am. Chem. Soc.*, **110**, 5687.

19. de Silva, A.P., Gunaratne, H.Q.N., and McCoy, C.P. (1993) *Nature*, **364**, 42.

20. (a) de Silva, A.P. (2005) *Nat. Mater.*, **4**, 15; (b) Credi, A. (2007) *Angew. Chem. Int. Ed.*, **46**, 5472.

21. (a) Langford, S.J. and Yann, T. (2003) *J. Am. Chem. Soc.*, **125**, 11198; (b) Szaciłowski, K. (2004) *Chem. Eur. J.*, **10**, 2520; (c) Margulies, D., Melman, G., and Shanzer, A. (2005) *Nat. Mater.*, **4**, 768; (d) Margulies, D., Melman, G., and Shanzer, A. (2006) *J. Am. Chem. Soc.*, **128**, 4865.

22. Amelia, M., Baroncini, M., and Credi, A. (2008) *Angew. Chem. Int. Ed.*, **47**, 6240.

23. Ceroni, P., Bergamini, G., and Balzani, V. (2009) *Angew. Chem. Int. Ed.*, **48**, 8516.

24. de Silva, A.P. and McClenaghan, N.D. (2002) *Chem. Eur. J.*, **8**, 4935.

25. For an example of irreversible reconfiguration in a molecule-based electronic device, see: Collier, C.P., Wong, E.W., Behloradsky, M., Raymo, F.M., Stoddart, J.F., Kuekes, P.J., Williams, R.S., and Heath, J.R. (1999) *Science*, **285**, 391.

26. Andréasson, J., Pischel, U., Straight, S.D., Moore, T.A., Moore, A.L., and Gust, D. (2011) *J. Am. Chem. Soc.*, **133**, 11641.

27. Balzani, V., Credi, A., Langford, S.J., and Stoddart, J.F. (1997) *J. Am. Chem. Soc.*, **119**, 2679.

28. (a) Ball, P. (1997) *New Sci.*, **155** (2093), 32; (b) Leigh D.A. and Murphy, A. (1999) *Chem. Ind.*, **5**, 178; (c) Freemantle, M. (2000) *Chem. Eng. News*, **78** (18), 12; (d) Ball, P. (2000) *Nature*, **406**, 118.

29. Semeraro, M. and Credi, A. (2010) *J. Phys. Chem. C*, **114**, 3209.

30. For examples of molecular half-subtractors, see [21a] and: (a) Coskun, A., Deniz, E., and Akkaya, E.U. (2005) *Org. Lett.*, **7**, 5187; (b) Perez-Inestrosa, E., Montenegro, J.M., Collado, D., Suau, R., and Casado, J. (2007) *J. Phys. Chem. C*, **111**, 6409; (c) Guo, Z., Zhao, P., Zhu, W., Huang, X., Xie, Y., and Tian, H. (2008) *J. Phys. Chem. C*, **112**, 7047.

31. Landauer, R. (1961) *IBM J. Res. Dev.*, **5**, 183.

32. Remon, P., Ferreira, R., Montenegro, J.M., Suau, R., Perez-Inestrosa, E., and Pischel, U. (2009) *ChemPhysChem*, **10**, 2004.

33. Clemente-Leon, M., Pasquini, C., Hebbe-Viton, V., Lacour, J., Dalla Cort, A., and Credi, A. (2006) *Eur. J. Org. Chem.*, **1**, 105.

34. Andréasson, J., Straight, S.D., Bandyopadhyay, S., Mitchell, R.H., Moore, T.A., Moore, A.L., and Gust, D. (2007) *Angew. Chem. Int. Ed.*, **46**, 958.
35. Andréasson, J., Straight, S.D., Bandyopadhyay, S., Mitchell, R.H., Moore, T.A., Moore, A.L., and Gust, D. (2007) *J. Phys. Chem. C*, **111**, 14274.
36. Perez-Inestrosa, E., Montenegro, J.M., Collado, D., and Suau, R. (2008) *Chem. Commun.*, 1085.
37. Andréasson, J., Straight, S.D., Moore, T.A., Moore, A.L., and Gust, D. (2008) *J. Am. Chem. Soc.*, **130**, 11122.
38. Giansante, C., Ceroni, P., Venturi, M., Sakamoto, J., and Schluter, A.D. (2009) *ChemPhysChem*, **10**, 495.
39. Juris, A., Balzani, V., Barigelletti, F., Campagna, S., Belser, P., and von Zelewsky, A. (1988) *Coord. Chem. Rev.*, **84**, 85.
40. Tokel, N.E. and Bard, A.J. (1972) *J. Am. Chem. Soc.*, **94**, 2862.
41. Seelig, G., Soloveichik, D., Zhang, D.Y., and Winfree, E. (2006) *Science*, **314**, 1585.
42. (a) Lukas, A.S., Bushhard, P.J., and Wasielewski, M.R. (2001) *J. Am. Chem. Soc.*, **123**, 2440; (b) Giordani, S. and Raymo, F.M. (2002) *Proc. Natl. Acad. Sci. U.S.A.*, **99**, 4941; (c) Kuznetz, O., Salman, H., Shakkour, N., Eichen, Y., and Speiser, S. (2008) *Chem. Phys. Lett.*, **451**, 63; (d) Keirstead, A.E., Bridgewater, J.W., Terazono, Y., Kodis, G., Straight, S., Liddell, P.A., Moore, A.L., Moore, T.A., and Gust, D. (2010) *J. Am. Chem. Soc.*, **18**, 6588.
43. Silvi, S., Constable, E.C., Housecroft, C.E., Beves, J.E., Dunphy, E.L., Tomasulo, M., Raymo, F.M., and Credi, A. (2009) *Chem. Eur. J.*, **15**, 178.
44. Bai, Y.C., Zhang, C., Fang, C.J., and Yan, C.H. (2010) *Chem. Asian J.*, **5**, 1870.
45. Raymo, F.M., Giordani, S., White, A.J.P., and Williams, D.J. (2003) *J. Org. Chem.*, **68**, 4158.
46. Constable, E.C., Housecroft, C.E., Thompson, A.C., Passaniti, P., Silvi, S., Maestri, M., and Credi, A. (2007) *Inorg. Chim. Acta*, **360**, 1102.
47. Silvi, S., Arduini, A., Pochini, A., Secchi, A., Tomasulo, M., Raymo, F.M., Baroncini, M., and Credi, A. (2007) *J. Am. Chem. Soc.*, **129**, 13378.
48. Silvi, S., Constable, E.C., Housecroft, C.E., Beves, J.E., Dunphy, E.L., Tomasulo, M., Raymo, F.M., and Credi, A. (2009) *Chem. Commun.*, 1484.
49. (a) Raymo, F.M., Alvarado, R.J., Giordani, S., and Cejas, M.E. (2003) *J. Am. Chem. Soc.*, **125**, 2361; (b) Guo, Z.Q., Zhu, W.H., Shen, L.J., and Tian, H. (2007) *Angew. Chem. Int. Ed.*, **46**, 5549; (c) Sun, W., Zhou, C., Xu, C.H., Fang, C.J., Zhang, C., Li, Z.X., and Yan, C.H. (2008) *Chem. Eur. J.*, **14**, 6342; (d) Andréasson, J., Straight, S.D., Moore, T.A., Moore, A.L., and Gust, D. (2009) *Chem. Eur. J.*, **15**, 3936; (e) de Ruiter, G., Tartakovsky, E., Oded, N., and van der Boom, M.E. (2010) *Angew. Chem. Int. Ed.*, **49**, 169; (f) de Ruiter, G., Motiei, L., Choudhury, J., Oded, N., and van der Boom, M.E. (2010) *Angew. Chem. Int. Ed.*, **49**, 4780.
50. Margulies, D., Felder, C.E., Melman, G., and Shanzer, A. (2007) *J. Am. Chem. Soc.*, **129**, 347.
51. Margulies, D., Melman, G., Felder, C.E., Arad-Yellin, R., and Shanzer, A. (2004) *J. Am. Chem. Soc.*, **126**, 15400.
52. Dietrich-Buchecker, C.O. and Sauvage, J.P. (1987) *Chem. Rev.*, **87**, 795.
53. Champin, B., Mobian, P., and Sauvage, J.P. (2007) *Chem. Soc. Rev.*, **36**, 358.
54. Létinois-Halbes, U., Hanss, D., Beierle, J.M., Collin, J.P., and Sauvage, J.P. (2005) *Org. Lett.*, **7**, 5753.
55. Periyasamy, G., Collin, J.P., Sauvage, J.P., Levine, R.D., and Remacle, F. (2009) *Chem. Eur. J.*, **15**, 1310.
56. Periyasamy, G., Sour, A., Collin, J.P., Sauvage, J.P., and Remacle, F. (2009) *J. Phys. Chem. B*, **113**, 6219.
57. Uchiyama, S., Iwai, K., and de Silva, A.P. (2008) *Angew. Chem. Int. Ed.*, **47**, 4667.
58. de Silva, A.P., James, D.Y., McKinney, B.O.F., Pears, D.A., and Weir, S.M. (2006) *Nat. Mater.*, **5**, 787.
59. (a) von Maltzahn, G., Harris, T.J., Park, J.H., Min, D.H., Schmidt, A.J., Sailor, M.J., and Bhatia, S.N. (2007) *J. Am. Chem. Soc.*, **129**, 6064; (b) Yashin, R., Rudchenko, S., and Stojanovic, M.N. (2007) *J. Am. Chem. Soc.*, **129**, 15581;

(c) Motornov, M., Zhou, J., Pita, M., Gopishetty, V., Tokarev, I., Katz, E., and Minko, S. (2008) *Nano Lett.*, **8**, 2993.
60. Kou, S., Lee, H.N., van Noort, D., Swamy, K.M.K., Kim, S.H., Soh, J.H., Lee, K.M., Nam, S.W., Yoon, J., and Park, S. (2008) *Angew. Chem. Int. Ed.*, **47**, 872.
61. Muramatsu, S., Kinbara, K., Taguchi, H., Ishii, N., and Aida, T. (2006) *J. Am. Chem. Soc.*, **128**, 3764.
62. Kikkeri, R., Grünstein, D., and Seeberger, P.H. (2010) *J. Am. Chem. Soc.*, **132**, 10230.
63. Konry, T. and Walt, D.R. (2009) *J. Am. Chem. Soc.*, **131**, 13232.
64. (a) Benenson, Y., Gil, B., Ben-Dor, U., Adar, R., and Shapiro, E. (2004) *Nature*, **429**, 423; (b) Amir, R.J., Popkov, M., Lerner, R.A., Barbas, C.F. III, and Shabat, D. (2005) *Angew. Chem. Int. Ed.*, **44**, 4378; (c) Ozlem, S. and Akkaya, E.U. (2009) *J. Am. Chem. Soc.*, **131**, 48.
65. Rinaudo, K., Bleris, L., Maddamsetti, R., Subramanian, S., Weiss, R., and Benenson, Y. (2007) *Nat. Biotechnol.*, **25**, 795.
66. Credi, A. (2008) *Nat. Nanotechnol.*, **3**, 529.
67. (a) Biancardo, M., Bignozzi, C., Doyle, H., and Redmond, G. (2005) *Chem. Commun.*, 3918; (b) Wen, G., Yan, J., Zhou, Y., Zhang, D., Mao, L., and Zhu, D. (2006) *Chem. Commun.*, 3016; (c) Nitahara, S., Terasaki, N., Akiyama, T., and Yamada, S. (2006) *Thin Solid Films*, **499**, 354; (d) Szaciłowski, K., Macyk, W., and Stochel, G. (2006) *J. Am. Chem. Soc.*, **128**, 4550; (e) Furtado, L.F.O., Alexiou, A.D.P., Gonçalves, L., Toma, H.E., and Araki, K. (2006) *Angew. Chem. Int. Ed.*, **45**, 3143; (f) Pita, M. and Katz, E. (2008) *J. Am. Chem. Soc.*, **130**, 36.
68. (a) Gupta, T. and van der Boom, M.E. (2008) *Angew. Chem. Int. Ed.*, **47**, 2260; (b) Gupta, T. and van der Boom, M.E. (2008) *Angew. Chem. Int. Ed.*, **47**, 5322; (c) de Ruiter, G. and van der Boom, M.E. (2011) *Acc. Chem. Res.*, **44** (8), 563, doi: 10.1021/ar200002v.

4
Photonically Switched Molecular Logic Devices
Joakim Andréasson and Devens Gust

4.1
Introduction

Organic molecules have been identified as potential switching elements in Boolean logic constructs. Replacement of the components traditionally used for these purposes (silicon-based semiconductor materials) by individual molecules or small ensembles would imply a giant step forward for miniaturization, minimal power consumption, and more. Following the pioneering example by de Silva in 1993 [1], a large number of molecules that perform the function of Boolean logic devices have been reported and summarized in several review articles [2–8]. The function of the majority of these molecular constructs relies on the on–off switching of an optically detectable property (output), such as absorption or emission, rather than the voltage switching used in traditional hardware. The stimulus (inputs) used to trigger the spectral changes are most often the addition of chemical species, for example, metal ions, protons, and so on. Repeated operation will therefore lead to buildup of chemical waste and/or dilution. Moreover, it imposes limitations on both the operational speed and the nature of the surrounding medium, as fluid solution is required. Molecular logic elements with optical inputs and outputs, such as the photochromic systems described in this chapter, do not require access for chemicals or wires, can in principle operate on a much faster timescale and in rigid media, and can be cycled numerous times with no buildup of byproducts. Furthermore, all-photonic operation allows remote control, and it formally overcomes the input–output heterogeneity barrier. In the following sections, we describe our approach to the realization of molecular logic devices using photochromic (supra)molecules. First, a number of simple one-input switches are presented. The function of these constructs relies on the on–off switching of energy and electron transfer reactions. Next, the function of a selection of two-input Boolean molecular logic gates is described. Finally, the use of the concept of *functional integration* [9] to design molecules that perform the function of small logic circuits, such as half-adders/half-subtractors (HAs/HSs), multiplexers/demultiplexers (MUXs/DMUXs), encoders/decoders, sequential logic devices, and other memory functions are illustrated.

4.2
Photochromic Molecules

Photochromic molecules are molecular photoswitches that can be isomerized between two more or less thermally stable forms using light [10]. On isomerization, the absorption spectrum changes substantially. Most photochromes demonstrate positive photochromism, that is, the thermally stable form absorbs almost exclusively in the UV region. On UV exposure, this form isomerizes to a visible-light absorbing species. Heat and/or visible light triggers isomerization back to the UV-absorbing form. Figure 4.1 shows the structures and the isomerization schemes for a selection of compounds from different photochromic families that have all been used in

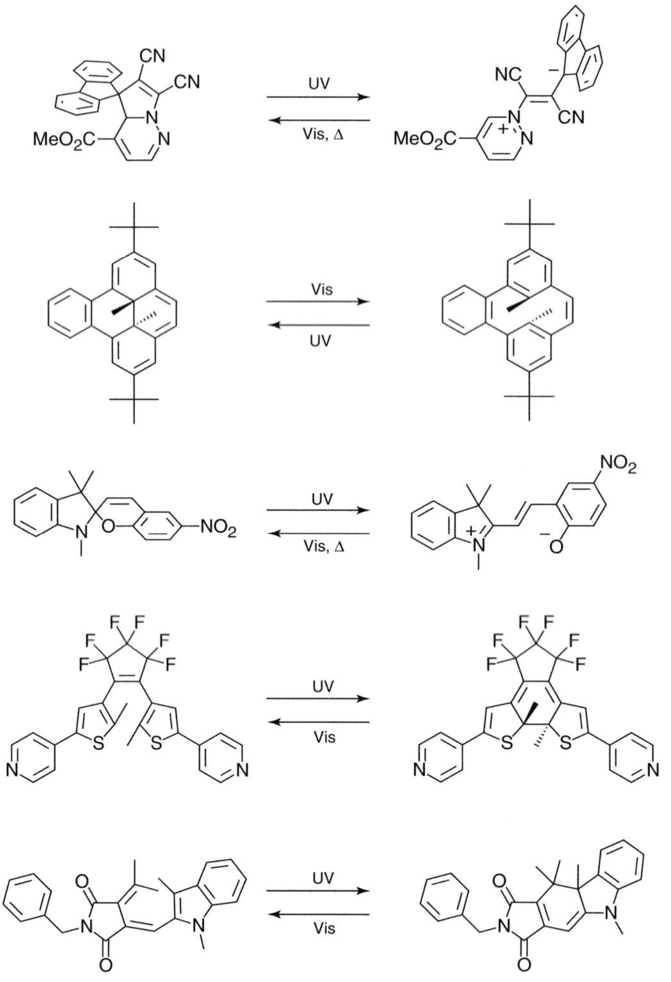

Figure 4.1 Structures and isomerization schemes of representative photochromic units used throughout the chapter.

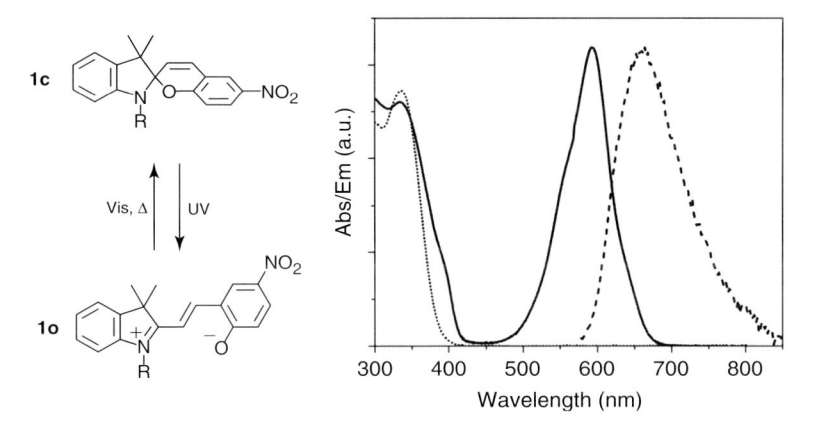

Figure 4.2 Structures and absorption spectra of a spiropyran unit. The closed spiro form **1c** (dotted line) is isomerized to the open merocyanine form **1o** (solid line) by exposure to UV light. The reverse reaction is triggered by visible light or heat. Shown is also the emission spectrum of **1o** (dashed line). R = $C_3H_6COOCH_3$.

our studies. The pronounced color change can be used for various applications; an example is photochromic eye wear. In addition to the color change, there are several other properties of the photoswitch that change on isomerization. The excited state properties, redox potentials, refractive index, and charge distribution are usually significantly different in the two forms. Figure 4.2 exemplifies both the color change and the structural rearrangement that follow isomerization of a photochromic molecule from the well-known spiropyran (SP) family (**1**). Also shown is the fluorescence spectrum of the open form **1o**. For digital logic applications, the absorbance at 590 nm as well as the emission intensity at 665 nm can be described in terms of 0 or 1 (*off* or *on*) for **1c** and **1o**, respectively. From this trivial example of on–off switching, it is obvious that photochromic molecules are potential candidates for optically controlled memory applications, which are discussed later in this chapter. The changes in the electronic properties that accompany the isomerization process of a photochromic unit may be used to switch also the emission intensity of an appended fluorescent chromophore. Most often, the fluorescence modulation relies on the on–off switching of energy and electron transfer reactions [11, 12].

4.3 Photonic Control of Energy and Electron Transfer Reactions

4.3.1 Energy Transfer

The emission of a fluorescent chromophore (energy donor) may be quenched by an energy acceptor if there is suitable spectral overlap between the donor emission and the acceptor absorption. As the absorption spectrum of most photochromic

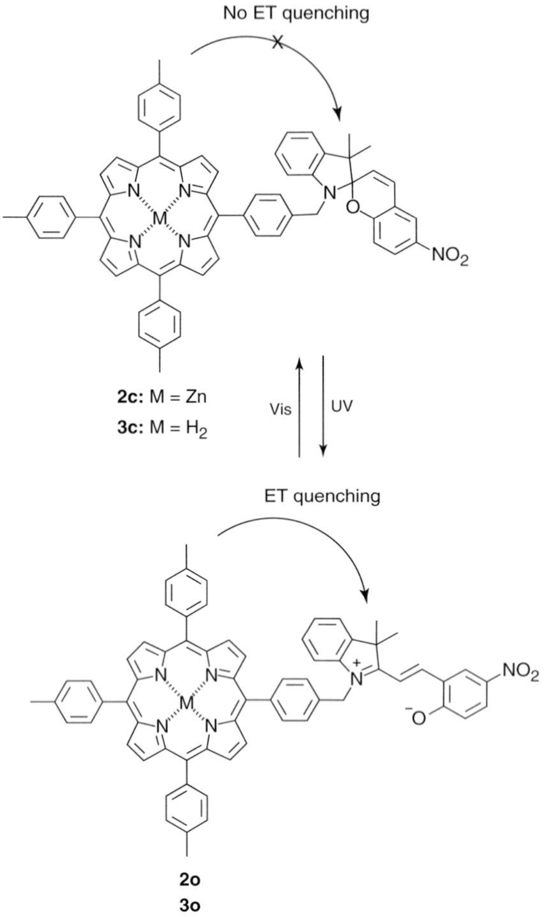

Figure 4.3 Structures of the dyads used for photochromic control of energy transfer reactions.

molecules changes dramatically on isomerization, the spectral overlap with the emission of an appended energy donor may experience pronounced changes. Given that one isomeric form displays a zero overlap, whereas the other form overlaps significantly, the energy transfer (ET) process is switched in a binary fashion, and the concomitant changes in the donor fluorescence intensity are conveniently monitored. This approach was used in the design of dyads **2** and **3** shown in Figure 4.3, whereby a photochromic SP (Figure 4.2) was attached to fluorescent tetraarylporphyrin reporter chromophores [13]. The emission spectra of the porphyrin units in **2** and **3** span the regions 580–740 nm and 625–770 nm, respectively. From Figure 4.2 it is obvious that only isomer **1o** displays absorption in these spectral regions and, as expected, the porphyrin emission was shown to be quenched in **2o** and **3o**, but remained virtually unaffected in **2c** and **3c**.

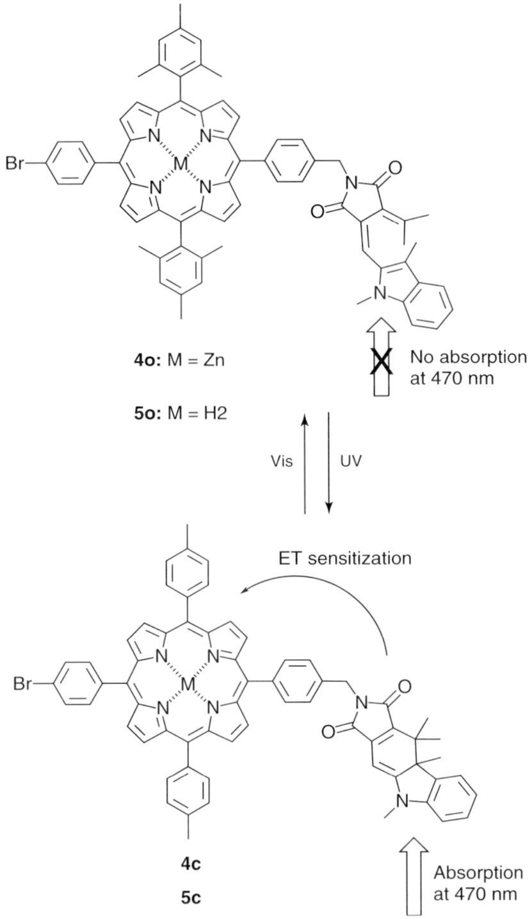

Figure 4.4 Structures of the photochromic dyads used for photoswitchable sensitization of the porphyrin excited state.

In the example described above, the modulation of the fluorescence intensity results from the isomerization of the SP between two forms with dramatically different energy acceptor properties. An alternative approach was taken in the design of dyads **4** and **5** shown in Figure 4.4, where a photochromic fulgimide (FG) was covalently attached to a fluorescent porphyrin reporter [14]. Here, light absorbed by the closed colored form of the FG in **4c** and **5c** is efficiently transferred to the porphyrin, that is, the porphyrin emission is sensitized by the excited state of the FG, which acts as the energy donor. Visible light exposure isomerizes the dyads to **4o** and **5o**, with the FG in the open, colorless form. With no absorption in the visible region, the FG can no longer act as an antenna for the 470 nm excitation light, and there is no sensitization of the porphyrin excited state through ET, resulting in a low emission intensity. The on–off switching of ET reactions

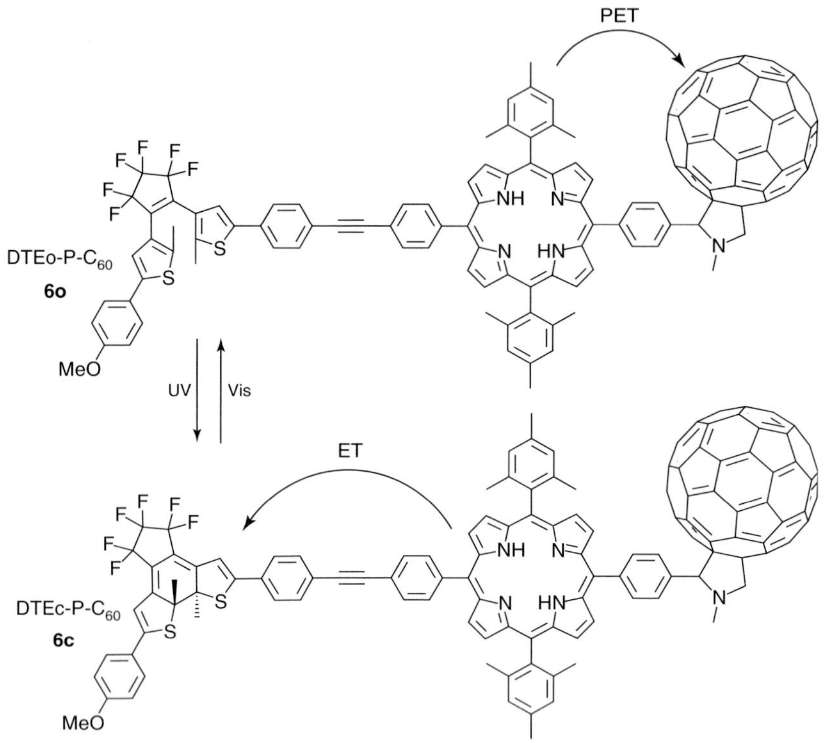

Figure 4.5 DTE-P-C_{60} triad used for photoinduced switching between energy- and electron transfer reactions.

can also be used to control the yield of a competing electron transfer reaction in the same supramolecule. Triad **6** shown in Figure 4.5 was synthesized for this purpose and consists of a photochromic dithienylethene (DTE) derivative linked to a porphyrin-C_{60} electron transfer motif (P-C_{60}) [15]. In the P-C_{60} dyad, the singlet excited state of P is efficiently quenched by photoinduced electron transfer (PET) to form $P^{\bullet +}$-$C_{60}^{\bullet -}$ with a rate constant of 25 ps and a yield of virtually unity. In the corresponding DTE-P dyad, the closed form (DTEc) of the photoswitch is the only quencher of the porphyrin excited state, as the absorption of the open form (DTEo) displays no spectral overlap with the emission from P. The quenching by DTEc is ascribed to ET with a rate constant of about 3 ps, that is, around a factor 10 faster compared to the PET reaction forming $P^{\bullet +}$-$C_{60}^{\bullet -}$. Consequently, in the DTEo-P-C_{60} triad **6o** the P excited state decays to form DTEo-$P^{\bullet +}$-$C_{60}^{\bullet -}$ with unity efficiency, whereas the dominating quenching mechanism in DTEc-P-C_{60} (**6c**) is ET to form ^{1}DTEc-P-C_{60}, resulting in a substantial reduction in the formation of the charge separated state. Interestingly, the same functional principle was also used in the design of a molecular pentad where the yield of PET was self-regulated by the visible light intensity [16].

4.3.2
Electron Transfer

PET reactions can also be directly controlled by using the changes in redox energies that follow isomerization of a photochromic unit. Essentially, one of the two isomeric forms of the photoswitch should be thermodynamically competent to accept/donate an electron to form the charge separated state, whereas the corresponding electron transfer reaction in the other isomeric form should be energetically unfavorable. Figure 4.6 shows the structure of dyad **7** that was designed to fulfill the above-mentioned criteria [17]. The dyad consists of a porphyrin chromophore (P) attached to a dihydroindolizine (DHI) photoswitch. When DHI is in its closed spirocyclic form in **7c**, unquenched emission from P is observed upon excitation at 650 nm. UV irradiation converts the DHI unit to the open betaine form shown in **7o**, which is significantly easier to reduce because of its higher reduction potential. Excitation of P in this isomeric form of the dyad results in efficient PET to form $P^{\bullet+}$-DHIo$^{\bullet-}$ with concomitant quenching of the P emission. Visible light exposure triggers the isomerization to **7c**, and the P emission intensity is restored to the initial unquenched level again.

Photoswitches from the dihydropyrene family experience similar isomerization-induced changes in the oxidation potential. The colored DHP form is more easily oxidized than the colorless cyclophanediene (CPD) form, and the two forms are conveniently interconverted by the use of UV and visible light. Triad **8** shown in Figure 4.7 incorporates a dimethyl dihydropyrene derivative and the porphyrin-C_{60} electron transfer motif (P-C_{60}) also used in **6**. As mentioned above, the P singlet excited state in the P-C_{60} dyad decays in ∼30 ps to form the $P^{\bullet+}$-$C_{60}^{\bullet-}$ charge separated state. This state returns to the ground state again by charge recombination in a few nanoseconds. This behavior is preserved in the CPD-P-C_{60} form of the triad (**8o**), as the oxidation potential of CPD is too high to take part in electron transfer with either the excited state of P or the $P^{\bullet+}$-$C_{60}^{\bullet-}$ charge separated state.

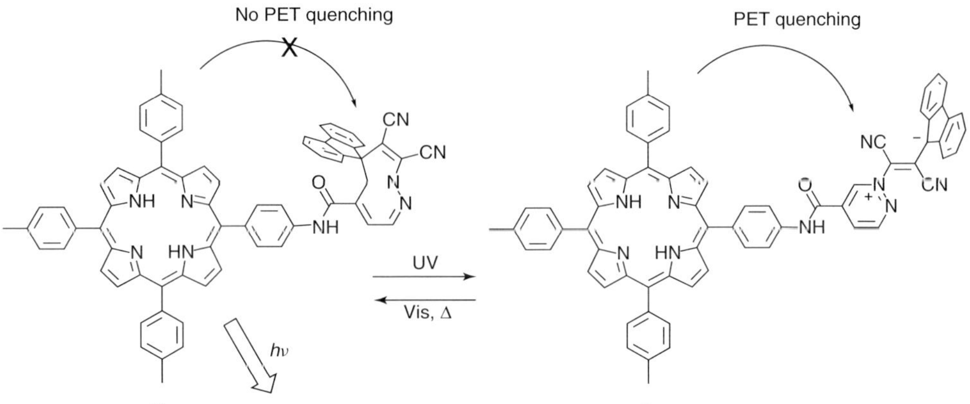

Figure 4.6 P-DHI dyad used for photonic control of electron transfer.

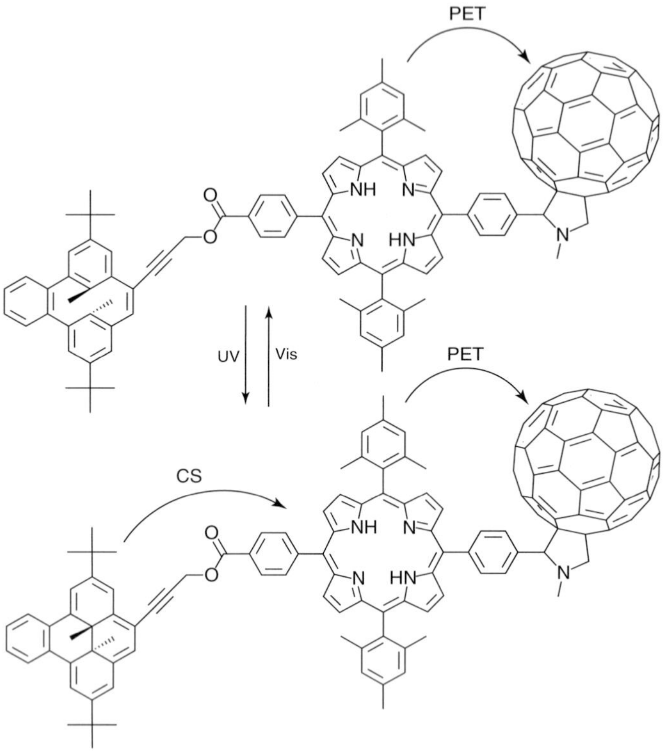

Figure 4.7 DHP-P-C_{60} triad used for the photocontrolled formation of a long-lived charge separated state.

UV light isomerizes the triad to the DHP-P-C_{60} form (**8c**). Excitation of the P unit leads again to efficient formation of the DHP-P$^{\bullet+}$-$C_{60}^{\bullet-}$ state in ~30 ps. As opposed to the CPD-P$^{\bullet+}$-$C_{60}^{\bullet-}$ charge separated state, which collapses directly back to the ground state, the DHP unit will donate an electron to P$^{\bullet+}$ in a charge shift (CS) reaction to form DHP$^{\bullet+}$-P-$C_{60}^{\bullet-}$ in 96% yield. This state lives for 2 µs before charge recombination repopulates the ground state. The isomerization process is reversible, as visible light converts the triad to the CPD-P-C_{60} form again. Hence, UV light and visible light are used to toggle the triad between two forms that display a long-lived and a short-lived charge separated state, respectively [18].

Such isomerization-induced redox changes can also be employed to direct PET down either of two pathways in a "left-or-right" manner. Devices possessing this feature are commonly referred to as *double-throw switches*, and triad **9** was designed to display this behavior on the molecular scale [19]. The triad shown in Figure 4.8 comprises the photochromic DHI redox switch, employed also in **7**, covalently linked to the porphyrin in a P-C_{60} electron transfer motif. In **9c**, with the photoswitch in the closed form with the lower reduction potential, excitation of the porphyrin unit is followed by PET to C_{60} with a ~2 ns time constant to give the DHIc-P$^{\bullet+}$-$C_{60}^{\bullet-}$ charge separated state. PET from P to DHIc would be

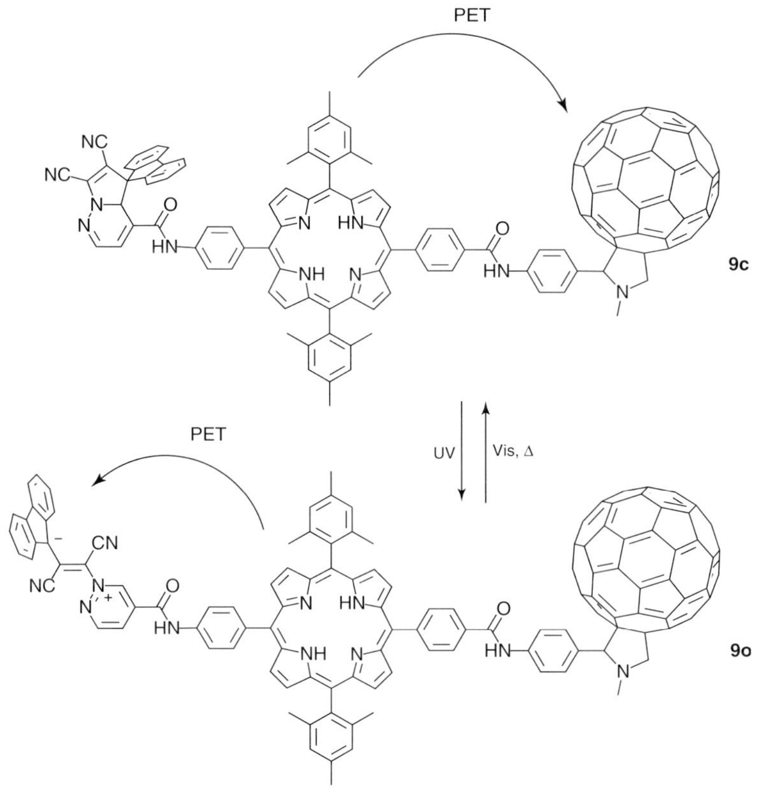

Figure 4.8 Structures and electron transfer patterns of the DHI-P-C_{60} triad designed to perform as a double-throw switch. UV and visible light are used to toggle between electron transfer to the C_{60} or to the DHI unit.

an endergonic process, and does not occur. UV exposure isomerizes the triad to DHIo-P-C_{60} (**9o**), with DHI in the easily reduced form. Excitation of the porphyrin unit is now followed by a very efficient (56 ps time constant) PET reaction to yield mainly the DHIo$^{\bullet-}$-P$^{\bullet+}$-C_{60} charge separated state. Hence, the electron is transferred "to the right" when the DHI redox switch is in the closed form, but "to the left" when the switch is isomerized to the open form by UV light exposure. The yields of the PET reactions are 82 and 99%, respectively, and in either form, no more than 1% of the electrons are transferred down the wrong branch.

4.4
Boolean Logic Gates

The functions of the various on–off switches described above rely on the switching of photochromic supramolecular constructs between two isomeric forms. Although the function of these molecules can be abstracted into one-input Boolean logic gates, the realization of two-input logic gates using molecular photoswitches is a more interesting (and complex) task [20]. Typically, two or more individually addressable photochromic units have to be incorporated into one and the same molecular construct, offering at least four different accessible isomeric forms. Often, each of these four isomeric forms display unique spectral features (e.g., absorption and emission), allowing for the parallel performance of several different logic gate functions by multiple wavelength readout (output reconfiguration). This feature will be described in the section about Advanced Logic Functions, whereas this chapter focuses on how to design molecules capable of performing the function of individual logic gates.

The first example to illustrate this idea is triad **10** comprising a fluorescent porphyrin reporter (P) covalently linked to both a dihydroindolizine and a dihydropyrene photochromic unit [21]. Both photoswitches exist in two isomeric forms, giving rise to a total of four isomers of the triad. The structures are shown in Figure 4.9. The dihydroindolizine in the open DHIo form as well as the dihydropyrene in the closed DHP form quench the emission from the covalently linked porphyrin by electron transfer processes. Hence, the porphyrin emission intensity in the DHP-P-DHIo form of the triad is low. In order to observe intense, unquenched fluorescence, DHIo must be isomerized to the nonquenching DHIc isomer, *and* DHP must be isomerized to the nonquenching CPD isomer. Hence, the only form of the triad that displays intense porphyrin emission is CPD-P-DHIc. With the triad initially set in the DHP-P-DHIo form, it is seen from Figure 4.9 that both heat (defined as Input 1) *and* red light (defined as Input 2) must be applied to observe the intense porphyrin emission (defined as the output) from CPD-P-DHIc. This behavior is described by the truth table for the logic AND gate shown in Table 4.1, where it is seen that both inputs must be switched to the *on* (1) state in order for the output to switch *on* (1). All other input combinations set the output in the *off* (0) state. The triad is reset to the initial state by UV exposure, which allows for recycling of the AND logic operation. It is worth mentioning that the heat input (that could be supplied by a pulsed IR laser) has no effect on the DHP isomer, which is the thermally stable form of the dihydropyrene switch. Furthermore, DHIo is not significantly affected by the dose of red light employed, as the isomerization yield DHIo $\rightarrow$ DHIc is relatively low.

Triad **10** can also be reconfigured to function as an INHIBIT (INH) gate by choosing DHP-P-DHIc as the initial state and using red light and UV light as Inputs 1 and 2, respectively. The truth table for the INH gate is shown in Table 4.2. The INH gate has, just like the AND gate, two inputs and one output. With both inputs in the *off* state, the output is *off*. Switching *on* Input 1 sets the output *on*, whereas the application of Input 2 does not switch the output *on*. Finally, switching both inputs *on* leaves the output in the *off* position. Hence, Input 2 inhibits the gate output response to Input 1. The application of red light (Input 1) isomerizes the nonfluorescent form DHP-P-DHIc to CPD-P-DHIc, and the fluorescence output switches *on*. If instead UV light is applied (Input 2), the triad isomerizes to DHP-P-DHIo, where again the porphyrin emission is weak, and the

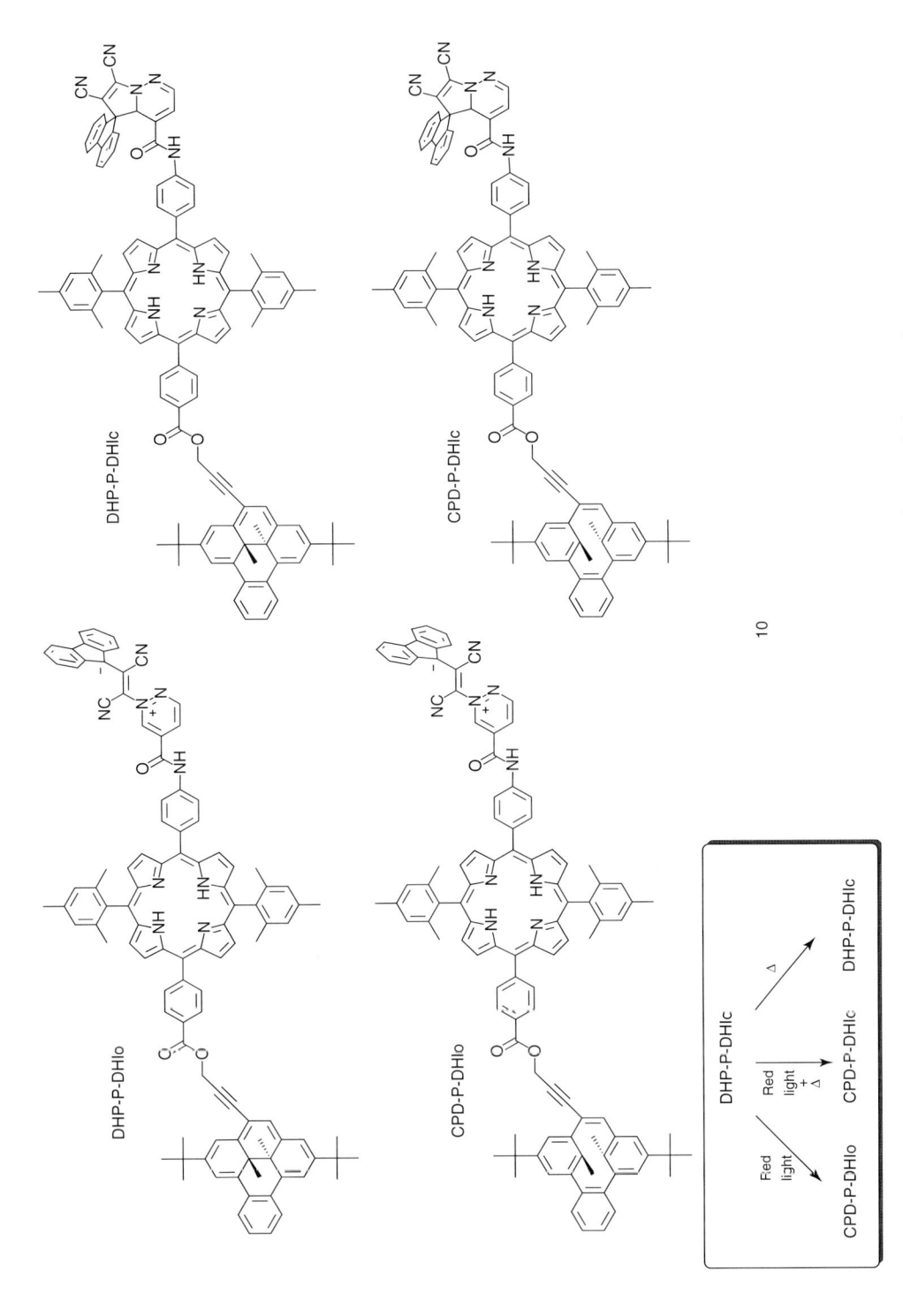

Figure 4.9 Structures of the four isomeric forms of the DHP-P-DHI dyad used as photonically controlled AND and INH gates.

Table 4.1 Truth table for the AND gate.

In1 (heat or IR)	In2 (red light)	Output (P fluorescence)
0	0	0
0	1	0
1	0	0
1	1	1

Table 4.2 Truth table for the INHIBIT gate.

In1 (red light)	In2 (UV light)	Output (P fluorescence)
0	0	0
0	1	0
1	0	1
1	1	0

output remains *off*. Application of both inputs generates the CPD-P-DHIo form, and the gate output is *off*. This concludes the truth table for the INH gate.

The examples above illustrate our main approach to the design of molecular switches and other logic devices – the photocontrolled on–off switching of excited state communication between chromophores in supramolecular constructs and the monitoring of the concomitant changes in the fluorescence intensity as the output [22]. We have designed several other photochromic constructs that perform the function of individual logic gates using the same underlying principles, together with examples where the function relies on reading other optical properties (such as the absorbance or the electric linear dichroism) as the outputs [23, 24]. A selection of these will be briefly described in the next section, together with several examples of how functional integration has been used in the performance of advanced logic functions.

4.5
Advanced Logic Functions

The realization of Boolean logic gates described above implies that one molecule is designed to function as one logic gate – typically processing the information contained in two inputs to deliver a one-output response. In order for molecules to perform arithmetic operations and other more advanced logic functions, another level of complexity is introduced. For example, the function of a 4:2 encoder is described by the circuit shown in Figure 4.10. This logic device receives information from four different inputs (In0–In3). Before the result is delivered by the two output

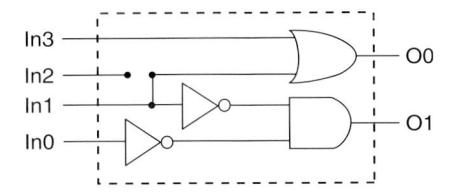

Figure 4.10 Logic circuit for the 4:2 encoder, expressed in conventional electronic form.

signals (O0 and O1), information has been processed by several individual logic gates, where the output from one gate may serve as the input for another. There are two principal approaches to implement these (and similar) logic functions on the molecular scale. One approach, referred to as *physical integration*, implies that separate molecules represent each logic gate in the circuit and that the individual molecular gates must be "wired" (concatenated) in order for output signals to serve as input signals to downstream gates. Although less common, there are examples where molecules have been designed and shown to execute logic operations according to these functional principles [7, 25–28]. The problems associated with the concatenation of molecules (e.g., input–output heterogeneity) form the main barrier to be overcome. An alternative and more frequently used approach is *functional integration* [9], where the function of a complex logic circuit is loaded onto one (supra)molecular species, rather than having separate molecules representing each logic gate in the circuit. The logical communication between gates thus occurs within the molecule. We have adopted this method in the design of molecules capable of performing the function of a large selection of advanced logic devices, individually described below.

4.5.1
Half-Adders and Half-Subtractors

An HA and an HS perform addition and subtraction, respectively, of two binary digits. Addition of two binary digits is the most trivial arithmetic operation. Using base-10, it corresponds to the following operations: $0 + 0 = 0, 0 + 1 = 1, 1 + 0 = 1$, and $1 + 1 = 2$. If binary numbers are used instead, the operations translate into $0 + 0 = 0, 0 + 1 = 1, 1 + 0 = 1$, and $1 + 1 = 10$. Hence, the HA consists of two binary outputs in addition to the two binary inputs. The inputs represent the two binary digits to be added (augend and addend), and the outputs represent the two least significant bits of the binary sum – the carry digit (C) and the sum digit (S), respectively. This is summarized in Table 4.3, where the truth table for the HA is shown. Here, it is seen that the HA can be implemented with an XOR gate and an AND gate with shared input signals, In1 and In2.

The truth table for the HS is shown in Table 4.4. The corresponding four operations executed by the HS in 10-base are $0 - 0 = 0, 1 - 0 = 1, 0 - 1 = 1$, and $1 - 1 = 0$. This is represented by the following operations in the binary system: $0 - 0 = 0, 1 - 0 = 1, 0 - 1 = 11$, and $1 - 1 = 0$. In terms of logic gates, these results are obtained by letting the borrow digit (B) and the difference digit (D)

Table 4.3 Truth table for the half-adder.

In1	In2	O1 AND (C)	O2 XOR (S)	Binary sum	10-base sum
0	0	0	0	00	0
0	1	0	1	01	1
1	0	0	1	01	1
1	1	1	0	10	2

Table 4.4 Truth table for the half-subtractor.

In1	In2	O1 INH (B)	O2 XOR (D)	Binary difference	10-base difference
0	0	0	0	00	0
0	1	0	1	01	1
1	0	1	1	11	−1
1	1	0	0	00	0

be represented by the outputs from an INH gate and an XOR gate, respectively. A large number of molecular approaches to the HA and the HS are reported in the literature and several of these have been described in comprehensive reviews [6, 8]. Inspired by the first pioneering reports on the molecular HA [29–31], our attempt to the first example of an all-photonic version started out by realizing that the combination of two previously used photochromic supramolecules (dyad **7** and triad **8**) and a nonlinear crystal could generate the required AND and XOR gates responses [32]. The input signals In1 and In2 are represented by 1064 nm and 532 nm light delivered by Nd:YAG lasers. Application of both inputs will generate 355 nm UV light, as the input beams are combined in a third harmonic generating (THG) crystal before hitting the sample. When triad **8** is used as the AND gate, it is initially in the CPD-P-C_{60} form and the output is read as the long-lived transient absorption signal of DHP$^{\bullet+}$-P-$C_{60}^{\bullet-}$ at 1000 nm [18]. The CPD-P-C_{60} form displays no long-lived transient signal, and irradiation of this isomer with either 1064 nm (In1) or 532 nm (In2) does not cause any isomerization. However, simultaneous irradiation by both input lasers generates 355 nm light in the THG, switching the triad to the DHP-P-C_{60} form. This isomer shows long-lived absorption at 1000 nm from DHP$^{\bullet+}$-P-$C_{60}^{\bullet-}$, and the output is switched *on*. The prerequisites for the AND gate are fulfilled. The XOR gate is implemented by dyad **7** in the P-DHIo initial state, and the output is taken as P emission at 720 nm [17]. In order for P to display intense fluorescence, the dyad must be isomerized to the P-DHIc form. This can be accomplished by either heat (equivalent to IR light) or visible light. Hence, both In1 and In2 alone switch the output to the *on* state. Simultaneous application of the inputs, however, generates 355 nm UV light in the THG and no net isomerization

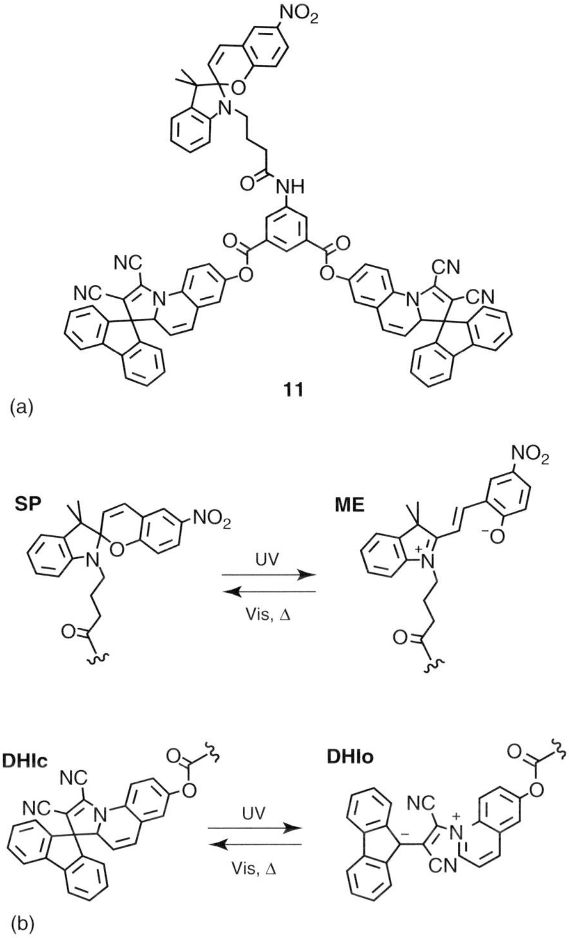

Figure 4.11 Structure of the all-closed form of triad **11** used as an all-photonic half-adder (a). Shown also are the isomerization schemes of the individual photochromic units (b).

is induced. The P emission is quenched by DHIo, and the output remains *off*. This is the function of the XOR gate.

Triad **11** shown in Figure 4.11, consisting of one SP and two quinoline derived dihydroindolizines, is another example of a molecular HA reported by us [33]. Here, the function relies on a delicate balance between thermal (decolorization) and UV-induced (colorization) isomerizations. The colored open merocyanine (ME) form of the SP displays red fluorescence. This fluorescence, however, is quenched in an ET reaction by the colored open DHIo form of the dihydroindolizine in the triad. Hence, both the absorption and the emission intensity in the visible depend on the intensity of the UV light used for colorization. The AND gate was realized by monitoring the total absorption at 581 nm, whereas the XOR gate was read as the fluorescence from ME at 690 nm. UV light at 355 nm was used to represent both (degenerate) inputs. Briefly, the absorbance at 581 nm increases with increasing UV intensity (switching *on* the output from the AND gate only when both UV inputs are applied), whereas the ME emission intensity describes an *off–on–off* response, which is required for the XOR gate function. The knowledge gained from the study of systems such as those described above has now allowed the design and synthesis of molecules capable of performing both addition and subtraction, rather than using one molecular species for each operation. These molecules are referred to as *moleculators* (molecular calculators) after Shanzer's pioneering work [34]. Later on in this chapter, an example of such an advanced device is discussed, which enables not only addition and subtraction but also the majority of all other logic operations that have been realized on the molecular scale [35]. Information transmission and compression are both examples of such operations, and are discussed below in the sections about multiplexing/demultiplexing and encoding/decoding.

4.5.2
Multiplexers and Demultiplexers

The function of a MUX is to direct each of two or more inputs to a common transmission medium (output) by switching the state of a selector input, Sel. Thus, digital MUXs are analogous to mechanical rotary switches that connect any of several inputs to the output. When Sel is in the *off* (0) state, the output O reports the state of input In1 and ignores the state of input In2. After Sel is switched *on* (1), the output instead reports only the state of In2. In 2007, we reported the first molecular 2 : 1 MUX based on triad **10** shown in Figure 4.9 [36]. Since then, other groups have interpreted the same function using cleverly designed molecules [37, 38]. As mentioned above, the porphyrin (P) emission is quenched by both the DHIo and the DHP isomers. Thus, CPD-P-DHIc is the only isomeric form displaying intense fluorescence from the porphyrin reporter, which is read as the MUX output. Using the thermally stable DHP-P-DHIc form as the initial state, red light as In1, heat (which can be supplied by an IR laser) as In2, and green light as Sel, the triad functions as follows (see Figure 4.12 for the relevant isomerization scheme): heat has no effect on the DHP-P-DHIc form as both photoswitches are in the thermally stable form, whereas red light isomerizes the triad to the fluorescent CPD-P-DHIc form and the output is switched on. Hence, with Sel *off*, the output takes the

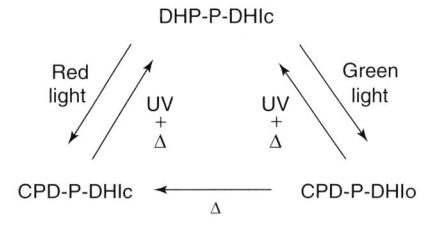

Figure 4.12 Isomerization scheme for the three relevant forms of triad **9** when operated as a 2 : 1 multiplexer.

value of In1. If instead Sel is switched on, the triad is isomerized mainly to the CPD-P-DHIo form, where the P emission is quenched by DHIo. In order to switch the fluorescence output to the *on* state, DHIo must be isomerized to DHIc, while leaving CPD unaffected. Heat has this effect on the isomeric distribution, as the thermal isomerization BT → DHI is efficient, whereas the corresponding process CPD → DHP is very slow. Red light exposure has little effect on the DHIc/DHIo distribution, as the light-induced isomerization DHIo → DHIc is inefficient and the CPD isomer is colorless and does not absorb the red light. Thus, In1 does not have any effect on the output after application of the selector input Sel, but the output takes the value of In2, which concludes the function of the MUX.

Once the data inputs have been combined into a common transmission line (the MUX output) they have to be disentangled again by directing the data from a common input signal to separate outputs. This function is performed by the DMUX, by switching the state of an address input, Ad. When Ad is set *off*, output O1 reports the state of the input In and O2 remains *off*, whereas when Ad is switched on, O2 reports the state of In. The first example of a molecular 1:2 DMUX was reported by us in 2007 using triad **10** together with a THG crystal [39], and other examples have followed [37, 38, 40]. We chose 532 nm light as the input signal In, whereas 1064 nm light was used as the Ad input. The absorbance in the visible (where DHIo is the main absorber) and the P emission were read as the outputs. Again, application of both inputs will generate 355 nm UV light, as the input beams are combined in a THG crystal before hitting the sample. As 532 or 355 nm light selectively isomerizes one photoswitch each, the input combinations (1,0) and (1,1) switch *on* one output each and the function of the DMUX is implemented.

4.5.3
Encoders and Decoders

An encoder converts data from one form to another by receiving data from 2^n input lines and delivering the data to n output lines. This is typically done for the purposes of compression, processing speed, or security. One example is the single bit 4:2 encoder that converts 0, 1, 2, and 3 in base-10 to the corresponding 2-bit binary number. The input lines In0–In3 represent 0–3 in base-10. The encoder delivers the outputs to O0 and O1, representing the least and the second least significant bits of a binary number, respectively. If, for example, In2 is switched on, the resulting output combination is O0 = 0 and O1 = 1, as 10 in the binary equals 2 in 10-base. The results of all input combinations (only one input applied at any given time) are collected in the truth table for the encoder shown in Table 4.5.

Triad **12** was designed to perform the function of both the 4:2 encoder and the 2:4 decoder [41]. The triad consists of two identical photochromic FG units and one photochromic DTE unit (Figure 4.13). Each photoswitch exists is an open, colorless form (FGo and DTEo) and a closed form (FGc and DTEc) which is colored in the visible. The function of the encoder relies on the fact that each of the four isomers above has absorption maxima in separate spectral regions. As a result, all four isomers can be independently isomerized by the use of one or two inputs

Table 4.5 Truth table for the 4 : 2 encoder.

In0 (Green light)	In1 (397 nm light)	In2 (302 nm light)	In3 (366 nm light)	O1 (A_{625})	O0 (A_{475})
1	0	0	0	0	0
0	1	0	0	0	1
0	0	1	0	1	0
0	0	0	1	1	1

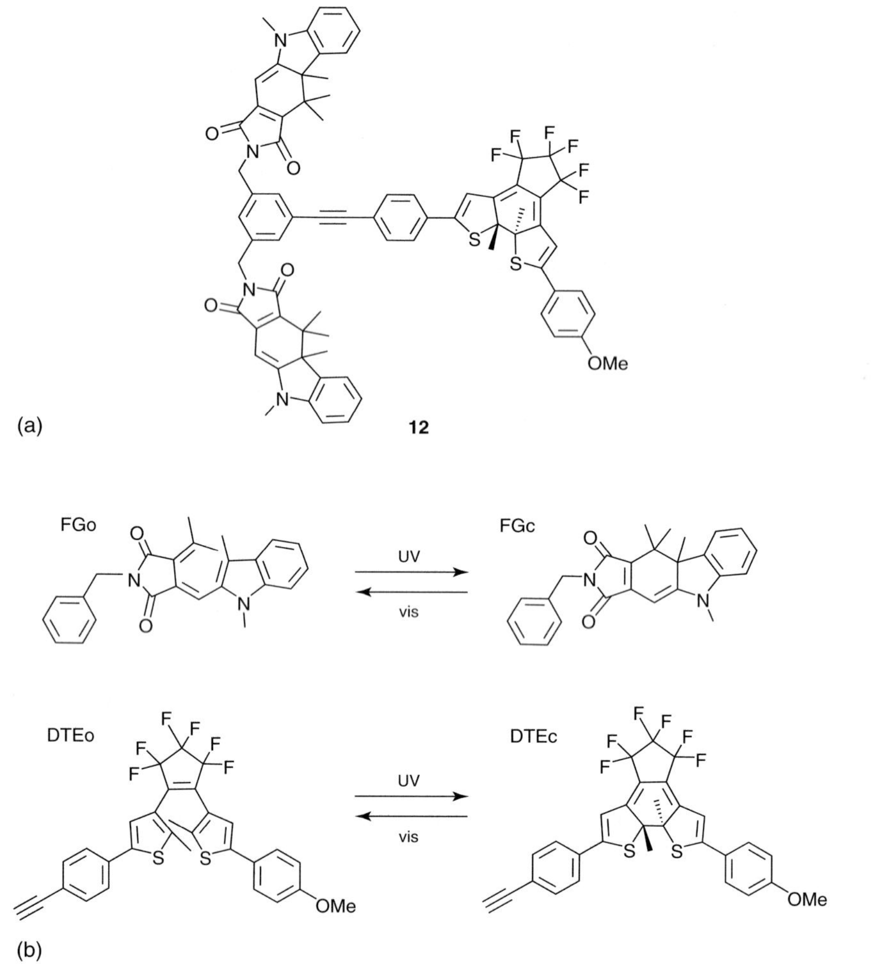

Figure 4.13 Structure of the all-closed form of triad **12** used as an all-photonic 4 : 2 encoder and 2 : 4 decoder (a). Shown also are the isomerization schemes of the individual photochromic units (b).

without causing significant net isomerization of the other forms. Furthermore, the outputs are conveniently chosen at wavelengths where the absorption is ascribed mainly due to one of the four isomers. The initial state is FGo-DTEo (no distinction is made between the two individual FG units), and the inputs are UV light at 302, 366, and 397 nm and green light. The 302 nm input triggers the DTEo → DTEc isomerization. The absorption at 625 is ascribed exclusively to DTEc, representing O1. Likewise, the 397 nm input isomerizes exclusively FGo to FGc. FGc is the main absorber at 475 nm, where O0 is read. The 366 nm light colorizes both the FG and the DTE photoswitches, converting the triad to the FGc-DTEc isomer absorbing strongly at both 625 and 475 nm. Finally, green light has no effect on the FGo-DTEo form, with no absorption in the visible. In this way, the absorption at 475 and 625 nm contains, in encoded form, the information entered into the system by the optical input signals in line with the encoder truth table.

A decoder reverses the operation of the encoder, that is, it converts binary information from n input lines to a maximum of 2^n output lines. The truth table for the 2:4 decoder is shown in Table 4.6. When operated as the decoder, the triad is again set to the FGo-DTEo initial state. Hence, the triad does not have to be chemically reconfigured when the operation is switched from encoder to decoder mode and vice versa, which was shown also for the $[Ru(bpy)_3]^{2+}$ encoder/decoder reported by Balzani and coworkers [42]. This feature (and others) will be further emphasized in a later section concerning the triad as an all-photonic multifunctional logic device. The inputs and the outputs used in the decoder mode are shown in Table 4.6. The use of 302 and 397 nm UV as input signals allows the triad to be isomerized among all four isomeric forms by the use of the various input combinations. As each of the four isomeric forms has at least one unique spectral feature, which together constitute the set of output signals, the decoder function is conveniently implemented.

4.5.4
Sequential Logic Devices

All functions described so far are so-called combinational logic functions. These functions have in common that the state of the output is independent on the order of the input application. For example, the output of a regular two-input AND gate

Table 4.6 Truth table for the 2:4 decoder.

In1 (302 nm light)	In0 (397 nm light)	O0 (T_{535})	O1 (Em_{624})	O2 (A_{393})	O3 (A_{535})
0	0	1	0	0	0
0	1	0	1	0	0
1	0	0	0	1	0
1	1	0	0	0	1

switches to the *on* state given that both inputs are switched on, irrespective of the order by which In1 and In2 are applied. This is not true for sequential logic devices, where the state of the output depends on the sequence by which the inputs are delivered. Hence, a sequential logic circuit is said to have a memory function, whereas a combinational one does not. The sequential logic device that has received most attention by the molecular logic community is the keypad lock. Inspired by Shanzer's first demonstration of a molecular version, several research groups have designed various molecules that perform the same function [27, 35, 43–48]. Three of these come from our laboratories [35, 43, 46]. The first example was realized using triad FG-P-DTE (triad **13**) that was originally designed to perform XOR and NOR logic functions [49]. The triad consists of a central fluorescent porphyrin reporter (P) covalently linked to photochromic FG and DTE units [43]. These are the same photochromic units that constitute triad **12**, and as described above, both FG and DTE may exist in an open form and a closed form giving rise to a total of four isomeric states of triad **13**. By the proper combination of UV and visible light, it is possible to prepare solutions that are greatly enriched in each isomeric state. In operation of the keypad lock, only three isomers are required. The structures and the isomerization scheme of these isomers are shown in Figure 4.14. As for the fluorescence properties of the triad, it is only the FGc-P-DTEo form that displays strong porphyrin emission on excitation at 470 nm. This behavior is rationalized by the following: first, FGc is the major absorber at 470 nm so that FGc is required for sufficient light absorption. FGc is also intrinsically fluorescent. In the triad, however, the fluorescence overlaps significantly with the P absorption, resulting is efficient ET and sensitization of the P singlet excited state. The porphyrin emission is in turn quenched by ET to DTEc, because of the pronounced spectral overlap. However, DTEo displays no such overlap, and no ET quenching is therefore observed with DTE in the open DTEo form. Thus, FGc is required to act as an antenna for the 470 nm light, and DTE must be in the open form DTEo not to quench the P emission. When used as a keypad lock, the triad is initially set in the FGo-P-DTEo form. In1 is UV light at 366 nm, and In2 is red light. The output is read as intense P emission. As is evident from Figure 4.14, the only input sequence that isomerizes FGo-P-DTEo to the fluorescent form FGc-P-DTEo is 366 nm (In1) followed by red light (In2). Hence, in this mode the triad properly describes the sequential logic behavior required for the keypad lock.

In our second interpretation of a molecular keypad lock, we designed three photochromic triads, each consisting of a fluorescent reporter chromophore (FRC = aminonaphthalimide, dansyl, and perylene) covalently linked to a SP photoswitch to constitute the FRC-SP dyads **14** with the generic structure shown in Figure 4.15 [46]. SPs exist in two constitutionally isomeric forms – the colorless SP form and the colored ME form. The ME form is readily protonated by the addition of acid to form MEH^+, with an absorption maximum in the spectral region between SP and ME. The interconversion pathways among the three different forms are shown in Figure 4.15. The triads were designed so that the FRC fluorescence has a substantial spectral overlap with the absorption of the ME and MEH^+ forms, but no corresponding overlap with the SP form. Hence, the fluorescence from the

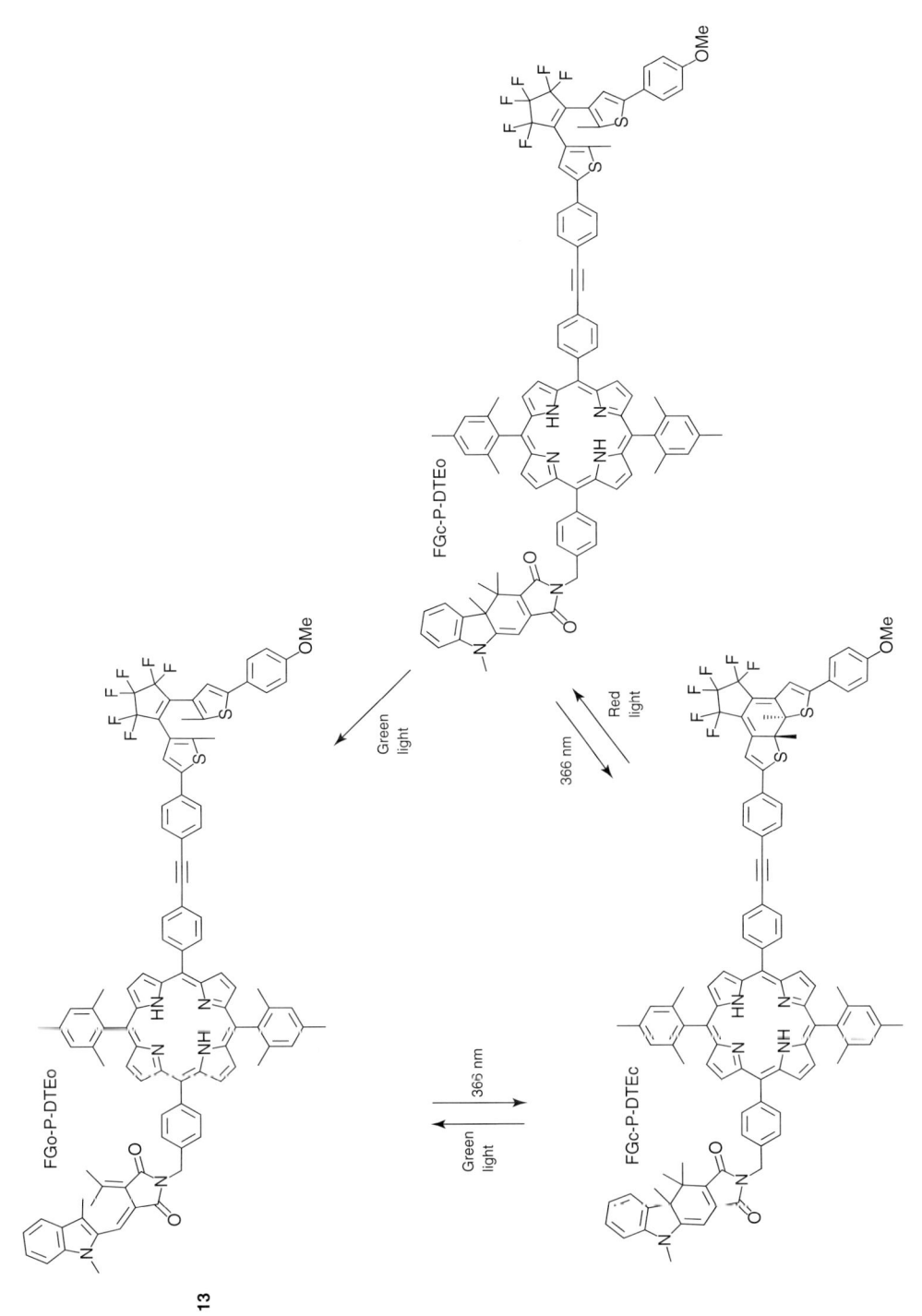

Figure 4.14 Structures and isomerization scheme for the three relevant forms of the FG-P-DTE triad when operated as an all-photonic keypad lock.

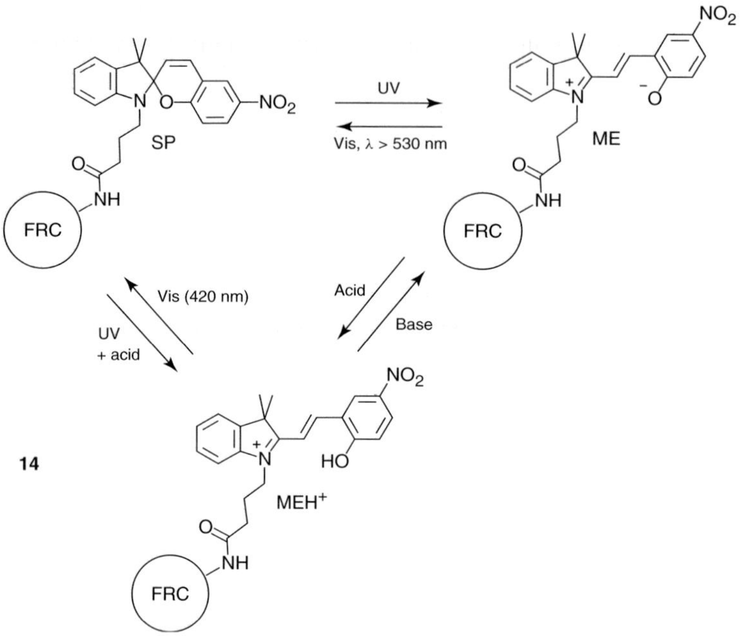

Figure 4.15 Structures and isomerization scheme for the photochromic dyads used as a keypad lock. Aminonaphthalimide, dansyl, and perylene were used as FRCs (fluorescent reporter chromophores).

FRC is quenched in the two former forms, but not in the latter. In operation as a keypad lock, the dyads are initially set in the FRC-MEH$^+$ form. In1 is base and In2 is visible light at $\lambda > 530$ nm. The output is read as intense FRC emission. From Figure 4.15 it is seen that the only input sequence that isomerizes FRC-MEH$^+$ to the fluorescent form FRC-SP is base (In1) followed by light at $\lambda > 530$ nm (In 2). Applying the opposite input sequence will lead to formation of the FRC-ME form displaying low fluorescence intensity, as light at $\lambda > 530$ nm is not absorbed by MEH$^+$ and, as an obvious consequence, does not lead to any isomerization.

Our third example of a keypad lock is discussed in a later section where we describe how the FG-DTE triad **12** can be used to implement a so far unprecedented number of logic operations [35]. Other logic circuits with a memory function, such as the S-R latch, have also been implemented on the molecular scale [50–55]. The S-R latch can be regarded as an on–off switch with two binary input signals – the Set input (S) and the Reset input (R). A regular binary on–off switch, such as the buffer gate, comes with only one input. If the input is set to 1, the output is set to 1, whereas if the input is set to 0, the output is set to 0. In the S-R latch, the output is switched to 1 by momentarily setting S to 1. When S subsequently is set to 0 by removing the active input, the output still remains at 1 until R momentarily is set to 1, which resets the latch output to 0. The latch output is set to 1 again by setting S to 1. Hence, the S-R latch has a built-in memory function, which in conventional electronics is guaranteed by feedback loops. Any photochromic molecule that displays good thermal stability of both isomeric forms performs the function of the S-R latch *per se*. We demonstrated the straightforward implementation of the S-R latch using a DTE monomer, where UV light and visible light were used as the S and the R inputs, respectively, and the output was represented by the UV-induced absorption of DTEc in the visible region [56].

4.5.5
An All-Photonic Multifunctional Molecular Logic Device

In the sections above, a large selection of molecules capable of performing the functions of different logic devices have been presented. In the majority of these cases, one molecule is designed to perform as one logic device. An extreme example of functional integration would be if *one* molecule could perform *all* these functions with one common initial state. The FG-DTE triad **12** is an example of such a molecule. More specifically, the triad performs the functions of the following devices using optical inputs and outputs: AND, XOR, INH, HA, HS, MUX, DMUX, encoder, decoder, keypad lock, and logically reversible transfer gates. Including also the individual gates that constitute the logic circuits, the triad can be operated as any of 13 logic devices [35]. This level of complexity is unprecedented. The functions rely on the ability to control the excited state communication between the FG and the DTE photoswitches, as well as the selective isomerization of the individual photoswitches by the use of different wavelengths. Furthermore, the outputs can be read as the absorption and the emission properties at different wavelengths, allowing for parallel performance of several logic functions. Finally, the triad has one common initial state for all the logic operations, which means that the reset operation (green light exposure) does not depend on the previous input combination/logic function.

4.6
Conclusion

In the sections above, we have shown that photochromic molecules are able to perform a large number of logic operations. The complexity of these operations varies from simple binary on–off switches, which in principle can be realized by any bistable system, to advanced functions such as addition/subtraction, multiplexing/demultiplexing, and encoding/decoding. The realization of the latter devices requires the design and the nontrivial synthesis of multiphotochromic systems where the excited state communication between the different chromophores has to be carefully tuned. The realization of even more complex functions would be facilitated if the concatenation problem could be overcome. All-photonic systems such as the ones described above overcome (at least formally) the input–output heterogeneity barrier, which assists in the concatenation of individual molecular logic units. While the concept of molecular computing has been the major driving force in the design of "smart molecules," it is becoming apparent that the replacement of already functional silicon-based predecessors with molecular analogs may not lead to the earliest practical applications. Instead, fields are identified where semiconductor material cannot go, for example, the delivery/activation of drugs and clinical diagnostics. Many more applications are yet to be uncovered, and will show up as this research field develops.

References

1. de Silva, A.P., Gunaratne, H.Q.N., and McCoy, C.P. (1993) A molecular photoionic and gate based on fluorescent signaling. *Nature*, **364**, 42–44.
2. Andréasson, J. and Pischel, U. (2010) Smart molecules at work-mimicking advanced logic operations. *Chem. Soc. Rev.*, **39**, 174–188.
3. Balzani, V., Credi, A., and Venturi, M. (2003) Molecular logic circuits. *Chemphyschem*, **4**, 49–59.
4. de Silva, A.P. and Uchiyama, S. (2007) Molecular logic and computing. *Nat. Nanotechnol.*, **2**, 399–410.
5. Katz, E. and Privman, V. (2010) Enzyme-based logic systems for information processing. *Chem. Soc. Rev.*, **39**, 1835–1857.
6. Pischel, U. (2007) Chemical approaches to molecular logic elements for addition and subtraction. *Angew. Chem. Int. Ed.*, **46**, 4026–4040.
7. Raymo, F.M. (2002) Digital processing and communication with molecular switches. *Adv. Mater.*, **14**, 401–414.
8. Szacilowski, K. (2008) Digital information processing in molecular systems. *Chem. Rev.*, **108**, 3481–3548.
9. Tian, H. (2010) Data processing on a unimolecular platform. *Angew. Chem. Int. Ed.*, **49**, 4710–4712.
10. Crano., J.C. and Guglielmetti., R.J. (eds) (1998) *Organic Photochromic and Thermochromic Compounds*, Plenum Press, New York.
11. Raymo, F.M. and Tomasulo, M. (2005) Fluorescence modulation with photochromic switches. *J. Phys. Chem. A*, **109**, 7343–7352.
12. Raymo, F.M. and Tomasulo, M. (2005) Electron and energy transfer modulation with photochromic switches. *Chem. Soc. Rev.*, **34**, 327–336.
13. Bahr, J.L., Kodis, G., de la Garza, L., Lin, S., Moore, A.L., Moore, T.A., and Gust, D. (2001) Photoswitched singlet energy transfer in a porphyrin-spiropyran dyad. *J. Am. Chem. Soc.*, **123**, 7124–7133.
14. Straight, S.D., Terazono, Y., Kodis, G., Moore, T.A., Moore, A.L., and Gust, D. (2006) Photoswitchable sensitization of porphyrin excited states. *Aust. J. Chem.*, **59**, 170–174.
15. Liddell, P.A., Kodis, G., Moore, A.L., Moore, T.A., and Gust, D. (2002) Photonic switching of photoinduced electron transfer in a dithienylethene-porphyrin-fullerene triad molecule. *J. Am. Chem. Soc.*, **124**, 7668–7669.
16. Straight, S.D., Kodis, G., Terazono, Y., Hambourger, M., Moore, T.A., Moore, A.L., and Gust, D. (2008) Self-regulation of photoinduced electron transfer by a molecular nonlinear transducer. *Nat. Nanotechnol.*, **3**, 280–283.
17. Terazono, Y., Kodis, G., Andréasson, J., Jeong, G.J., Brune, A., Hartmann, T., Durr, H., Moore, A.L., Moore, T.A., and Gust, D. (2004) Photonic control of photoinduced electron transfer via switching of redox potentials in a photochromic moiety. *J. Phys. Chem. B*, **108**, 1812–1814.
18. Liddell, P.A., Kodis, G., Andréasson, J., de la Garza, L., Bandyopadhyay, S., Mitchell, R.H., Moore, T.A., Moore, A.L., and Gust, D. (2004) Photonic switching of photoinduced electron transfer in a dihydropyrene-porphyrin-fullerene molecular triad. *J. Am. Chem. Soc.*, **126**, 4803–4811.
19. Straight, S.D., Andréasson, J., Kodis, G., Moore, A.L., Moore, T.A., and Gust, D.

20. de Silva, A.P. and McClenaghan, N.D. (2004) Molecular-scale logic gates. *Chem. Eur. J.*, **10**, 574–586.
21. Straight, S.D., Andréasson, J., Kodis, G., Bandyopadhyay, S., Mitchell, R.H., Moore, T.A., Moore, A.L., and Gust, D. (2005) Molecular AND and INHIBIT gates based on control of porphyrin fluorescence by photochromes. *J. Am. Chem. Soc.*, **127**, 9403–9409.
22. Gust, D., Moore, T.A., and Moore, A.L. (2006) Molecular switches controlled by light. *Chem. Commun.*, 1169–1178.
23. Andréasson, J., Terazono, Y., Albinsson, B., Moore, T.A., Moore, A.L., and Gust, D. (2005) Molecular AND logic gate based on electric dichroism of a photochromic dihydroindolizine. *Angew. Chem. Int. Ed.*, **44**, 7591–7594.
24. Andréasson, J., Terazono, Y., Eng, M.P., Moore, A.L., Moore, T.A., and Gust, D. (2011) A dihydroindolizine-porphyrin dyad as molecule-based all-photonic AND and NAND gates. *Dyes Pigm.*, **89**, 284–289.
25. Frezza, B.M., Cockroft, S.L., and Ghadiri, M.R. (2007) Modular multi-level circuits from immobilized DNA-Based logic gates. *J. Am. Chem. Soc.*, **129**, 14875–14879.
26. Niazov, T., Baron, R., Katz, E., Lioubashevski, O., and Willner, I. (2006) Concatenated logic gates using four coupled biocatalysts operating in series. *Proc. Natl. Acad. Sci. U.S.A.*, **103**, 17160–17163.
27. Strack, G., Ornatska, M., Pita, M., and Katz, E. (2008) Biocomputing security system: concatenated enzyme-based logic gates operating as a biomolecular keypad lock. *J. Am. Chem. Soc.*, **130**, 4234–4235.
28. Yashin, R., Rudchenko, S., and Stojanovic, M.N. (2007) Networking particles over distance using oligonucleotide-based devices. *J. Am. Chem. Soc.*, **129**, 15581–15584.
29. de Silva, A.P. and McClenaghan, N.D. (2000) Proof-of-principle of molecular-scale arithmetic. *J. Am. Chem. Soc.*, **122**, 3965–3966.
30. Guo, X.F., Zhang, D.Q., Zhang, G.X., and Zhu, D.B. (2004) Monomolecular logic: ''Half-adder'' based on multistate/multifunctional photochromic spiropyrans. *J. Phys. Chem. B*, **108**, 11942–11945.
31. Stojanovic, M.N. and Stefanovic, D. (2003) Deoxyribozyme-based half-adder. *J. Am. Chem. Soc.*, **125**, 6673–6676.
32. Andréasson, J., Kodis, G., Terazono, Y., Liddell, P.A., Bandyopadhyay, S., Mitchell, R.H., Moore, T.A., Moore, A.L., and Gust, D. (2004) Molecule-based photonically switched half-adder. *J. Am. Chem. Soc.*, **126**, 15926–15927.
33. Andréasson, J., Straight, S.D., Kodis, G., Park, C.D., Hambourger, M., Gervaldo, M., Albinsson, B., Moore, T.A., Moore, A.L., and Gust, D. (2006) All-photonic molecular half-adder. *J. Am. Chem. Soc.*, **128**, 16259–16265.
34. Margulies, D., Melman, G., and Shanzer, A. (2006) A molecular full-adder and full-subtractor, an additional step toward a moleculator. *J. Am. Chem. Soc.*, **128**, 4865–4871.
35. Andréasson, J., Pischel, U., Straight, S.D., Moore, T.A., Moore, A.L., and Gust, D. (2011) All-photonic multifunctional molecular logic device. *J. Am. Chem. Soc.*, **133**, 11641–11648.
36. Andréasson, J., Straight, S.D., Bandyopadhyay, S., Mitchell, R.H., Moore, T.A., Moore, A.L., and Gust, D. (2007) Molecular 2:1 digital multiplexer. *Angew. Chem. Int. Ed.*, **46**, 958–961.
37. Amelia, M., Baroncini, M., and Credi, A. (2008) A simple unimolecular multiplexer/demultiplexer. *Angew. Chem. Int. Ed.*, **47**, 6240–6243.
38. Arugula, M.A., Bocharova, V., Halamek, J., Pita, M., and Katz, E. (2010) Enzyme-based multiplexer and demultiplexer. *J. Phys. Chem. B*, **114**, 5222–5226.
39. Andréasson, J., Straight, S.D., Bandyopadhyay, S., Mitchell, R.H., Moore, T.A., Moore, A.L., and Gust, D. (2007) A molecule-based 1:2 digital demultiplexer. *J. Phys. Chem. C*, **111**, 14274–14278.

40. Perez-Inestrosa, E., Montenegro, J.M., Collado, D., and Suau, R. (2008) A molecular 1:2 demultiplexer. *Chem. Commun.*, 1085–1087.
41. Andréasson, J., Straight, S.D., Moore, T.A., Moore, A.L., and Gust, D. (2008) Molecular all-photonic encoder-decoder. *J. Am. Chem. Soc.*, **130**, 11122–11128.
42. Ceroni, P., Bergamini, G., and Balzani, V. (2009) Old molecules, new concepts: [Ru(bpy)$_3$]$^{2+}$ as a molecular encoder-decoder. *Angew. Chem. Int. Ed.*, **48**, 8516–8518.
43. Andréasson, J., Straight, S.D., Moore, T.A., Moore, A.L., and Gust, D. (2009) An all-photonic molecular keypad lock. *Chem. Eur. J.*, **15**, 3936–3939.
44. Halamek, J., Tam, T.K., Strack, G., Bocharova, V., Pita, M., and Katz, E. (2010) Self-powered biomolecular keypad lock security system based on a biofuel cell. *Chem. Commun.*, **46**, 2405–2407.
45. Margulies, D., Felder, C.E., Melman, G., and Shanzer, A. (2007) A molecular keypad lock: a photochemical device capable of authorizing password entries. *J. Am. Chem. Soc.*, **129**, 347–354.
46. Remon, P., Hammarson, M., Li, S.M., Kahnt, A., Pischel, U., and Andréasson, J. (2011) Molecular implementation of sequential and reversible logic through photochromic energy transfer switching. *Chem. Eur. J.*, **17**, 6492–6500.
47. Sun, W., Zhou, C., Xu, C.H., Fang, C.J., Zhang, C., Li, Z.X., and Yan, C.H. (2008) A fluorescent-switch-based computing platform in defending information risk. *Chem. Eur. J.*, **14**, 6342–6351.
48. Suresh, M., Ghosh, A., and Das, A. (2008) A simple chemosensor for Hg2+ and Cu2+ that works as a molecular keypad lock. *Chem. Commun.*, 3906–3908.
49. Straight, S.D., Liddell, P.A., Terazono, Y., Moore, T.A., Moore, A.L., and Gust, D. (2007) All-photonic molecular XOR and NOR logic gates based on photochemical control of fluorescence in a fulgimide-porphyrin-dithienylethene triad. *Adv. Funct. Mater.*, **17**, 777–785.
50. Baron, R., Onopriyenko, A., Katz, E., Lioubashevski, O., Willner, I., Sheng, W., and Tian, H. (2006) An electrochemical/photochemical information processing system using a monolayer-functionalized electrode. *Chem. Commun.*, 2147–2149.
51. de Ruiter, G., Tartakovsky, E., Oded, N., and van der Boom, M.E. (2010) Sequential logic operations with surface-confined polypyridyl complexes displaying molecular random access memory features. *Angew. Chem. Int. Ed.*, **49**, 169–172.
52. Elbaz, J., Moshe, M., and Willner, I. (2009) Coherent activation of DNA tweezers: a "SET-RESET" logic system. *Angew. Chem. Int. Ed.*, **48**, 3834–3837.
53. Periyasamy, G., Collin, J.P., Sauvage, J.P., Levine, R.D., and Remacle, F. (2009) Electrochemically driven sequential machines: an implementation of copper rotaxanes. *Chem. Eur. J.*, **15**, 1310–1313.
54. Pischel, U. (2010) Advanced molecular logic with memory function. *Angew. Chem. Int. Ed.*, **49**, 1356–1358.
55. Pita, M., Strack, G., MacVittie, K., Zhou, J., and Katz, E. (2009) Set-reset flip-flop memory based on enzyme reactions: toward memory systems controlled by biochemical pathways. *J. Phys. Chem. B*, **113**, 16071–16076.
56. Pischel, U. and Andréasson, J. (2010) A simplicity-guided approach toward molecular set-reset memories. *New J. Chem.*, **34**, 2701–2703.

5
Engineering Luminescent Molecules with Sensing and Logic Capabilities
David C. Magri

5.1
Introduction

The design of multifunctional molecular devices is an expanding frontier area of research encompassing the fields of materials science, nanotechnology, and supramolecular chemistry [1–3]. A current theme is to design molecules with individual components that operate independently of one another, yet work cooperatively as a team to produce devices with greater function than the sum of their individual parts [4–6]. Molecules with the capacity to perform a variety of stimuli-responsive functions based on built-in algorithms can, therefore, independently process information, and then of their own accord, provide a composite result autonomously [7–17]. Molecules with the ability to perform Boolean algebraic operations are the focus of this chapter.

Intelligent molecular systems designed to be activated by some external stimulus by changing a physical or chemical state can be viewed as switches, and therefore, may be useful for the transfer, processing, and storage of information. Dozens of research laboratories around the world have embraced the design, synthesis, characterization, and testing of chemical systems that mimic the basic operations of electrical circuitry [18]. The general engineering approach to designing information-processing molecules is based on a modular strategy with specific units of structural and functional importance [19]. Earlier reviews described in detail the various strategies for designing luminescence sensing devices: these ideas range in scope from ways of covalently arranging modular units to the exploitation of self-assembly [20–22].

This multidisciplinary research field demands a number of skill sets from a background in organic synthesis for the strategic design and molecular engineering of novel chemical systems, to analytical chemistry for the quantification of analyte concentrations and the determination of binding constants, as well as other thermodynamic and kinetic parameters, and Boolean algebra and binary computation for expressing observed chemical phenomena (analog information) as easy-to-interpret digital information. Then, there are also the skills and tools required for anchoring these molecular devices at interfaces and surfaces.

Chemical sensors are based on ion-binding equilibria of populations of molecules that produce a sigmoidal response function. Logic gates are digital devices with a defined threshold limit so that the continuum of data is given a binary code value of either "**1**" or "**0**." By viewing sensors as a two-state device rather than as an analog device, our point of view changes – a sensor becomes a digital one-input logic gate. A common mode of output has been fluorescence, or a related form of luminescence, while the earliest examples used inorganic chemical species as inputs, although now larger molecules and physical inputs are also useful for specific applications [3–9].

This chapter gives an overview of some recent contributions to the area of molecular logic gates and molecular computation resulting from the contributing author's personal research experiences while working at Queen's University of Belfast, and afterwards while working independently in Canada. A brief introduction is provided regarding some considerations to take into account when creating luminescent molecular logic gates (and more advanced circuitry analogs) with a modular construction paradigm based on the competition between photoinduced electron transfer (PET) and fluorescence. The reader is encouraged to refer to Chapter 2, which outlines the basic design principles of luminescent PET sensors and also includes truth tables, notably for two-input and three-input AND logic, which will be discussed in this chapter. Specific examples of molecular logic gates operating in solution, including future potential applications as "lab-on-a-molecule" systems for medical diagnostics [23] and redox-fluorescent logic gates for environmental monitoring [24], among others, are covered in some detail.

5.2
Engineering Luminescent Molecules

To the general public, engineering is a discipline associated with building large structures such as towers, bridges, and cars. Chemistry is also a discipline associated with building structures, but at the subnanometer realm. Hence, the chemist who builds molecules with specific functions is in essence an engineer. Organic chemistry is central to the overall enterprise as the design, synthesis, and characterization of carbon-containing molecules precedes the demonstration of luminescence switching and information processing at the molecular level. As molecular engineers, chemists not only perform chemical transformations but also spend considerable time planning and designing intelligent molecules – and afterwards, considerable time testing them too. The discipline has now progressed to a level where there is an assortment of beautiful proof-of-concept examples of intelligently engineered molecules with multifunctional capabilities, summarized nicely in many recent reviews [7–13]. Some specific favorites include the concepts of molecular calculators [25, 26], combinatory identification tags for tiny objects [27], security keypad locks (sequential memory logic) [28, 29], and a selective photodynamic therapy agent [30].

How do you go about engineering such intelligent molecules for digital information-processing applications? Foremost, you need some manner of communication both for the inputs and the outputs channels: chemical, photonic, and electrochemical approaches are the most commonly used [6]. Systems based on optical outputs have been the forerunner at the molecular level since the emergence of the field [31], motivated by the simplicity of the techniques used to the measure the output signals. Changes in color can be seen by the "naked eye" and recorded by absorption or fluorescence spectroscopy. However, understanding these changes is not always simple as a number of excited state phenomena provide the inventor with a selection of mechanisms including PET, internal charge transfer (ICT), electronic energy transfer (EET), excimer and exciplex formation, and excited state proton transfers. Although this chapter focuses on examples based on the PET mechanism, the importance of ICT and EET for more advanced logic operations such as arithmetic is worth mentioning [11].

ICT systems paired with PET systems provide direct conjugation between a receptor and chromophore, which causes the molecular orbitals of the modular components to overlap. These conjugated pi-systems [32, 33] are necessary components for more complex logic systems such as half-subtractor [34, 35] and half-adder [36, 37] mimics. ICT systems are purposely designed such that one end of the molecule is electron rich while the other end is electron poor. A common strategy has been to use changes in pH to control a wavelength shift, which often accompanies a dramatic color change between the complexed and uncomplexed species [35]. Spectral shifts are also useful for ratio analysis of two separated wavelength intensities in intracellular sensing [38].

The "fluorophore-spacer-receptor" modular configuration, based on the competition between two photophysical processes, namely, intramolecular PET and fluorescence, is an established design principle [21–23]. Its general applicability comes from a rational model that allows for various components, in particular the receptor sites, to operate in parallel in terms of substrate binding, yet collectively in terms of quenching the fluorophore emission [24]. The engineering of luminescent molecular logic gates begins from the selective choice of a fluorophore with a reasonable quantum yield of fluorescence [39]. Fluorescent polyaromatic hydrocarbons, such as anthracene **1** (Scheme 5.1), satisfy the elementary PASS 1 logic gate function [40]. In other words, independent of the absence or presence of an input, represented by either **0** or **1**, respectively, the molecule always emits a light signal, which corresponds to an output of **1**. Although it may seem trivial, the spacer and receptor must satisfy the elementary PASS 0 function: they need to be independent of the input (whether **0** or **1**) and never emit a light signal, which corresponds to an output of **0**. Once the three basic components (the fluorophore, spacer, and receptor) are connected together, a two-state sensor results. Single-input PET systems can operate as "off–on" switches, otherwise known in digital terms as YES logic gates. In the "off" state, the absence of an input **0** does not yield a light signal so the output is **0**; in the "on" state, the presence of an input **1** yields a light signal so the output is **1**. Two early examples with chemical inputs are **2** for protons [41] and **3** for sodium ions (Scheme 5.1) [42]. By appropriate selection of the receptor

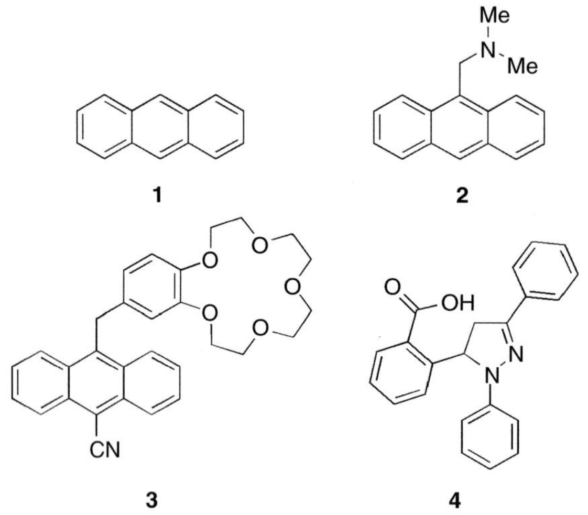

Scheme 5.1

and consideration of the thermodynamics, it is also possible to design single-input "on–off" PET systems, otherwise known as *NOT logic gates* [43]. In this case, in the "on" state, the absence of an input **0** results in a light signal so the output is **1**; in the "off" state, the presence of an input **1** shuts off the emission so the output is a **0**. The pyrazoline molecule **4** is in the "off" state when the carboxylic acid is protonated (Scheme 5.1).

A common strategy is to construct the logic gates by a convergent synthesis in which the final reaction involves a reductive animation or "click" chemistry to bring the components together, while at the same time introducing a spacer. In PET systems, the spacer is customarily kept at a length of one or two carbon bonds. This has the effect of maximizing the switching efficiency (signal-to-noise ratio) by allowing the rate for PET to be as fast as possible. Thus, a short spacer such as the methylene unit is quite common, as it is well known that increasing the spacer length results in an exponential decrease in the rate of electron transfer, and hence a lower switching ability. Virtual spacers using sterically induced orthogonal geometries provide an alternative design for keeping spacer distances to a minimal. The receptor is chosen based on the analyte to be detected and the average analyte concentration to be sensed. Thought has to be given to the selectivity of the receptor against matrix effects when considering real-life applications and the possibility of intramolecular interactions for larger supramolecular assemblies.

The theory for predicting the feasibility of PET is well established [44]. Because of this, the designer has considerable forecasting power for predicting the feasibility of PET by using the Weller equation as given in Eq. (5.1) [45].

$$\Delta G_{PET} = E_{OX} - E_{RED} - E_S - e^2/\varepsilon r \tag{5.1}$$

In addition to knowledge of the redox potentials where E_{OX} is the standard oxidation potential and E_{RED} the standard reduction potential, ΔG_{ET} is dependent on knowledge of the singlet energy of the fluorophore, E_S, and the Coulomb term $e^2/\varepsilon r$, which accounts for electrostatic interactions. In polar solvents, the contribution from the Coulomb term is generally negligible. Although the Weller equation has a high success rate for accurately predicting the feasibility of PET in the "fluorophore-spacer-receptor" format, some care is still needed. The designer must not forget, among other things, about the importance of kinetics and the paradox of the Marcus equation [46]. Regiostereochemical effects due to a repulsive photogenerated electric field or a frontier orbital node at the point of attachment to the fluorophore are other exceptions to be aware of as discovered in 4-aminonaphthalimides [47]. Homoconjugation of a heteroatom to the fluorophore is another issue to avert, because even when the receptor fully binds the analyte, the lone pairs of the heteroatom deactivate the excited state by a charge transfer mechanism [48].

5.3
Logic Gates with the Same Modules in Different Arrangements

As PET-based luminescent systems are modular, it was only a matter of time before the field progressed to sets of sensors with more than one receptor. Hence, the next leap forward resulted from the cumulative addition of two YES gates connected in parallel, but with only a single fluorophore, the result being the first example of a molecular AND logic gate, **5** (Scheme 5.2), in 1993 by de Silva and coworkers [31]. The molecule was designed on the premise of a "fluorophore-spacer$_1$-receptor$_1$-spacer$_2$-receptor$_2$" format by exploiting earlier work on proton [49] and sodium sensors [42] (**2** and **3**, Scheme 5.1). The "programed" algorithm is based on the requirement of two chemical inputs exceeding the binding thresholds of each selective receptor to yield a light output. Rather than sensing for two analytes independently, **5** communicates a cooperative fluorescence enhancement when both analytes are detected together above threshold levels. The tertiary amine receptor captures protons and the benzo-15-crown-5 ether sodium ions. In digital terms, an enhanced fluorescence signal (output **1**) is observed only in the presence of both sodium and protons (inputs **1,1**). The absence of either (inputs **0,1** or **1,0**) or both (inputs **0,0**) chemicals results in a weak emission (output **0**). Hence, blocking of only one of the two PET channels still results in quenching the fluorescence emission, thus turning it "off." The reader is encouraged to refer to Table 2.1 in the previous chapter by de Silva for details.

Rearrangement of the modular components to yield version **6** (Scheme 5.2) was an improvement on the original design [50]. This second AND logic gate was constructed with the fluorophore in the middle of the molecular device rather than at the end according to a "receptor$_1$-spacer$_1$-fluorophore-spacer$_2$-receptor$_2$." The outcome was a molecule with a greater fluorescence enhancement resulting from shortening the separation distances between the fluorophore and both receptors so that the rate of PET was even more competitive with fluorescence, hence, lowering the minimum fluorescence quantum yields in the absence of one or both analytes.

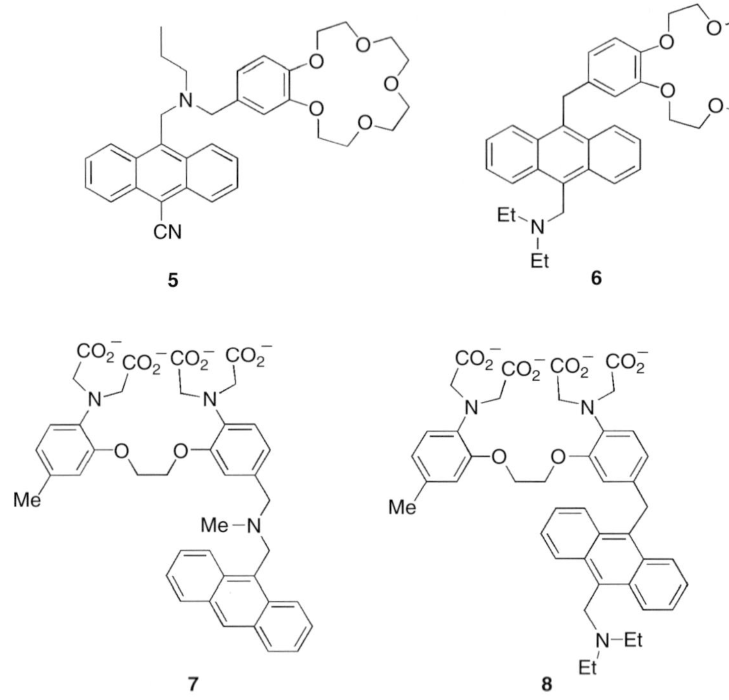

Scheme 5.2

In another example, rearrangement of modular components not only results in improved quantum yields but can also bring about some interesting capabilities. Molecules **7** and **8** (Scheme 5.2) are constructed using the same receptors, spacers, and fluorophore: these being a tertiary amine for binding H^+, a BAPTA ligand for binding Ca^{2+}, and anthracene as the fluorophore [51]. In the "receptor$_1$-spacer$_1$-fluorophore-spacer$_2$- receptor$_2$" format, **8** operates as an AND logic gate. In the "fluorophore-spacer$_1$-receptor$_1$-spacer$_2$-receptor$_2$" arrangement, **7** not only operates as an AND logic gate, but from a tertiary perspective, it operates as a H^+-driven "off–on–off" system with the Ca^{2+} acting as an enabling switch that suppresses the PET reaction. In this arrangement at low pH the carboxylic acid aligns with the pi-cloud orbitals of the anthracene fluorophore and deactivates the excited state.

5.4
Consolidating AND Logic

Sensors and logic gates with three PET pathways are a new avenue in the field of luminescent sensing and molecular computation [52]. One of the earlier examples of this century, **9**, shown in Scheme 5.3, combined two different receptor sites for protons, a pyridine and a tertiary amine, along with

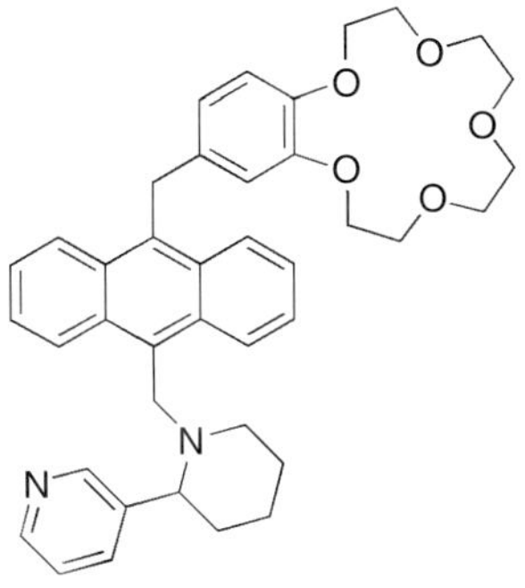

Scheme 5.3

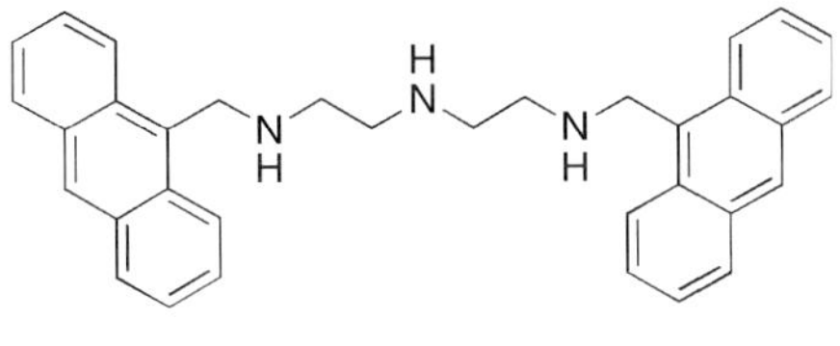

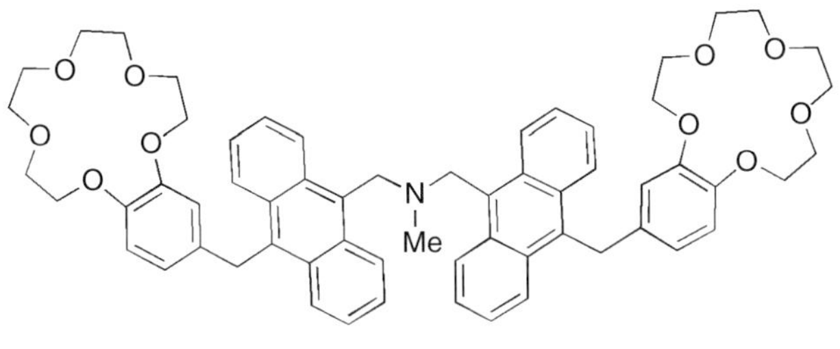

a benzo-15-crown-5 ether for sodium [53]. The molecule based on a "receptor$_1$-spacer$_1$-fluorophore-spacer$_2$-receptor$_2$-spacer$_3$-receptor$_3$" module arrangement behaves as a sodium-enabled "off–on–off" pH switch. The three complexed states, **9-Na$^+$**, **9-Na$^+$-H$^+$**, and **9-Na$^+$-2H$^+$** represent the fluorescence "off", "on," and "off" states, respectively. In the presence of sodium ions and protons (pH 6), which only results in protonation of the tertiary amine, **9** behaves as an AND logic gate. However, increasing the proton concentration to pH 2 protonates the pyridine and initiates a PET process between the excited state anthracene and the pyridinium, which quenches the emission of **9-Na$^+$-2H$^+$**.

Another example is a pH-controlled polyamine with a diethylenetriamine chain between two anthracene fluorophores **10** (Scheme 5.3), which behaves as a two-channel reconfigurable logic gate by exploiting both the monomer and excimer emission spectra [54, 55]. Depending on the transition metal input, various logic gate functions are observed when monitoring two output channels at 416 and 520 nm, corresponding to monomer and excimer emission, respectively. The switching function in water is rationalized by the formation of an intramolecular charge transfer complex in the ground state via pi-stacking interactions between the two fluorophores. At pH 2, the monomer emission at 416 nm is bright. In contrast, at a pH greater than 9, only the excimer emission at 520 nm is observed albeit the excimer emission being restricted to only 12% of the maximum monomer emission.

Our first success at a three-receptor logic gate **11** (Scheme 5.3) combined two benzo-15-crown-5 ethers and a tertiary amine [56]. This novel "receptor$_1$-spacer$_1$-fluorophore-spacer$_2$-receptor$_2$-spacer$_2$-fluorophore-spacer$_1$-receptor$_1$" format contains two fluorophores and three receptor sites. The molecule has a plane of symmetry with the crown ethers on opposite ends, anthracene fluorophores on the insides, and a tertiary amine right in the center. Figure 5.1A illustrates the fluorescence output response of **11** in the presence of H$^+$ and Na$^+$ concentrations according to AND logic. A high sodium ion concentration corresponds to the addition of 0.1 M sodium methanesulfonate and a high proton concentration to 0.015 M methanesulfonic acid. Low sodium levels were maintained by the absence of the sodium salt and a low proton levels by the addition of tetramethylammonium hydroxide. The maximum quantum yield of fluorescence in methanol was found to be 0.22, which is identical to that observed with previous YES [42] and AND [50] logic gate models. In the "on" state, the stoichiometry of **11** to Na$^+$ to H$^+$ is one-to-one-to-two as confirmed by electrospray ionization mass spectroscopy. The second sodium ion acts as the "trigger" that switches "on" the fluorescence.

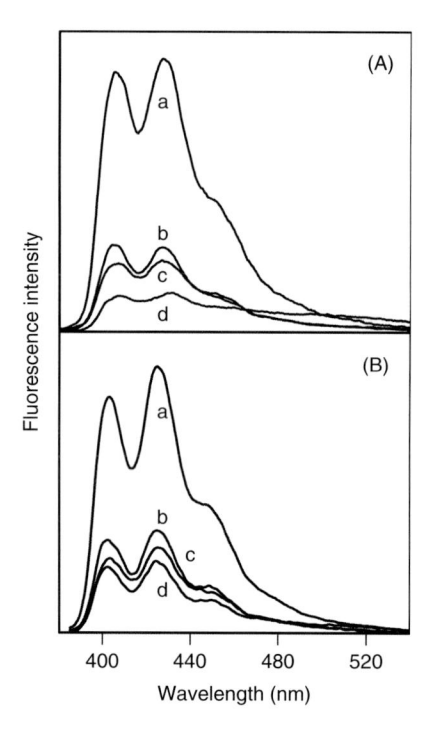

Figure 5.1 Fluorescence emission spectra of AND gate **11** in methanol excited at 379 nm under four input conditions in the presence of (A) (a) 0.1 M CH$_3$SO$_3$Na and 0.015 M CH$_3$SO$_3$H, (b) 0.1 M CH$_3$SO$_3$Na, (c) 0.015 M CH$_3$SO$_3$H, and (d) 0.015 M (CH$_3$)$_4$NOH and (B) (a) 0.1 M CH$_3$SO$_3$Cs and 0.015 M CH$_3$SO$_3$H, (b) 0.1 M CH$_3$SO$_3$Cs, (c) 0.015 M CH$_3$SO$_3$H, and (d) 0.015 M (CH$_3$)$_4$NOH. (Adapted from Ref. [56], reproduced with permission from Elsevier.)

As shown in Figure 5.1B, **11** was also found to obey AND logic in response to H^+ and Cs^+. The additional spacer-receptor unit, as compared to **6** (Scheme 5.2), adds extra functionality as the molecule can also compute with Cs^+. The flexibility about the single bonds of the spacers allows for a conformation that can capture a cesium ion, which is normally too large for one benzo-15-crown-5 ether alone. Sandwich-type systems with two crown ethers have previously been reported for complexing potassium or rubidium ions [57].

The anthracene emission of **11** in methanol is similar in acidic solution whether the metal ion is sodium or cesium with peaks centered at 408 and 429 nm. Contrary to what might be expected, no excimer formation was observed on complexation of Cs^+, suggesting that the anthracene units do not align in a stacking orientation. Excimer formation was anticipated based on the Hirayama rule [58], given the conformational flexibility around the central tertiary amine. In comparison, an excimer was observed for a dual anthracene system with an oxygen-containing linkage between the aromatic rings [59]. Excimer emissions tend to show a preference in nonpolar solvents such as cyclohexane rather than in methanol or acetonitrile [60]. An alternative possibility for the absence of excimer formation in **11** is that the fluorophore excited state lifetime in methanol is short, so PET quenches the excited state before the excimer can form.

5.5
"Lab-on-a-Molecule" Systems

The first example of a third-generation logic system with three PET processes and three distinct inputs within a single molecule **12** (Scheme 5.4) was reported in 2006 [23]. The detection of three chemical species has opened up the field of molecular computation to handle a congregation of entities. This unprecedented "receptor$_1$-spacer$_1$-fluorophore-spacer$_2$-receptor$_2$-spacer$_3$-receptor$_3$" "supermolecule" allows for PET from each receptor to quench the fluorescence unless all three sites are blocked with the appropriate analyte, these being H^+, Na^+, or Zn^{2+}. Figure 5.2, curve a, shows a threefold fluorescence emission (output **1**) on detection of all three analytes (inputs **1,1,1**). Refer to table 2.3 in chapter 2 for details. Hence, this three-input AND logic gate performs three separate tests and provides one composite result. An earlier report of a three-input AND gate was based on two chemical species and a light input as the third [61].

Our "lab-on-a-molecule" prototype **12** is an example of how data from several sensory channels can be processed on-board a single molecular platform with a hard-wired algorithm to make "yes–no" decisions by combining sensing, computing, and functionality all within a single molecule. The current PET sensor technology measures the blood analytes of ions such as Na^+, K^+, and Ca^{2+} using one molecule per analyte [62–64]. Multi-analyte logic gates could assist busy doctors by having an array of molecules analyze many samples, as is done nowadays, with the added function of the molecules being able to do diagnosis. Rather than

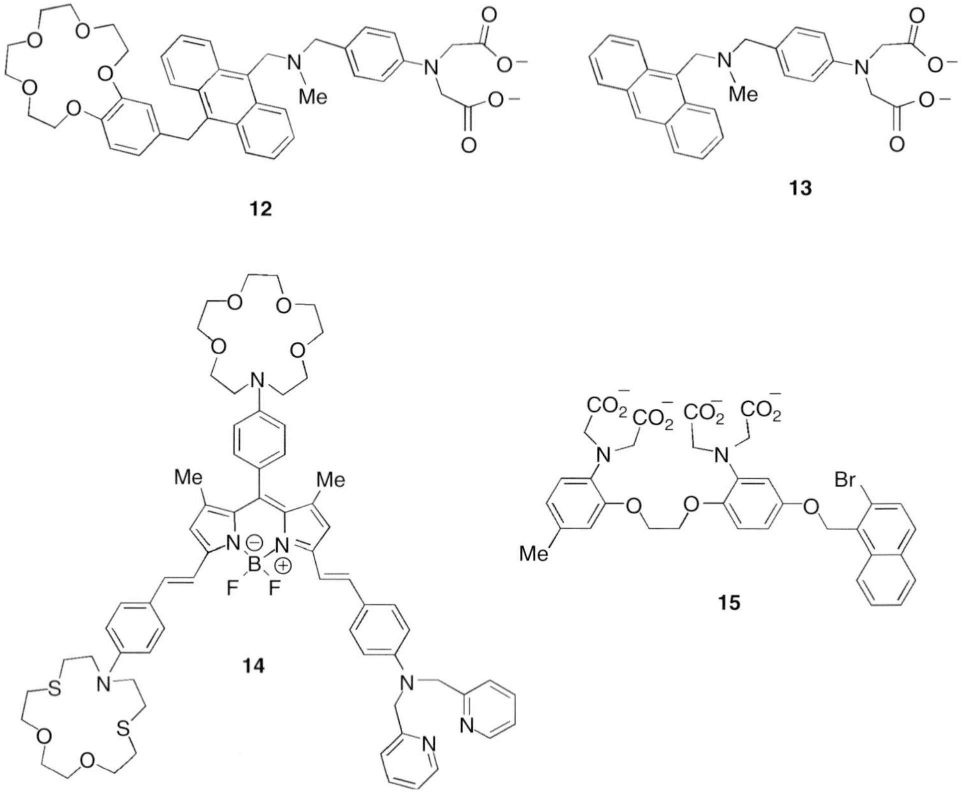

Scheme 5.4

detecting for individual analytes, we could have intelligent molecules with a bit of computing ability able to detect for disease conditions [65].

An accomplishment with our "lab-on-a-molecule" prototype is that it operates in aqueous solution. However, it was quite surprising that the quantum yield of fluorescence when all three analytes were bound was a modest 0.02 – an order of magnitude lower than expected. After amassing a collection of data from related one-input [41, 42] and two-input logic gates [50], the model compound **13** (Scheme 5.4) was synthesized to fill the knowledge void and resolve the quantum yield discrepancy [66]. Depending on the output wavelength chosen, **13** can be configured as an AND gate or INHIBIT gate for protons and zinc cations. The proton and zinc binding constants in water were found to be identical to those of yield **12**, and the quantum yield 0.19 almost on par with anthracene **1**. Thus, the lower fluorescence output observed for **12** can be attributed to the low stability constant for binding of the sodium ion to the benzo-15-crown-5-ether in aqueous solution, which does not effectively deactivate the PET pathway. In retrospect, this finding is not surprising considering that the binding constant for Na^+ by benzo-15-crown-5 ether in **12** is −0.3 in water, 3 orders of magnitude lower than the

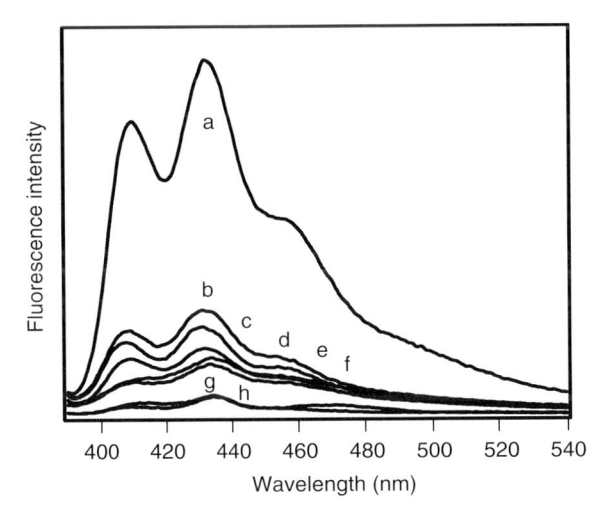

Figure 5.2 The eight fluorescence output spectra of 1×10^{-5} M of **12** in water corresponding to three-input AND logic: (a) high Na^+, H^+, Zn^{2+}, (b) high Na^+, H^+, low Zn^{2+}, (c) high Na^+, Zn^{2+}, low H^+, (d) high Na^+, low H^+, Zn^{2+}, (e) high H^+, Zn^{2+}, low Na^+, (f) high Zn^{2+}, low Na^+, H^+, (g) high H^+, low Na^+, Zn^{2+}, and (h) low Na^+, H^+, Zn^{2+}. High and low sodium levels correspond to 5 M sodium methanesulfonate and no added sodium salt. High and low proton levels correspond to $10^{-6.0}$ and $10^{-9.5}$ M protons adjusted with methanesulfonic acid and tetramethylammonium hydroxide. High and low zinc levels were maintained with 9.1 mM zinc sulfate in the presence of 10 mM iminodiacetic acid at the corresponding pH values. Low input levels (curves (g) and (h)) were maintained by scavenging trace heavy metals with 1 mM ethylenediaminetetraacetic acid. (Adapted from Ref. [23] with permission. Copyright American Chemical Society, 2006.)

binding constant for similar systems in methanol. Measurements in 50% aqueous methanol supported this hypothesis, as the binding constant was found to be greater by 1.2 log units than in aqueous solution alone [23]. Alternatively, besides altering the solvent composite, another strategy for improving the quantum yield of **12** would be to replace the inferior benzo-15-crown-5 ether with a stronger binding receptor for sodium ions such as N-(o-methoxyphenyl)aza-15-crown-5 ether [62].

The Akkaya group [67] has successfully demonstrated a second example of a "lab on a molecule," **14** (Scheme 5.4), with three selective receptors for the simultaneous detection of Ca^{2+}, Hg^{2+}, and Zn^{2+}. On the basis of a boron-dipyrromethene (BODIPY) dye, the excitation wavelength is 620 nm and the output fluorescence is 656 nm. The quantum yield in the presence of elevated cation concentrations is 0.27 in acetonitrile, and the output enhancement in the "on" state is a respectable factor of 3. The molecule operates by a combination of ICT and PET processes. The use of the BODIPY fluorophore provides some noticeable advantages such as versatility in synthetic design and a long excitation wavelength for biological applications. Notwithstanding that the study was done in acetonitrile, because of poor solubility in water, and the laborious effort required to synthesize the molecule, it is a great example of multifunctional engineering in a single molecule.

It is noteworthy to point out that in biological systems not only can too much of an analyte be life-threatening, but the same is also true of too little. Hence, from a clinical diagnostic point of view, three-input AND logic coincides with a condition where all three analytes are detected above set threshold levels (inputs **1,1,1**; output **1**). For diagnostic purposes, we could also just as easily consider three-input INHIBIT logic, which is witnessed when two analytes are above threshold levels and the third is below (inputs **1,1,0**; output **1**). The concept was first illustrated with **15** (Scheme 5.4), consisting of a 2-bromonaphthalene phosphor linked to a BAPTA receptor for binding Ca^{2+} [68]. The bromonaphthalene emits a light signal when (i) the BAPTA receptor is occupied by Ca^{2+}, (ii) the bromonaphthalene unit is complexed with β-cyclodextrin, and (iii) the solution is free of oxygen. In other words, when the concentrations of Ca^{2+} and β-cyclodextrin are both high and the concentration of oxygen is low, the light output is high. The other seven possible truth table entries (inputs **0,0,0**; **1,0,0**; **0,1,0**; **0,0,1**; **0,1,1**; **1,0,1**; and **1,1,1**) all result in a low output luminescence (output **0**). The Ca^{2+} input is required to prevent the PET pathway between the BAPTA receptor to the bromonaphthalene and the β-cyclodextrin to maximize the phosphorescence output. No signal is observed, whatever the concentration of either Ca^{2+} or β-cyclodextrin, when the oxygen concentration is high, as it quenches the triplet state of the bromonaphthalene-β-cyclodextrin complex.

5.6
Redox-Fluorescent Logic Gates

Molecular switches responsive to individual physicochemical inputs including light, temperature, and redox potential have been known for quite some time [4, 69]. However, only recently are molecular engineers looking at ways of incorporating both physical and chemical stimuli-responsive components. First-generation examples of molecules responsive to an electrochemical input and a luminescent output, developed by Fabbrizzi and coworkers, were based on the redox-switching of a coordinated transition metal [70, 71]. Stemming from an earlier example by Lehn and coworkers [72], the past decade has brought about the emergence of a new generation of redox-active switches based on the oxidation (or reduction) of an organic component that result in fluorescence emission [73–81]. Commonly used redox units include ferrocene, tetrathiafulvalene (TTF), and quinones connected to aromatic hydrocarbons such as anthracene, pyrene, and perylene. Nowadays, these types of sensors are commonly referred to as *redox-fluorescent switches*.

Multi-input molecular logic gates with an electrochemical switching input function have also emerged. The bulk of the two-input redox-fluorescent examples are limited to NOR, AND, and INHIBIT logic gates [82–86]. An elegant example based on an absorbance output, an AND logic gate based on a two-component molecular-level machine responsive to redox and proton inputs, entailed a macrocyclic polyether and a π-deficient cyclophane [83]. A simpler approach, **16** (Scheme 5.5), contains a ferrocene moiety as the redox unit and anthracene

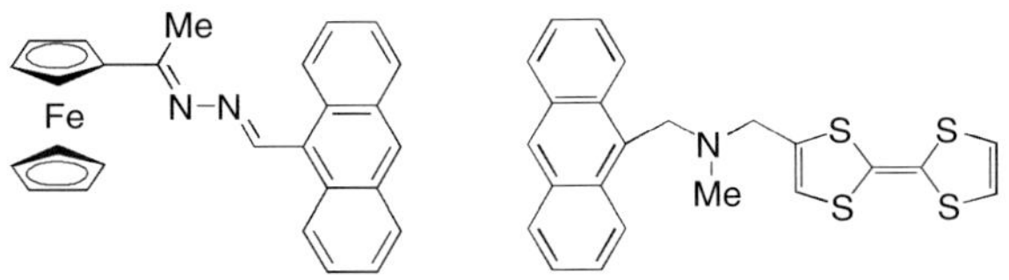

Scheme 5.5

as the signaling unit [86]. The ferrocene moiety can be reversibly converted to the ferrocenium ion both chemically and electrochemically. In the presence of 20 mM trifluoroacetic acid and 0.1 mM $(NH_4)_2Ce(NO_3)_6$ oxidant in tetrahydrofuran solution, the ferrocene unit is oxidized to its radical cation, resulting in a threefold fluorescent enhancement. Oxidation of the ferrocene moiety prevents electron transfer to the excited state anthracene resulting in an increase in the fluorescence emission. As there is some overlap between the absorption and emission spectra of ferrocene and anthracene, the quenching of the fluorescence may result from an energy transfer as well as a PET mechanism. Nonetheless, in an acidic environment and after oxidation of the ferrocene component, an AND logic function is observed.

The author has recently demonstrated a redox-fluorescent AND logic gate **17** (Scheme 5.5) that also simultaneously processes information regarding the acid concentration and oxidizing ability in solution by communicating a fluorescent light signal [24]. The conceptual design, based on the modular PET mechanism, has a fluorophore-spacer$_1$-receptor$_1$-spacer$_2$-redox donor format. A tertiary amine is the receptor for protons, and a TTF unit acts as the redox-activate component. As discussed later, profound changes were observed upon characterization by cyclic voltammetry and UV–visible absorbance spectroscopy and demonstration of AND logic by fluorescence spectroscopy.

Cyclic voltammetry is a versatile analytical technique [87]. The technique is used primarily for the determination of standard reduction potentials; however, in conjunction with other techniques, it can be used to determine a host of other pertinent thermodynamic, kinetic, as well as mechanistic information [88]. This is nicely illustrated with **17**, which was studied by voltammetry in acetonitrile solution as shown in Figure 5.3A, curves (a) and (b), and Figure 5.3B, curves (c) and (d). For comparison, voltammograms of authentic samples of TTF (Figure 5.3C, curve (e)) and anthracene (Figure 5.3C, curve (f)) are also shown. TTF has two reversible redox couples centered at 0.64 and 0.27 V corresponding to the radical dication and cation, respectively, and anthracene has a sharp, irreversible peak at 1.5 V. The irreversible nature of the anthracene peak is due to the follow-up chemistry that causes the anthracene radical cation to undergo another reactivity (e.g., dimerization) before reduction at the electrode occurs at the conventional scan rates employed. Figure 5.3C also illustrates that the presence of 10 mM methanesulfonic acid causes no changes in either of the voltammograms of TTF or anthracene, and hence confirms that the standard potentials of the individual

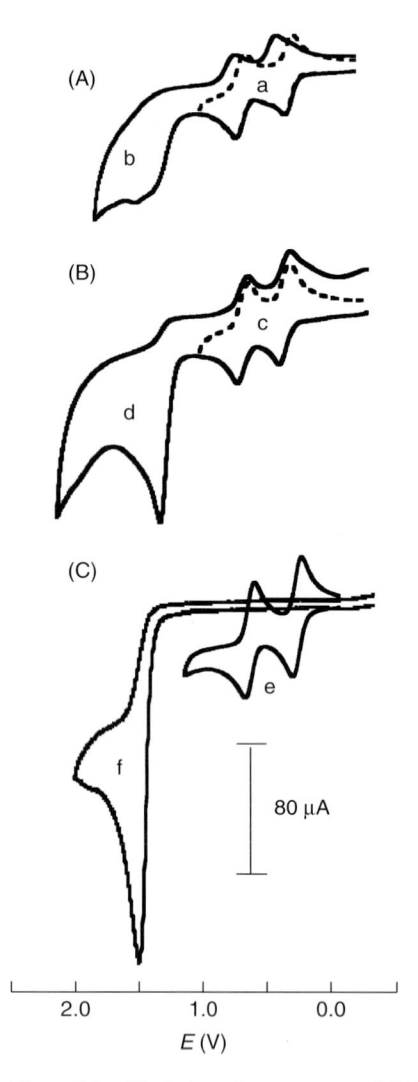

Figure 5.3 (A) Cyclic voltammograms (CVs) of 1.1 mM **17** in acetonitrile containing 0.10 M tetrabutylammonium perchlorate at a 3 mm glassy carbon electrode at 0.1 V s^{-1}; (B) CVs of 1.0 mM **17** in the presence of 10 mM methanesulfonic acid; (C) CVs of (e) 1.2 mM tetrathiafulvalene and (f) 1.3 mM anthracene. Curves (a), (c), and (e) were scanned between −0.2 and 1.2 V. Curves (b), (d), and (f) were scanned between −0.2 and 2.0 V. (Adapted from Ref. [24], reproduced with permission from the Royal Society of Chemistry (RSC) for the National de la Recherche Scientific (CNRS) and the RSC.)

components in both their neutral and oxidized states are robust in the presence of a weak acid.

Figure 5.3A shows that on scanning to 1.2 V (curve (a)) the oxidation of TTF behaves similarly to the authentic TTF sample as observed in Figure 5.3C, curve (e). However, scanning out to 2.0 V (curve (b)) reveals a broad, irreversible wave of lesser height at about 1.5 V, ascribed to the oxidation of the anthracene. On the return scan, the peak widths of the TTF redox couples no longer have a Nernstian peak separation of 59 mV. By repetitively scanning multiple times between 0.2 and 1.2 V – after initially scanning to 2.0 V – the presence of four redox couples was confirmed, the additional two standard potentials being at 0.97 and 0.52 V [24]. Hence, scanning out to 2.0 V results in the formation of a new redox-active product that consumes the anthracene radical cation. Additional mechanistic insight is provided on addition of 10 mM methanesulfonic acid (Figure 5.3B), which causes the redox couples at 0.97 and 0.52 V to disappear and the peak current at a potential of 1.4 V to increase by a factor of 3. From these observations, it can be concluded that excess acid prevents the competing reaction pathway associated with the anthracene radical cation and that the 70 mV anodic shift in the TTF peaks in Figure 5.3B is due to the electron-withdrawing effects resulting from protonation of the tertiary amine.

The UV–visible spectra of **17** in the absence (curve (a)) and presence (curve (b)) of 50 mM methanesulfonic acid are illustrated in Figure 5.4. In the absence of acid, upon controlled addition of 1 F mol^{-1} of charge at an applied potential of 0.60 V, a vivid green-colored solution results, with concomitant formation of two new absorbance bands centered at 444 and 593 nm, which are characteristic of the TTF radical cation. Similarly, the electrochemical oxidation of the redox unit in the presence of 50 mM methanesulfonic acid results in absorbance bands at 435 and 580 nm (Figure 5.4, curves (c–f)) [20]. Further oxidation of the radical cation to the dication after controlled addition of another equivalent

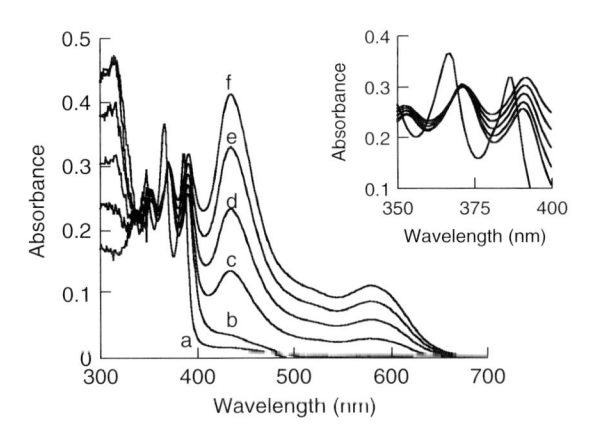

Figure 5.4 UV–visible absorption spectra of 3×10^{-5} M **17** in acetonitrile. Curve (a) corresponds to the absence of both an applied potential and methanesulfonic acid. On addition of 50 mM methanesulfonic acid and an applied potential at 0.60 V versus SCE, new bands are observed at 435 and 580 nm according to (b) 0, (c) 13, (d) 26, (e) 39, and (f) 52 mC of charge to oxidation of TTF to its radical cation state. The inset highlights an isosbestic point at 369 nm. (Ref. [24], reproduced with permission from The Royal Society of Chemistry (RSC) for the National de la Recherche Scientific (CNRS) and the RSC.)

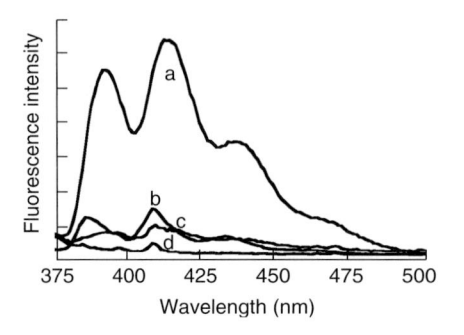

Figure 5.5 Fluorescence spectra of 10^{-5} M **17** in acetonitrile excited at the isosbestic point of 369 nm according to two-input AND logic: (a) high voltage, high H$^+$; (b) low voltage, high H$^+$; (c) high voltage, low H$^+$; and (d) low voltage, low H$^+$. High voltage levels at an applied voltage of 0.90 V using a 12 mm rotating disk glassy carbon electrode after addition of 2 F mol^{-1} of charge. Low voltage level: no applied voltage added. High proton level: 10 mM methanesulfonic acid. Low proton level: no acid added. (Ref. [24], reproduced with permission from The Royal Society of Chemistry (RSC) for the National de la Recherche Scientific (CNRS) and the RSC.)

of charge at an applied potential of 0.90 V recovers a clear, colorless solution with concomitant disappearance of the longer wavelength absorbance bands. The "off–on–off" switching of TTF from its neutral state to the radical cation and to the dication is also observed on addition of subsequent stoichiometric additions of Fe(ClO$_4$)$_3$.

Figure 5.5 illustrates the fluorescence output of **17** in acetonitrile according to the four experimental input conditions for AND logic after removal of two electrons and the addition of acid, which results in a threefold fluorescence enhancement. The acid input was 10 mM methanesulfonic acid, and for maximum switching enhancement the oxidation of the redox unit was controlled electrochemically. Compound **17** provides a high fluorescence signal (output **1**, curve (a)) upon the presence of protons and a positive applied voltage (inputs **1,1**). The absence of protons or the absence of an applied voltage (inputs **0,1** or **1,0**) or both (inputs **0,0**) result in a low fluorescence signal (output **0**, curves (b–d). AND logic is also observed when Fe(ClO$_4$)$_3$ is introduced into the solution in place of the applied voltage, although the maximum fluorescence output is only half that achieved by electrochemical control. Furthermore, the addition of more than 1 equiv. of Fe(ClO$_4$)$_3$ has the detrimental effect of quenching the anthracene fluorescence.

An unexpected result from the study was that the fluorescence quantum yield was a conservative 0.015 (Figure 5.5, curve (a)), which is an order of magnitude lower than expected from the anthracene fluorophore. A possible explanation is that a competing PET from the excited state anthracene to the TTF dication quenches a majority of the emission output. Although the driving force for PET is favorable at −1.2 V as calculated by Eq. (5.1), such a highly exothermic value suggests that the PET reaction occurs in the Marcus inverted region [46]. Residual PET reactions in the "on" state have been observed from crown ethers to fluorophores from the inability of metal cations to significantly reduce the HOMO energy of the receptor [23, 66]. This example illustrates a system with a residual PET in the "on" state that occurs in the opposite direction from the fluorophore to the dication.

5.7
Summary and Perspectives

The molecular engineer with the ability to rationally plan and construct molecules capable of performing intelligent functions, and with the ability, skill, and resources to study them, will be in demand in our growing knowledge-based economy. Multi-analyte sensors based on logic and computation have great application potential for information processing and intelligent medical diagnostics [65]. The concept of "lab-on-a-molecule" systems that can detect for at least three analytes is a premier example that demonstrates the cross-fertilization of synthetic skill with the design of a molecule with parallel processing capability. The current paradigm in our hospitals is for a doctor to order blood tests for a host of analytes and then examine the results for parameters that overshoot and undershoot normal analyte levels. "Lab-on-a-molecule" technology could assist doctors, not only by measuring the parameters, but also by making an autonomous intelligent decision on behalf (or in consultation) of the doctor. With the growing crisis in government health care costs, the creation of intelligent molecules for the simultaneous detection of a congregation of analytes may not only assist doctors diagnose patients but also help to alleviate clinical testing budget expenses too.

The emerging subfield of redox-fluorescent logic gates combines chemical and redox inputs with a fluorescent output. The relationship between the standard potentials and proton concentrations of metals in aqueous solutions was recognized by Pourbaix, who summarized the stability of numerous metal species as a function of the pE and pH [89]. With any new concept, it is difficult to forecast what applications may become possible in the future. Some possibilities may be in the areas of molecular electronics, water quality monitoring, and heavy metal analysis in tailing ponds about mining sites.

Presently, the current models for predicting the driving force for electron transfer are reasonably well established. On the contrary, however, predicting the quantum yield of fluorescence based on these calculations still comes with a considerable amount of uncertainty as discovered with **12** and **17**. However, by reflecting on past knowledge of simple sensor and logic gate models, we can consolidate this information and gain even more predictive power with regards to the fluorescence quantum yields [66]. By looking at sets of related molecules more holistically, we gain further insight into the rational design of ever smarter molecules with built-in orthogonality and cooperativity for ever more advanced logic operations [7–10].

References

1. Kay, E.R., Leigh, D.A., and Zerbetto, F. (2007) *Angew. Chem. Int. Ed.*, **46**, 72.
2. Northrop, B.H., Yang, H.B., and Stang, P.J. (2008) *Chem. Commun.*, 5896.
3. de Silva, A.P., Uchiyama, S., Vance, T.P., and Wannalerse, B. (2007) *Coord. Chem. Rev.*, **251**, 1623.
4. Balzani, V., Venturi, M., and Credi, A. (2003) *Molecular Devices and Machines*, Wiley-VCH Verlag GmbH, Weinheim.
5. Ballardini, R., Ceroni, P., Credi, A., Gandolfi, M.T., Maestri, M., Semarano, M., Venturi, M., and Balzani, V. (2007) *Adv. Funct. Mater.*, **17**, 740.
6. Balzani, V., Credi, A., and Venturi, M. (2008) *Chem. Eur. J.*, **14**, 26.
7. Pischel, U. (2010) *Aust. J. Chem.*, **63**, 148.
8. Andréasson, J. and Pischel, U. (2010) *Chem. Soc. Rev.*, **39**, 174.
9. Szacilowski, K. (2008) *Chem. Rev.*, **108**, 3481.
10. de Silva, A.P. and Uchiyama, S. (2007) *Nat. Nanotechnol.*, **2**, 399.
11. Pischel, U. (2007) *Angew. Chem. Int. Ed.*, **46**, 4026.
12. Credi, A. (2007) *Angew. Chem. Int. Ed.*, **46**, 5472.
13. Magri, D.C., Vance, T.P., and de Silva, A.P. (2007) *Inorg. Chim. Acta*, **360**, 751.
14. Gust, D., Moore, T.A., and Moore, A.L. (2006) *Chem. Commun.*, 1169.
15. de Silva, A.P. and McClenaghan, N.D. (2004) *Chem. Eur. J.*, **10**, 574.
16. Balzani, V., Credi, A., and Venturi, M. (2003) *ChemPhysChem*, **4**, 49.
17. Raymo, F.M. (2002) *Adv. Mater.*, **14**, 401.
18. de Silva, A.P. (2011) *Chem. Asian. J.*, **6**, 750.
19. de Silva, A.P., Vance, T.P., West, M.E.S., and Wright, G.D. (2008) *Org. Bio. Chem.*, **6**, 2468.
20. de Silva, A.P., Gunaratne, H.Q.N., Gunnlaugsson, T., Huxley, A.J.M., McCoy, C.P., Rademacher, J.T., and Rice, T.E. (1997) *Chem. Rev.*, **97**, 1515.
21. Bissell, R.A., de Silva, A.P., Gunaratne, H.Q.N., Lynch, P.L.M., Maguire, G.E.M., McCoy, C.P., and Sandanayake, K.R.A.S. (1993) *Top. Curr. Chem.*, **168**, 223.
22. Bissell, R.A., de Silva, A.P., Gunaratne, H.Q.N., Lynch, P.L.M., Maguire, G.E.M., and Sandanayake, K.R.A.S. (1992) *Chem. Soc. Rev.*, **21**, 187.
23. Magri, D.C., Brown, G.J., McClean, G.D., and de Silva, A.P. (2006) *J. Am. Chem. Soc.*, **128**, 4950.
24. Magri, D.C. (2009) *New. J. Chem.*, **33**, 457.
25. Margulies, D., Melman, G., and Shanzer, A. (2006) *J. Am. Chem. Soc.*, **128**, 4865.
26. Margulies, D., Melman, G., and Shanzer, A. (2005) *Nat. Mater.*, **4**, 768.
27. de Silva, A.P., James, M.P., McKinney, B.O.F., Pears, D.A., and Weir, S.M. (2006) *Nat. Mater.*, **5**, 787.
28. Margulies, D., Felder, C.E., Melman, G., and Shanzer, A. (2007) *J. Am. Chem. Soc.*, **129**, 347.
29. Andréasson, J., Straight, S.D., Moore, T.A., Moore, A.L., and Gust, D. (2009) *Chem. Eur. J.*, **15**, 3936.
30. Ozlem, S. and Akkaya, E.U. (2009) *J. Am. Chem. Soc.*, **131**, 48.
31. de Silva, A.P., Gunaratne, H.Q.N., and McCoy, C.P. (1993) *Nature*, **364**, 42.
32. de Silva, A.P. and McClenaghan, N.D. (2002) *Chem. Eur. J.*, **8**, 4935.
33. de Silva, A.P. and Magri, D.C. (2005) *Chimia*, **59**, 218.
34. Langford, S. and Yann, T. (2003) *J. Am. Chem. Soc.*, **125**, 11198.
35. Coskun, A., Deniz, E., and Akkaya, E.U. (2005) *Org. Lett.*, **7**, 5187.
36. de Silva, A.P. and McClenaghan, N.D. (2000) *J. Am. Chem. Soc.*, **122**, 3965.
37. Andréasson, J., Kodis, G., Terazono, Y., Liddell, P.A., Bandyo-padhyay, S., Mitchell, R.H., Moore, T.A., Moore, A.L., and Gust, D. (2004) *J. Am. Chem. Soc.*, **126**, 15926.
38. Valeur, B. (2003) *Molecular Fluorescence*, Wiley-VCH Verlag GmbH, Weinheim.
39. Murov, S.L., Carmichael, I., and Hug, G.L. (1993) *Handbook of Photochemistry*, Marcel Dekker, Inc., New York.
40. Berlman, I.B. (1971) *Handbook of Fluorescence Spectra of Aromatic Molecules*, 2nd edn, Academic Press, New York.
41. Griener, G. and Maier, I. (2002) *J. Chem. Soc., Perkin Trans.*, **2**, 1005.

42. de Silva, A.P. and Sandanayake, K.R.A.S. (1991) *Tetrahedron Lett.*, **32**, 421.
43. de Silva, A.P., de Silva, S.A., Dissanayake, A.S., and Sandanayake, K.R.A.S. (1989) *J. Chem. Soc. Chem. Commun.*, 1054.
44. Kavarnos, G.J. (1993) *Fundamentals of Photoinduced Electron Transfer*, Wiley-VCH Verlag GmbH, Weinheim, New York.
45. Weller, A. (1968) *Pure Appl. Chem.*, **16**, 115.
46. Marcus, R.A. and Sutin, N. (1985) *Biochim. Biophys. Acta*, **811**, 265.
47. Gao, Y.Q. and Marcus, R.A. (2002) *J. Phys. Chem. A*, **106**, 1956.
48. Magri, D.C., Callan, J.F., de Silva, A.P., Fox, D.B., McClenaghan, N.D., and Sandanayake, K.R.A.S. (2005) *J. Fluoresc.*, **15**, 769.
49. de Silva, A.P. and Rupasinghe, R.A.D.D. (1985) *J. Chem. Soc. Chem. Commun.*, 1669.
50. de Silva, A.P., Gunaratne, H.Q.N., and McCoy, C.P. (1997) *J. Am. Chem. Soc.*, **119**, 7891.
51. Callan, J.F., de Silva, A.P., and McClenaghan, N.D. (2004) *Chem. Commun.*, 2048.
52. Callan, J.F., de Silva, A.P., and Magri, D.C. (2005) *Tetrahedron*, **61**, 8551.
53. de Silva, S.A., Amorelli, B., Isidor, D.C., Loo, K.C., Crooker, K.E., and Pena, Y.E. (2002) *Chem. Commun.*, 1360.
54. Shiraishi, Y., Tokitoh, Y., Nishimura, G., and Hirai, T. (2005) *Org. Lett.*, **7**, 2611.
55. Shiraishi, Y., Tokitoh, Y., and Hirai, T. (2005) *Chem. Commun.*, 5316.
56. Magri, D.C., Coen, G.D., Boyd, R.L., and de Silva, A.P. (2006) *Anal. Chim. Acta*, **568**, 156.
57. Takeshita, M. and Irie, M. (1998) *J. Org. Chem.*, **63**, 6643.
58. Hirayama, F. (1965) *J. Chem. Phys.*, **42**, 3163.
59. Desvergne, J.P., Bouas-Laurent, H., Lahmani, F., and Sepiol, J. (1992) *J. Phys. Chem.*, **B 96**, 10616.
60. Lahlou, S., Bitit, N., and Desvergne, J.P. (1998) *J. Chem. Res. (S)*, (6), 302.
61. Guo, X.F., Zhang, D.Q., and Zhu, D.B. (2004) *Adv. Mater.*, **16**, 125.
62. He, H., Mortellaro, M., Leiner, M.J.P., Fraatz, R.J., and Tusa, J. (2003) *J. Am. Chem. Soc.*, **125**, 1468.
63. He, H., Mortellaro, M.A., Leiner, M.J.P., Young, S.T., Fraatz, R.J., and Tusa, J.K. (2003) *Anal. Chem.*, **75**, 549.
64. Tusa, J.K. and He, H. (2005) *J. Mater. Chem.*, **15**, 2640.
65. Konry, T. and Walt, D.R. (2009) *J. Am. Chem. Soc.*, **131**, 13232.
66. Magri, D.C. and de Silva, A.P. (2010) *New J. Chem.*, **34**, 476.
67. Bozdemir, O.A., Guliyev, R., Buyukcakir, O., Selcuk, S., Koleman, S., Gulseren, G., Nalbantoglu, T., Boyaci, H., and Akkaya, E.U. (2010) *J. Am. Chem. Soc.*, **132**, 8029.
68. de Silva, A.P., Dixon, I.M., Gunaratne, H.Q.N., Gunnlaugsson, T., Maxwell, P.R.S., and Rice, T.E. (1999) *J. Am. Chem. Soc.*, **121**, 1393.
69. Feringa, B.L. (ed.) (2001) *Molecular Switches*, Wiley-VCH Verlag GmbH, Weinheim.
70. Amendola, V., Fabbrizzi, L., Foti, F., Licchelli, M., Mangano, C., Pallavicini, P., Poggi, A., Sacchi, D., and Taglietti, A. (2006) *Coord. Chem. Rev.*, **250**, 273.
71. Fabbrizzi, L., Licchelli, M., and Pallavicini, P. (1999) *Acc. Chem. Res.*, **32**, 846.
72. Goulle, V., Harriman, A., and Lehn, J.M. (1993) *J. Chem. Soc. Chem. Commun.*, 1034.
73. Zhang, R., Wang, Z., Wu, Y., Fu, H., and Yao, J. (2008) *Org. Lett.*, **10**, 3065.
74. Benniston, A.C., Copley, G., Elliott, K.J., Harrington, R.W., and Clegg, W. (2008) *Eur. J. Org. Chem.*, (16), 2705.
75. Martínez, R., Ratera, I., Tárraga, A., Molina, P., and Veciana, J. (2006) *Chem. Commun.*, 3809.
76. Xiao, X.W., Xu, W., Zhang, D.Q., Xu, H., Liu, L., and Zhu, D.B. (2005) *New. J. Chem.*, **29**, 1291.
77. Leroy-Lhez, S., Baffreau, J., Perrin, L., Levillain, E., Allain, M., Blesa, M.J., and Hudhomme, P. (2005) *J. Org. Chem.*, **70**, 6313.
78. Zhang, G.X., Zhang, D.Q., Guo, X.F., and Zhu, D.B. (2004) *Org. Lett.*, **6**, 1209.
79. Gorodetsky, B., Samachetty, H.D., Donkers, R.L., Workentin, M.S., and

Branda, N.R. (2004) *Angew. Chem. Int. Ed.*, **43**, 2812.
80. Li, H., Jeppesen, J.O., Levillain, E., and Becher, J. (2003) *Chem. Commun.*, 846.
81. Arounaguiri, S. and Maiya, B.G. (1999) *Inorg. Chem.*, **38**, 842.
82. Biancardo, M., Bignozzi, C., Doyle, H., and Redmond, G. (2005) *Chem. Commun.*, 3918.
83. Ashton, P.R., Baldoni, V., Balzani, V., Credi, A., Hoffmann, H.D.A., Martínez-Díaz, M.V., Raymo, F.M., Stoddart, J.F., and Venturi, M. (2001) *Chem. Eur. J.*, **7**, 3482.
84. Fang, C.J., Zhu, Z., Sun, W., Xu, C.H., and Yan, C.H. (2007) *New J. Chem.*, **31**, 580.
85. Komura, T., Niu, G.Y., Yamaguchi, T., and Asano, M. (2003) *Electrochim. Acta.*, **48**, 631.
86. Fang, C.J., Li, C.Y., Fu, X.F., Yue, Y.F., and Yan, C.H. (2008) *Chin. J. Inorg. Chem.*, **24**, 1832.
87. Bard, A.J. and Faulkner, L.R. (2001) *Electrochemical Methods Fundamentals and Applications*, John Wiley & Sons, Inc.
88. Magri, D.C. and Workentin, M.S. (2008) *Chem. Eur. J*, **14**, 1698.
89. Pourbaix, M. (1966) *Atlas of Electrochemical Equilibria in Aqueous Solutions*, Pergamon Press, Oxford.

6
Supramolecular Assemblies for Information Processing
Cátia Parente Carvalho and Uwe Pischel

6.1
Introduction

The idea of using chemical systems to mimic logic operations at the molecular level continues to receive wide attention among scientists [1–8]. This book is a witness for this ongoing development and the actual diversification of the research efforts in many more directions than the initially outlined (but still rather distant) possibility to construct a competitive molecular computer. Indeed, for many applications, the implementation of just one or at most a few logic gate operations is sufficient to combine the imagination of chemical design with Boolean logic for the benefit of developing novel switchable intelligent nanomaterials, vehicles for drug delivery, approaches for prodrug activation, and molecular systems for clinical diagnostics and actuation [9–19].

Since the early days of molecular logic, the design of systems in the form of molecular entities, where all actuating parts are linked by covalent bonds, has been in the focus of the field. A survey of the published literature reveals that in fact not many supramolecular logic gates are known. This may come as a surprise, because the noncovalent and reversible assembly of functional units, containing chemical input information, offers advantages in terms of (i) resetting the logic device to a predefined initial state and (ii) reconfiguring the logic operation by variation of chemical inputs and/or photonic inputs. These are two essential points in advanced molecular logic design.

In this chapter, a brief compilation of supramolecular assemblies, which have been reported as logic devices or which can be potentially interpreted as such (although not published in that context), will be presented. Photophysical engineering of the excited state properties of supramolecular assemblies occupies a rather large space in the interest spectrum related to their functional exploitation. Hence, it should not be surprising that most of the following examples make use of photochemical and/or photophysical principles and rely partially (chemical input signals, photonic output signal) or completely (all-photonic operation) on optical signaling. In order to pinpoint the subject of this chapter as clearly as possible,

Molecular and Supramolecular Information Processing: From Molecular Switches to Logic Systems,
First Edition. Edited by Evgeny Katz.
© 2012 Wiley-VCH Verlag GmbH & Co. KGaA. Published 2012 by Wiley-VCH Verlag GmbH & Co. KGaA.

it should be noted that "biomolecular logic systems" relying on supramolecular recognition of substrates by enzymes are not discussed herein.

6.2 Recognition of Metal Ion Inputs by Crown Ethers

Among the various binding motifs used for the complexation of metal ions as chemical inputs of molecular logic gates, crown ethers have enjoyed certain preference (see also Chapters 2 and 5). The first molecular logic device is a two-input AND gate (**1**, Scheme 6.1), which was published by de Silva *et al.* in 1993 [20]. It uses two receptors, which quench the fluorescence of an appended anthracene signaling unit via photoinduced electron transfer (PET). On simultaneous recognition of the corresponding ionic input signals, PET is switched off and a high fluorescence output is observed. One of the receptors is benzo-15-crown-5, fitted to bind Na^+ ions with high selectivity. The same crown ether motif later played a central role in the design of sophisticated AND logic gates, which are based on the same PET fluorescence switching principle. Compound **2** is one among them (Scheme 6.1), which is equipped with a C_8 alkyl side chain to facilitate its structural integration in tetramethylammonium dodecyl sulfate micelles [21]. The use of micelles to confine molecular logic gates in nanospaces of a defined architecture is another example for the exploitation of supramolecular principles [22], which, however, is not covered in more detail in this chapter. The Bharadwaj group [23] has reported another example of an anthracene logic switch, in this case with two macrocyclic receptors (an azacrown ether and a cryptand). On the basis of the general principles

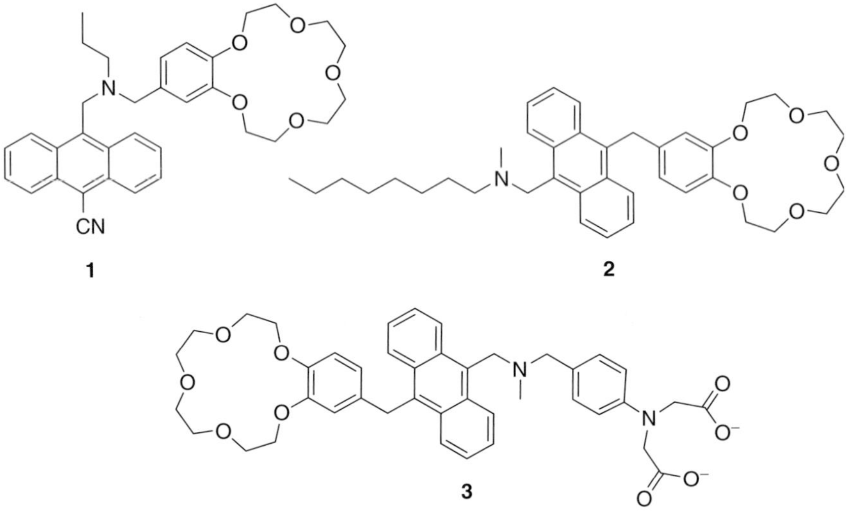

Scheme 6.1 Examples of members of the "anthracene family" of molecular AND logic switches with crown ether receptors.

of PET switching by metal ion complexation, AND logic is accessible. Compound **3** (Scheme 6.1), another member of the family of anthracene-based fluorescent logic switches, relies on a consequent extension of the above-mentioned PET design principle [11]. It responds to three chemical signals, which are again connected by an AND logic operation to yield the highest fluorescence output in the simultaneous presence of all inputs. Once more, Na^+ ions are one input, which is recognized by the benzo-15-crown-5 macrocycle.

Akkaya and coworkers reported in 2009 a difluoroboron dipyrromethene (BODIPY) dye **4** with a benzo-15-crown-5 as one receptor (Scheme 6.2). The combined action of Na^+ ions and protons (captured by a conjugated pyridine receptor) enables the activation of the BODIPY excited triplet state and the

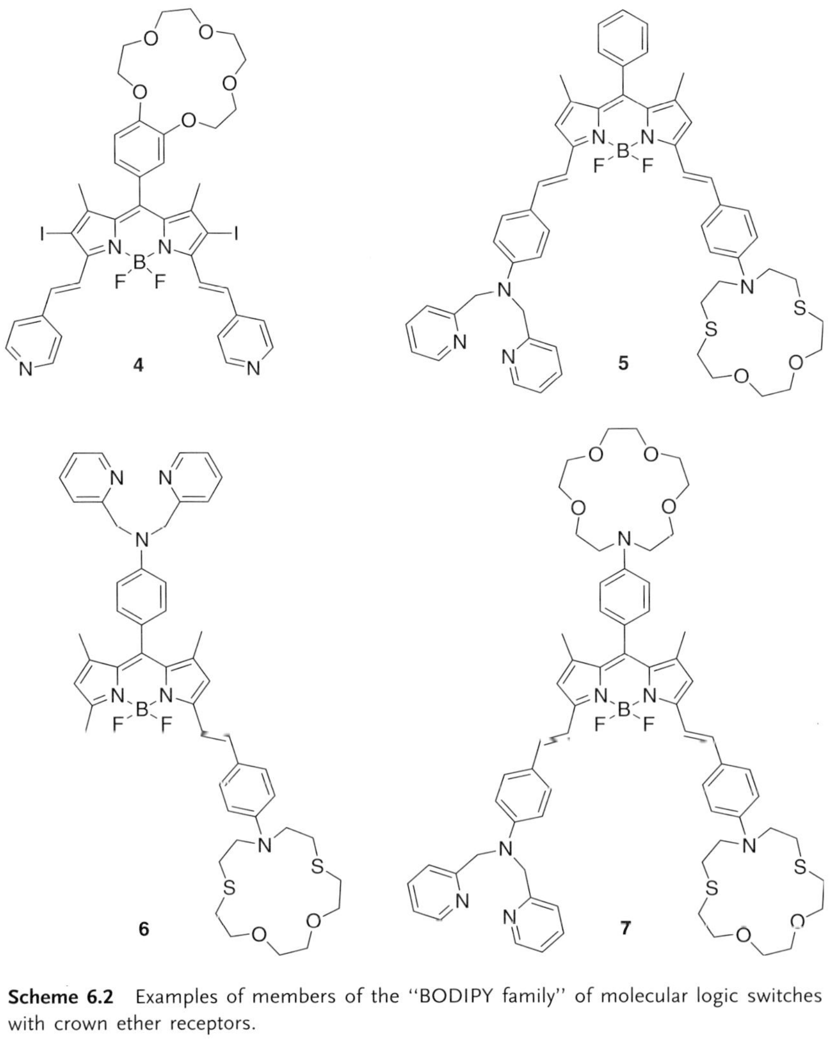

Scheme 6.2 Examples of members of the "BODIPY family" of molecular logic switches with crown ether receptors.

production of singlet oxygen. This utility output makes this device a promising example for the logically mediated switching of an actuation mechanism (for photodynamic therapy in this concrete case) [15]. In 2010 the same group published a report on a series of derivatized BODIPY dyes (5–7; see structures in Scheme 6.2), where they demonstrate the general usefulness of this fluorophore motif in molecular logic [24]. An effective control of its spectral characteristics (absorption and fluorescence) is possible by manipulation of excited state processes such as PET and internal charge transfer (ICT) in function of ionic input information. Thiaazacrown ethers are employed to bind Hg^{2+} metal ions, which in combination with Zn^{2+} ions (complexed by dipicolylamine-derived receptors) yields molecular equivalents for a half-adder (5), which is an XOR/AND combination, and a two-input AND gate (6). The combination of a BODIPY signaling unit with two crown ethers (for Hg^{2+} and Ca^{2+}) and one Zn^{2+} receptor (in compound 7) enables the realization of a three-input AND logic gate. These few examples illustrate the use of crown ethers as metal ion receptors. The resetting of logic devices with ionic inputs can be achieved by the addition of stronger competing ligands for the metal ions. However, it should be kept in mind that this will build up undesired waste products, which may compromise the functionality of the logic devices. For molecular computing, which requires recycling and long-term stability, this is a drawback. However, for other applications in the area of life sciences resetting is not necessarily a precondition.

6.3
Hydrogen-Bonded Supramolecular Assemblies as Logic Devices

In 2000, the Akkaya group reported an interesting example that relies on the hydrogen bonding of the complementary deoxynucleotides dAMP and dTMP (Scheme 6.3) [25]. The signaling unit is established by the known fluorescent DNA probe DAPI (4′,6-diamidino-2-phenylindole, 8) which is able to form hydrogen bonds through its two amidinium groups. By a clever choice of the fluorescence emission wavelength for reading the response in function of the presence of

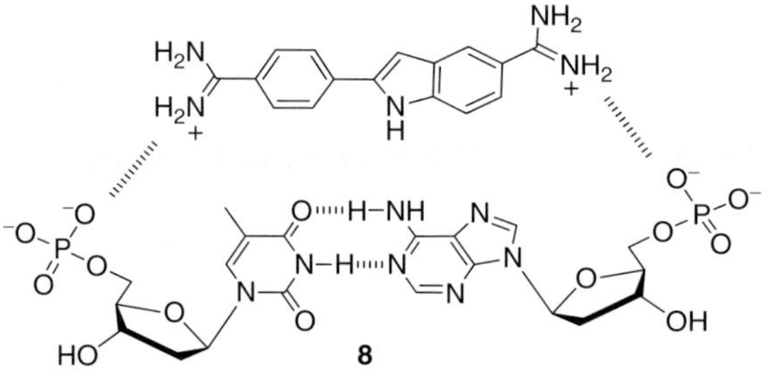

Scheme 6.3 Supramolecular ternary complex composed of dTMP, dAMP, and the dye **8** (DAPI) for the implementation of molecular NAND and TRANSFER gates with fluorescence output.

dAMP, dTMP, or both, the operation of a NAND logic gate ($\lambda_{obs} = 455$ nm) is mimicked. In the absence of nucleotide inputs or in presence of only dAMP, the emission has a maximum at about 475 nm. However, when dTMP is added to DAPI, a blue shift of the emission with an isoemissive point at about 455 nm is observed. When both inputs are present, a supramolecular assembly in the form of a ternary complex is postulated, where the nucleobases form two hydrogen bonds between them and each is held to compound **8** by interaction with the amidinium groups of the dye. This is accompanied by a further blue shift and significant fluorescence quenching. This observation is believed to be at least partially related to a complexation-induced pK_a shift, resulting in DAPI deprotonation. Hence, the fluorescence signal at 455 nm is always high (binary 1), except for the presence of both nucleotides. This is compatible with NAND logic. The change of the emission observation wavelength to 411.5 nm leads to a reconfiguration of the logic operation, yielding a TRANSFER gate.

In 2001, Rotello and coworkers [26] reported the reversible switching of a hydrogen-bonded supramolecular assembly (**9**) with orthogonal electrochemical and photonic inputs (Scheme 6.4). In brief, a molecular scaffold, which is modified with a donor−acceptor−donor hydrogen-bonding motif and a photoswitchable azobenzene unit, is used to form an assembly with a naphthalenediimide derivative (NDI). The latter introduces a convenient opportunity for electrochemical switching via one- or two-electron reduction. As shown in Scheme 6.4, the binding constant of NDI depends critically on two factors: (i) its own oxidation state, with a weaker binding for its reduced form and (ii) a favorable π-π stacking interaction with one phenyl of the azobenzene, which is only given for the trans-isomer. Hence, the binding constant of NDI is highest when the assembly is in state I (Scheme 6.4). The network of the four shown states would allow the construction of various logic gates with the output defined as the binding strength of NDI. For example, if state I is chosen for this purpose and UV light (360 nm) and the one-electron reduction (slightly dependent on the cis/trans configuration of the photochromic part) are the two gate inputs, then a NOR gate can be envisioned. It should be emphasized that in the original publication the general utility of molecular switches for the implementation of logic gates is briefly mentioned, but no concrete gate is discussed. However, the reported experimental results support the herein made assignment.

6.4
Molecular Logic Gates with [2]Pseudorotaxane- and [2]Rotaxane-Based Switches

The implementation of switching functions with pseudorotaxanes and rotaxanes is a central topic of supramolecular chemistry [27]. In most cases, bistable scenarios are created by application of an external stimulus (e.g., pH changes, redox reactions, and light-induced processes). A couple of systems, which lead to logic operations such as exclusive OR (XOR), exclusive NOR (XNOR), INHIBIT, reversible logic, a half-adder (combination of AND and XOR gates), and a half-subtractor (combination of INHIBIT and XOR gates), have been designed; see below.

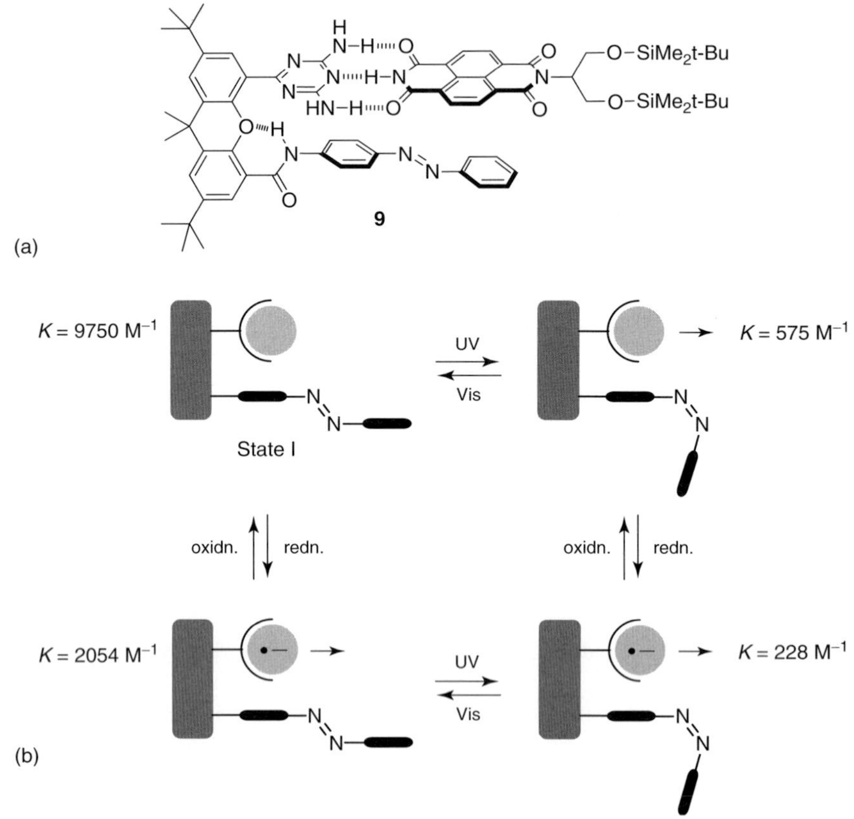

Scheme 6.4 Structure of the hydrogen-bonded assembly **9** in its neutral trans form (a). Orthogonal photo- and electrochemical switching of **9** (b). Note that only the one-electron reduction switching is shown. The original work [26] includes two-electron reduction processes also.

In 1997, Credi et al. used the [2]pseudorotaxane **10** (Scheme 6.5) for the first molecular implementation of the XOR gate [28]. Assembly **10** is formed by charge transfer (CT) interaction between the electron-donating 2,3-dinaphtho-30-crown-10 ring and the electron-accepting 2,7-dibenzyl-diazapyrenium dication as thread. As a consequence of this interaction, the naphthalene fluorescence of the ring at $\lambda_{max} = 343$ nm is quenched. The addition of protons (triflic acid) as one input signal and competitor of the guest is followed by the release of the thread, which is accompanied by naphthalene fluorescence recovery. The same effect is observed on addition of a base input (tri-n-butylamine), which forms a stronger complex with the thread than the ring does. The simultaneous presence of both inputs has no effect on the stability of the supramolecular assembly **10**, because acid–base neutralization is observed. Hence, the naphthalene fluorescence signal remains quenched. These characteristics, that is, a high output signal in the presence of either input, but a low output signal for their simultaneous absence or presence

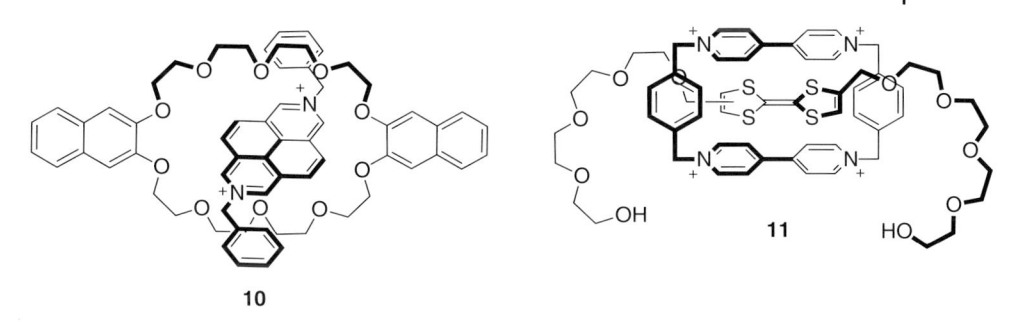

Scheme 6.5 Structures of [2]pseudorotaxanes **10** and **11** for the realization of molecular XOR and XNOR gates, respectively.

is identified as XOR operation. More recently, the enhanced potential of the assembly of a diazapyrenium guest and the 2,3-dinaphtho-30-crown-10 macrocycle was disclosed in a report by the Credi group [29]. They were able to implement features such as a bidirectional half-subtractor and a reversible logic device (see Chapter 3 for a detailed discussion).

The inversion of the XOR output leads to the complementary logic operation known as XNOR. This logic gate type was mimicked in 1997 with the [2]pseudorotaxane **11** (see structure in Scheme 6.5) [30]. The switching is, as described above for the XOR gate, again based on threading/dethreading. However, this time electrochemical stimuli are used. The electron-donating tetrathiafulvalene (TTF) thread forms a CT complex with a paraquat-derived electron-accepting cyclophane ring, which is signaled by a red-shifted absorption band at about 830 nm. The electrolysis of an acetonitrile solution of **11** at −0.3 V leads to the reduction of the cyclophane, which causes decomplexation of the thread and the concomitant disappearance of the CT absorption band (output). The oxidation of the TTF thread (electrolysis at a potential of +0.5 V) weakens the CT interaction in **11** as well, leading to dethreading with the same consequences for the CT band as observed for the reduction of the cyclophane. The simultaneous application of both potentials would hardly reduce/oxidize the components of the same complex, but macroscopically the result is an unchanged solution with its CT absorption band present. The described behavior, namely, the exclusive observation of the CT absorption band at about 830 nm in the absence of inputs or their simultaneous presence, is coincident with the function of an XNOR logic gate.

In 1997, the Balzani group [31] presented a photochemical version of a switchable [2]pseudorotaxane (see structure **12** in Scheme 6.6). In brief, the fluorescence of the naphthalene thread is quenched in the CT complex with a paraquat-derived tetracationic ring. On dethreading, which is induced by PET from an anthracene photosensitizer to the ring and consequent weakening of the CT interaction, the fluorescence of the thread is restored. The process is inverted by reoxidation of the reduced ring by oxygen. The oxidized photosensitizer can be recovered by reaction with a sacrificial electron donor (triethanolamine). In terms of the logic interpretation of the system, the oxygen input is the disabling input and the

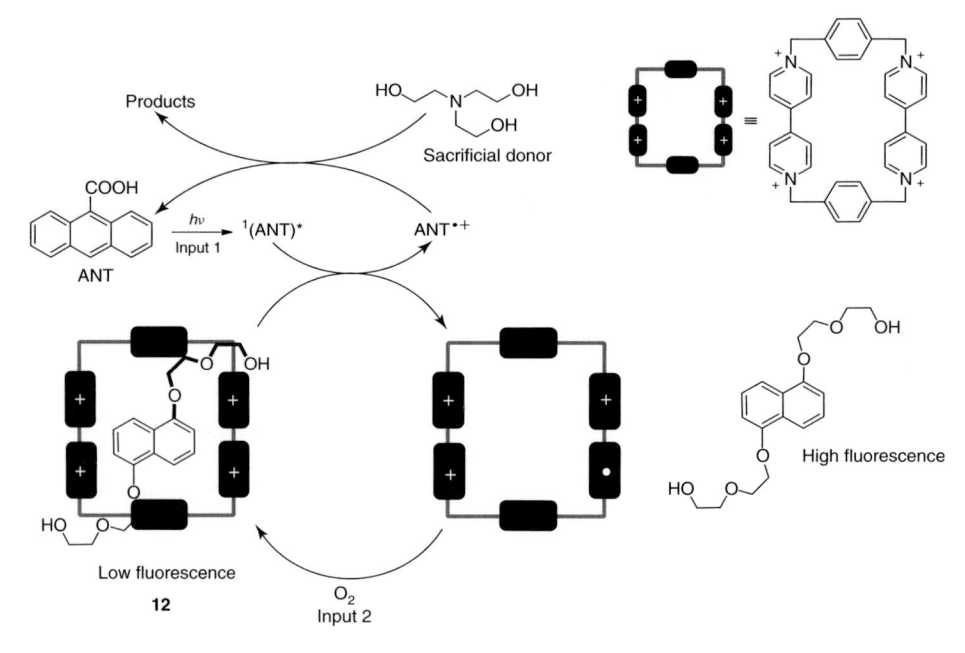

Scheme 6.6 Photoinduced switching of [2]pseudorotaxene **12**.

UV light needed to activate the photosensitizer constitutes the other input. The fluorescence of the system is only read as high when the light input is on and oxygen is absent. This is compatible with an INHIBIT gate behavior. Later on, second-generation versions of this switching principle with the photosensitizer as integral part of the ring (**13**) or the thread (**14**) were presented as well (Scheme 6.7) [32, 33].

Scheme 6.7 Structures of integrated photosensitizer-[2]pseudorotaxane architectures **13** and **14**.

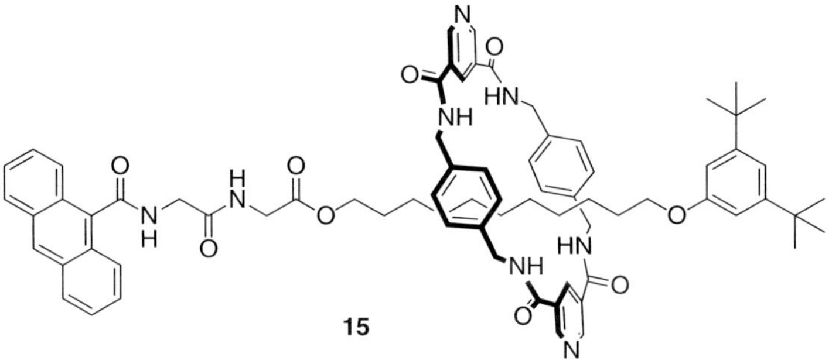

Scheme 6.8 Structure of the proton- and solvent-addressable [2]rotaxane **15**.

In 2005, Leigh et al. reported the fluorescence switching of the [2]rotaxane **15** (Scheme 6.8) for harnessing the function of an INHIBIT logic gate [34]. The thread of rotaxane **15** is composed of a glycylglycine dipeptide station with a nearby anthracene stopper and a long alkyl chain as solvophobic station. The ring is a tetramide cycle with integrated protonable pyridine units. The logic function can be realized in solution as well as in polymer films, by using trifluoroacetic acid (TFA) and dimethylsulfoxide (DMSO) as chemical input information. The fluorescence of the anthracene stopper is high for the unprotonated ring, which is not implied in emission quenching processes. The submolecular motion of the ring can be controlled by the addition of DMSO, which moves the ring from the hydrogen-bonded situation (for example, in dichloromethane or chloroform), involving the dipeptide station, to the solvophobic part of the thread. However, when the pyridines of the ring get protonated on addition of TFA, PET fluorescence quenching of the anthracene may potentially occur, depending on the relative position of the ring with respect to the fluorophore stopper. When the protonated ring is close, that is, when located over the dipeptide station, PET is operative and a low fluorescence is observed. However, when DMSO is applied simultaneously with the acid, the ring moves away from the anthracene, and PET is much less competitive as fluorescence quenching pathway. Thus, when quenching of the fluorescence is considered as output, an INHIBIT gate results. It must be considered a surplus that the supramolecular logic gate can be operated in polymer films. However, a poor reversibility was reported, because heating of the films in order to remove the TFA and DMSO is accompanied by significant decomposition of the assembly.

Tian and coworkers [35] concentrated on the design of [2]rotaxanes with two orthogonally photoswitchable stations (see examples in Scheme 6.9). In 2005, this group reported system **16**, which is composed of a thread with azobenzene and stilbene stations, two electronically different fluorescent naphthalimide stoppers, and α-cyclodextrin (α-CD) as ring component. The two stations are $E \rightleftarrows Z$ configurational switches, which can be addressed via reversible photoinduced processes (see central part of Scheme 6.9). The E-form of the azobenzene can be addressed by irradiation with 380 nm light, and the generated Z-form is back-isomerized by 450 nm light. The *E*-stilbene station is converted to the *Z*-form by irradiation with 313 nm light, while 280 nm light induces the inverse process. The diverse isomerizations are accompanied by significant changes in the UV spectral range of the absorption spectrum, that is, in comparison to the *E*/*E* form the other three isomeric forms (*E*/*Z*, *Z*/*E*, and *Z*/*Z*; azobenzene/stilbene) have a higher absorption around 270 nm and a decreased absorption around 350 nm. In the *E*/*E* form of the thread, the ring shuttles between both stations. However, on photoinduced *E* → *Z* isomerization of one of the stations, the ring binds preferentially to the other station, which remains in its *E*-form. On simultaneous isomerization of both stations to their respective *Z*-form, the ring is found to reside at the center of the thread between both stations. The aminonaphthalimide stopper in proximity to the azobenzene emits with a maximum at about 520 nm, while the naphthalimide close to the stilbene station fluoresces at about 395 nm. A significant increase in the fluorescence intensity of the respective stopper next to the station, which remains in the fitting *E*-configuration for accommodating the α-CD ring, is observed. It was speculated that this is the result of a rigidification effect. By defining light of the two irradiation wavelengths for photoisomerization, that is, 313 and 380 nm, as inputs and by reading the above-discussed absorbance and fluorescence changes, two logic gates can be implemented: AND and XOR. The parallel operation of both leads to a half-adder system, which can add binary digits. The output of the XOR gate defines the sum digit and the AND gate output stands for the carry digit. A slight inconvenience is that the fluorescence change as XOR gate output has to be read for different optical channels (395 nm for the *Z*/*E* form and 520 nm for the *E*/*Z* form). However, the discussed system is an early example for all-photonic logic switches, where input as well as output signals are of photonic nature. This type of switching based on photochromic compounds is nowadays very popular (see also Chapter 4) and has been used, for example, to implement a multifunctional platform with advanced logic behavior [36].

A structurally related [2]rotaxane **17** (see structure in Scheme 6.9), where one of the naphthalimide stoppers of **16** (the one without 4-amino substitution) is exchanged for a dicarboxylic acid motif, was used by the Tian group [37] in 2006 to realize INHIBIT gates. These can be addressed either by two photonic inputs (the same as described above for the half-adder) or one photonic input (313 nm) and one chemical input signal (protons). In both cases, the output was recorded as fluorescence intensity change at 520 nm. The initial state of the [2]rotaxane is described by the *E*-configuration of both stations and the dicarboxylic acid in its deprotonated form (as sodium salt). Hence, only when the assembly is irradiated with 313 nm light (but not with 380 nm light or with both wavelengths) and the *E*/*Z* (azobenzene/stilbene) form is generated, the fluorescence signal of the aminonaphthalimide stopper is significantly enhanced (see discussion above for rotaxane **16**). When the inputs are protons and 313 nm light, the absence of any input enables shuttling of the α-CD ring between both stations in their *E*-form. When protons are added, the ring is locked by hydrogen bonding to the dicarboxylic acid motif. The only situation, in accordance with INHIBIT logic, where a high output signal is obtained, is the irradiation at 313 nm in the absence of protons.

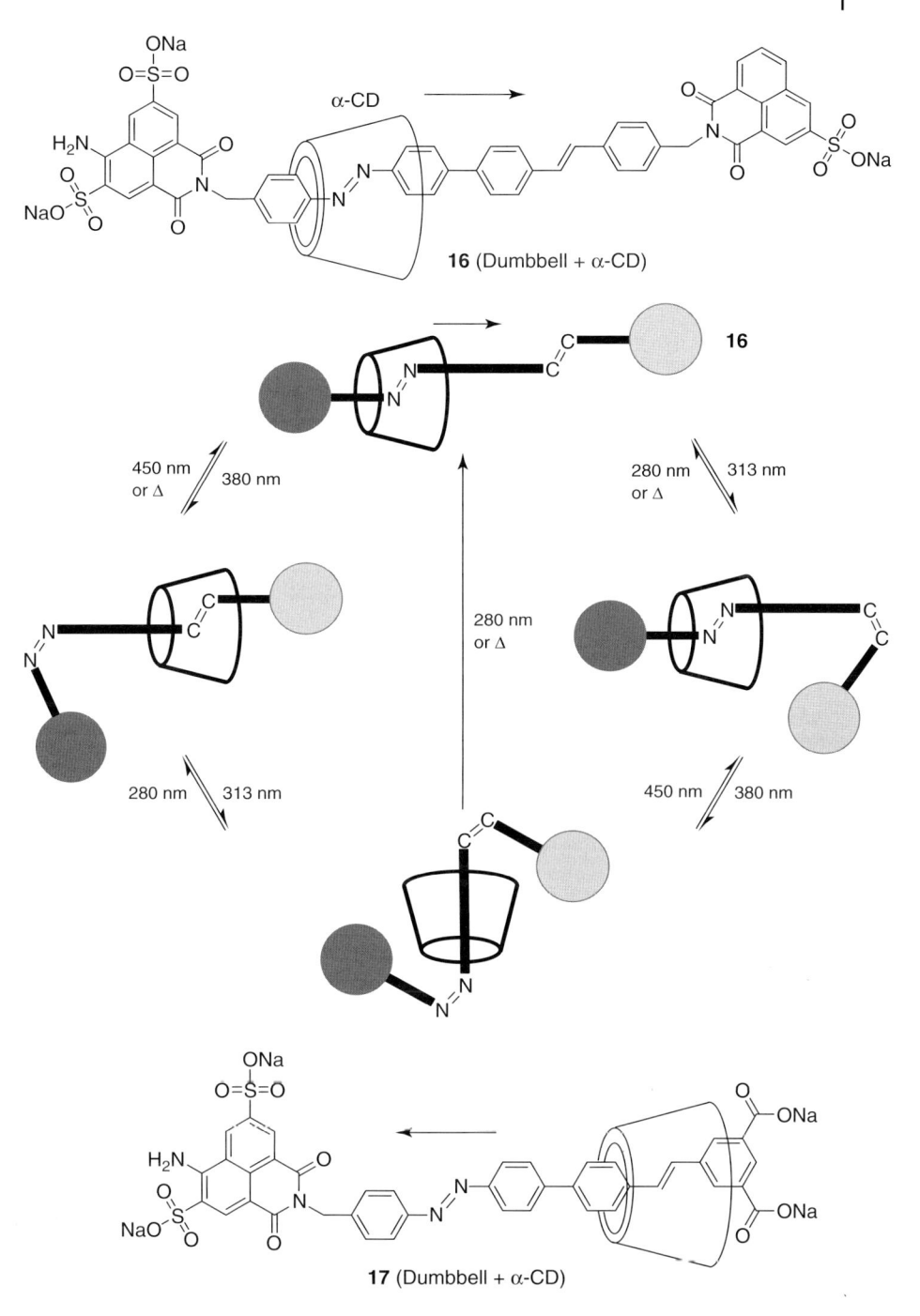

Scheme 6.9 Structures of [2]rotaxanes **16** and **17**. The all-photonic switching of **16** is shown in the central part.

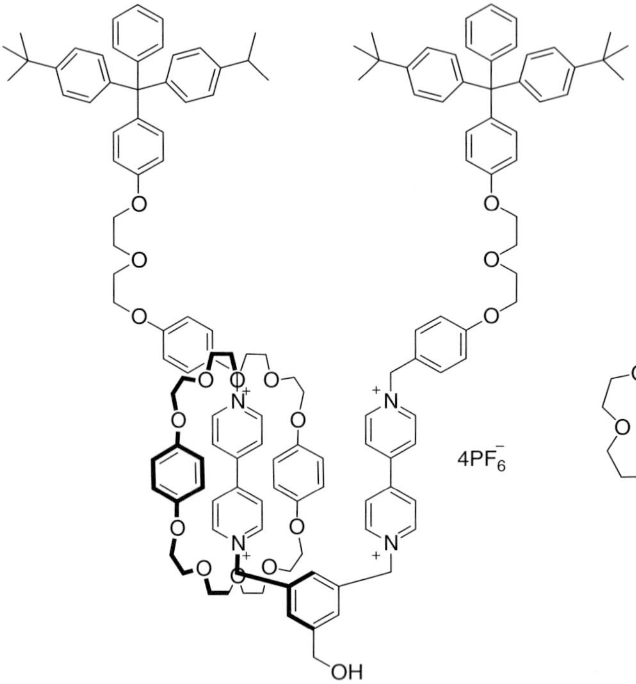

Scheme 6.10 Structures of [2]rotaxane **18** and [2]catenane **19**.

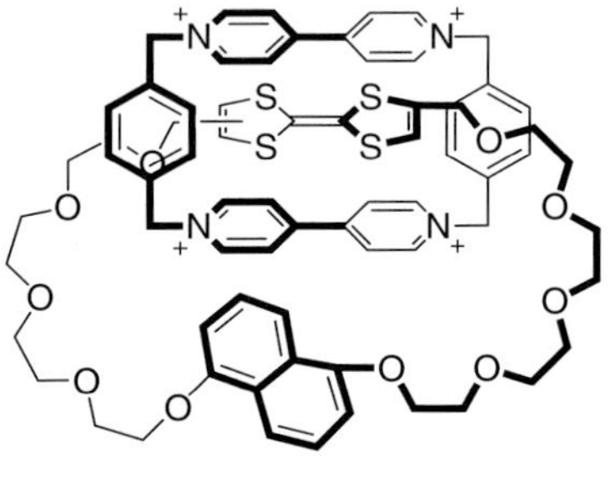

Under these circumstances, the E/Z form is generated and the ring is located at the azobenzene station in its E-form. Consequently, the fluorescence of the aminonaphthalimide stopper is enhanced.

Electrochemical switching of [2]rotaxanes and also [2]catenanes has been applied in a series of reports from the Stoddart and Heath groups [38–41]. They were able to integrate these organic supramolecular architectures (see, for example, **18** and **19** in Scheme 6.10) on electrode surfaces and realize molecular electronic devices with AND, XOR, and OR functions or working as Set–Reset memories. Remacle and coworkers showed in 2009 that the electrochemical switching of a copper [2]rotaxane can be harnessed for the realization of a Set–Reset memory (see Chapter 3 for a more detailed description) [42–45].

6.5
Supramolecular Host-Guest Complexes with Cyclodextrins and Cucurbiturils

In the early days of supramolecular chemistry, CDs and calixarenes were among the most frequently used macrocyclic hosts [46, 47]. More recently, cucurbiturils (CBs) have received increased attention [48]. Despite the popularity of such molecular containers for the photophysical engineering of the properties of fluorescent guests

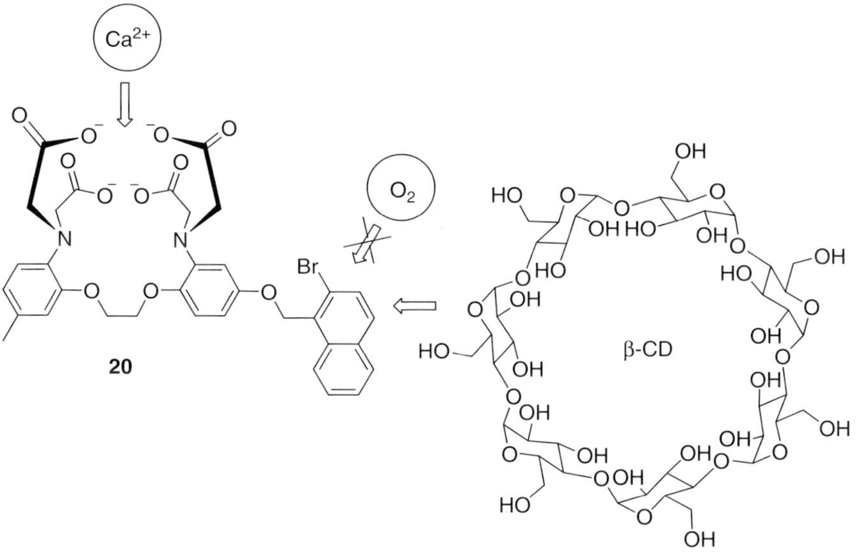

Scheme 6.11 Functional integration of a molecular three-input INHIBIT logic gate with compound **20** and Ca^{2+}, O_2, and β-CD as chemical inputs.

[49], relatively little attention has been paid to their utility for the demonstration of molecular logic. One of the few examples, which applies a CD as chemical input signal, was published by the de Silva group in 1999 [50]. It is precisely in the context of delivering proof of principle for the possibility of functional integration of logic operations that require the actuation of more than one logic gate (for example, the INHIBIT gate as a combination of AND gate and INVERTER) that β-CD was used as the input stimulus for phosphorescence switching of compound **20** (Scheme 6.11). In brief, the phosphorescence of the naphthalene-derived signaling unit can be activated by the hindrance of PET from a receptor unit upon Ca^{2+} ion binding. However, the long-lived emission signal is only observed for the additional presence of β-CD and for the simultaneous absence of oxygen. The macrocycle shields the naphthalene phosphor from deactivation by triplet–triplet annihilation, while oxygen is an efficient triplet quencher (even for the naphthalene/β-CD complex), which needs to be removed. All other input signal combinations lead to low phosphorescence signals, which is compatible with a three-input INHIBIT logic gate.

CBs are water-soluble macrocycles with n glycoluril units ($n = 5, 6, 7, 8, 10$) linked via methylene bridges (see CB6 and CB7 in Scheme 6.12) [51–53]. They have two carbonyl-lined electron-rich portals, which explain the preference of these macrocycles for charged guests (e.g., pyridinium cations) or guests that can be protonated (e.g., aliphatic and aromatic amines and benzimidazoles). An often observed phenomenon, which accompanies the CB-complexation of guests with protonable functions, is the host-assisted guest protonation [54–56]. This is expressed in a shift of the protonation constant toward a higher pK_a value

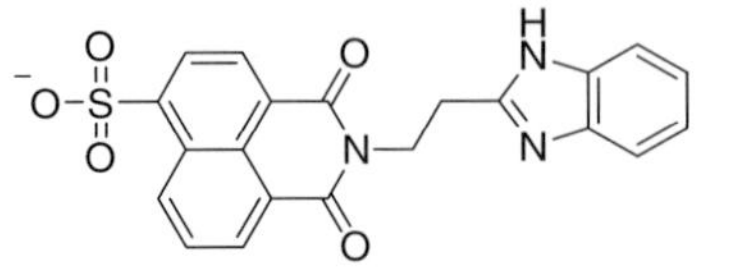

21

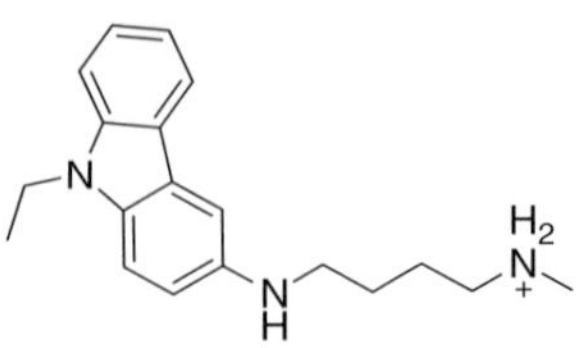

22

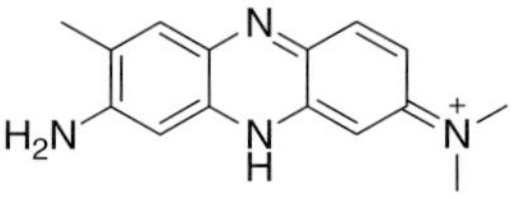

23

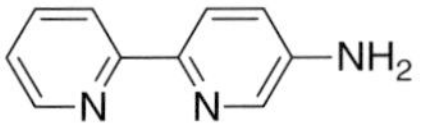

24

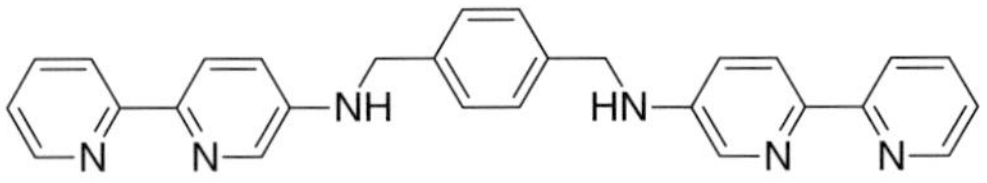

25

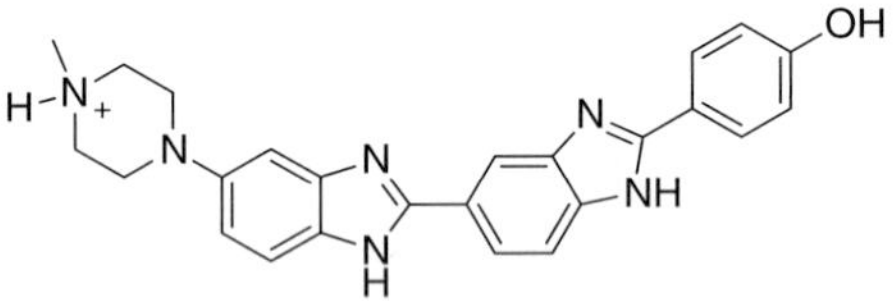

26

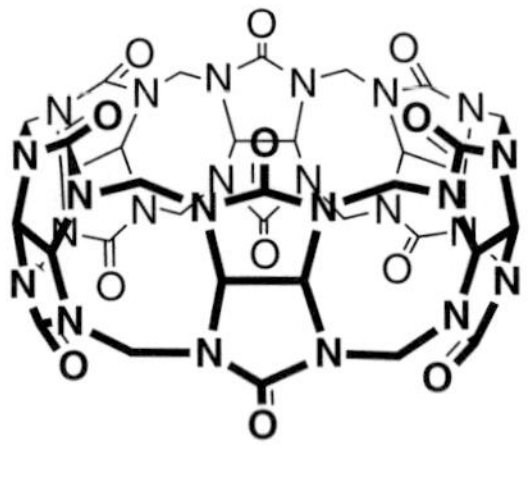

CB6

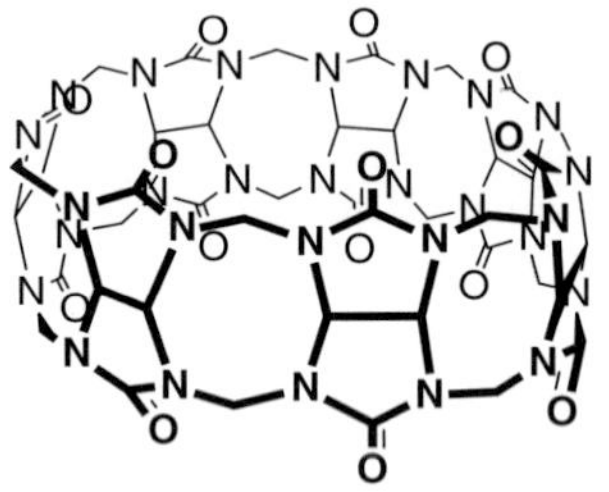

CB7

Scheme 6.12 Structures of guests **21**, **23**, and **26** for CB7 complexation and guests **22**, **24**, and **25** for CB6 complexation. In the bottom part the structures of the cucurbiturils CB6 and CB7 are shown.

(e.g., facilitated protonation). In other words, guests that are unprotonated in their free form may get protonated upon inclusion in the CB cavity. Many fluorescence phenomena are dependent on ICT or PET involving amine functions. Hence, the complexation by CBs and the accompanying protonation effects can be potentially used to switch these processes and the fluorescence output signal [57].

In 2010, the Pischel and Nau groups [58] used this approach for the design of the fluorophore-anchor dyad **21** (see structure in Scheme 6.12), which can be complexed by CB7 through binding of the benzimidazole anchor unit. In its uncomplexed form, dye **21** shows fluorescence quenching of the appended naphthalimide through PET from the electron-donating benzimidazole. However, the complexation-induced protonation of the latter (pK_a shift by 2.1 pH units) blocks PET, and the fluorescence of the naphthalimide is recovered. The pK_a shift can be used to construct various molecular logic gates by using the CB7 macrocycle and pH variations (addition of base or acid) as chemical inputs. Setting the initial state to the uncomplexed dye **21** at pH 9, a low fluorescence is observed because of the active PET process. Lowering the pH to 7 (addition of acid) is not sufficient to protonate the benzimidazole ($pK_a = 6.0$), and the fluorescence signal remains at a low level. The same is observed for the addition of CB7, which hardly leads to a host–guest complex at pH 9 because of the low binding constant [$K_b = (2.5 \pm 0.8) \times 10^3$ M^{-1}]. However, lowering the pH to 7 and adding CB7 yields a strong host–guest complex [$K_b = (1.3 \pm 0.1) \times 10^5$ M^{-1}], where the benzimidazole is protonated ($pK_a = 8.1$ in the complex) and therefore inactive as an electron donor. Now a strong emission signal is observed, that is, only the simultaneous addition of acid and CB7 yields a high fluorescence output in conformity with AND logic behavior. The initial conditions can be modified, and other logic gates are obtained. For example, setting pH 7 as the initial state and defining base (addition until reaching pH 9) as the input besides CB7 results in INHIBIT logic. For the same initial state, the application of a base input (hydroxide ions) and cadaverine as a strong competitor results in the disassembly of the host–guest complex for each input and also their combination. This is expressed as a drop in the fluorescence, in agreement with NOR logic: only the absence of any input yields a high output signal.

Another interesting example for host-induced fluorescence switching of a fluorophore-anchor dyad has been reported by the Nau group in 2008 for the carbazole derivative **22** as guest dye (see structure in Scheme 6.12) and CB6 as host [54]. Compound **22** contains a putrescine-derived anchor for CB6 binding, with one of the amino nitrogens as an integral part of the 3-amino-9-ethylcarbazole fluorophore. The addition of CB6 at pH 7 yields the complexation of the anchor ($K_b = (2.2 \pm 0.2) \times 10^7$ M^{-1} in 10 mM ammonium acetate buffer at pH 7) and the protonation of the aromatic amine nitrogen. Consequently, the ICT state of the fluorophore is destabilized, and the quenching of the ICT emission band at 458 nm is accompanied by the buildup of a new blue-shifted emission at 375 nm (assigned to the locally excited state). The observed pK_a shift between free and complexed **22** reaches 4.5 pH units, which is among the most accentuated shifts observed for a CB complex. Although no logic operation was interpreted in the original article, several considerations in this direction can be made. For example, defining cadaverine, a strongly competing guest of CB6 ($K_b = (9.5 \pm 1.1) \times 10^9$ M^{-1} in 10 mM ammonium acetate buffer at pH 7), as one chemical input and CB6 itself as the other input and setting the initial state as the free dye **22** at pH 7 yields the implementation of two logic gates. These can be reconfigured by choosing between 375 nm (INHIBIT gate) and 458 nm emissions (IMPLICATION gate) as the output signal. The emission intensity changes of the two output channels are coupled inversely to each other, and thus, complementary logic operations are obtained.

As inferred from the two examples discussed above, for the realization of molecular logic gates with CB complexes, chemical inputs in the form of pH changes, competing guests (e.g., diamines or polyamines), and the CB macrocycles themselves can be considered. The competing guest input is not necessarily always a species that gets immersed into the cavity of the CB; it may be a metal cation that has affinity to the carbonyl-lined portals of the host. An interesting example that can be potentially reinterpreted in terms of Boolean logic was reported in 2008 by the Pal group [59]. They demonstrated that the complex between the dye neutral red (**23**; see structure in Scheme 6.12) and CB7 is affected not only by the pH but also by a salt effect (addition of NaCl). The addition of salt enabled the tuning of the complex stability, which is also reflected in a gradual pK_a shift of the dye. Defining the initial state of the system as a dye–CB7 complex at pH 8 and monitoring the absorbance at 535 nm, corresponding to the protonated dye, allows the interpretation of molecular NOR logic with base (e.g., hydroxide ions) and sodium ion inputs. The absorbance at 535 nm is high for the initial state. However, an increase in the pH (addition of base until pH 9.5), addition of sodium ions (0.5 M), or both simultaneously lead to deprotonation of the dye and possibly decomplexation from the macrocycle cavity. This is accompanied by a drop in the 535 nm absorbance. Hence, all combinations of the inputs, except the simultaneous absence of both, yield a low output signal (binary 0) in accordance with a NOR logic function. The list of CB-derived systems with logic gate functionality on application of chemical input signals could be surely extended. For example, a recent work by the Keinan group [60], in which chemically switchable CB6-bipyridine beacons (see 4-aminobipyridine structures **24** and **25** in Scheme 6.12) are demonstrated, bears potential in this direction. However, the herein discussed few case studies are meant to just illustrate the general idea.

The switching of CB assemblies via light-induced processes is less common. In 2011, the Pischel group [61] reported an example where a photoinduced relay mechanism led to the switching of the fluorescence and ultimately the release of a guest from the CB7 cavity. The complex between CB7 and the Hoechst 33258 dye (**26**; see structure in Scheme 6.12) was studied in detail. The pH modulation of the complex stability is evident: $K_b = (1.7 \pm 0.4) \times 10^6$ M^{-1} at pH 7 versus $(2.8 \pm 0.2) \times 10^4$ M^{-1} at pH 9. The pH change, as in the cases discussed above, can be effectuated by manual addition of base. However, a photoinduced reaction could also be used for this purpose. The photolysis (300 nm light) of malachite green leucohydroxide yields a pronounced pH jump triggered by the release of hydroxide ions [62]. In accordance with the pH-dependent binding, the guest is released on irradiation: 2% free dye at pH 7 versus 55% free dye at pH 9, under the experimental conditions of the work [61]. The event of release can be followed conveniently by fluorescence measurements, because as opposed to the high fluorescence of **26** in the CB7 complex ($\Phi_f = 0.74$ at pH 7), the free dye is only very weakly fluorescent ($\Phi_f = 0.01$ at pH 7). The photoinduced guest release and the accompanying fluorescence switching itself do not constitute a complex logic operation. It would be just an INVERTER logic gate: no light impulse – high fluorescence, light impulse – low fluorescence. However, the conjugation of the phototriggered relay mechanism with other inputs would make this system an approach for controlled and "intelligent" drug delivery. Although not discussed in the original publication [61], it can be easily imagined that the addition of high sodium ion concentrations or a strongly competitive amine (such as cadaverine or spermine) would also lead to the release of dye **26** without the need to actuate the light pulse. Reading the fluorescence of **26** as the output leads straightforwardly to a NOR logic operation. The presence of either input (UV light or competitor) or both inputs simultaneously would yield guest release accompanied by a strongly diminished fluorescence. The photoinduced pH jump is in principle reversible; however, it was found that repeated recycling is difficult because of the likely involvement of secondary photodegradation of the malachite green leucohydroxide. Noteworthy, if the approach is considered for the targeted delivery of a drug, recycling is not an important issue. In general, the switching of the complexation/decomplexation of a discrete host–guest assembly by light-triggered pH changes is scarcely reported in the literature. The groups of Credi and Raymo presented an example in 2007, where the proton release upon photochromic processes in a spiropyran/merocyanine system (see also Chapter 3) is used to trigger the complexation of a pyridine-derived guest by a modified calix[6]arene [63]. The assembly/disassembly of supramolecular complexes can also be triggered by photoisomerization either of the guest [64, 65] or of the host [66]. However, despite their potential interest for molecular logic, none of these examples have been discussed along these lines.

An example where a light input and pH changes were used to trigger the release of a drug under the regime of an AND-controlled logic operation was provided by the groups of Stoddart and Zink in 2009 [12]. They used mechanized mesoporous silica nanoparticles [67] (**27**, Scheme 6.13), which in this concrete case were modified with nanovalves and nanoimpellers. The nanovalves are constituted by pH-switchable [2]pseudorotaxanes with CB6 as ring. While CB6 is threaded onto bisammonium stalks at the outer surface of the nanoparticles, the valve is closed and any loaded content in the mesopores is retained. On the other hand, the deprotonation of the ammonium groups by addition of base opens the nanovalve. The nanoimpellers are light-addressable azobenzenes tethered to the inner surface of the nanoparticle pores. The irradiation with visible light (448 nm) creates a dynamic wagging motion, which favors the guest release. As both concepts are combined in the same nanoparticle system, visible light irradiation has to be applied together with the addition of base to remove the CB6 macrocycles. Hence, the condition of an AND gate is fulfilled: the guest is only released if both inputs are applied. Manipulating only the nanovalve or the nanoimpeller function does not lead to guest release from the pores. If the guest is fluorescent, the correct functioning of the AND gate can be also followed by fluorescence spectroscopy. This last example shows in an instructive manner how molecular logic and principles of supramolecular chemistry can be merged to obtain "intelligent" functional materials.

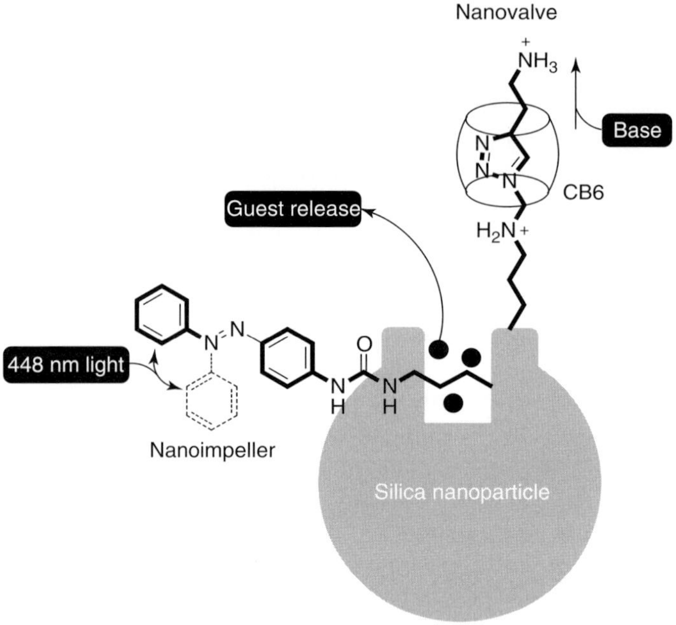

Scheme 6.13 AND-controlled drug delivery with mechanized silica nanoparticles (**27**).

6.6
Summary

On the basis of selected examples, the application of principles of supramolecular chemistry for the realization of simple molecular logic functions was briefly reviewed. Special emphasis was given to the implication of host–guest assemblies involving macrocyclic components. The bandwidth of systems ranges from classic crown ethers via pseudorotaxanes and rotaxanes to host–guest complexes with CBs, a class of recently much applied macrocyclic nanocontainers. While some of the systems discussed were explicitly reported in the literature as molecular logic gates, others were reinterpreted for the first time as logic devices. The discussion of supramolecular interactions for the design of logic functionality was completed by consideration of some systems based on hydrogen bonding.

Supramolecular assemblies have found relatively little attention in molecular logic. This is more surprising when taking into account that resetting and recycling of the systems may benefit from the inherently reversible nature of supramolecular interactions. Furthermore, the combination of the logically controlled thermodynamic stability of supramolecular assemblies and their potential applications, for example in drug delivery, [56, 61, 67–69] may provide an additional layer of functionality. It is to be hoped that future activities in the field will profit more from such synergies.

Acknowledgments

The financial support by the Spanish Ministry of Economy and Competitiveness (MINECO), Madrid (grant CTQ2011-28390 for U.P.) and the Portuguese Foundation for Science and Technology (FCT), Lisbon (PhD fellowship SFRH/BD/81628/2011 for C.P.C.) is gratefully acknowledged.

References

1. Balzani, V., Credi, A., and Venturi, M. (2003) *ChemPhysChem*, **4**, 49–59.
2. de Silva, A.P. and McClenaghan, N.D. (2004) *Chem. Eur. J.*, **10**, 574–586.
3. Pischel, U. (2007) *Angew. Chem. Int. Ed.*, **46**, 4026–4040.
4. Szaciłowski, K. (2008) *Chem. Rev.*, **108**, 3481–3548.
5. Andréasson, J. and Pischel, U. (2010) *Chem. Soc. Rev.*, **39**, 174–188.
6. Katz, E. and Privman, V. (2010) *Chem. Soc. Rev.*, **39**, 1835–1857.
7. Pischel, U. (2010) *Aust. J. Chem.*, **63**, 148–164.
8. de Silva, A.P. (2011) *Chem. Asian J.*, **6**, 750–766.
9. Amir, R.J., Popkov, M., Lerner, R.A., Barbas, C.F. III, and Shabat, D. (2005) *Angew. Chem. Int. Ed.*, **44**, 4378–4381.
10. de Silva, A.P., James, M.R., McKinney, B.O.F., Pears, D.A., and Weir, S.M. (2006) *Nat. Mater.*, **5**, 787–790.
11. Magri, D.C., Brown, G.J., McClean, G.D., and de Silva, A.P. (2006) *J. Am. Chem. Soc.*, **128**, 4950–4951.
12. Angelos, S., Yang, Y.W., Khashab, N.M., Stoddart, J.F., and Zink, J.I. (2009) *J. Am. Chem. Soc.*, **131**, 11344–11346.
13. Konry, T. and Walt, D.R. (2009) *J. Am. Chem. Soc.*, **131**, 13232–13233.
14. Margulies, D. and Hamilton, A.D. (2009) *J. Am. Chem. Soc.*, **131**, 9142–9143.
15. Ozlem, S. and Akkaya, E.U. (2009) *J. Am. Chem. Soc.*, **131**, 48–49.
16. Tokarev, I., Gopishetty, V., Zhou, J., Pita, M., Motornov, M., Katz, E., and Minko, S. (2009) *ACS Appl. Mater. Interfaces*, **1**, 532–536.
17. Hammarson, M., Andersson, J., Li, S., Lincoln, P., and Andréasson, J. (2010) *Chem. Commun.*, **46**, 7130–7132.
18. Privman, M., Tam, T.K., Bocharova, V., Halámek, J., Wang, J., and Katz, E. (2011) *ACS Appl. Mater. Interfaces*, **3**, 1620–1623.
19. Xie, Z., Wroblewska, L., Prochazka, L., Weiss, R., and Benenson, Y. (2011) *Science*, **333**, 1307–1311.
20. de Silva, A.P., Gunaratne, H.Q.N., and McCoy, C.P. (1993) *Nature*, **364**, 42–44.
21. Uchiyama, S., McClean, G.D., Iwai, K., and de Silva, A.P. (2005) *J. Am. Chem. Soc.*, **127**, 8920–8921.
22. de Silva, A.P., Dobbin, C.M., Vance, T.P., and Wannalerse, B. (2009) *Chem. Commun.*, 1386–1388.
23. Bag, B. and Bharadwaj, P.K. (2005) *Chem. Commun.*, 513–515.
24. Bozdemir, O.A., Guliyev, R., Buyukcakir, O., Selcuk, S., Kolemen, S., Gulseren, G., Nalbantoglu, T., Boyaci, H., and Akkaya, E.U. (2010) *J. Am. Chem. Soc.*, **132**, 8029–8036.
25. Baytekin, H.T. and Akkaya, E.U. (2000) *Org. Lett.*, **2**, 1725–1727.
26. Goodman, A., Breinlinger, E., Ober, M., and Rotello, V.M. (2001) *J. Am. Chem. Soc.*, **123**, 6213–6214.
27. Balzani, V., Credi, A., and Venturi, M. (2008) *Molecular Devices and Machines – Concepts and Perspectives for the Nanoworld*, 2nd edn, Wiley-VCH Verlag GmbH, Weinheim.
28. Credi, A., Balzani, V., Langford, S.J., and Stoddart, J.F. (1997) *J. Am. Chem. Soc.*, **119**, 2679–2681.
29. Semeraro, M. and Credi, A. (2010) *J. Phys. Chem. C*, **114**, 3209–3214.
30. Asakawa, M., Ashton, P.R., Balzani, V., Credi, A., Mattersteig, G., Matthews, O.A., Montalti, M., Spencer, N., Stoddart, J.F., and Venturi, M. (1997) *Chem. Eur. J.*, **3**, 1992–1996.

31. Ashton, P.R., Ballardini, R., Balzani, V., Boyd, S.E., Credi, A., Gandolfi, M.T., Gómez-López, M., Iqbal, S., Philp, D., Preece, J.A., Prodi, L., Ricketts, H.G., Stoddart, J.F., Tolley, M.S., Venturi, M., White, A.J.P., and Williams, D.J. (1997) *Chem. Eur. J.*, **3**, 152–170.

32. Ashton, P.R., Ballardini, R., Balzani, V., Constable, E.C., Credi, A., Kocian, O., Langford, S.J., Preece, J.A., Prodi, L., Schofield, E.R., Spencer, N., Stoddart, J.F., and Wenger, S. (1998) *Chem. Eur. J.*, **4**, 2413–2422.

33. Ashton, P.R., Balzani, V., Kocian, O., Prodi, L., Spencer, N., and Stoddart, J.F. (1998) *J. Am. Chem. Soc.*, **120**, 11190–11191.

34. Leigh, D.A., Morales, M.A.F., Pérez, E.M., Wong, J.K.Y., Saiz, C.G., Slawin, A.M.Z., Carmichael, A.J., Haddleton, D.M., Brouwer, A.M., Buma, W.J., Wurpel, G.W.H., León, S., and Zerbetto, F. (2005) *Angew. Chem. Int. Ed.*, **44**, 3062–3067.

35. Qu, D.H., Wang, Q.C., and Tian, H. (2005) *Angew. Chem. Int. Ed.*, **44**, 5296–5299.

36. Andréasson, J., Pischel, U., Straight, S.D., Moore, T.A., Moore, A.L., and Gust, D. (2011) *J. Am. Chem. Soc.*, **133**, 11641–11648.

37. Qu, D.H., Ji, F.Y., Wang, Q.C., and Tian, H. (2006) *Adv. Mater.*, **18**, 2035–2038.

38. Collier, C.P., Wong, E.W., Belohradský, M., Raymo, F.M., Stoddart, J.F., Kuekes, P.J., Williams, R.S., and Heath, J.R. (1999) *Science*, **285**, 391–394.

39. Collier, C.P., Mattersteig, G., Wong, E.W., Luo, Y., Beverly, K., Sampaio, J., Raymo, F.M., Stoddart, J.F., and Heath, J.R. (2000) *Science*, **289**, 1172–1175.

40. Luo, Y., Collier, C.P., Jeppesen, J.O., Nielsen, K.A., DeIonno, E., Ho, G., Perkins, J., Tseng, H.R., Yamamoto, T., Stoddart, J.F., and Heath, J.R. (2002) *ChemPhysChem*, **3**, 519–525.

41. Green, J.E., Choi, J.W., Boukai, A., Bunimovich, Y., Johnston-Halperin, E., DeIonno, E., Luo, Y., Sheriff, B.A., Xu, K., Shin, Y.S., Tseng, H.R., Stoddart, J.F., and Heath, J.R. (2007) *Nature*, **445**, 414–417.

42. Periyasamy, G., Collin, J.P., Sauvage, J.P., Levine, R.D., and Remacle, F. (2009) *Chem. Eur. J.*, **15**, 1310–1313.

43. de Ruiter, G., Tartakovsky, E., Oded, N., and van der Boom, M.E. (2010) *Angew. Chem. Int. Ed.*, **49**, 169–172.

44. Pischel, U. (2010) *Angew. Chem. Int. Ed.*, **49**, 1356–1358.

45. Pischel, U. and Andréasson, J. (2010) *New J. Chem.*, **34**, 2701–2703.

46. Vögtle, F. (1993) *Supramolecular Chemistry: An Introduction*, John Wiley & Sons, Ltd, Chichester.

47. Lehn, J.M. (1995) *Supramolecular Chemistry – Concepts and Perspectives*, Wiley-VCH Verlag GmbH, Weinheim.

48. Nau, W.M. and Scherman, O.A. (2011) *Isr. J. Chem.*, **51**, 492–494. (thematic issue about cucurbiturils).

49. Dsouza, R.N., Pischel, U., and Nau, W.M. *Chem. Rev.*, **111**, 7941–7980; doi: 10.1021/cr200213s

50. de Silva, A.P., Dixon, I.M., Gunaratne, H.Q.N., Gunnlaugsson, T., Maxwell, P.R.S., and Rice, T.E. (1999) *J. Am. Chem. Soc.*, **121**, 1393–1394.

51. Lagona, J., Mukhopadhyay, P., Chakrabarti, S., and Isaacs, L. (2005) *Angew. Chem. Int. Ed.*, **44**, 4844–4870.

52. Liu, S., Ruspic, C., Mukhopadhyay, P., Chakrabarti, S., Zavalij, P.Y., and Isaacs, L. (2005) *J. Am. Chem. Soc.*, **127**, 15959–15967.

53. Isaacs, L. (2009) *Chem. Commun.*, 619–629.

54. Praetorius, A., Bailey, D.M., Schwarzlose, T., and Nau, W.M. (2008) *Org. Lett.*, **10**, 4089–4092.

55. Saleh, N., Koner, A.L., and Nau, W.M. (2008) *Angew. Chem. Int. Ed.*, **47**, 5398–5401.

56. Macartney, D.H. (2011) *Isr. J. Chem.*, **51**, 600–615.

57. Bhasikuttan, A.C., Pal, H., and Mohanty, J. (2011) *Chem. Commun.*, **47**, 9959–9971.

58. Pischel, U., Uzunova, V.D., Remón, P., and Nau, W.M. (2010) *Chem. Commun.*, **46**, 2635–2637.

59. Shaikh, M., Mohanty, J., Bhasikuttan, A.C., Uzunova, V.D.,

Nau, W.M., and Pal, H. (2008) *Chem. Commun.*, 3681–3683.
60. Sinha, M.K., Reany, O., Parvari, G., Karmakar, A., and Keinan, E. (2010) *Chem. Eur. J.*, **16**, 9056–9067.
61. Parente Carvalho, C., Uzunova, V.D., Da Silva, J.P., Nau, W.M., and Pischel, U. (2011) *Chem. Commun.*, **47**, 8793–8795.
62. Irie, M. (1983) *J. Am. Chem. Soc.*, **105**, 2078–2079.
63. Silvi, S., Arduini, A., Pochini, A., Secchi, A., Tomasulo, M., Raymo, F.M., Baroncini, M., and Credi, A. (2007) *J. Am. Chem. Soc.*, **129**, 13378–13379.
64. Dube, H., Ajami, D., and Rebek, J. Jr. (2010) *Angew. Chem. Int. Ed.*, **49**, 3192–3195.
65. Kim, Y., Ko, Y.H., Jung, M., Selvapalam, N., and Kim, K. (2011) *Photochem. Photobiol. Sci.*, **10**, 1415–1419.
66. Molard, Y., Bassani, D.M., Desvergne, J.P., Horton, P.N., Hursthouse, M.B., and Tucker, J.H.R. (2005) *Angew. Chem. Int. Ed.*, **44**, 1072–1075.
67. Ambrogio, M.W., Thomas, C.R., Zhao, Y.L., Zink, J.I., and Stoddart, J.F. *Acc. Chem. Res.*, **44**, 903–913; doi: 10.1021/ar200018x
68. Meng, H., Xue, M., Xia, T., Zhao, Y.L., Tamanoi, F., Stoddart, J.F., Zink, J.I., and Nel, A.E. (2010) *J. Am. Chem. Soc.*, **132**, 12690–12697.
69. Walker, S., Oun, R., McInnes, F.J., and Wheate, N.J. (2011) *Isr. J. Chem.*, **51**, 616–624.

7
Hybrid Semiconducting Materials: New Perspectives for Molecular-Scale Information Processing

Sylwia Gawęda, Remigiusz Kowalik, Przemysław Kwolek, Wojciech Macyk, Justyna Mech, Marek Oszajca, Agnieszka Podborska, and Konrad Szaciłowski

7.1
Introduction

Omnipresent revolution in information technologies will not last forever. Numerous fundamental and technological obstacles will, in the near future, slow down and finally stop the development of classic, silicon-based electronics [1–6]. Therefore, new technological platforms are being developed [5]. Application of single molecules seems to be a feasible alternative to the classic silicon-based devices [7–16]. Among the molecular logic devices, various biomolecules seem to be especially appealing [17–19] since almost every biochemical process at the cellular level can be understood in terms of information processing [19]. On the other hand, application of molecules and information carriers in fluid media leads to enormous increase in chemical complexity, which in consequence impairs operation of more complex devices [20]. Therefore, an intermediate approach involving combination of molecular species with semiconductor devices may be more reasonable [21–24]. Immobilization of molecular species onto semiconducting surfaces first of all increases the stability of the systems and provides the tool to increase spatial complexity of molecular-scale logic devices. Combination with various chemical processes controlled by diffusion and kinetics may lead to future generation of information-processing devices [25–27]. Increased geometrical complexity combined with chemical simplicity can in principle result in neuromorphic systems – large arrays of simple nodes with hierarchical structures [28, 29].

Molecular-scale engineered semiconducting materials offer not only new emergent properties resulting from mutual interactions of the components but also a great structural diversity of molecular species combined with robustness and electronic properties (i.e., band structure) of solids. Bulk, thin-layer, and nanoparticulate semiconductors can be used for preparation of these materials, but the properties of molecule–semiconductor hybrids are strongly affected not only by the chemical composition of the counterparts but also by the physical form of the semiconducting material [30–32]. In the case of nanoparticulate systems, quantum confinement and the absence of band bending (BB) are crucial factors influencing properties of the material, while BB of dipole–dipole coupling is indispensable in the case of bulk and thin-layer materials. The latter effects are especially strongly pronounced in the systems where the thickness of the space-charge layer is comparable with the thickness of the semiconducting thin film [33]. Furthermore, these systems can be regarded as an interface between molecular nanoworld and macroscopic world of everyday life [34, 35].

One and the most intriguing emergent property of molecule–semiconductor systems is the photoelectrochemical photocurrent switching (PEPS) effect. This effect, initially regarded as a scientific curiosity, has become a field of intense research [36]. Classical p-n, p-i-n, or Schottky junctions generate photocurrent on illumination, and changes in bias potential or incident photon energy usually affect the energetic and/or quantum efficiency of photocurrent generation [37]. In the case of PEPS effect, the polarity of photocurrent can also be easily changed from anodic to cathodic and vice versa. Numerous materials exhibit this kind of behavior, but the most typical ones are wide-band-gap semiconductors (TiO_2, CdS) with surfaces modified with appropriate molecular species. Surface binding, redox, and optical properties of these surface species, as well as the degree of electronic coupling between the surface and bulk of the hybrid materials, are key factors influencing the performance of PEPS-based devices.

This review summarizes our extensive research in the field of the PEPS effect, including synthesis of new materials, mechanistic studies, and practical applications.

7.2
Synthesis of Semiconducting Thin Layers and Nanoparticles

Many techniques provide deposition of thin-film semiconductors, that is, bath deposition, electrochemical deposition, physical or chemical vapor deposition, spray pyrolysis, laser-induced synthesis, molecular beam epitaxy, or sputtering [38, 39]. Screen printing, ink-jet printing, spin coating, and cast coating of nanoparticle suspensions also yield semiconducting films of reasonable quality [40, 41]. For the application as parts of electronic devices, bath and electrochemical deposition methods have an advantage in their practical and economical nature. These methods require neither advanced apparatus nor high temperature and pressure conditions. Inexpensive and simple methods are viable to obtain thin layers with sufficient morphological quality. These fabrication methods yield films that are closer to the equilibrium as compared with the sputtered films and do not involve the use of toxic organometallic compounds. These features make bath and electrochemical deposition desired methods for thin film synthesis. Besides, electrodeposition is the leading method in industrial and scientific thin film preparation, especially in the case of binary and ternary chalcogenides. Thus synthesis mechanisms are well known, enabling a precise process control. In the case of nanoparticles, microwave synthesis seems to be the method of choice. It allows rapid and uniform heating of the reaction media and requires reasonably short reaction times [42].

Another emerging field of preparation of layered nanostructures encompasses utilization of the directed self-assembly of nanoparticles. This phenomenon is based on various stimuli, such as temperature, light, pH, solvent polarity, and van der Waals forces, by which nanoparticles organize spontaneously. By careful selection and construction of the building blocks it is possible to obtain specific structures of the self-assembled nanoparticles.

7.2.1
Microwave Synthesis of Nanoparticles

One method used in the synthesis of semiconductor nanoparticles is microwave-assisted preparation. Microwave irradiation (MWI) as a heating method has found a number of applications in chemistry since 1986. The effect of heating is created by the interaction of the dipole moment of the molecules with the high-frequency electromagnetic radiation (2.45 GHz) [43–45]. Compared with conventional methods, microwave synthesis has many advantages, among others: short reaction time, small particle size, narrow particle size distribution (about 5–10%), high product yields, and purity due to suppression of unwanted side reactions [46, 47].

The microwave reaction can be divided into three reaction stages: (i) initiation of nucleation, (ii) the growth regime, and (iii) reaction termination [45]. The first step, the nucleation process, is very important because the quality of the final product depends on the initiation of the reaction. Rapidly increasing temperature guarantees heating the whole volume of the reacting mixture. During the second step, the growth process, the power is reduced to maintain a controlled growth stage. The temperature is held constant by active air cooling of the reaction vessel to allow power and temperature to be controlled independently. By changing the reaction time, the growth of semiconductor nanocrystals can be controlled. For example, during the synthesis of ZnS small spherical particles were produced in 30–60 s; in 1–2 min rods were formed; and in 3 min only wires were obtained [48]. In the last step, the microwave power is turned off to terminate the reaction, and the solution is rapidly cooled by the air flow. Owing to reaction quenching, the size distribution of nanoparticles can be significantly minimized.

The method based on MWI allows control of the size and shape of semiconductor nanostructures not only by reaction time and temperature but also by relative concentrations of the organic surfactants used during synthesis [49]. Complexing agents play a very important role during the microwave reaction. They can slow down the reaction and cause the nanoparticles to be smaller [43]. Moreover, using different organic surfactants it is possible to modify the shape of semiconductor nanoparticles.

Microwave-assisted synthesis is usually used for the preparation of semiconductors type II–VI (sulfides: CdS, ZnS [44, 48], Bi_2S_3 [47]; selenides: CdSe [43, 45], PbSe, CuSe [43], ZnSe [48]; and tellurides: Cu_{2-x}Te, HgTe [50]) or III–V such as InP, InGaP [45]. Taking into account all the factors described above, one can design and obtain high-quality semiconductor nanomaterials that can be used in nanoelectronic or optoelectronic devices.

7.2.2
Chemical Bath Deposition

The simplest method of preparation of binary chalcogenides involves rapid mixing of appropriate salts. This method, however, yields powders of variable crystallinity and wide particle size distribution. Therefore, the precursor ions must be present in solution in low concentrations and/or should be gradually generated during the deposition process. Thus, metal ions must be complexed with appropriate ligands, while chalcogenide ions can be generated by decomposition of soluble precursors. Cadmium sulfide was selected as an example to demonstrate the principles of chemical bath deposition of semiconducting films.

7.2.2.1 Sulfide Ion Precursors

The most widely used sulfide ion precursor in basic solutions is thiourea, which decomposes in basic solutions probably according to the scheme Eqs. (7.1) and (7.2) [51]:

$$SC(NH_2)_2 + OH^- \rightarrow HS^- + NH_2CN + H_2O \tag{7.1}$$

$$HS^- + OH^- \rightarrow S^{2-} + H_2O \tag{7.2}$$

Increase in temperature and/or pH results in acceleration of thiourea decomposition [52]. Sometimes dimethylthiourea, thioacetamide, or thiosulfate is employed instead of thiourea [53].

7.2.2.2 Commonly Used Ligand

While the most popular ligand in the synthesis of CdS is ammonia, other complexing agents are also possible, such as various aliphatic amines, for example, ethylenediamine [53], sodium citrate alone or together with ammonia [54, 55], potassium nitrilotriacetate [51], nitrilotriacetic acid together with hydrazine monohydrate [56], ethylenediaminetetraacetic acid together with ammonia [57] or tartaric acid [58].

The simplest deposition mechanism is based on reaction of sulfide ions with cadmium ions according to the scheme Eq. (7.3):

$$Cd^{2+} + S^{2-} \rightarrow CdS \tag{7.3}$$

If the concentration of the complexing agent is too low to prevent $Cd(OH)_2$ formation, these particles serve as nucleation centers and cluster mechanism operates (Eqs. (7.4) and (7.5)):

$$Cd^{2+} + 2OH^- \rightarrow Cd(OH)_2 \tag{7.4}$$

$$Cd(OH)_2 + S^{2-} \rightarrow CdS + 2OH^- \tag{7.5}$$

Growth rate is the parameter that strongly affects thin film properties. Band-gap energy is particularly sensitive to the deposition rate. For CdS, the bulk value of the band gap amounts to 2.42 eV [59]. The faster the reaction, the wider the band gap, because dimensions of the grains decrease [60, 61]. This phenomenon is derived from the quantum size effect. Typically, values of band-gap energies of thin films are within 2.5 and 3.2 eV [54].

Film thickness depends both on the reaction rate and time of synthesis. The faster the reaction, the thinner the film that is formed (even less than 20 nm) [61]; longer time results in thicker film till the self-passivation point. Typically, terminal thickness is between 0.1 and 0.2 µm and varies with bath composition [62].

A good method to synthesize adherent and smooth films with low porosity is to enrich the bath composition with surfactants [63, 64]. The main disadvantage of chemical bath deposition is usually very poor yield. Only about 2% of starting material is used for the substrate coverage [65].

7.3
Electrochemical Deposition

Electrochemical deposition of metals with formation of metallic alloys or compounds has been known for a long time and is applied for over 150 years [66]. An increasing interest in semiconductor electrodeposition is caused by a wide range of unique features of this method. The most important reasons of such interest are

- processes occurring during electrochemical deposition are close to equilibrium (unlike in the case of high-temperature methods) therefore allowing formation of materials characterized by various states of stresses;
- precise control of film thickness, uniformity, and deposition rate;
- possibility of controlling the processes taking place during film deposition and therefore control of their composition, morphology, and structure;
- possibility of large areas coating (even porous or with complex shape);
- good adherence of the deposit to the substrate, which opens a possibility to obtain multilayer periodical nanostructures;
- low-temperature process slows down diffusion of atoms between particular layers, as well as between the substrate and the film;
- deposition of multidimensional structures with the use of templates;
- no use of toxic organometallic compounds;
- low costs of processes because of avoiding vacuum or high temperatures;
- simple instrumentation and low energy consumption.

There are several review articles and books on the electrodeposition of semiconductors [63–65, 67–70]. Almost all semiconductors can be deposited by the electrochemical method, for example, elemental semiconductors such as silicon [71, 72] or germanium [73], binaries (III–V and II–VI) [74, 75], and ternaries (chalcopyrite-like materials, for example, $CuInSe_2$) [76]. Recently, a rapid development of electrodeposition of oxide semiconductors was presented in the literature (ZnO, Cu_2O) [77, 78]. The material properties depend on a number of parameters, namely, type of electrolyte (aqueous, organic, ionic liquids), ionic concentrations, pH, stirring rate, temperature, applied potential, or current density. The optimization of these parameters makes it possible to produce semiconducting layers with the desired structural, electronic, and optical properties. Depending on the electrochemical properties of the deposited materials, different types of electrolytes can be applied. Most often used are aqueous and organic solutions, which are very convenient because of the low temperature of operation. High-temperature molten salts are used mostly for the deposition of elements that cannot be deposited from aqueous solution (e.g., silicon) because of their very negative reduction potential. Currently, high-temperature molten salts are replaced by ionic liquids [79–81]. Key advantages that enable them to overcome the limits imposed by common aqueous or organic media are their wide electrochemical windows and low melting points.

Currently, much effort is being devoted to building electronic and optical devices/sensors by systematically putting together individual nanoscale or submicron structures. Electrodeposition gives the possibility of controlling the process of growth at the nanometer scale. Template-assisted electrodeposition is an important technique for synthesizing metallic nanomaterials with controlled shapes and sizes [82, 83]. Arrays of nanostructured materials with specific arrangements can be prepared by this method, employing either an active or restrictive template as a cathode in an electrochemical cell.

Besides classic electrodeposition, a very promising possibility has appeared for semiconductor compounds synthesis, namely, electrochemical atomic layer epitaxy (ECALE) proposed by Stickney [84]. The ECALE method exploits surface-limited electrochemical reactions, such as underpotential deposition (upd), to form alternate atomic layers of the elements making up compound layers of compounds. Each deposition cycle forms a monolayer (ML) of the compound, and the number of cycles determines how thick the deposit will be.

Electrochemical deposition of the individual elements from the electrolyte begins after the electrode has been brought to a potential more negative than the equilibrium potential characteristic of the redox potential of the reaction shown in Eq. (7.6):

$$SiCl_4 + 4e^- \rightarrow Si + 4Cl^- \tag{7.6}$$

The synthesis of semiconductor compounds is more complex, and either an anodic or a cathodic method can be applied. The anodic route is possible during spontaneous corrosion of metals in a chalcogenide environment or by applying a positive potential at the metal electrode [85–87]. Major disadvantages of anodic and electroless synthesis routes are self-limitation of the film growth and its difficult control.

On the other hand, cathodic synthesis is a more flexible method. When semiconductors are deposited, for example, II–VI or III–V, both precursors have to be dissolved in the solution. If there are two different ions in the solution, namely, A^{m+} and B^{n+}, one of them will always be relatively nobler. By applying a more negative potential value than the equilibrium potential of the less noble species, both metals can be deposited on the electrode surface. If, additionally, the semiconducting phase A_xB_y in the binary A-B system exists, there is a driving force between the deposited elements, which permits the reaction shown in Eq. (7.7)

$$xA + yB \rightarrow A_xB_y \tag{7.7}$$

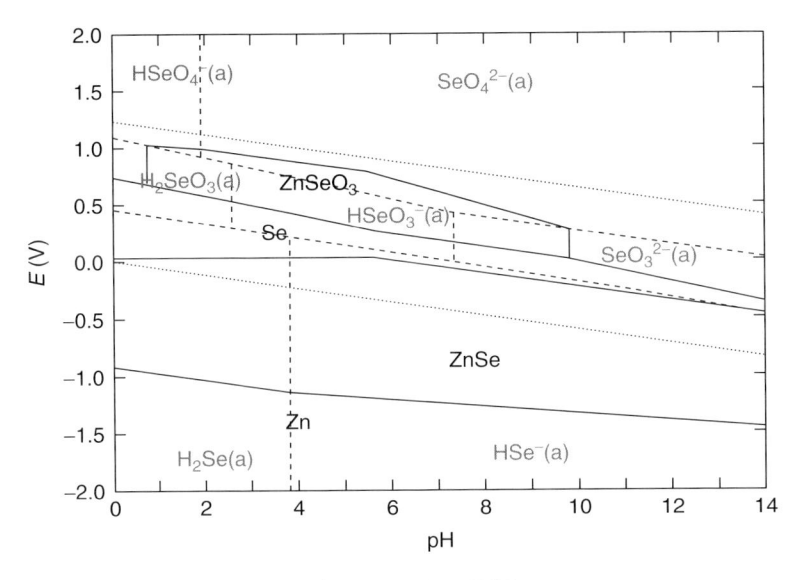

Figure 7.1 Pourbaix diagram of Zn-Se-H$_2$O at 25 °C.

to occur in the solid state. Thus, the electrochemical process may yield the semiconducting phase as a deposit on the electrode surface. This simple picture becomes much more complex if one considers electrodeposition of ZnSe, for example. Inspection of the Pourbaix diagram for the Zn-Se-H$_2$O system [88] shows that selenium species in aqueous solutions are neutral (Figure 7.1).

According to Kröger's theory, it is possible to deposit both elements simultaneously without exceeding the equilibrium potential of one of them [89]. This phenomenon is called *underpotential deposition*, and should be observed during deposition of semiconductor compounds. Applying this theory to the ZnSe deposition process, one must assume that two electrochemical reactions take place between the solution and the electrode (Eqs. (7.8) and (7.9)):

$$Zn^{2+} + 2e^- \rightarrow Zn_{(s)} \tag{7.8}$$

$$H_2SeO_3 + 4e^- + 4H^+ \rightarrow Se_{(s)} + 3H_2O \tag{7.9}$$

The equilibrium potentials of these two reacting species can be described by the following equations (Eqs. (7.10) and (7.11)).

$$E_{(Zn^{2+}/Zn)} = E^o_{(Zn^{2+}/Zn)} + \frac{RT}{2F} \ln \left(\frac{a_{Zn^{2+}}}{a_{Zn}} \right) \tag{7.10}$$

with $E^o_{(Zn^{2+}/Zn)} = -0.76$V and

$$E_{(H_2SeO_3/Se)} = E^o_{(H_2SeO_3/Se)} + \frac{RT}{4F} \ln \left(\frac{a_{H_2SeO_3} a^4_{H^+}}{a_{Se}} \right) \tag{7.11}$$

with $E^o_{(H_2SeO_3/Se)} = 0.74$V.

The activities of a_{Se} and a_{Zn} in the deposit are determined by the equilibrium constant of the reaction (Eq. (7.12))

$$Zn_{(s)} + Se_{(s)} \rightleftarrows ZnSe_{(s)} \tag{7.12}$$

which yields Eq. (7.13):

$$(a_{Se}a_{Zn}) = \exp\left(\frac{\Delta G^o_{ZnSe}}{RT}\right) \tag{7.13}$$

This yields, at the Zn/ZnSe phase boundary ($a_{Zn} = 1$), Eq. (7.14)

$$\frac{RT}{F} \ln\left(\frac{\sqrt{a_{Zn^{2+}}}}{\sqrt[4]{a_{Se^{4+}}}}\right) - \frac{RT}{F} \ln(a_{H^+})$$
$$= \left(E^o_{H_2SeO_3/Se} - E^o_{Zn^{2+}/Zn}\right) - \frac{\Delta G^o_{ZnSe}}{4F} \tag{7.14}$$

and Eq. (7.15)

$$\frac{RT}{F} \ln\left(\frac{\sqrt{a_{Zn^{2+}}}}{\sqrt[4]{a_{Se^{4+}}}}\right) - \frac{RT}{F} \ln(a_{H^+})$$
$$= \left(E^o_{H_2SeO_3/Se} - E^o_{Zn^{2+}/Zn}\right) + \frac{\Delta G^o_{ZnSe}}{2F} \tag{7.15}$$

at the ZnSe/Se phase boundary ($a_{Se} = 1$), respectively.

In the absence of selenium species in the solution, application of the potentials higher than -0.76 V will not cause zinc deposition. If selenium is present, for potentials between -0.76 and $+0.74$ V, selenium will deposit, while zinc still remains in the solution. However, individual potentials will be affected if the interaction between different atoms in the solid phase is present. It can be easily demonstrated that the potential given by Eq. (7.8) will change at the Se/ZnSe phase boundary (that is, on "the other side of the phase") by the term associated with ΔG^o_{ZnSe} (Eq. (7.16)):

$$E_{(Zn^{2+}/Zn)} = E^o_{(Zn^{2+}/Zn)} + \left(-\frac{\Delta G^o_{ZnSe}}{2F}\right) \tag{7.16}$$

Since in this phase diagram region $a_{Se} = 1$, one obtains from Eq. (7.13) $\ln(a_{Zn}) = \frac{\Delta G^o_{ZnSe}}{RT}$. This yields the value of the potential $E_{(Zn^{2+}/Zn)} = +0.04$ V, if the activity of Zn^{2+} species in the solution is assumed to be unity. The change in zinc potential with the activity is shown in Figure 7.2. From this figure, it is clear that bringing the potential below $E^o_{(H_2SeO_3/Se)}$ will cause selenium to deposit. If, however, the potential becomes more negative and passes $+0.04$ V, ZnSe will appear. Because of the interaction in the deposit, electrodeposition of selenium also enables deposition of zinc at the same potential range.

The range of deposition conditions, involving applied potential and hydrodynamic properties, also affects the process of electrolysis and must be well defined, since the film composition strictly depends on these parameters [75, 88]. For example, at the potential of -0.6 V versus SCE only four-electron process of Se reduction can be observed. In turn, at -0.7 V versus SCE, the efficiency of ZnSe electrodeposition

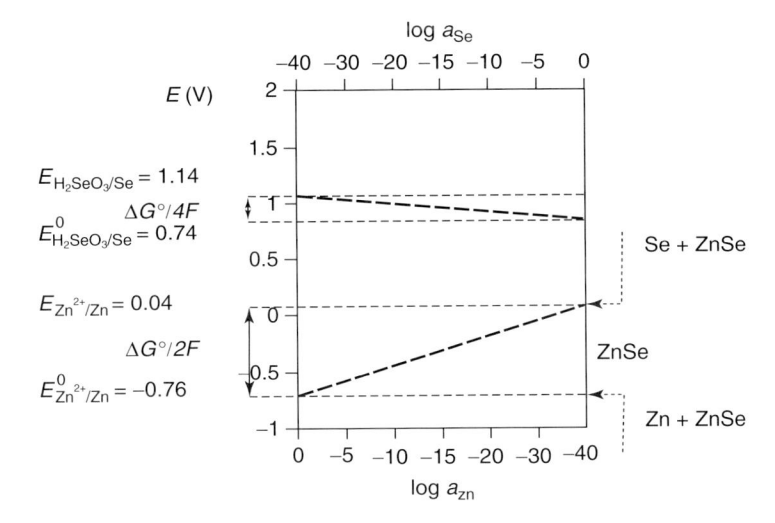

Figure 7.2 Equilibrium potentials of Zn^{2+}/Zn and H_2SeO_3/Se as a function of the a_{Zn} and a_{Se} activities relative to an electrolyte with $a_{Zn^{2+}} = 1$ and $a_{H_2SeO_3} = 1$.

decreases because of hydrogen evolution, and additionally, the deposited selenium is reduced to H_2Se according to Eq. (7.17):

$$Se + 2H^+ + 2e^- \rightarrow H_2Se \tag{7.17}$$

The selenious acid can be also reduced to H_2Se (Eq. (7.18)):

$$H_2SeO_3 + 6H^+ + 6e^- \rightarrow H_2Se + 3H_2O \tag{7.18}$$

In turn, H_2Se reacts with zinc ions, and zinc selenide may be formed according to Eq. (7.19):

$$H_2Se + Zn^{2+} \rightarrow ZnSe + 2H^+ \tag{7.19}$$

The mutual relationship between process parameters and the composition and morphology of the products must be found experimentally.

The above description shows briefly the idea of the electrodeposition process of semiconductor compounds for the ZnSe example. Basic analysis of the synthesis of any II–VI compounds required knowledge of (i) the equilibrium potentials of the components, (ii) the interaction of the components when forming the alloy or the compound, and (iii) the values of activities of ionic species in the electrolyte at the solid–electrolyte interface during deposition. However, additional knowledge is required to describe the process of the electrosynthesis in detail: composition of the electrolyte (e.g., presence of other ions and complexing agents in the electrolyte), pH, temperature, or other competitive reactions. Moreover, the kinetics of the process has to be known; for example, the exchange current of components between the electrolyte and the electrode, and the rate constants of partial processes are required to properly model the process of electrodeposition.

The electrochemical synthesis of metal sulfides is not very well understood because of a much more complex chemistry of sulfur. Different schemes are proposed for synthesis of sulfides, depending on the source of sulfur. Chemical processes can stimulate the film formation, for example, decomposition of sulfur-containing precursors in the presence of metal ions. An insoluble compound is formed, and a film is deposited by crystallization on a substrate. The general model assumes disproportionation of the thiosulfate ions (Eq. (7.20)), which is assisted by electrochemical reaction with metal ions, according to the overall scheme (Eqs. (7.21) and (7.22)) [90].

$$S_2O_3^{2-} \rightarrow S + SO_3^{2-} \tag{7.20}$$

$$M^{2+} + S_2O_3^{2-} + 2e^- \rightarrow MS + SO_3^{2-} \tag{7.21}$$

$$2M^{2+} + S_2O_3^{2-} + 6H^+ + 8e^- \rightarrow 2MS + 3H_2O \tag{7.22}$$

In addition, the following electrode reactions may occur (Eqs. (7.23) and (7.24)):

$$S_2O_3^{2-} + 6H^+ + 4e^- \rightarrow 2S + 3H_2O \tag{7.23}$$

$$S + 2H^+ + 2e^- \rightarrow H_2S \tag{7.24}$$

Formed H_2S can subsequently react with metal ions:

$$M^{2+} + H_2S \rightarrow MS + 2H^+ \tag{7.25}$$

However, these models are still under dispute in the literature [91, 92], and the process of synthesis of sulfides requires further studies.

Cathodic electrodeposition of thin films of various metal oxides is also possible [64, 78]. Films of SnO_2 can be formed from nitrate solutions. The process of deposition consists of a few steps. The aim of the first one is to decrease the local pH value close to the surface of the electrode. The reaction described by Eq. (7.26) occurs, resulting in local pH elevation [77, 93]:

$$NO_3^- + 2H^+ + 2e^- \rightarrow NO_2^- + H_2O \tag{7.26}$$

Next steps involve precipitation of tin hydroxides and oxides, according to Eqs. (7.27) and (7.28):

$$Sn^{4+} + 4H_2O \rightarrow Sn(OH)_4 + 4H^+ \tag{7.27}$$

$$Sn(OH)_4 \rightarrow SnO_2 + 2H_2O \tag{7.28}$$

Similarly, ZnO characterized by various structures and morphologies can be deposited by an analogous method [64]. Its cathodic electrodeposition from aqueous $Zn(NO_3)_2$ solution is possible according to the net Eq. (7.29):

$$Zn^{2+} + NO_3^- + 2e^- \rightarrow ZnO + NO_2^- \tag{7.29}$$

Copper oxide deposition can be realized from alkaline solution containing lactate [94, 95]. During the electrochemical process, soluble Cu^{2+} ions are oxidized to insoluble Cu^+ species, namely, copper(I) oxide (Eq. (7.30)):

$$2Cu^{2+} + 2e^- + H_2O \rightarrow Cu_2O + 2H^+ \tag{7.30}$$

Ternary compounds with a chalcopyrite structure are very important materials because of their significance in photovoltaics. One-step electrodeposition of copper indium gallium selenide (CIGS) has been studied extensively by many groups [70]. The electrosynthesis of these compounds is very complex because of specific electrochemical properties of all components of the semiconductor. The mechanism of codeposition of selenium and copper strongly depends on substrates used for the synthesis [96].

The electrochemical method is very convenient for deposition of so-called multilayers or sandwich structures. The layer-by-layer deposition enables periodic modulation of structure or composition of coatings. The actual deposition of the multilayered composite can be carried out by different routes. The modulation of potential during deposition from a single bath was applied during synthesis of CdSe-ZnSe [97] or PbSe-PbTe [98] systems. An interesting example of the synthesis of superstructures from a single bath was proposed by Switzer [99]. The self-organizing phenomenon appeared during electrodeposition of Cu_2O-Cu system. The oscillating cathodic processes were responsible for the formation of ordered structures. The self-oscillations strongly depend on pH and temperature and induce the alternate deposition of Cu_2O and a composite of Cu and Cu_2O. Spontaneous potential oscillations are observed when applying a constant cathodic current density (Figure 7.3) [100].

Otherwise electrodeposition of multilayers from different electrolytes seems to be more accurate. The flow system was adapted by Wei to the electrochemical cell to manipulate precisely the composition of multilayers [101].

A special type of semiconductor film can be obtained by the electrochemical atomic layer deposition (ECALD) method, which was suggested by Stickney in the 1990s and is now being intensively developed by his team [102]. The method relies on the phenomenon of underpotential deposition of elements. Metals or nonmetals can be deposited within potentials more positive than their reversible equilibrium potentials. The process of film creation is surface limited and results from the influence of the material of electrode on the element being reduced. Therefore, it is possible to create a ML of a given element on the electrode surface by a precise control of the deposition process. The whole process of a semiconductor synthesis with the use of the ECALD method is based on cyclic alternate deposition of one element ML followed by the second one, which subsequently leads to semiconductor compound synthesis. The effect of underpotential deposition in this case results from a negative value of free energy of synthesized compounds creation. The negative value of ΔG shifts the deposition potential of each element and enables the process of underpotential deposition. Appropriate electrode selection allowing deposition of the first ML under underpotential conditions is a key problem of this process. Subsequent layers are deposited based on the mutual influence of formerly deposited layers on the next components of the semiconductor compound being deposited. The process is conducted in a flow reactor where the electrolyte is

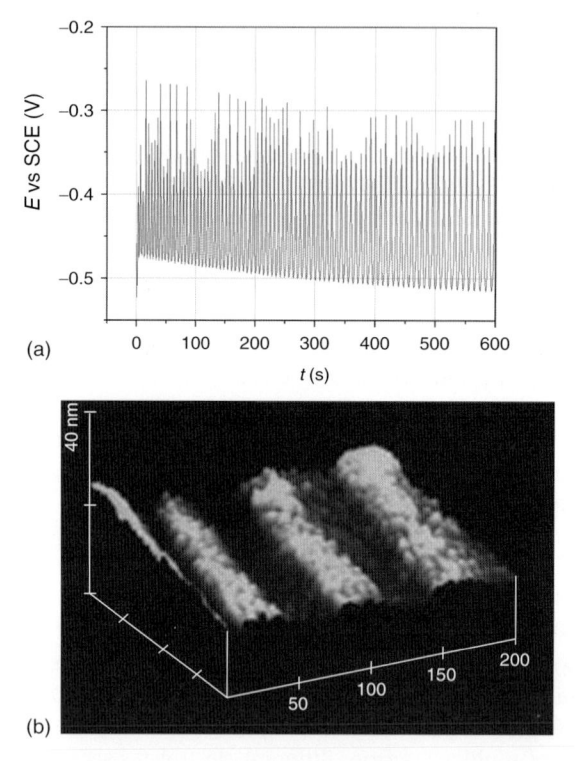

Figure 7.3 Potential oscillations observed on application of a cathodic current density of 0.5 mA cm^{-2} (solution containing 0.4 M CuSO$_4$, 1.4 M lactate at pH = 8, temperature 20 °C) and cross-sectional scanning tunneling microscopic (STM) image of film after etching with dilute nitric acid. (Partially reproduced from Ref. [100] with permission. Copyright American Chemical Society 1998.)

periodically exchanged depending on the element being deposited. The deposition conditions (electrolyte composition, deposition potential) are selected in order not to allow uncontrolled overpotential deposition and to avoid dissolving of formerly deposited layers. The film thickness depends on the number of performed cycles.

ECALD allows the synthesis of very complex structures, with the deposition process controlled at the atomic level. Consequently, a precise control of the semiconductor layer thickness at such a level can be applied for precise manipulation of its band-gap energy. The layers may consist of a few atomic layers of one semiconductor, and on its surface another semiconductor can be deposited. Such a sequence can be repeated until the formation of a layer of desired thickness and structure. Manipulation of both the synthesized semiconductor type and thickness of particular layers enables fine-tuning of the semiconductor layer properties depending on its future applications. Furthermore, in thus prepared layers one should expect quantum size effects – a very evident dependence of electronic and optical properties on the layer thickness. This is associated with a one-dimensional quantum confinement of charge carriers and efficient electronic coupling between layers in semiconducting heterostructures. Moreover, under these conditions, resonance effects associated with interaction of free electrons in metallic support and confined charge carriers in thin semiconducting layer should be expected.

7.3.1
Nanoheterostructure Preparation

Synthesis of various composite semiconductor nanoclusters is another step to obtain more advanced and useful materials. Linking two or more nanocrystals results in new materials with unique properties [32, 103–106]. Composite semiconductor nanoclusters can be classified into two categories, namely, capped (core–shell) and coupled (dumbbell) heterostructures (Figure 7.4).

The core–shell systems are much more useful because of a better and more reproducible interaction between two materials. Furthermore, in this case, the inner component is effectively isolated from the environment, and the undesired surface processes are thus avoided. The most interesting electronic properties are represented by core–shell nanostructures, which could be composed of (i) two different semiconductors (CdSe-ZnS), (ii) two different metals (Ag-Au), or (iii) a metal and a semiconductor (Au-CdSe). These nanostructures can be prepared via multistep syntheses where presynthesized nanocrystals are used as seeds for nucleation and growth of other inorganic phases. In core–shell nanostructures, depending on the materials used for core and shell, the confinement of electron and hole wave functions can be controlled, which, in turn, determines optical and electronic properties.

In semiconductor–semiconductor core–shell heterostructures, five different band alignment types can be distinguished, depending on the band-gap energies and relative band-edge positions. This, in turn, provides different spatial confinement modes. Type I heterostructures (Figure 7.5a) combine one narrow-band-gap and one wide-band-gap semiconductor [32, 104]. If a narrow gap semiconductor is covered by a shell of semiconductor with wide band gap, both electron and hole wave functions will be strongly confined to the core (Figure 7.5a), which provides the lowest energy states for both electrons and holes. Moreover, the excitons in the core are protected from interaction with the surface and the environment. This type of heterostructure can be fabricated in CdSe/ZnS, InAs/ZnS, and InP/ZnS systems. Confinement of the exciton to the inner sphere results in high quantum efficiencies and stability of photo- and electroluminescence.

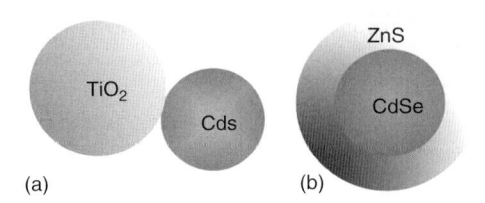

Figure 7.4 Coupled semiconductors (TiO$_2$/CdS, (a)) and capped semiconductors (CdSe/ZnS, (b)).

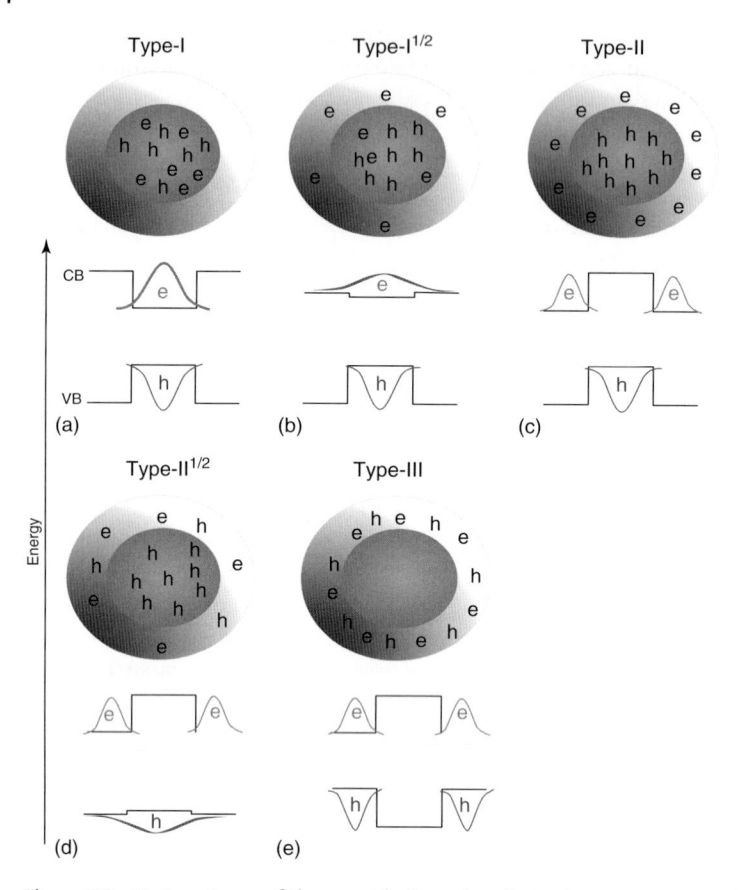

Figure 7.5 Various types of the core/shell semiconductor heterostructures with schematic representation of charge carrier confinement and the radial probability functions for the lowest energy electron (e) and hole (h) wave functions.

In some cases, for example, in the CdSe/CdS core–shell nanostructures, the band gap of the shell material is not wide enough to confine both electron and hole wave functions. In this case, the electron can easily move between CdSe and CdS phases, whereas the hole is confined to the CdSe core because of its large effective mass and substantial offset of the valence band (VB) energies (Figure 7.5b). This system is called *type I$^{1/2}$* (also known as *quasi type II*) [32, 104].

In type II heterostructures, the energies of the conduction band (CB) and the VB of the shell are either both higher or both lower than those of the core. As a result, one of the carriers is mostly confined to the core of the heterostructures, while the other one remains in the shell (Figure 7.5c). Their effective band gaps are heavily affected by the band offsets of the cores and shells. In the lowest excited state, the electron and hole wave functions are spatially separated, which results in the electron wavefunction mainly residing in one semiconductor, whereas the hole wavefunction is localized more in the second semiconductor.

The $e^−$–h^+ recombination energy is smaller than the band gap of either of the constituent material components. Therefore, radiative recombination of the $e^−$–h^+ pair produces a red-shifted emission, which would not be available with a single semiconductor. For example, the CdTe/CdSe core/shell heterostructures emit infrared radiation, which cannot be achieved in CdTe or CdSe materials. Type II nanocrystals are characterized by low quantum yields and enhanced sensitivity to the local environment. A small oscillator strength of the radiative transition caused by reduced overlap of the electron and hole wavefunctions results also in longer lifetimes of the excited state. This property of type II heterostructures causes them to be highly attractive for applications in photovoltaics, where charge separation in space domain is desirable.

In the previous examples, the narrow-band-gap semiconductor was covered with a layer of semiconductor with a wider band gap. The opposite situation is, however, also possible and leads to interesting phenomena. Here, one can distinguish two additional types of confined heterostructures. The most typical case can be observed when the VB edges are of comparable energies but the CB edge of the core semiconductor is higher than the shell one. Such a system is called *type II$^{1/2}$* (or *"inverted" type II*), and it was observed in ZnSe/CdSe nanocrystals [107] (Figure 7.5d). These heterostructures can support only partial spatial separation between electrons and holes. As a result, electrons are confined in the shell region while holes are delocalized over the whole volume of the heterostructure. Such a situation leads to reduced (but nonzero) overlap between the electron and hole wavefunctions. The ZnSe/CdSe heterostructure is characterized by a high quantum yield of light emission, reaching even 80–90% [107].

It is also possible to consider a system in which both the electron and hole are localized in the shell of the heterostructure (Figure 7.5e). This example can be called the *type III heterostructure*. In principle, such heterostrucures should behave just like empty shells. An additional effect may, however, result from plasmonic effects, as the plasmon oscillation of an empty shell and a core–shell system will be very different.

7.3.2
Nanoparticles Directed Self-Assembly

From the traditional point of view, the key issue in nanoparticle synthesis and morphology control is the stabilization of colloid, which avoids aggregation of the nanoparticles. Since the introduction of the bottom-up approach in nanofabrication of the functional materials, the nanoparticles need to be properly modified to be able to aggregate into desired structures and functionalities. Nevertheless, this requires a high level of control of interactions between molecules localized at nanoparticle interfaces for directing the aggregation process. The most usual stimuli in this regard include temperature, light, pH, and solvent polarity (e.g., Langmuir-Blodgett or surfactant-assisted assembly). In addition, adjusting proper electrostatic interactions between nanoparticles leads them to self-assemble into crystalline or noncrystalline arrays. It can be achieved through introduction of a variety of charged, capping groups. However, these interactions must not be too strong since random aggregation might emerge. Therefore, a delicate balance must be maintained between the entropically favored close-packed arrays and electrostatically driven non-close-packed structures. Thus, it is convenient to introduce amphoteric capping agents, which can change the surface charge with varying pH. Readers more interested in this topic can find more information in these reviews [108, 109].

7.4
Organic Semiconductors–toward Hybrid Organic/Inorganic Materials

Acenes are the family of linearly fused polycyclic hydrocarbons. This group of compounds is recognized as rather "unfriendly" since they are toxic and potentially carcinogenic. The simplest member of this family is benzene, while the largest, unsubstituted stable acene is pentacene (Figure 7.6). Higher acenes (more than five aromatic rings) are extremely unstable [110]. For example, heptacene, octacene, or nonacene cannot be isolated, but there is a possibility for their preparation and studies *in situ* in cold matrices [111–113]. Despite extremely low stability and difficulties in processing, higher homologs exhibit extraordinary semiconducting properties, such as pentacene, which is the benchmark of performance for thin-film devices [114].

Stabilization of higher acenes, leading to the formation of a large group of acene-like structures, is implemented by incorporation of heteroatoms [114, 115] or alkyl, aryl, and thioaryl chains into the aromatic core [116, 117] (Figure 7.7). In

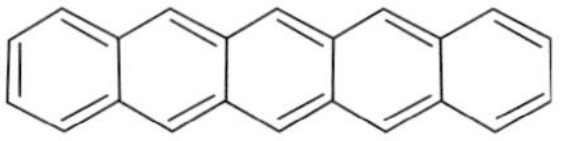

Figure 7.6 Molecular structure of pentacene.

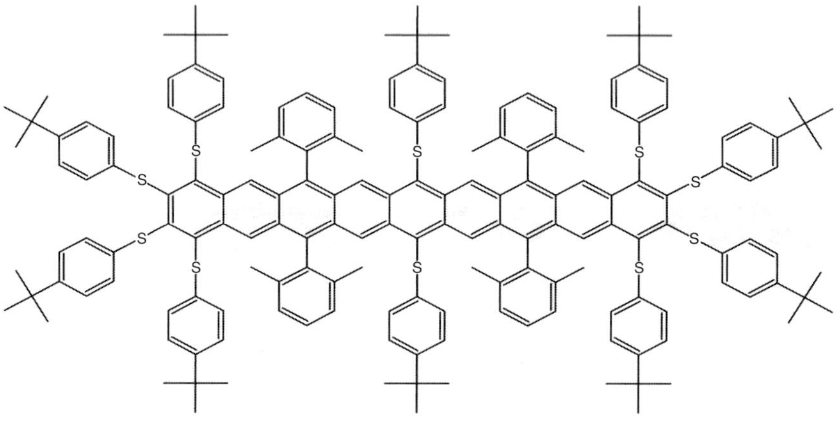

Figure 7.7 The largest known stable acene derivative [116].

7.4 Organic Semiconductors–toward Hybrid Organic/Inorganic Materials

the case of lower homologs, for example, tetracene, introduction of substituents may improve their electrical features [114].

Acenes and acene-like structures are the group of compounds that are exploited mainly in electronic devices such as field-effect transistors (FETs) and organic light-emitting diodes (OLEDs). There is a very strong correlation between the structural order of these materials and device performance [114]. Therefore, investigating and understanding the way of organizing of acenes is not only interesting but also important and useful scientific task.

7.4.1
Self-Organization Motifs Exhibited by Acenes and Acene-Like Structures

Crystallization of acenes in general results in two packing motifs. The first structure is the so-called herringbone, where the molecules are ordered in edge-to-face, two-dimensional layers. The second way of ordering is stacking them face to face (π-stacked motif). Similar to the herringbone motif, the two-dimensional structure is also preserved. Since the π-stacked morphology exhibits a strong electronic coupling between neighboring molecules, further interactions with adjacent stacks may result in two-dimensional electronic coupling in the solid state [114, 118].

In the case of the "herringbone" structure, molecules in the undistorted stack are rotated around their long axes by a certain angle, while molecules from adjacent stack are rotated by the opposite angle. The last step is the vertical translation of alternating stacks by a half of the unit cell length (Figure 7.8). Since the angular distortion destroys the overlapping between π-orbitals of neighboring molecules from the same column almost completely, the largest interaction between molecules from adjacent stacks is observed. This effect is called *diagonal interaction* [119].

The starting point for the π-stacked structure is the same cofacial stack as for the herringbone one. The difference lies in the angle of rotation. Currently, molecules rotate around their short axes. Similar to the former case, adjacent stacks rotate in opposite directions. Then the alternate stacks are moved by half length of the stacking distance (Figure 7.9). Unlike in the herringbone structure,

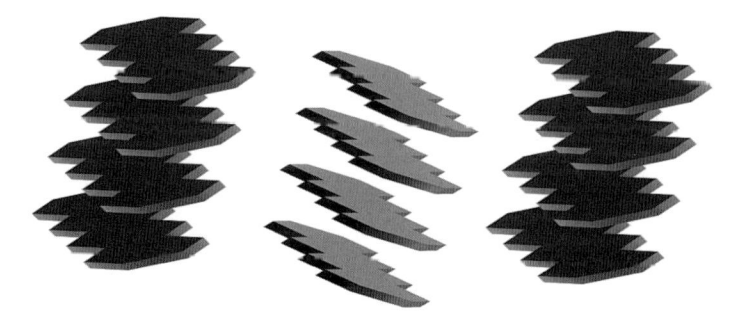

Figure 7.8 Scheme of the herringbone structure. The picture does not represent real ratios of distance to dimension of molecules.

Figure 7.9 Sscheme of the π-stacked structure. The picture does not represent real ratios of distance to dimension of molecules.

in the π-stacked packing motif an angular distortion does not destroy the π-overlap between adjacent molecules from the same column [119].

From the foregoing description of the types of ordering of the acenes in solid state stems, it can be seen that they are quite similar. Both of them increase packing density in comparison with undistorted cofacial stacks. The primary distinction, which induces differences in electronic structure of the stacks, is the direction of the angular distortion [119].

The crucial factor influencing the packing motif of the acenes is the presence and nature of side chains. In the case of their absence, acenes usually adopt the herringbone structure. For acenes substituted by, for example, linear alkyl side chains, π-stacked (lamellar) structure is observed [119].

Consideration of packing motifs of acenes is essential from the point of view of carrier transfer phenomenon. Charge transport in such structures generally occurs by carrier hopping from the lower to the upper molecule in the stack. Since the π-stacked structure is characterized by higher π-orbitals overlapping than in the herringbone one, acenes that crystallized in the former order should exhibit better charge carrier transport properties. First of all, higher π-orbital overlapping causes the larger valence bandwidth (in the case of p-type semiconductors), which increases hole mobility. In addition, the hopping rate rises with intermolecular orbital overlapping and falls with the hop distance [119].

Besides the crystalline form, acene-like structures can occur as discotic liquid crystals, which are currently considered as a new generation of organic semiconductors [120]. They are half way between crystalline and amorphous phases. Discotic

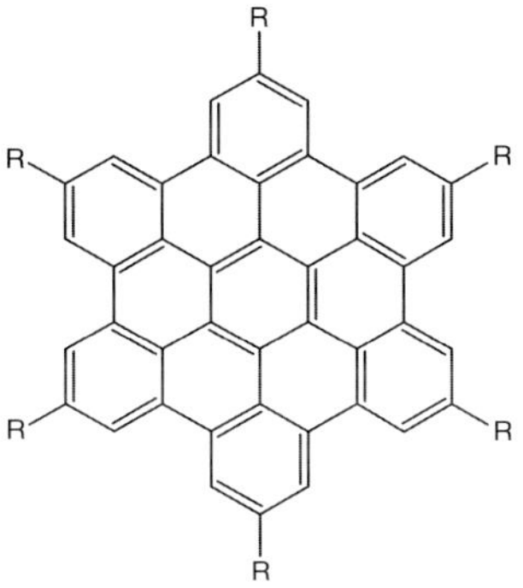

Figure 7.10 Hexa-*peri*-hexabenzocoronene.

liquid crystals consist of disordered columns of disc-shaped molecules organized in a two-dimensional array [121]. These molecules are comprised of flat, rigid aromatic cores substituted with aliphatic chains, which may strongly affect their structures. The most common discotics have large, conjugated cores, for example, porphyrins, phtalocyanides, triphenylenes, hexaazatriphenylenes, perylenes, macrocycles, and hexa-peri-hexabenzocoronenes (Figure 7.10) [120].

Very elegant examples of the dependence of side chains on the columnar organization of acene-like structures are polycyclic aromatic hydrocarbons (PAHs) substituted with alkyl side chains. PAHs are organized in columnar order, which is based on interactions of π-orbitals of neighboring aromatic cores in the stack. The distance between two adjacent cores is about 3.5 Å, while the distance between columns depends on the length of the side chains. The larger the aromatic core, the higher is the columnar stability of the structure observed [120].

Introducing alkyl side chains into rigid aromatic cores of PAHs may tune their intermolecular organization from the crystalline phase through the liquid crystal to the disordered structure. The length of side chain, degree of branching, and position of attachment of side chains to the aromatic core are important factors governing the organization of molecules [120]. In the crystalline phase, which exhibits the highest order, molecules possess a strongly reduced ability to move. Owing to optimization of electronic interactions between molecules in the stacks, they are tilted, which also results in a higher density of the structure (see above) [119, 120]. In the case of liquid crystal structure, molecules have the ability to perform lateral, longitudinal, or rotary movements. They are organized in a columnar order (undistorted cofacial stack). At elevated temperatures, aliphatic side chains, which are longer than before, behave similar to fluid. Incorporation of long, branched alkyls into aromatic cores destroys intermolecular interactions, and a columnar, disordered liquid crystal phase occurs [120]. The longer and more branched the side chains, the lower is the temperature of phase transition between crystalline and liquid crystal. On the other hand, the bigger the aromatic core, the higher is the temperature of this transition [120].

Both the character of the side chains and the presence of ionic species introduced into PAHs affect their structures. Self-assembled nanostructures can be formed in solution of 9-phenyl-benzol[1,2]quinolizino[3,4,5,6-*fed*]phenanthidinylium (PQP)-substituted salts that contain alkyl sulfonate or disulfonate anions. Their spontaneous organization in solution leads to two-dimensional nanostructures such as twisted belts or twisted belt-like nanostructures, one-dimensional rods, and puckered two-dimensional belts or planks. Since self-organized PQPs may serve as channels for transport of dissolved ionic species, improving their ionic conductivity via ordering morphology of nanostructures is crucial [122].

In addition, centrally charged PQPs with alkyl chains form different morphologies depending on the type of counter ion and the chain length. Increasing the latter parameter results in transition from fibers to ribbons, while changing the counter ion from Cl^- to BF_4^- alters the nanostructure from nanoribbons to helicenes and tubes [123].

The charge carrier transport in ordered discotic liquid crystals is mainly one dimensional [121]. It stems from the much stronger electronic interactions of molecules within the stack in comparison with molecules from adjacent stacks. Moreover, the columnar structure resembles an insulated wire (aromatic cores create a conducting pathway that is surrounded by insulating alkyl chains) [120]. Since the liquid crystal phase is intermediate between crystalline and amorphous phases with a relatively low viscosity, it exhibits self-healing properties [120, 121]. The consequence of this ability of disappearing of defects is insignificant deep trapping of charge carriers. They can move along the columns, subjected to time-dependent potentials, resulting from motions of adjacent aromatic cores, while columnar fluctuations remove structural traps [121]. The comparison of charge carrier mobility for aligned liquid crystal and amorphous organic phases indicates that for the former, hole mobility along the column is 2 orders of magnitude higher than for the latter [121].

In the case of low intrinsic charge carrier density, the conductivity may be increased by doping, that is, by incorporation of electron acceptors (e.g., $AlCl_3$ or $NOBF_4$) into a hydrocarbon chain matrix. Acceptors extract electron from the aromatic core, resulting in the creation of a radical cation. Such materials exhibit charge transport perpendicular to the columns. It is possible only at dislocations or with the assistance of "defect bridges." At elevated temperatures, the charge carrier performs random-walk travel scattered from insulating side chains as long as it reaches an adjacent column. Intercolumnar transport can also occur because of ionic exchanging processes [121]. Another possibility of three-dimensional charge carrier migration is hexa-peri-hexabenzocoronene asymmetrically decorated with three alkoxy chains (only on one side of the core). In such a configuration, conducting aromatic core is not completely surrounded by insulating mantel, and charge carriers have the possibility of hopping between adjacent columns as well as along the stack [120].

Besides the self-assemblies of acene and acene-like structures, self-assembled MLs on semiconducting or dielectric surfaces are also of great importance. An elegant example of such a structure is self-assembled ML of tetracenodiol (which behaves as a p-type semiconductor) onto Al_2O_3 surface, which is the common material for gates in transistors. To ensure binding of the organic semiconductor to aluminum oxide, the terminal ring of tetracene is substituted by two hydroxyl groups; thus, it interacts with the surface in manner similar to that of catechol with metal oxides. Similar to adsorption of catechol onto titanium dioxide, bidentate chelating structure is observed. Organic molecules on Al_2O_3 stand upright, and the structure is tightly packed, almost as tight as in bulk crystal. Self-assembly of tetracenodiol may be controlled by tuning crystallographic properties of the aluminum oxide surface [124].

7.4.2
Applications of Acenes in Organic Electronic Devices

The most common electronic devices using acenes are FETs and OLEDs. They are also applied in sensors or photovoltaics. Employing these organic compounds in solar cells reduces manufacturing costs and enables printing devices over a large area. Organic solar cells exhibit efficiencies over 4% and thus may be a good alternative for silicon-based ones. Acenes also give a possibility of replacing inorganic liquid crystals in emissive layers for less power consumption and may be applied in lower performance devices such as radio frequency identification (RFID) tags or flat-panel displays [15].

The key parameter that determines the performance of FETs or OLEDs is the charge carrier mobility. In the case of acenes, holes are dominant charge carriers. To compete effectively with amorphous silicon-based electronics, the value of hole mobility should be in the order of $0.5\ cm^2\ V^{-1}\ s^{-1}$ [114]. It depends on device configuration and measurement conditions, but the most relevant parameter is the spatial organization of acene molecules in the solid state. Thus, a precise control of molecular order is crucial [114]. Using the example of FETs, one may see how the packing motif of employed acenes influences the hole mobility value and in consequence performance of the whole device.

Precise control over the geometry of the lattice is critical for optimal device performance. Thin-layer devices show optimal performance with materials with two-dimensional ordering (i.e., which prefer the herringbone arrangement), while the highest charge carrier mobilities in single-crystal devices were observed for highly π-stacked compounds, with rubrene being the best example [114, 125].

Fiber-oriented acenes can create 3-D networks showing solvent trapping. Organogels thus obtained can be used in energy transfer processes [126].

Rubrene (Figure 7.11) is a very good example of a material that is utilized for single-crystal FETs. In fact, it exhibits so high value of charge carrier mobility that it is considered as a performance standard for this type of electronic devices. Rubrene has a tetracene structure substituted by four phenyl groups, which leads to a strongly π-stacked order in solid state. So far, the highest observed mobility value is $20\ cm^2\ V^{-1}\ s^{-1}$ at room temperature. Contrary to its superb properties in the single-crystal form, rubrene used in thin-film devices yields poor performance [114].

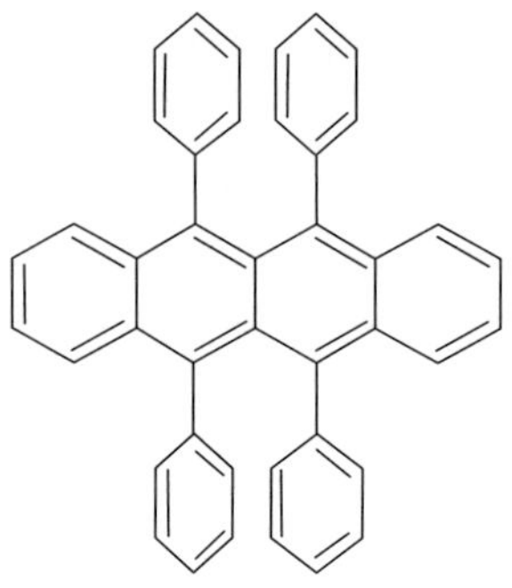

Figure 7.11 Rubrene (5,6,11,12-tetraphenyltetracene).

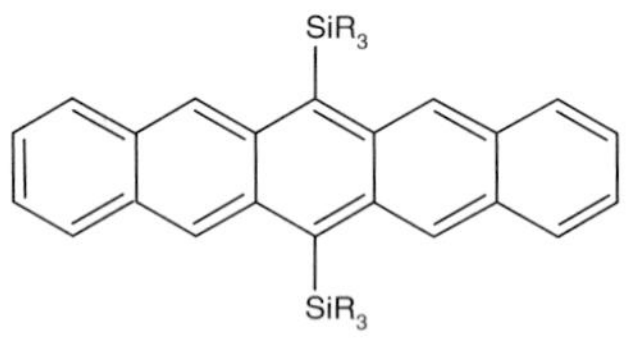

Figure 7.12 The example of alkyne-functionalized pentacene: triisopropyl silyl pentacene derivative (R = isopropyl) [114].

Like rubrene for single-crystal devices, pentacene may afford a supreme example of acene employed in thin-film transistors (Figure 7.6).

Similar to all unsubstituted acenes, it crystallizes into the herringbone structure. Mobility values characterizing thin-film FET devices reach up to 5 cm^2 V^{-1} s^{-1}. Another application of pentacene is organic solar cells with power conversion efficiency of about 2.7%, where it is used as a p-type material together with C$_{60}$ as the n-type material.

Another elegant example of influence of the acene order on the performance of thin-film FETs are alkyne-functionalized pentacenes. The structure is controlled by changing the alkyne substituent dimensions. For the substituent diameter less than half length of the acene, one-dimensional π-stacking interactions occur. For the diameter roughly the same as half length, two-dimensional π-stacking interactions are observed (e.g., triisopropyl silyl pentacene derivative – TIPS pentacene, Figure 7.12). For higher values of substituent diameter, herringbone structure emerges. Among these three types of acene ordering in the solid state, only the second reveals a high-quality thin-film FET device with hole mobility equal to 0.4 cm^2 V^{-1} s^{-1} (TIPS-pentacene) [114].

7.5
Mechanisms of Photocurrent Switching Phenomena

The effect of photocurrent switching depends on various parameters, which must be considered in order to understand the nature of this phenomenon. These parameters involve the redox properties of a semiconductor surface and its bulk, availability of electron donors and acceptors in the electrolyte solution, and certainly applied potential and color (i.e., energy) of incident light. The direction of photogenerated net currents is a result of competition between various redox processes: anodic photocurrents are observed when oxidation reactions prevail at the working electrode, while cathodic photocurrents require good efficiencies of reduction processes.

7.5.1
Neat Semiconductor

In the case of film consisting of separated nanocrystals, no BB takes place. The effect of BB starts to play a role when larger crystals or aggregates with a high degree of percolation cover the photoelectrode. In the latter case, the direction of BB depends on the Fermi level and charge accumulated at the film surface, leading to the formation of accumulation, depletion, or inversion layers.

Photocurrent generation occurs on excitation of the semiconductor film with light of sufficient photon energy. Let us consider an electrode covered with the n-type semiconductor. Electrons generated in the CB can be transferred to the electrode whenever its potential is higher than the potential of trapped electrons. In the presence of electron donors and acceptors, interfacial electron transfer between the semiconductor and the electrolyte solution may occur. Anodic photocurrents require that an electron donor is easily oxidized by photogenerated holes, and the electrode potential enables electron transfer from the CB of the semiconductor to the electrode (Figure 7.13a). Several processes compete with those responsible for anodic photocurrent generation: recombination within the semiconductor particles (radiative and nonradiative), recombination of holes with electrons from the electrode, as well as reduction of electron acceptors in the electrolyte solution. The evolution of anodic photocurrents follows the kinetics presented in Figure 7.13c: the rate of electrode charging and discharging (depending on the electrolyte content,

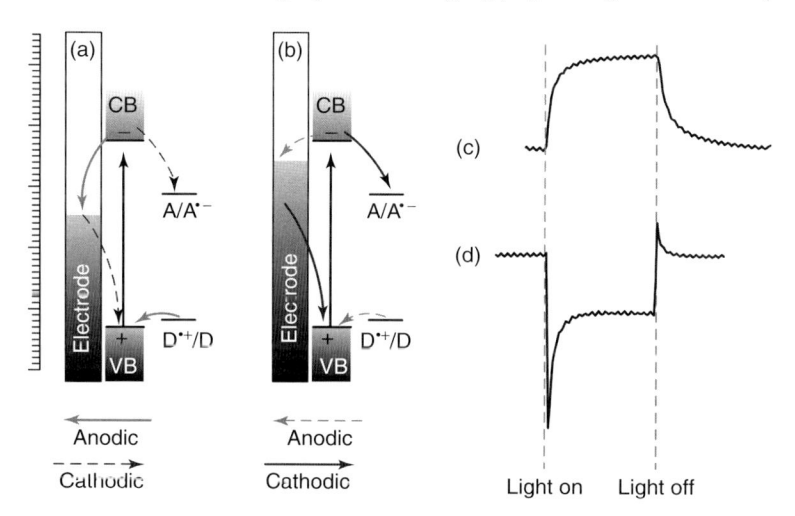

Figure 7.13 Mechanisms of anodic (a) and cathodic (b) photocurrent generation at the electrode covered with an n-semiconductor. Kinetics of anodic (c) and cathodic (d) photocurrent evolution.

film morphology, diffusion, and light intensity) reflects in the curvature of $i(t)$ graph recorded on shutter opening and closing.

Cathodic photocurrents occur when reduction of the electron acceptor by electrons from the CB and consumption of holes are more efficient than the mechanisms responsible for anodic photocurrent generation (Figure 7.13b). When the electrode potential facilitates reduction of the semiconductor, dark cathodic currents appear, leading to consumption of the electron acceptor and decrease in the net photocurrent. The kinetics of cathodic photocurrent generation is presented in Figure 7.13d.

7.5.2
Composite Semiconductor Materials

The photocurrent switching effect can be observed in neat semiconducting materials; however, the choice of materials and conditions are strongly restricted because of fixed band-edge potentials (within the BB limit, see below). On the other hand, composite materials offer almost unlimited variability of electrical and optical properties, which in turn should facilitate design of semiconductors with predetermined properties. The structure of composites is based on at least two different constituent materials, which retain their chemical identities. In the case of composites, the interference of optical or electrical properties is obvious and can be designed at the synthesis stage by a proper selection of components. Bulk heterojunction materials consist of at least two n- and p-type materials, distinguishable by their chemical or physical features, with outspread interphase surface. The class of p-n bulk heterojunction composite materials demonstrates unusual optical and electrical properties such as nonlinear optical effects, photoluminescence, selective absorption, or catalytic effects [127, 128]. There are two distinct cases of photocurrent switching: potential induced and light induced. In both cases, two criteria must be strictly fulfilled: optical and electronic. Optical criteria impose different values of band-gap energies of constituent materials (i.e., p- and n-type semiconductors). In an optical switching regime, semiconductors can be excited selectively, which results in anodic or cathodic photocurrents, depending on the characteristics of the photoexcited material. Electronic criteria, which impose limits on band-edge positions, are especially important in potential-induced photocurrent switching and prevent parasitic electron transfer between particles of different conductivity types. Figure 7.14 presents schematically various modes of photocurrent switching.

When the Fermi level of electrode lies between the bottom of a CB of n-type semiconductor and the peak of the VB of p-type semiconductor, both polarizations of photocurrent are possible, depending on the energy of incident light (Figure 7.14a). Photocurrent switching (switching between anodic and cathodic photocurrent) occurs upon changing the wavelength of incident light. When the light has $h\nu_1$ photon energy, the n-type semiconductor is excited and photogenerated electrons are transferred to the conducting support. As a result, anodic photocurrent is observed.

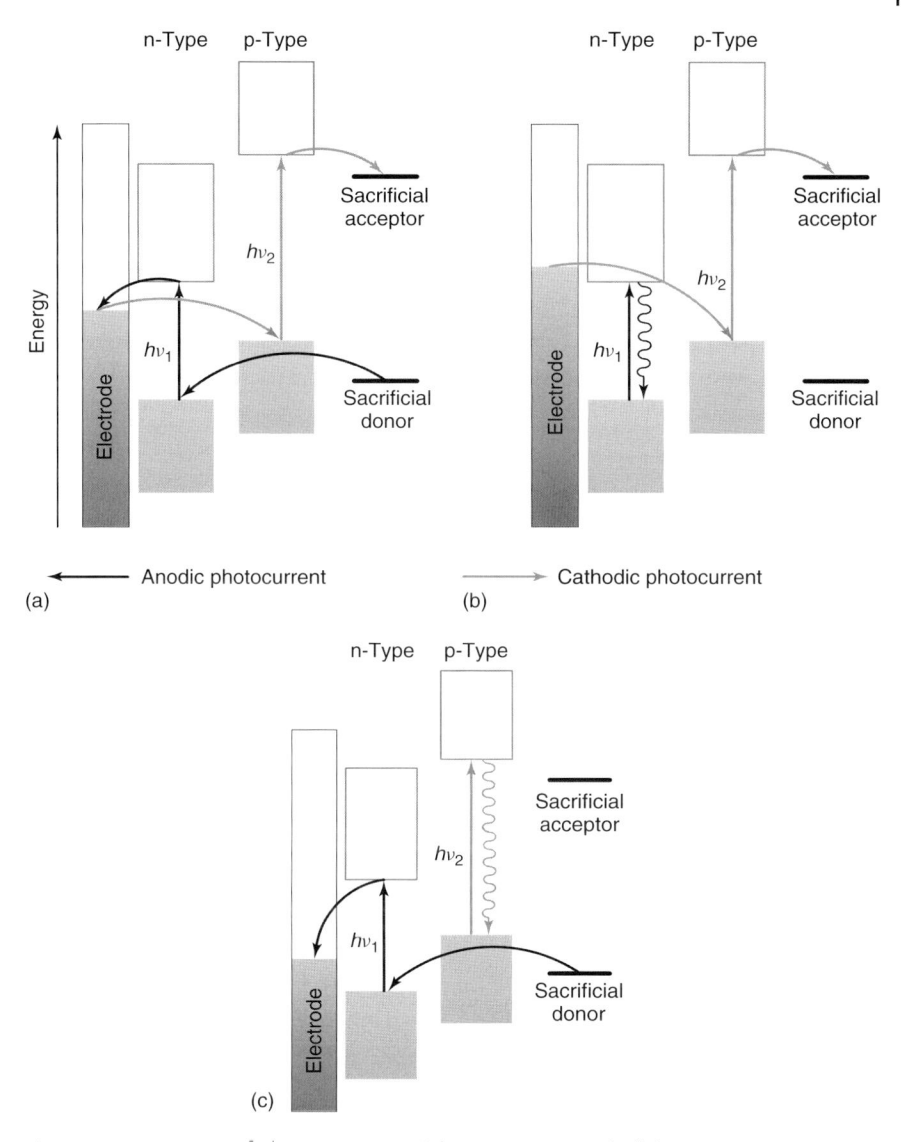

Figure 7.14 Diagram of photocurrents switching processes in bulk heterojunction composite at no external bias (a) and under cathodic (b) and anodic (c) polarization.

External polarization of the electrode determines the direction of photocurrent flow, independent of incident light wavelength. Cathodic polarization with bias potential lower than the Fermi level of the n-type semiconductor results exclusively in the generation of cathodic photocurrent (Figure 7.14b). The opposite effect is observed in the case of cathodic polarization with potentials higher than the Fermi level of the p-type component (Figure 7.14c). Composite materials with

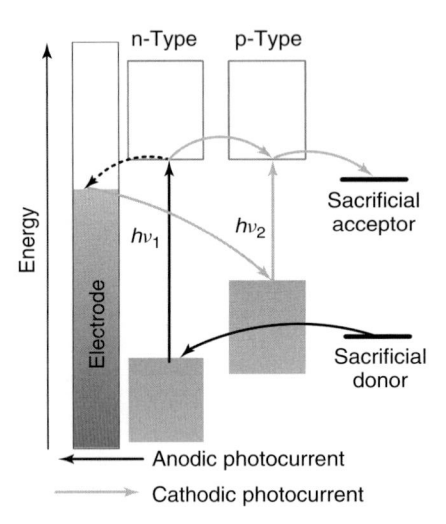

Figure 7.15 Diagram of photocurrent when electrical criterion is not fulfilled.

reported photocurrent switching following the above mechanism include p-n bulk structures based on n-TiO$_2$-N/p-CuI [129], n-BiVO$_4$/p-Co$_3$O$_4$ [130], n-BiVO$_4$/p-CuO [130], n-TiO$_2$/p-Se [131, 132], and n-CdS/p-CdTe [133].

If the optical criterion fails, the selective photoexcitation of semiconducting components is no longer possible. In the case when the electronic criterion fails (Figure 7.15), selective photoexcitation does not result in photocurrent switching, because electrons can easily migrate between n- and p-type components and only one photocurrent direction is preferred. This was observed for the n-TiO$_2$/p-CuO material [130].

Similar switching characteristics were reported for quasi-planar p-n junction structures. Combination of organic–inorganic thin-film composites leads to very good photoelectrochemical properties. As an example of such a structure, polybithiophene/TiO$_2$ (PBT) can be considered [134]. PBT deposited onto TiO$_2$ behaves like a p-type semiconductor of 1.8 eV band gap. Application of an external potential results in the switching of electrical properties of PBT from conducting (oxidized state) at high potentials to insulating or semiconducting (reduced state) at low potentials. At 0 V (vs SCE) potential, only cathodic photocurrents can be observed. Anodic polarization of the composite (0.5 V vs SCE) leads to both anodic and cathodic photocurrents depending on the wavelength. As a result, the photocurrent switching effect occurs at 400 nm wavelength: at $\lambda < 400$ nm anodic photocurrent is generated, while at longer wavelengths cathodic photocurrents can be observed [134]. Similar effects were observed in the case of a polythiophene/TiO$_2$ (PTh) core–shell nanocomposite [135].

Composites of a TiO$_2$ nanoparticle core with a PTh shell have been investigated regarding their energetic structures and photoelectrochemical properties [135]. The real structure of obtained material shown by transmission electron microscopy reveals anatase particles coated by a 2–3 nm thick PTh layer. For electrochemical

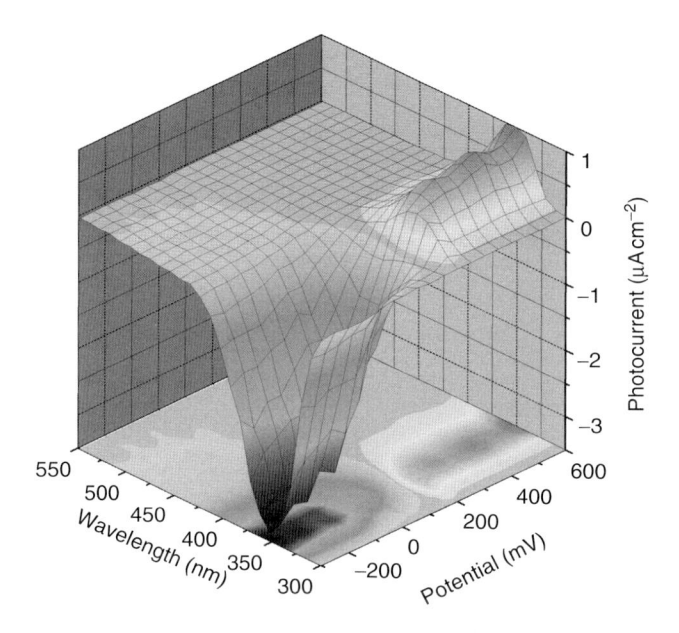

Figure 7.16 The photocurrent switching conditions for TiO_2–Prussian blue composite. Potential measured versus Ag/AgCl electrode. (Adapted from Ref. [136].)

and optical measurements, a PTh/TiO_2 electrode was prepared by electrophoretic deposition method. This fabricated photoelectrode is characterized by high porosity. Photocurrent action spectra are very similar to those of PBT/TiO_2 thin layer [134]. Cathodic photocurrents result from PTh excitation ($\lambda_{max} = 530$ nm), whereas anodic photocurrents originate from excitation of TiO_2 support.

Photocurrent switching in Prussian blue/TiO_2 composite is controlled by external potential (Figure 7.16). The mechanism responsible for this effect is based on a quasi-reversible redox process, observed in the neat polymer as well. Prussian blue is stable only at potentials 0.2–0.9 V versus Ag/AgCl. At lower potential values, Prussian blue is reduced to Prussian white, and at potentials higher than 0.9 V, the polymer is oxidized to Prussian yellow. The interaction between TiO_2 particles and the Prussian blue polymer does not involve any significant modification of electronic structures of components; only a small band gap increase is observed. The composites containing Fe^{III} centers behave like a neat TiO_2 material and, on illumination, generate only anodic photocurrents. The situation is totally different at potentials below 0.3 V, where reduction of Fe^{III} centers modifies the energetics of the electron transfer and cathodic photocurrent is observed. Furthermore, weak photosensitization due to Fe^{II}-Ti^{IV} electronic coupling can be noticed (see above) [136].

The examples presented above illustrate three different regimes of photocurrent switching phenomena related to composite materials. In the case of a composite consisting of two inorganic semiconductors, the switching process relies exclusively on the variable energetics of different electron transfer pathways. Composites based on semiconducting polymers offer an additional control via the electrochemical doping/dedoping of the polymer, which results in metal–semiconductor–insulator transitions. In both cases, photocurrent generation is observed within the whole absorption spectrum of the composite. Finally, composites based on cyanometallate polymers offer photocurrent switching based on tuning the polymer's properties, which in turn can influence the electron transfer pathways. Photocurrent generation is observed within the absorption spectrum of semiconducting particles, while the optical properties of the polymer matrix have only slight influence on photocurrent action spectra. The latter materials are thus closely related to surface-modified semiconductors, which are discussed in detail in the following section of this review.

7.5.3
Semiconductor–Adsorbate Interactions

The PEPS effect originates from interaction between a wide band gap semiconductor and a surface-modifying molecule. In order to give a broader view on this field, we feel that a summary of the knowledge about electronic structures of interfaces is most appropriate. The electronic structure of a semiconductor consists of energy bands (Figure 7.17), of which the most important are the VB and the CB, separated by an energy band gap (E_g). At the semiconductor surface, the periodicity of the lattice is broken, which automatically results in perturbation of the electronic structure of the semiconductor. The VB and CB wavefunctions cannot terminate abruptly at the interface, but they tail from the surface into the vacuum because the wavefunctions are, by definition, continuous, differentiable, and square integrable. Tailing of the wavefunctions makes the vacuum side of the interface negatively charged, while the deficiency of electron density (with respect to the charge of the nuclei) within the crystal renders the inner part of the interface positively charged (Figure 7.17) [137, 138].

This process generates a dipole layer and electrostatic potential drop at the surface, which is called the *surface potential* (V_s). Electrostatic interaction of this dipole with VB and CB electrons results in BB; its energy can be expressed as

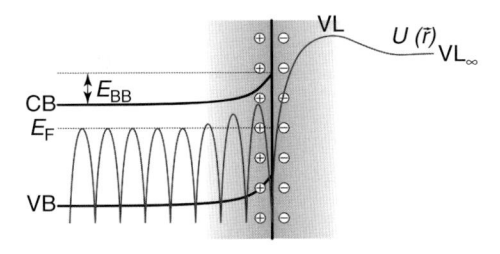

Figure 7.17 Generation of surface potential and band bending due to wavefunction tailing. $U(\vec{r})$ is the crystal potential, which is the measure of interaction of electron with ions and electrons within the solid as well as charges external to the solid [138].

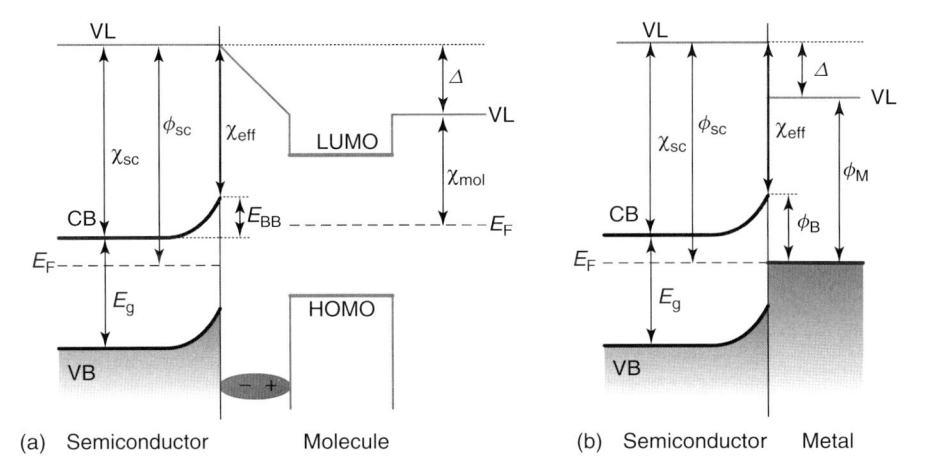

Figure 7.18 Energy level alignment during the formation of semiconductor–molecule (a) and semiconductor–metal (b) interfaces.

shown in Eq. (7.31):

$$E_{BB} = eV_s \tag{7.31}$$

where e is the charge of an electron, χ_{SC} (Figure 7.18a) is the energy needed to bring the electron from vacuum level (VL) to the bottom of CB (Figure 7.18a). χ_{SC} is a theoretical property measured for bulk, whereas an effective electron affinity (χ_{eff}) is measured on the surface. The energy needed to bring an electron from the bulk to the VL is called the *work function* (ϕ_{SC}). VL is the potential above which the electron can escape from the bulk. The molecular layer deposited atop the semiconductor cannot support the same wavefunction tailing; therefore, the VLs for different parts of the device are different. The VL should not be confused with the VL for an electron resting in vacuum (VL$_\infty$) [137]. E_F is the Fermi energy and is equivalent to the level which divides occupied states from unoccupied states at 0 K.

When a molecule comes into contact with a semiconductor, often an additional dipole layer forms at the interface. Various reasons, such as charge transfer across the interface, redistribution of electron cloud, interfacial chemical reaction, and other forms of rearrangement, lead to reaching electrical equilibrium at which the Fermi levels would be at the same energy. In the context of electronic devices, the interfacial electron transfer involving surface molecules appears to be the most important process. The adsorbed surface molecule can form an ion either by accepting or donating one electron. The probability of such a process is given by the average of its molecular electron affinity, E_a, and its ionization potential, I_p, the so-called Mulliken electronegativity coefficient of the adsorbed molecule, χ_{mol} (Eq. (7.32)):

$$\chi_{mol} = \frac{E_a + I_p}{2} \tag{7.32}$$

or, according to the Koopmans theorem [139], (Eq. (7.33)):

$$\chi_{mol} = \frac{E_{HOMO} + E_{LUMO}}{2} \quad (7.33)$$

On the other hand, the ability of a semiconductor to donate or accept electrons is uniquely related to the energy of the electron at its Fermi level, E_F. The transfer of a fractional charge δe can be viewed as a redox equilibrium between the surface molecule and the electronic continuum of the semiconductor, in which the roles of the electron donor and the electron acceptor are relative and governed by the difference between χ_{mol} and the Fermi energy E_F, respectively (Eq. (7.34)) [140].

$$\delta = \xi \left(E_F - \chi_{mol} \right) \quad (7.34)$$

In the case of $|\delta| < 1$ the surface charge transfer complex is formed, while in the case of $|\delta| = 1$ the interfacial redox reaction occurs [141]. The proportionality factor ξ is related to the global softness of the molecule (Eq. (7.35)) [142]:

$$S = -\left(\frac{\partial \mu}{\partial N} \right)_V^{-1} \approx \frac{2}{E_{LUMO} - E_{HOMO}} \quad (7.35)$$

In more complex systems, when a molecule may interact with the semiconductor in several different binding modes, localized indices should be used, that is, the Fukui function (Eq. (7.36)) and local softness (Eq. (7.37)) [143, 144]:

$$f_{\pm}(\vec{r}) = \left(\frac{\partial \rho(\vec{r})}{\partial N} \right)_V^{\pm} \quad (7.36)$$

$$s_{\pm}(\vec{r}) = \left(\frac{\partial \rho(\vec{r})}{\partial \mu} \right)_V^{\pm} = S \cdot f_{\pm}(\vec{r}) \quad (7.37)$$

where "+" and "−" denote the right-sided and left-sided derivatives, respectively. These functions determine the response of the electron density $\rho(\vec{r})$ of the molecular system to the change in the number of electrons at the constant external potential and the response of the electron density of the system to the change in chemical potential at the constant external potential, respectively [145]. In the first approximation, the Fukui functions can be derived from appropriate frontier orbitals (Eqs. (7.38–7.39)):

$$f_+(\vec{r}) = |\psi_{LUMO}(\vec{r})|^2 \quad (7.38)$$

$$f_-(\vec{r}) = |\psi_{HOMO}(\vec{r})|^2 \quad (7.39)$$

In the case of degenerated frontier molecular orbitals, appropriate symmetry correction must be applied [146]. The f_+ and f_- functions indicate the most probable sites of nucleophilic and electrophilic attack, respectively. In general, the Fukui function represents the propensity of the electron density to deform at a given location within a molecule, indicating the most reactive sites [145].

The dipole layer induces a shift of VL at the interface (shown in Figure 7.18a as Δ). The value of Δ depends on the magnitude of the molecular dipole of the surface-modifying molecule by Eq. (7.40):

$$\Delta = \frac{N\mu \cos\theta}{\varepsilon\varepsilon_0} \qquad (7.40)$$

where N is the density of dipoles (m^{-2}), μ is the molecular dipole oriented at an average angle θ relative to the normal to the surface; ε is the dielectric constant of the absorbed layer, and ε_0 is the permittivity of free space. The molecular density and θ angle are controlled by molecular packing on the surface. Therefore BB is strongly influenced by the type of the surface-adsorbed molecule and bonding. Moreover, a recent article by Paska and Haick [147] gives evidence that modifying the molecular dipole via tuning the intermolecular interactions within just one type of adsorbed molecules can give changes of 0.12 eV in χ_{eff} while keeping N, θ, and ε constant. Other reports [148, 149] show that the presence of electron-withdrawing groups, such as –CF$_3$ and –CN, in molecules adsorbed at the semiconductor surface increases the current in metal–semiconductor junctions, whereas donating groups, such as –OCH$_3$ and –CH$_3$, decrease this current. Different substituents in the molecule change its molecular dipole and therefore change χ_{eff} when adsorbed at surface. The importance of these phenomena can be understood when compared with Schottky theory for metal–semiconductor junctions, where the barrier for electron transport at the interface should be determined by the difference between χ_{eff} and metal work function (ϕ_M) as shown in Eq. (7.41):

$$\phi_b = \phi_M + \Delta - \chi_{eff} \qquad (7.41)$$

where Δ is the difference between work functions of metal and semiconductor (Figure 7.18b), but in practice, the variation of the Schottky barrier with the metal work function is much weaker [138]. On the other hand, molecular-scale modification of semiconductor surfaces offers tools for wide range tuning of semiconductor work function [21–23, 140, 148, 150].

When increasing degrees of freedom by tuning the remaining parameters from Eq. (7.41), one needs to distinguish other possible interactions between adsorbates, which can occur, for example, with increasing density of surface dipoles. The decreasing heat of adsorption with increasing coverage is a frequent observation [151]. Direct interaction via overlapping wave functions, leading to Pauli repulsion, is one of the reasons (Figure 7.19a). Rarely, when the energy states are close to the Fermi level, direct interaction between molecules can cause an attraction by shifting states through the Fermi level [152].

Furthermore, when a molecule is adsorbed on a surface, it brings about a significant perturbation of electronic structures of both counterparts [153]. The commonly described effect is a downshifting of the d states of the neighboring transition metal atoms at the solid surface. Another adsorbate would then bind more weakly to the surface, and eventually this indirect interaction would change the adsorption energy (Figure 7.19b) [154]. Adsorbing molecules can also lead to local distortions of the surface lattice. Other adsorbates would feel it as a

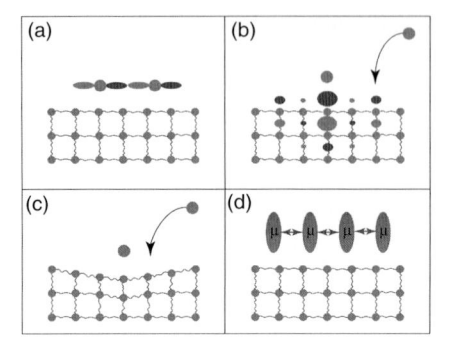

Figure 7.19 Usual reasons of interactions between adsorbates: (a) Pauli repulsion/attraction via electronic interactions between adsorbate molecules, (b) apparent repulsion resulting from perturbations of electronic structure at the surface, (c) surface lattice distortions causing repulsion, and (d) higher order interactions.

repulsion, due to an increased potential barrier needed to reach the binding site (Figure 7.19c). Higher order intermolecular interactions between adsorbates, such as van der Waals forces, are also significant, since they usually limit the surface coverage (Figure 7.19d).

7.5.4
Surface-Modified Semiconductor

Semiconductor surfaces can interact with diverse molecules, both organic and inorganic. The nature of interactions, as described above, may influence the optical and electronic properties of semiconductor particles, as well as the particle-surrounding interface. The weaker molecule-particle interaction, the less influence on the spectroscopic properties can be observed; however, photoelectrochemical properties may change significantly also in the case of weak interactions. A good example of this can be observed for a series of iron(II) complexes deposited onto TiO_2 surface: ferrocene (Fc), ferrocenylboronic acid (FcB), and hexacyanoferrate(II) [155]. Ferrocene is not bound chemically to the surface, while FcB and hexacyanoferrate(II) are. The latter anions bind very efficiently to the titania surface, forming surface charge transfer complexes. As a result of these modifications, the electronic spectrum of Fc@TiO_2 material resembles the sum of the spectra of both components. However, photocurrents developed at the photoelectrode covered with Fc@TiO_2 are significantly higher than those recorded for the components alone. Also, photocurrent switching appears at different potentials: at about -250 and $100\,mV$ versus Ag/AgCl for TiO_2 and Fc@TiO_2, respectively. In the case of a strong modifier binding to the semiconductor surface, as in $[Fe(CN)_6]^{4-}$@TiO_2 material, both optical and electrochemical properties differ from those of the material constituents. Photocurrent switching phenomenon by $[Fe(CN)_6]^{4-}$@TiO_2 is described in detail below.

Covalent binding of the adsorbate to the semiconductor surface results in a strong modification of the interface properties. This type of bonding is a good platform for strong electronic coupling between surface species and the electronic continuum of a semiconductor [156, 157]. Such interactions may be responsible, for instance, for efficient electron injection from the exited states of the molecule (photosensitization effect) to the semiconductor particle. This effect can be used for construction of dye-sensitized solar cells or visible light active photocatalysts. In the case of physisorption of molecules onto the semiconductor surface, much weaker interactions influencing the mechanism and efficiency of the charge exchange between the components of the system are observed [155]. Interactions between semiconductor surfaces and molecular species are essential for a photosensitization effect in that they not only modify the electronic properties of semiconductors but also influence spectroscopic and electrochemical properties of surface-bound molecules. Particularly important are interactions between redox-active species.

Photosensitization of a wide-band-gap semiconductor with surface-bound photosensitizers requires a photoinduced electron (with photons of sub-band-gap energy) or hole injection into CB or VB, respectively. As an example, photosensitization of titanium dioxide can be considered. The efficiency of this process depends on electronic interaction between the photosensitizer moiety (surface complex) and the TiO_2 particle. It is possible to distinguish between at least two types of the charge injection mechanisms: in the first, the photogenerated charge is transferred from the excited state of the sensitizer molecule to the CB or VB (Figure 7.20a), while the second mechanism involves a direct molecule-to-band charge transfer (MBCT, Figure 7.20b). The MBCT process can be realized by surface titanium(IV) complexes formed by coordination of various organic ligands or inorganic moieties (e.g., another complex bound via a bridging ligand) [158]. The excited state of the surface complex can be considered as Ti^{III} moiety, formally identical with the electron in the CB. Therefore, the efficiency of electron injection to the CB in the case of direct photosensitization is very high; however, the excited state may also

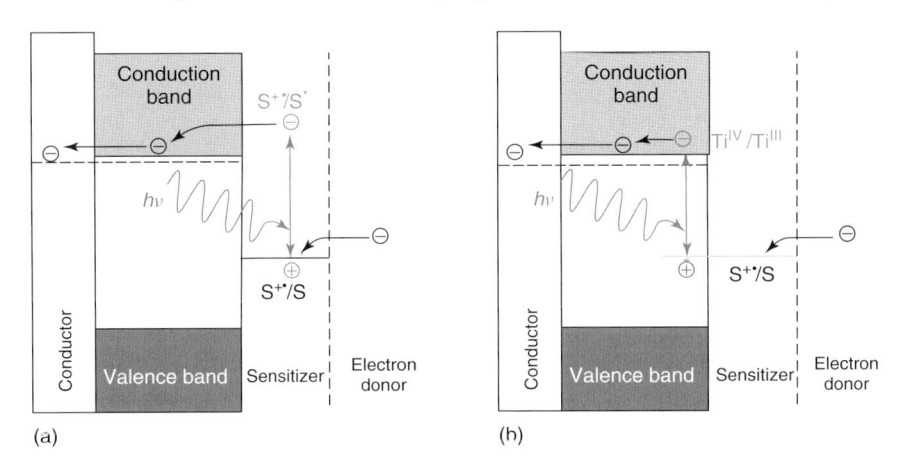

Figure 7.20 The mechanism of (a) indirect photosensitization (photoinduced electron transfer, for example, Ru^{II} complexes attached to TiO_2 via carboxylated bipy ligands) and (b) direct photosensitization (optical ET, for example, Ti^{IV} surface complexes).

undergo an efficient relaxation, resulting in the loss of electrons from CB. Efficient direct photosensitization can therefore be achieved only by surface titanium(IV) complexes undergoing a possibly slow back electron transfer. Also, a high mobility of electrons within the semiconductor particle, as well as an efficient oxidation of an electron donor, should prevent fast deactivation of the excited state. In the case of indirect photosensitization, charge injection from the excited state of the sensitizer to the semiconductor particle competes with the relaxation processes.

While spectroscopic and redox properties of surface complexes influence the photoelectrochemical behavior of modified materials, the chemical nature (coordination modes, hydrogen bonds, etc.) of the modification determines adsorption/desorption properties and therefore the stability of these systems. The surface complexes can be formed as a result of coordination of various ligands. Those suitable for titanium dioxide modification should possess carboxylic, phosphoric, boronic groups, and so on, or even better, more than one of them; an especially high affinity to the TiO_2 surface has been observed in the case of catechols, salicylates, phthalates, and so on (Figure 7.21).

The strength and mode of complexation of the modifier depends on the accessible titanium(IV) sites (planes, edges, corners) and pH, since TiO_2 surface undergoes protonation and deprotonation reactions [158, 159].

A particularly interesting system can be composed of the coordinated catechol (1,2-dihydroxybenzene) or its derivatives. Catechol usually binds to the surface of titanium dioxide as catecholate substituting surface hydroxyl groups [160, 161]. Two main structures of surface complexes have been confirmed: bidentate chelating

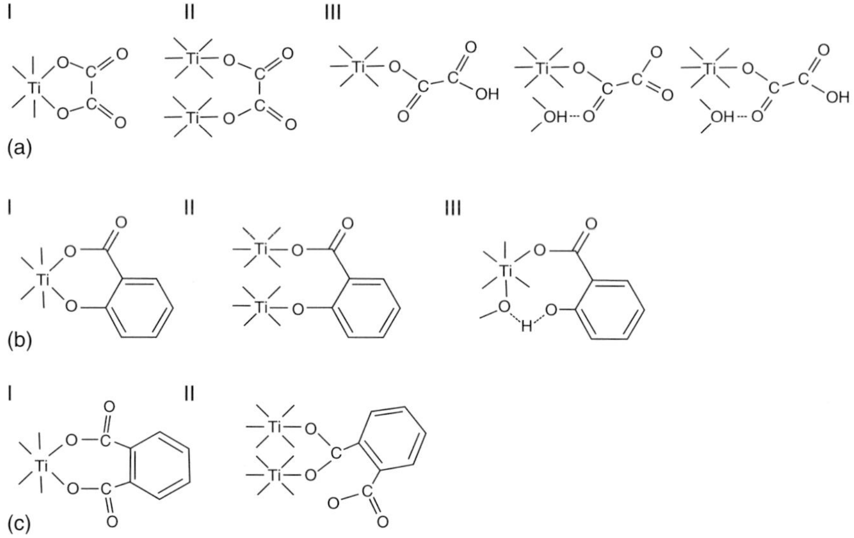

Figure 7.21 Coordination modes for selected chelating ligands: (a) oxalate; (b) salicylate; and (c) phthalate. Structures: bidentate chelating (I), bidentate bridging (II), monodentate with possible stabilization by hydrogen bonds (III).

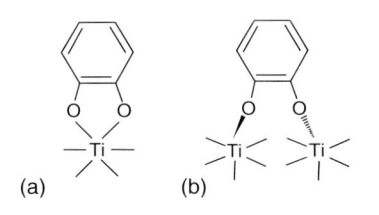

Figure 7.22 Bidentate chelating (a) and bidentate bridging (b) surface [$L_n Ti^{IV}$(catecholate)] complexes.

and bidentate bridging [160, 162]. In the case of catechol adsorption onto rutile (110) surface [160] or anatase [163] bidentate bridging complexes prevail because of a similar distance between catechol groups and adjacent surface Ti^{IV} centers (Figure 7.22). The formed structures undergo visible-light-driven ligand-to-metal charge transfer excitation, called also as a *molecule-to-band charge transfer*. Recent studies revealed formation of intra-band-gap electronic states located about 1 eV above the VB upper edge [164]. These new donor states decrease the band-gap energy enabling electron excitation to the CB upon <3 eV photons absorption. However, only catecholate adsorbed at the surface planes induces this effect. In contrast, complexes formed at the surface steps or at isolated points do not introduce new electronic states within the band gap. Since bidentate chelating structures should be favored at the defect sites and edges, while bidentate bridging complexes should prevail at planes, the intra-band-gap states are related to bridged complexation. The new states are related to the formation of ordered catecholate arrays forming π-stacks. This process leads to Davydov splitting [165] associated with formation of new bandlike structures of organic component. On the other hand, spectroscopic studies and chemical modeling [142] for mononuclear Ti^{IV} complexes (i.e., bidentate chelating structure) with catechol revealed visible light absorption due to charge transfer complex formation. The difference between abilities to form charge transfer complexes in the case of two modes of catechol binding to dinuclear titanium(IV) complex are presented in Figure 7.23. Lower energy transitions ($\lambda > 350$ nm) appear only in the case of the bidentate chelating structure. Therefore, the nature of excitation involving intra-band-gap states and MBCT must be different. In the case of titanium dioxide nanoparticles, the ratio of planar to edged titanium(IV) decreases, diminishing the role of intra-band-gap states and amplifying the role of MBCT excitation.

Another interesting observation involves the size of acene molecule bound to the surface (compare Section 7.3). Again, ligand-to-metal charge transfer process can be considered by comparing shapes of the highest occupied molecular orbital (HOMO) and the lowest unoccupied molecular orbital (LUMO) (Figure 7.24) [166]. The bidentate chelating structures of [Ti(OH)$_2$(acene)] (acene = catechol analog) are characterized by HOMO orbitals localized at ligands and LUMOs at titanium(IV) sites, but only for acenes constituted of up to six rings. For bigger structures LUMOs are distributed along the whole complex. A similar effect was observed for other acenes and derivatives of salicylic or phthalic acids; however, in these cases, distribution of LUMOs along the whole complex appeared already at smaller systems, constituted of four rings.

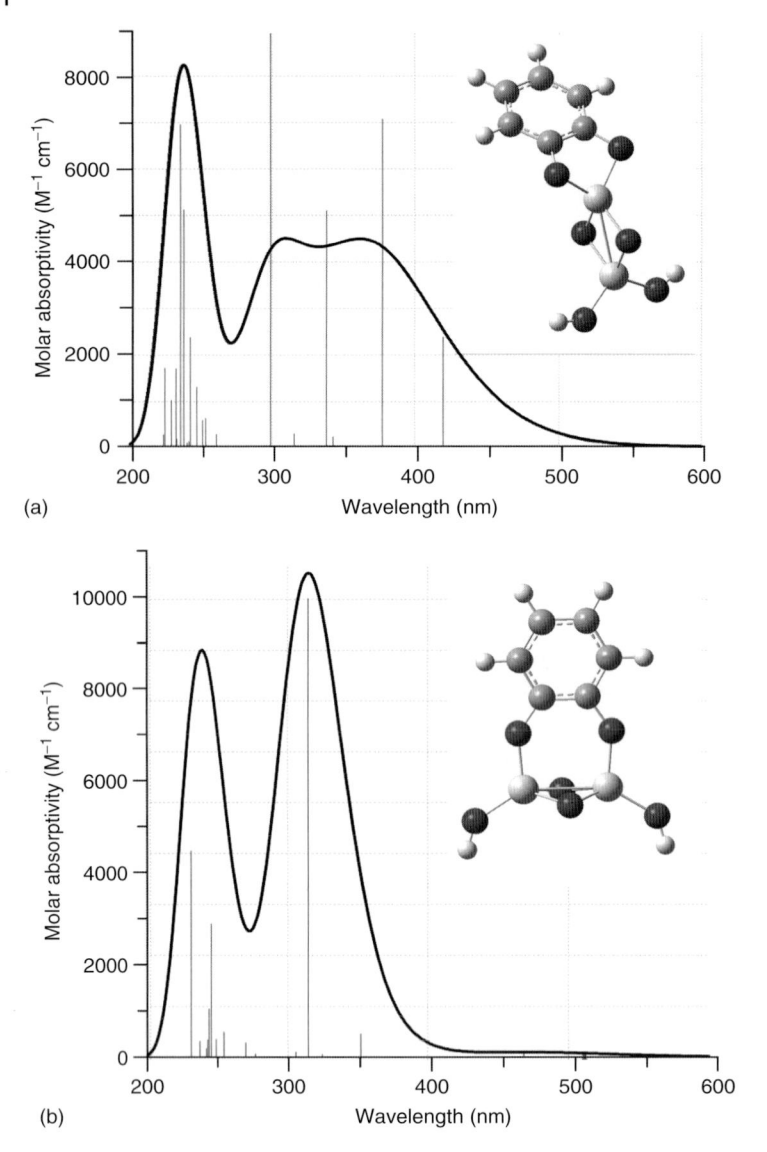

Figure 7.23 Bidentate chelating (a) and bidentate bridging (b) complexation of catechol and their simulated UV–vis spectra. Both models have been constructed of [Ti(μ-O)$_2$Ti] unit with coordinated catecholate and two OH$^-$ ions.

Adsorption of molecules onto the surface of titanium dioxide is, in general, a very complex issue. Binding of a monocarboxylic anion illustrates this problem, since several structures can be distinguished in this case (Figure 7.25): (i) an electrostatic interaction between the positively charged surface and a negatively charged anion; (ii) a monodentate structure with the organic ligand occupying one coordination site of the titanium(IV) center; (iii) a bidentate chelating structure

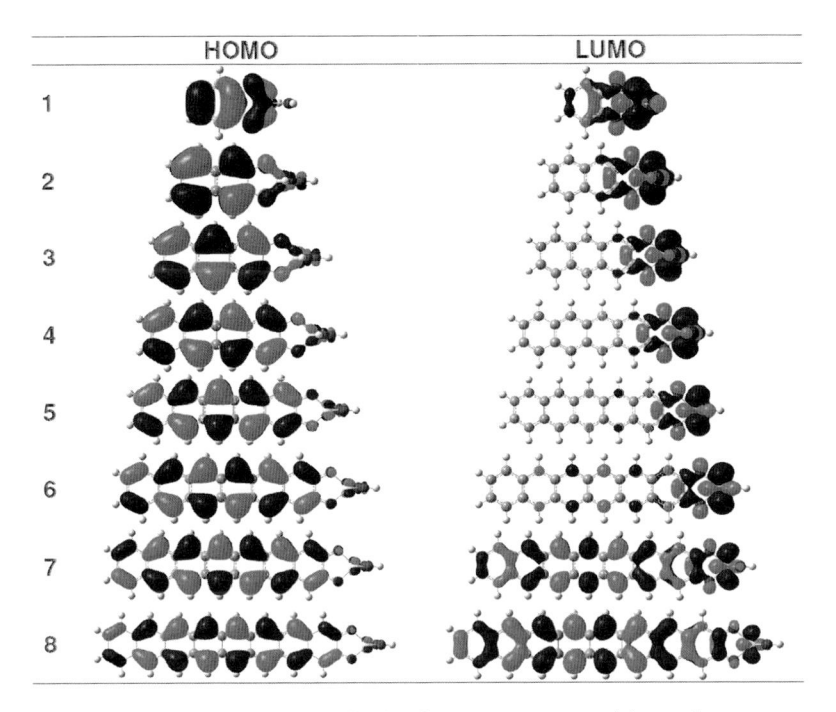

Figure 7.24 Frontier molecular orbitals of [Ti(OH)$_2$acene] model complexes (acene = catechol analog). (Adapted from Ref. [166].)

with the ligand occupying two coordination sites; (iv) a bidentate bridging structure composed of chelating ligand bound to two neighboring Ti^{IV} centers; and adsorption involving a single (v) and multiple (vi) hydrogen bonds [158, 167]. Knowing the possible structures of surface complexes and their stabilities, it is possible to predict (to some extent) the adsorption properties of various organic modifiers. For instance, the surface binding of carminic acid (CA) should strongly resemble the binding of salicylic acid to the TiO$_2$ surfaces, involving formation of chelating structures [159, 168, 169]. However, CA adsorbed onto the surface of titanium dioxide from DMF solution follows the Langmuir isotherm, while salicylic acid adsorption is described by a high affinity multisite isotherm. The Lineweaver–Burk equation fitted to the experimental data allows calculation of maximal coverage and adsorption equilibrium constants. The Langmuir binding constant of CA is relatively large ($K = 2.5 \times 10^{-2}$), while salicylic acid is bound relatively weakly ($K = 2.3 \times 10^{-4}$). On the other hand, the maximum coverage (Γ_{max}) found for CA is 0.65 ± 0.02 µmol m^{-2}, which is lower than that of salicylic acid (~ 14.7 µmol m^{-2}). These results correspond to the average packing of 0.4 and 8.4 molecules per square nanometer for CA and salicylic acid, respectively. The size of molecules can explain these differences [159, 170].

A particularly interesting system consists of iron complexes bound to the surface of titanium dioxide via a CN$^-$ bridging ligand (Figure 7.26). Cyanoferrates may exist in their oxidized or reduced forms, depending on the electrode potential and

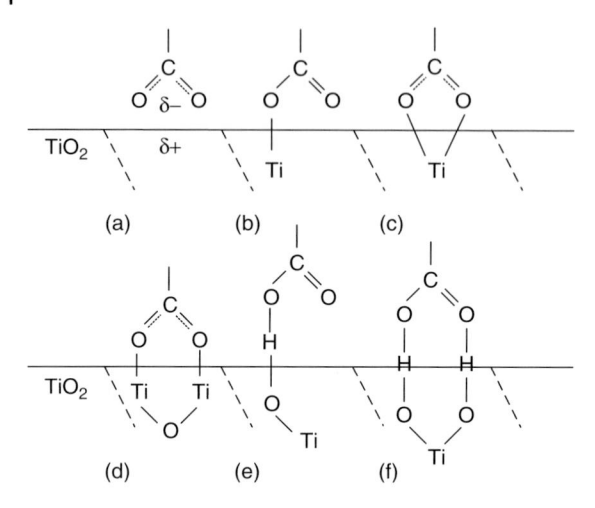

Figure 7.25 Modes of carboxylic group binding to the surface of TiO$_2$ [175]. For a detailed description see text.

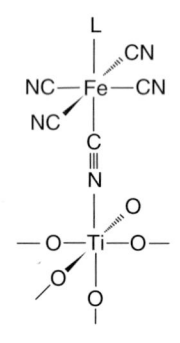

Figure 7.26 Dinuclear surface complex formed at the surface of titanium dioxide on cyanoferrate bonding through one CN$^-$ ligand. "L" denotes a ligand with C-, O-, N-, or S-donor atom.

reduction potential of the iron center, which may be tuned by ligand exchange [171, 172]. At high potentials, the Fe^{III} form prevails. Photocurrent generation can be realized only on excitation of the semiconductor support, that is, on UV light excitation. The mechanism of anodic photocurrent development (Figure 7.27) is therefore the same as presented in Figure 7.13a. The situation changes at lower potentials at which the reduced form of iron, Fe^{II}, appears. Excitation within the MBCT (Fe^{II} → Ti^{IV}) band results in direct electron transfer into the CB. This mechanism is in principle the same as that valid for TiO$_2$ photosensitization by multidentate surface complexes (see above, Figure 7.20b); however, in the presence of a suitable electron acceptor (e.g., oxygen molecule), cathodic photocurrents prevail. Excitation of TiO$_2$ support directly with ultraviolet light results also in the reduction of the CB, followed by electron transfer to the acceptor molecule. The direction of net photocurrent may be therefore switched by a change in the electrode potential around the redox potential of the surface iron species.

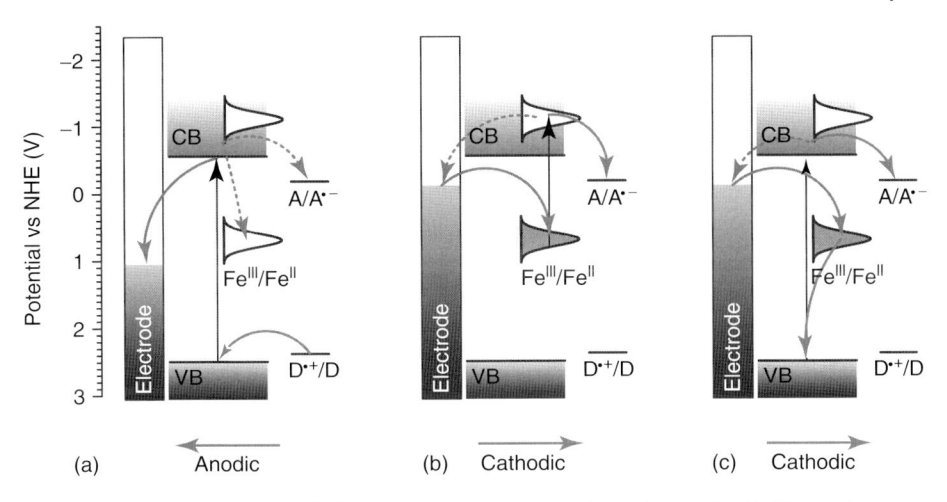

7.5 Mechanisms of Photocurrent Switching Phenomena | 159

Figure 7.27 The mechanism of photocurrent generation at electrodes made of TiO_2 modified with iron cyanide complexes on UV (a,c) and visible (b) light generation. High electrode potentials (a) ensure presence of Fe^{III} surface species, while Fe^{II} prevails at low potentials (b,c). (Adapted from Ref. [185].)

An interesting behavior of the $Fe^{II/III}@TiO_2$ system can be observed at potentials close to the redox potential of iron(II)/iron(III). Depending on the photon energy, that is, color of light, the surface complex or TiO_2 may be excited, thus inducing competitive photocurrents according to the mechanisms presented in Figure 7.27, respectively. In a relatively narrow range of potentials (tens to a few hundreds millivolts), photocurrent switching can be induced by a change of light color (Figure 7.28) [171, 173, 174]. The effect of net photocurrent direction change, induced by either changes of electrode potential or color of incident light, has been named the PEPS effect.

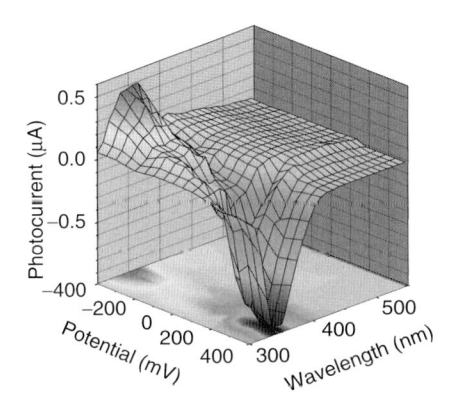

Figure 7.28 Photocurrent dependence on the electrode potential and color of incident light: anodic photocurrents are depicted as negative values, the cathodic as positive. Electrode covered with $[Fe(CN)_6]^{4-}@TiO_2$. (Adapted from Ref. [173].)

7.5.5
Optoelectronic Devices Based on Organic Molecules/Semiconductors

Modifications of semiconductor or metal surfaces by molecules with tunable properties can be used for control of electronic properties of hybrid organic/inorganic materials. Control over the electronic energy levels at the surfaces of semiconductors and metals is achieved by assembling on the solid surfaces poorly organized, partial MLs of molecules instead of the more commonly used ideal ones. In this way, electronic devices based on incorporation of molecules onto the semiconductor or metal support can be constructed. Gallium arsenide (GaAs)-based sensors, as well as Au-Si and Au-GaAs diodes, are good examples of such devices. Immobilization of molecules onto solid interfaces can be realized by a "soft" electric contact procedure (physisorption), which does not significantly change the structure of the immobilized molecules. There are only a few molecular limitations for this procedure, so it opens possibilities for the use of more complex and fragile molecules and biomolecules [23, 176].

Chemisorbed molecules can influence both the potential at interfaces (static effects) and charge transport across the interface (dynamic effects). Both effects are strongly influenced by the chemical interaction between the molecules and the support surface. Bound molecules can alter the net surface charge and therefore also the space-charge region, which can be regarded as an "internal dipole." Molecular adsorption can influence the effective electron affinity and/or the surface potential of the semiconductor. In general, the formed layer modifies the existing surface polarity and adds an external dipole layer. The formed hybrid systems, linking the collective properties of semiconductors (electron transport) with the functional flexibility of molecules, show a synergetic arrangement of the molecular and nonmolecular world. Molecular layers, as thin as 1 nm, can then redefine the electronic properties of the solid surface and can be used to tune the performance of solid-state devices such as sensors and Schottky diodes [23, 177].

Silicon as a substrate is of interest not only because of its role in complementary metal oxide semiconductor (CMOS) technology but also because adsorption of organic molecules onto Si is relatively simple and the systems formed show good stability. The semiconductor surface can act as a template for molecular adsorption [178]. The situation is different in the case of gold with chemisorbed thiols: the energy barrier between various chemisorption modes on Au (the corrugation energy) allows the bound thiolate to move at room temperature easily from one Au site to another, even without applying a voltage [21, 179]. By combining the rich functionality of biomolecules with Si technology, it is possible to extend the functionality of silicon-based electronics beyond what can be achieved without bound molecules. Formation of molecular MLs at polycrystalline semiconductors might improve their performance in various applications, such as flexible electronics, solar cells, and (bio)sensors. Moreover, the hybrid architectures may appear to be a step ahead toward purely molecular devices. In most current-carrying device structures, the electron transport is determined by interfaces. Good interfacial band alignment ensures better charge injection efficiency in light-emitting diodes

(LEDs) and higher open-circuit voltage of solar cells. This kind of appliance often involves metal/semiconductor interfaces, and the height of the transport barrier is controlled more by interfacial interactions than by fundamental properties of semiconductor or metal. Molecules can influence the charge transport directly (via molecular charge transfer processes) and via electrostatic effects involving energy band alignment [21].

One of the easiest ways to influence the reactivity of a specific adsorbate is to create a solvation shell around it, similar to solvation of molecules in solutions. The solvation of ions before their adsorption significantly influences the relative nucleophilicity and electronic properties of the surface, resulting in changes of charge transfer mechanisms. Such an effect was observed for III–V semiconductors with adsorbates containing halogen, sulfur, or metal atoms [145].

Immobilization of organic molecules can proceed as a result of various reactions, even those involving cycloaddition, as demonstrated at surfaces of silicon, germanium, and diamond [180]. Similar to titanium dioxide, surfaces of other semiconductors, such as CdTe, CdSe, InP, and GaAs, can also react with dicarboxylic acids [181]. The molecules are chemisorbed as MLs.

Two main factors govern the changes of surface electronic properties: the redox properties of adsorbed molecules and the matching of adsorbate redox potential with Fermi energy levels of the semiconductor support. The smaller the energy distance, the stronger the surface-molecule coupling and larger the induced changes in surface electronic properties that are observed [181]. A huge variety of organic molecules and possibilities to tune their HOMO and LUMO energy levels enables a choice of modifiers appropriate to fine-tune different semiconductor surfaces. Surface science has now evolved to the point where it can be applied to engineering of semiconductor properties suitable for fabrication of devices possessing unique physical and chemical properties.

7.6
Digital Devices Based on PEPS Effect

The PEPS effect provides a universal platform for construction of chemical switches and logic gates, which are the basic components of all electronic computing systems. The following section presents how simple chemical systems could be used for construction of switches and more sophisticated logic devices based on Boolean algebra. In all described examples, the output of photoelectrochemical devices can be controlled by selective stimulation with incident light and/or applied potentials (which are regarded as inputs of these devices).

The most elemental devices can be constructed from unmodified wide-band-gap semiconductors (e.g., ZnO, TiO_2, or CdS), which generate anodic photoelectrochemical responses during illumination with photons of sufficiently high energy. These devices are, in terms of Boolean algebra, one-input YES gates. Information fed to the input in the form of light pulses can be recovered at the output as electrical current pulses. In order to perform Boolean analysis of these systems,

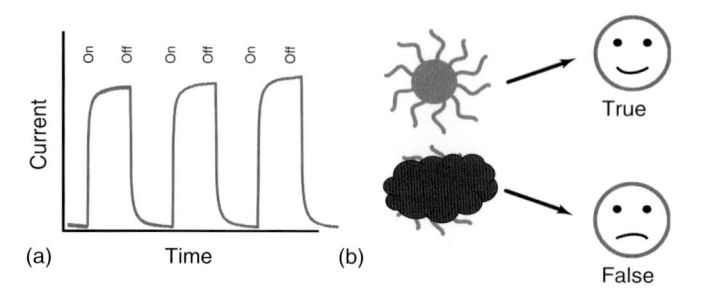

Figure 7.29 Photoresponse of nanocrystalline CdS during pulsed illumination (465 nm) (a) and the assignment of Boolean values (b).

appropriate assignments of stimuli and observables must be done. For simplicity, the "off" and "on" states of illumination are assigned to Boolean "false" (0) and "truth" (1), respectively. *Per analogiam* the null output photocurrent is assigned to Boolean "0" and anodic photocurrent to Boolean "1" (Figure 7.29).

Surface-engineered wide-band-gap semiconductors can be also regarded as more advanced logic devices based on the combination of a few simple logic gates. TiO_2 modified with redox-amphoteric organic chromophores such as folic acid (FA@TiO_2) [182], (CA@TiO_2) [170], or alizarin [183] are the best examples of such materials, which are capable of performing more complex information processing. In these cases, the photocurrent polarity strongly depends on photoelectrode potential. Anodic polarization results in an anodic photocurrent, while cathodic polarization results in a reversed photocurrent (see above). Therefore, one can assign positive and negative polarization of the photoelectrode to Boolean "0" and "1," respectively. In the case of photocurrent switching, when both polarities of photocurrent are allowed, it is convenient to split the electric output into two channels, one associated with anodic photocurrent and the other with the cathodic one. Defined in this way, the photoelectrochemical system follows the behavior of 1 : 2 demultiplexer (Table 7.1, Figure 7.30). Similar effects have been observed in the case of various composite materials, for example, Prussian blue/TiO_2 [136], ferrocene/TiO_2 [155], as well as Se/TiO_2 [131], n-TiO_2-N/p-CuI, n-$BiVO_4$/p-Co_3O_4 [130], and n-$BiVO_4$/p-CuO [130] bulk heterojunction composites.

Other materials have also been used for construction of optoelectronic 1 : 2 demultiplexers, for example, nanocrystalline S-doped CdS [184], ferrocene-titanium dioxide composite material [155], or Prussian blue-modified titanium dioxide [136]. The fact that, in the case of sulfur-doped cadmium sulfide, all the switching phenomena occur at the single nanoparticle level and do not require any cooperative interactions between nanoparticles deserves special attention. The presence of additional sulfur atoms in the structure of CdS (appearing as an extra energy level within the band gap) causes the generation of anodic and/or cathodic photocurrent, depending on incident light wavelength and applied bias. Therefore, single nanoparticle of this material can be regarded as the smallest semiconductor-based logic circuits ever reported (average diameter of 5–7 nm) [184].

Table 7.1 The truth table for the two-channel demultiplexer based on FA@TiO$_2$ hybrid material.

Light	Input 1	Photoelectrode potential	Input 2	Photocurrent	Output 1	Output 2
OFF	0	Positive	0	NO	0	0
ON	1	Positive	0	Cathodic	1	0
OFF	0	Negative	1	NO	0	0
ON	1	Negative	1	Anodic	0	1

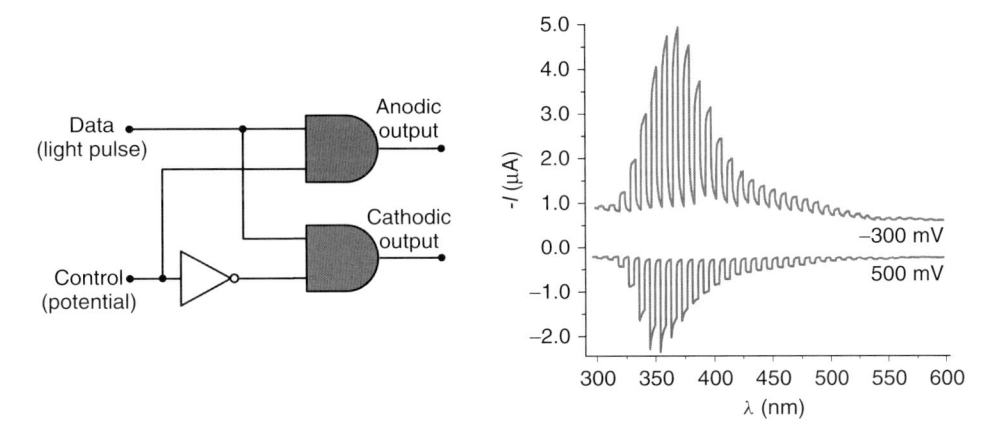

Figure 7.30 Electronic equivalent circuit of FA@TiO$_2$ photoelectrode working as a two-channel optoelectronic demultiplexer. (Adapted from Ref. [182].)

One of the most complex photoelectrochemical responses was recorded for pentacyanoferrate (PCF) and hexacyanoferrate (HCF) modified titanium dioxide materials [136, 155, 171–174, 185, 186]. This complex is bound to the surface of nanocrystals via formation of cyanobridged species, such as $(CN)_5Fe^{II}-C\equiv N-Ti^{IV}$ (titanium ion belongs to the surface of the nanocrystal). A combination of redox reactivity of the cyanoferrates and the tendency of Fe^{II} to form metal-to-metal charge-transfer complexes (MMCT; Fe^{II} → Ti^{IV}) with titanium ions results in a complex photocurrent switching processes (see below). Excitation of neat semiconductor or surface-modified semiconductor with oxidized surface species generates anodic photocurrents; cathodic photocurrent is generated on excitation of reduced surface states at sufficiently low photoelectrode potentials. Chemisorbed molecules should easily be oxidized/reduced in both the ground and excited states, so the system requires suitable redox properties [36]. Photocurrent dependence on both the incident light wavelength and the photoelectrode potential creates unique possibilities for construction of complex optoelectronic logic devices.

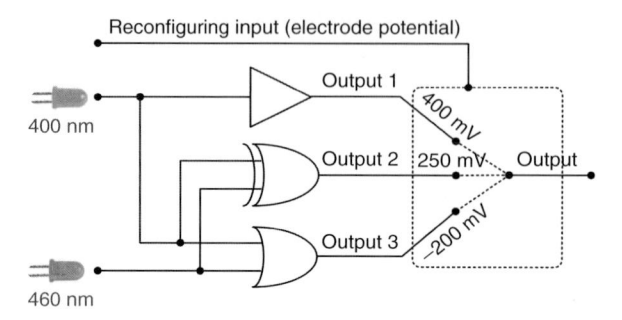

Figure 7.31 Electronic equivalent circuit for a reconfigurable logic system based on HCF@TiO$_2$. Output 1 follows the input 1 signal, output 2 computes the XOR function of input data, while output 3 corresponds to the logic sum (OR) of input data. Reconfiguring input and three-position switch represent the reconfigurability of the device through the photoelectrode potential. (Adapted from Refs [36, 186].)

To construct a logic system based on the HCF-modified titanium dioxide photoelectrodes, electroluminescent diodes (400 and/or 460 nm) were used as light sources. Boolean 0 and 1 states were assigned to "off" and "on" states of each diode, and two different wavelengths were assigned to two different device inputs. Analogously, Boolean 0 was assigned to null net photocurrent, while any nonzero photocurrent intensity was assigned Boolean 1, irrespective of its polarity. During pulsed irradiation with a violet diode (400 nm) at potentials that completely oxidize the surface species (+400 mV vs Ag/AgCl), anodic photocurrent is generated, but irradiation with a blue LED (460 nm) does not generate any photocurrent. Synchronized irradiation with both diodes gives the same effect as the violet diode alone (Figure 7.31, Table 7.2, output 1). Lower potentials cause electrochemical reduction of the surface species. Irradiation of this material with violet or blue diodes leads to generation of cathodic photocurrent. Simultaneous irradiation by both LEDs creates a photocurrent of much higher intensity. This behavior of the photoelectrode at −200 mV versus Ag/AgCl corresponds to the OR logic gate (Figure 7.31, Table 7.2, output 3). Partial oxidation of the surface complex causes a different effect. Violet light pulses generate anodic photocurrent, which is consistent with the excitation of the inner part of the semiconductor particles. Blue light pulses result in generation of cathodic photocurrent. Synchronized irradiation with both LEDs gives zero net current, as anodic and cathodic photocurrents effectively compensate each other (Figure 7.31, Table 7.2, output 2). This corresponds to the XOR function. This system represents the first example of a photoelectrochemical XOR logic gate with two optical inputs. Furthermore, the system is reconfigurable, and its logic properties can be changed via appropriate polarization of the photoelectrode [36, 186]. The rich chemistry of cyanoferrate complexes, combined with the reactivity of wide-band-gap semiconductors, creates possibilities for other, even more complex logic devices based on simple chemical systems. Information is supplied to the systems by means of light pulses, and processed information is retrieved in the form of current pulses. This behavior allows facile communication between various electronic devices and chemical logic systems [171, 174, 186].

7.6 Digital Devices Based on PEPS Effect

Table 7.2 The truth table for logic gates based on $[Fe(CN)_6]^{4-}$ modified titanium (HCF@TiO$_2$) dioxide photoelectrodes.

Input 1	Input 2	Output		
400 nm	460 nm	400 mV	250 mV	−200 mV
⎍⎍	⎍⎍	⎍⎍	⎍⎍	⎍⎍
0	0	0	0	0
0	1	0	1	1
1	0	1	1	1
1	1	1	0	1
Boolean function		YES	XOR	OR

Adapted from Refs [36, 186].

The photoelectrochemical properties of nanoparticulate CdS deposited on conducting substrates open the possibility of construction of diverse photoelectrochemical logic devices. A single indium tin oxide (ITO)/CdS junction operating in iodide-loaded ionic liquid electrolyte presents voltage–current characteristics similar to that of Schottky diodes [187]. The single Schottky junction was made by chemical bath deposition of a thin layer of CdS on the ITO surface or cast-coating the conductive support with thiourea-capped CdS nanoparicles. Combination of two such devices in series behaves in turn like AND and XOR gates, depending

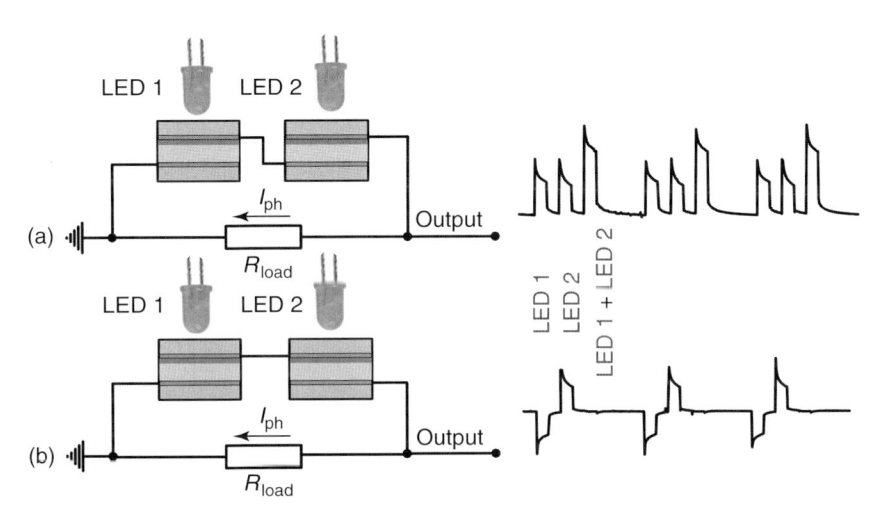

Figure 7.32 Schemes of optoelectronic AND (a) and XOR (b) logic gates together with corresponding photocurrent responses to pulsed illumination.

on connectivity (Figure 7.32a,b). The optoelectronic binary half-adder can be built from nanoparticle-based ITO/CdS Schottky junctions working in iodide-loaded semisolid ionic liquid electrolyte. At no external bias, illumination of the junction results in anodic photocurrent. Combination of two Schottky photodiodes with identical polarities results in an AND logic gate, since a high-intensity photocurrent can be recorded only on illumination of both junctions (Figure 7.32a). Opposite polarization of photodiodes naturally leads to the XOR gate, as illumination of any single junction generates net photocurrent, while concomitant illumination of both junctions results in compensation of photocurrents (Figure 7.32b).

An appropriate connection of these two optoelectronic systems (Figure 7.33) leads to construction of optoelectronic binary half-adder (Table 7.3). In this device, two input signals (light pulses) generate current pulses in two different circuits (corresponding to the AND and XOR gates) and thus yield the binary representation of the arithmetic sum of input signals [187]. These devices also do not require any interactions between individual particles, and therefore the whole arithmetic unit can be confined to the size of two small nanoparticles (about 40 nm^2).

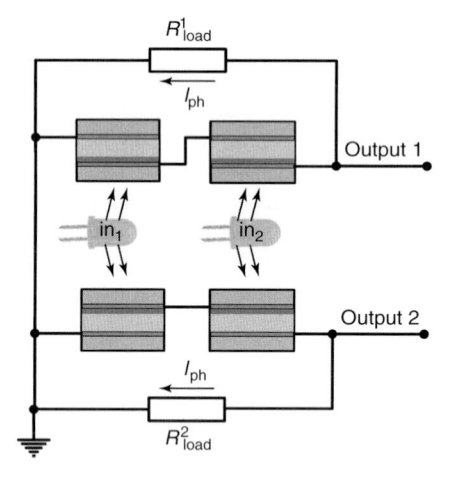

Figure 7.33 Connection diagram of binary half-adder based on CdS/ITO Schottky junctions. R^1_{load} and R^2_{load} denote internal resistances of two-channel potentiostat (both at 0 V).

Table 7.3 The truth table of the binary half-adder from Figure 7.33.

Input 1	Input 2	Output 1	Output 2	Input, decimal	Output, decimal
0	0	0	0	0 + 0	0
0	1	0	1	0 + 1	1
1	0	0	1	1 + 0	1
1	1	1	0	1 + 1	2

While the operation of the devices presented above does not require cooperation between the individual nanoparticles, all the results were obtained with photoelectrodes with large numbers of deposited nanoparticles. This approach can help to prove the principle of operation of novel optoelectronic devices and at the same time generates serious problems with the rate of switching. Under conditions enabling reversibility of switching, the fastest devices operate within millisecond time regime. It is a few orders of magnitude slower than typical silicon-based devices, but is comparable with the human nervous system. The electrochemical processes at the surface of nanoparticles associated with oxidation/reduction of surface molecules are of Faradaic character. Therefore, the increasing number of nanoparticles operating as a single switch increases the capacitance of the device and hence also the switching time. This is responsible for the very slow operation of the PEPS-effect-based devices, but the processes at the single nanoparticle level should be faster by many orders of magnitude.

7.7 Concluding Remarks

Wide-band-gap semiconductors (oxides and chalcogenides) are unique materials. While their electric performance is much poorer than that of silicon, their fabrication is much simpler and cheaper. Furthermore, instead of complex surface processing using multistep lithographic processes, polycrystalline or nanoparticulate layers of these materials can be easily engineered at the molecular scale, thus yielding hybrid semiconductors of tailored electrical and optical properties. These materials are especially suited for applications in various optoelectronic switches, since these applications do not require the highest energetic efficiency (as compared to solar cells). Information-processing characteristics of devices based on hybrid semiconductors usually are not the result of complex structural engineering (as in the case of silicon electronic devices) but rather molecular-scale tailoring and fine-tuning of electronic properties of interfaces. Their performance is defined by the nature of the chemical species adsorbed at the surface (or immobilized within the nanocrystals) and is a result of subtle interplay between the components of the system. Furthermore, their function can be modified by simple chemical methods, for example, oxidation, reduction, or covalent modification of immobilized molecules. The underlying physics is straightforward: the photocurrent switching phenomena are simply the result of thermodynamic competition between various pathways of photoinduced electron transfer. This simple principle offers an almost unlimited number of optoelectronically switchable systems. The real problem, however, is not to find the system but to perform the switching at the isolated nanoparticle and subsequently wire several nanoparticles within a circuit. At the moment, we can present a proof of the principle answering the question "Can it be made to work?" In the future, however, one should try to answer another one: "If it works, what can we use it for?" [188].

Acknowledgments

Financial support from the Ministry of Science and Higher Education (grants No. 1609/B/H03/2009/36, 0117/B/H03/2010/38, and 649/N-GDRE-GAMAS/2010/0), the National Centre for Research and Development (grant No. NCBiR/ENIAC-2009-1/1/2010), the European Nanoelectronics Initiative Advisory Council ENIAC (contract No. 120122), and the AGH University of Science and Technology (contract No. 11.11.180.509/11) is gratefully acknowledged. A.P. thanks the Foundation for Polish Science for the research grant within the VENTURES initiative. M.O. thanks the Foundation for Polish Science for the MPD Programme fellowship cofinanced by the EU European Regional Development Fund.

References

1. Meindl, J.D., Chen, Q., and Davis, J.A. (2001) *Science*, **293**, 2044–2049.
2. Lloyd, S. (2000) *Nature*, **406**, 1047–1054.
3. Pop, E. (2010) *Nano Res.*, **3**, 147–169.
4. Cavin, R.K. III and Zhirnov, V.V. (2006) *Solid State Electron.*, **50**, 520–526.
5. Welser, J.J., Bourianoff, G.I., Zhirnov, V.V., and Cavin, R.K. III (2008) *J. Nanopart. Res.*, **10**, 1–10.
6. Hutchby, J.A., Cavin, R., Zhirnov, V., Brewer, J.E., and Bourianoff, G. (2008) *Comput. Soc.*, **41**, 28–32.
7. Szaciłowski, K. (2008) *Chem. Rev.*, **108**, 3481–3548.
8. Andréasson, J. and Pischel, U. (2010) *Chem. Soc. Rev.*, **39**, 174–188.
9. Pischel, U. (2010) *Aust. J. Chem.*, **63**, 148–164.
10. Credi, A. (2007) *Angew. Chem. Int. Ed.*, **46**, 5472–5475.
11. Balzani, V., Credi, A., and Venturi, M. (2008) *Molecular Devices and Machines. Concepts and Perspectives for the Nanoworld*, Wiley-VCH Verlag GmbH, Weinheim.
12. Balzani, V., Credi, A., and Venturi, M. (2008) *Chem. Eur. J.*, **14**, 26–39.
13. Gust, D., Moore, T.A., and Moore, A.L. (2006) *Chem. Commun.*, 1169–1178.
14. Amelia, M., Zou, L., and Credi, A. (2010) *Coord. Chem. Rev.*, **254**, 2267–2280.
15. De Silva, A.P., Vance, T.P., West, M.E.S., and Wright, G.D. (2008) *Org. Biomol. Chem.*, **6**, 2468–2481.
16. De Silva, A.P. and Uchiyama, S. (2007) *Nat. Nanotechnol.*, **2**, 399–410.
17. Benenson, Y. (2009) *Mol. BioSyst.*, **5**, 675–685.
18. Katz, E. and Privman, V. (2010) *Chem. Soc. Rev.*, **39**, 1835–1857.
19. Mann, S. (2008) *Angew. Chem. Int. Ed.*, **47**, 5306–5320.
20. Privman, V., Strack, G., Solenov, D., Pita, M., and Katz, E. (2008) *J. Phys. Chem. B*, **112**, 11777–11784.
21. Vilan, A., Yaffe, O., Biller, A., Salomon, A., Kahn, A., and Cahen, D. (2010) *Adv. Mater.*, **22**, 140–159.
22. Ashkenasy, C., Cahen, D., Cohen, R., Shanzer, A., and Vilan, A. (2002) *Acc. Chem. Res.*, **35**, 121–128.
23. Vilan, A. and Cahen, D. (2002) *Trends Biotechnol.*, **20**, 22–29.
24. Natan, A., Kronik, L., Haick, H., and Tung, R.T. (2007) *Adv. Mater.*, **19**, 4103–4117.
25. Grzybowski, B.A. (2009) *Chemistry in Motion: Reaction – Diffusion Systems for Micro- and Nanotechnology*, John Wiley & Sons, Ltd, Chichester.
26. Sott, K., Lobovkina, T., Lizan, L., Tokarz, M., Bauer, B., Konkoli, Z., and Orwar, O. (2006) *Nano Lett.*, **6**, 209–214.
27. Czolkos, I., Hannestad, J.K., Jesorka, A., Kumar, R., Brown, T., Albinsson, B., and Orwar, O. (2009) *Nano Lett.*, **9**, 2482–2486.
28. George, D. and Jaros, B. (2007) *The HTM Learning Algorithms.*, Report, Numenta Inc.

29. Hawkins, J. and George, D. Hierarchical Temporal Memory. Concepts, Theory, and Terminology., (2006) Report, Numenta Inc.
30. Tao, F., Bernasek, S.L., and Xu, G.-Q. (2009) *Chem. Rev.*, **109**, 3991–4024.
31. Zabet-Khadousi, A. and Dhirani, A. (2008) *Chem. Rev.*, **108**, 4072–4124.
32. Talapin, D.V., Lee, J.-S., Kovalenko, M.V., and Shevchenko, E.V. (2010) *Chem. Rev.*, **110**, 389–458.
33. Hasegawa, S. and Grey, F. (2002) *Surf. Sci.*, **500**, 84–104.
34. Haick, H. and Cahen, D. (2008) *Progr. Surf. Sci.*, **83**, 217–261.
35. Adams, D.M., Brus, L., Chidsey, C.E.D., Creager, S., Creutz, C., Kagan, C.R., Kamat, P.V., Lieberman, M., Lindsay, S., Marcus, R.A., Metzger, R.M., Michel-Beyerle, M.E., Miller, J.R., Newton, M.D., Rolison, D.R., Sankey, O., Schanze, K.S., Yardley, J., and Zhu, X. (2003) *J. Phys. Chem. B*, **107**, 6668–6697.
36. Gawęda, S., Podborska, A., Macyk, W., and Szaciłowski, K. (2009) *Nanoscale*, **1**, 299–316.
37. Li, S.S. (2006) *Semiconductor Physical Electronics*, Springer Science+Business Media LLC, New York.
38. Bunshah, R.F. (1994) *Handbook of Deposition Technologies for Films and Coatings*, Noyes Publications, Park Ridge, NJ.
39. Seshan, K. (2002) *Handbook of Thin-Film Deposition Processes and Techniques*, Noyes Publications, Park Ridge, NJ.
40. Coleman, J.P., Lynch, A.T., Madhukar, P., and Wagenknecht, J.H. (1999) *Sol. Energy Mater. Sol. Cells*, **56**, 395–418.
41. Singh, M., Haverinen, H.M., Dhagat, P., and Jabbour, G.E. (2010) *Adv. Mater.*, **22**, 673–685.
42. Dahl, J.A., Maddux, B.L.S., and Hutchison, J.E. (2007) *Chem. Rev.*, **107**, 2228–2269.
43. Zhu, J., Palchik, O., Chen, S., and Gedanken, A. (2000) *J. Phys. Chem. B*, **104**, 7344–7347.
44. Zhu, J., Zhou, M., Xu, J., and Liao, X. (2001) *Mater. Lett.*, **47**, 25–29.
45. Gerbec, J., Magana, D., Washington, A., and Strouse, G. (2005) *J. Am. Chem. Soc.*, **127**, 15791–15800.
46. Dallinger, D. and Kappe, O. (2007) *Chem. Rev.*, **107**, 2563–2591.
47. Liao, X., Wang, H., Zhu, J., and Chen, H. (2001) *Mater. Res. Bull.*, **36**, 2339–2346.
48. Panda, A.B., Glaspell, G., and El-Shall, M.S. (2006) *J. Am. Chem. Soc.*, **128**, 2790–2791.
49. Panda, A.B., Glaspell, G., and El-Shall, M.S. (2007) *J. Phys. Chem. C*, **111**, 1861–1864.
50. Pachlik, O., Kerner, R., Zhu, Z., and Gedanken, A. (2000) *J. Solid State Chem.*, **154**, 530–534.
51. Nemec, I., Nahalkova, P., Nemcowa, Y., Trojanek, F., Maly, P., and Nemec, P. (2002) *Thin Solid Films*, **403–404**, 9–12.
52. Chu, S.S., Schultz, N., Wang, C., Wu, C.Q., and Chu, T.L. (1992) *J. Electrochem. Soc.*, **139**, 2443–2446.
53. Hodes, G. (2003) *Chemical Solution Deposition of Semiconductor Films*, Marcel Dekker, Inc., New York.
54. Lokhande, C.D. and Mane, R.S. (2000) *Mater. Chem. Phys.*, **65**, 1–31.
55. Popovici, E.J., Ladar, M., Pascu, L., Indrea, E., and Grecu, R. (2004) *J. Optoelectron. Adv. Mater.*, **6**, 127–132.
56. Oladeji, I.O., Chow, L., and Khallaf, H. (2008) *Thin Solid Films*, **516**, 5967–5973.
57. Miranda, M.A.R., Sasaki, J.M., Araújo-Silva, M.A., and Feitosa, A.V. (2004) *Braz. J. Phys.*, **34**, 656–658.
58. Srivastava, S.K. and Roy, P. (2006) *Mater. Chem. Phys.*, **95**, 235–241.
59. Podborska, A., Gaweł, B., Pietrzak, Ł., Szymańska, I.B., Jeszka, J.K., Łasocha, W., and Szaciłowski, K. (2009) *J. Phys. Chem. C*, **113**, 6774–6784.
60. Johnson, D.R., Sadeghi, M., Sivapathasundaram, D., Peter, L.M., Furlong, M.J., Goodlet, G., Shingleton, A., Lincut, D., Mokili, B., Vedel, J., and Ozsan, M.E. (1994) First World Conference on Photovoltaic Energy Conversion, Waikoloa.

61. Gümüş, C., Esen, R., and Çetinörgü, E. (2006) *Thin Solid Films*, **515**, 1688–1693.
62. Herrero, J. and Dona, J.M. (1997) *J. Electrochem. Soc.*, **144**, 4081–4091.
63. Dharmadasa, I.M. and Haigh, J. (2005) *J. Electrochem. Soc.*, **153**, G47–G52.
64. Lincot, D. (2005) *Thin Solid Films*, **487**, 40–48.
65. Pandey, R.K., Sahu, S.N., and Chandra, S. (1996) *Handbook of Semiconductor Electrodeposition*, Marcel Dekker, New York.
66. Brenner, A. (1963) *Electrodeposition of Alloys*, Academic Press, New York.
67. Rajeshwar, K. (1992) *Adv. Mater.*, **4**, 23–29.
68. Savadogo, O. (1998) *Sol. Energy Mater. Sol. Cells*, **52**, 361–388.
69. Schlesinger, M. and Paunovic, M. (2000) *Modern Electroplating*, John Wiley and Sons, Inc., New York.
70. Lincot, D., Guillemoles, J.F., Taunier, S., Guimard, D., Sicx-Kurdi, J., Chaumont, A., Roussel, O., Ramdani, O., Hubert, C., Fauvarque, J.P., Bodereau, N., Parissi, L., Panheleux, P., Fanouillere, P., Naghavi, N., Grand, P.P., Benfarah, M., Mogensen, P., and Kerrec, O. (2004) *Sol. Energy*, **77**, 725–737.
71. Agrawal, A.K. and Austin, A.E. (1981) *J. Electrochem. Soc.*, **128**, 2292–2296.
72. Zein El Abedin, S., Borissenko, N., and Endres, F. (2004) *Electrochem. Commun.*, **6**, 510–514.
73. Endres, F. (2001) *Phys. Chem. Chem. Phys.*, **3**, 3165–3174.
74. Yang, M.C., Landau, U., and Angus, J.C. (1992) *J. Electrochem. Soc.*, **139**, 3480–3488.
75. Kowalik, R., Żabiński, P., and Fitzner, K. (2008) *Electrochim. Acta*, **53**, 6184–6190.
76. Bhattacharya, R.N. and Rajeshwar, K. (1986) *Sol. Cells*, **16**, 237–243.
77. Therese, G.H.A. and Kamath, P.V. (2005) *Chem. Mater.*, **12**, 1195–1204.
78. Karuppuchamy, S., Nonomura, K., Yoshida, T., Sugiura, T., and Minoura, H. (2002) *Solid State Ionics*, **151**, 19–27.
79. Armand, M., Endres, F., MacFarlane, D.R., Ohno, H., and Scrosati, B. (2009) *Nat. Mater.*, **8**, 621–629.
80. El Abedin, S.Z., Borissenko, N., and Endres, F. (2004) *Electrochem. Commun.*, **6**, 510–514.
81. Freyland, W., Zell, C.A., Zein El Abedin, S., and Endres, F. (2003) *Electrochim. Acta*, **48**, 3053–3061.
82. Galatsis, K., Wang, K.L., Ozkan, M., Ozkan, C.S., Huang, Y., Chang, J.P., Monbouquette, H.G., Chen, Y., Nealey, P., and Botros, Y. (2010) *Adv. Mater.*, **22**, 769–778.
83. Huczko, A. (2000) *Appl. Phys. A*, **70**, 365–376.
84. Gregorya, B.W. and Stickney, J.L. (1991) *J. Electroanal. Chem.*, **300**, 543–561.
85. Ham, D., Mishra, K.K., Weiss, A., and Rajeshwar, K. (1989) *Chem. Mater.*, **1**, 619–625.
86. Ham, D., Mishra, K.K., and Rajeshwar, K. (1991) *J. Electrochem. Soc.*, **138**, 100–108.
87. Miller, B. and Heller, A. (1976) *Nature*, **262**, 680–681.
88. Kowalik, R. and Fitzner, K. (2004) *Metall. Foundry Eng.*, **30**, 129–141.
89. Kröger, F.A. (1978) *J. Electrochem. Soc.*, **125**, 2028–2034.
90. Dennison, S. (1993) *Electrochim. Acta*, **38**, 2395–2403.
91. Bouroushian, M. (2010) *Electrochemistry of Metal Chalcogenides*, Springer, Berlin.
92. Zarębska, K. and Skompska, M. (2011) *Electrochimica Acta*, **56**, 5731–5739.
93. Mech, K., Kowalik, R., and Fitzner, K. (2011) *Arch. Metall. Mater.*, **56**, 659–663.
94. Zhou, Y. and Switzer, J.A. (1998) *Scr. Mater.*, **38**, 1731–1738.
95. Bohannan, E.W., Schmusky, M.G., and Switzer, J.A. (1999) *Chem. Mater.*, **11**, 2289–2291.
96. Kemell, M., Ritala, M., and Leskela, M. (2005) *Crit. Rev. Solid State Mater. Sci.*, **30**, 1–31.
97. Krishnan, V., Ham, D., Mishro, K.K., and Rajeshwar, K. (1992) *J. Electrochem. Soc.*, **139**, 23–27.

98. Streltsov, E.A., Osipovicha, N.P., Ivashkevicha, L.S., and Lyakhova, A.S. (1998) *Electrochim. Acta*, **44**, 407–413.
99. Bohannan, E.W., Huang, L.-Y., Miller, F.S., Schumsky, M.G., and Switzer, J.A. (1999) *Langmuir*, **15**, 813–817.
100. Switzer, J.A., Hung, C.J., Huang, J.Y., Switzer, E.R., Kammler, D.R., Golden, T.D., and Bohannan, E.W. (1998) *J. Am. Chem. Soc.*, **120**, 3530–3531.
101. Wei, C. and Rajeshwar, K. (1992) *J. Electrochem. Soc.*, **139**, L40–L41.
102. Stickney, J.L. (2001) in *Advances in Electrochemical Science and Engineering*, vol. 7 (eds R.C. Alkire and D.M. Kolb), Wiley-VCH Verlag GmbH, pp. 1–105.
103. Alivisatos, A.P. (1996) *J. Phys. Chem.*, **100**, 13226–13239.
104. de Mello Donega, C. (2011) *Chem. Soc. Rev.*, **40**, 1512–1546.
105. Kaprov, S.V. and Mikushev, S.V. (2010) *Phys. Solid State*, **52**, 1750–1756.
106. Laheld, U.E.H., Pedersen, F.B., and Hemmer, P.C. (1995) *Phys. Rev. B*, **52**, 2697.
107. Ivanov, S., Piryatinski, A., Nanda, J., Tretiak, S., Zavadil, K., Wallace, W., Werder, D., and Klimov, V. (2007) *J. Am. Chem. Soc.*, **129**, 11708–11719.
108. Kinge, S., Crego-Calama, M., and Reinhoudt, D.N. (2008) *ChemPhysChem*, **9**, 20–42.
109. Grzelczak, M., Vermant, J., Furst, E.M., and Liz-Marzán, L.M. (2010) *ACS Nano*, **4**, 3591–3605.
110. Zade, S.S. and Bendikov, M. (2010) *Angew. Chem. Int. Ed.*, **49**, 4012–4015.
111. Mondal, R., Shah, B.K., and Neckers, D.C. (2006) *J. Am. Chem. Soc.*, **128**, 9612–9613.
112. Mondal, R., Tönshoff, C., Khon, D., Neckers, D.C., and Bettinger, H.F. (2009) *J. Am. Chem. Soc.*, **131**, 14281–14289.
113. Tönshoff, C. and Bettinger, H.F. (2010) *Angew. Chem. Int. Ed.*, **49**, 4125–4128.
114. Anthony, J.E. (2006) *Chem. Rev.*, **106**, 5028–5048.
115. Gao, B., Wang, M., Cheng, Y., Wang, L., Jing, Z., and Wang, F. (2008) *J. Am. Chem. Soc.*, **130**, 8297–8306.
116. Kaur, I., Jazdzyk, M., Stein, N.N., Prusevich, P., and Miller, G.P. (2010) *J. Am. Chem. Soc.*, **132**, 1261–1263.
117. Kaur, I., Stein, N.N., Kopreski, R.P., and Miller, G.P. (2009) *J. Am. Chem. Soc.*, **131**, 3424–3425.
118. Mas-Torrent, M. and Rovira, C. (2011) *Chem. Rev.*, **111**, 4833–4856.
119. Curtis, M.D., Cao, J., and Kampf, J.W. (2004) *J. Am. Chem. Soc.*, **126**, 4318–4328.
120. Pisula, W., Feng, X., and Müllen, K. (2010) *Adv. Mater.*, **22**, 3634–3649.
121. Boden, N., Bushby, R.J., Clements, J., Movaghar, B., Donovan, K.J., and Kreouzis, T. (1995) *Phys. Rev. B*, **52**, 13274–13280.
122. Wu, D., Liu, R., Pisula, W., Feng, X., and Müllen, K. (2011) *Angew. Chem. Int. Ed.*, **50**, 2791–2794.
123. Wu, D., Zhi, L., Bodwell, G.J., Cui, G., Tsao, N., and Müllen, K. (2007) *Angew. Chem. Int. Ed.*, **46**, 5417–5420.
124. Tulevski, G.S., Miao, Q., Fukuto, M., Abram, R., Ocko, B., Pindak, R., Steigerwald, M.L., Kagan, C.R., and Nuckolls, C. (2004) *J. Am. Chem. Soc.*, **126**, 15048–15050.
125. Gershenson, M.E., Podzorov, V., and Morpurgo, A.F. (2006) *Rev. Mod. Phys.*, **78**, 973–989.
126. Del Guerzo, A., Olive, A.G.L., Reichwagen, J., Hopf, H., and Desvergne, J.-P. (2005) *J. Am. Chem. Soc.*, **127**, 17984–17985.
127. Ricard, D., Roussignol, P., and Flytzanis, C. (1985) *Opt. Lett.*, **10**, 511–513.
128. Maeda, Y., Tsukamoto, N., Yazawa, Y., Kanemitsu, Y., and Masumoto, Y. (1991) *Appl. Phys. Lett.*, **59**, 3168–3170.
129. Beranek, R. and Kisch, H. (2008) *Angew. Chem. Int. Ed.*, **47**, 1320–1322.
130. Long, M., Beránek, R., Cai, W., and Kisch, H. (2008) *Electrochim. Acta*, **53**, 4621–4626.
131. de Tacconi, N.R., Chenthamarakshan, C.R., Rajeshwar, K., and Tacconi, E.J. (2005) *J. Phys. Chem. B*, **109**, 11953–11960.
132. Somasundaram, S., Chenthamarakshan, C.R., de Tacconi, N.R., Ming, Y., and Rajeshwar, K. (2004) *Chem. Mater.*, **16**, 3846–3852.
133. Agostinelli, G. and Dunlop, E.D. (2003) *Thin Solid Films*, **431–432**, 448–452.

134. Rammelt, U., Hebestreit, N., Fikus, A., and Plieth, W. (2001) *Electrochim. Acta*, **46**, 2363–2371.
135. Vu, Q.-T., Pavlik, M., Hebestreit, N., Rammelt, U., Plieth, W., and Pfleger, J. (2005) *React. Funct. Polym.*, **65**, 69–77.
136. Szaciłowski, K., Macyk, W., and Stochel, G. (2006) *J. Mater. Chem.*, **16**, 4603–4611.
137. Ishii, H., Sugiyama, K., Ito, E., and Seki, K. (1999) *Adv. Mater.*, **11**, 605–625.
138. Tung, R.T. (2001) *Mater. Sci. Eng.*, **R35**, 1–138.
139. Koopmans, T. (1934) *Physica*, **1**, 104–113.
140. Janata, J. and Josowicz, M. (1998) *Acc. Chem. Res.*, **31**, 241–248.
141. Janata, J. (1991) *Anal. Chem.*, **63**, 2546–2550.
142. Macyk, W., Szaciłowski, K., Stochel, G., Buchalska, M., Kuncewicz, J., and Łabuz, P. (2010) *Coord. Chem. Rev.*, **254**, 2687–2701.
143. Nguyen, L.T., De Proft, F., Amat, M.C., Van Lier, G., Fowler, P.W., and Geerlings, P. (2003) *J. Phys. Chem. A*, **107**, 6837–6842.
144. Yang, W. and Parr, R.G. (1985) *Proc. Natl. Acad. Sci.*, **82**, 6723–6726.
145. Lebedev, M.V. (2002) *Prog. Surf. Sci.*, **70**, 153–186.
146. Flores-Moreno, R. (2010) *J. Chem. Theory Comput.*, **6**, 48–54.
147. Paska, Y. and Haick, H. (2009) *J. Phys. Chem. C*, **113**, 1993–1997.
148. Vilan, A., Shanzer, A., and Cahen, D. (2000) *Nature*, **404**, 166–168.
149. Selzer, Y. and Cahen, D. (2001) *Adv. Mater.*, **13**, 508–511.
150. Potje-Kamloth, K. (2008) *Chem. Rev.*, **108**, 367–399.
151. Brown, W.A., Kose, R., and King, D.A. (1998) *Chem. Rev.*, **98**, 797–831.
152. Ge, Q., King, D.A., Lee, M.-H., White, J.A., and Payne, M.C. (1997) *J. Chem. Phys.*, **106**, 1210–1215.
153. Braun, S., Salaneck, W.R., and Fahlman, M. (2009) *Adv. Mater.*, **21**, 1450–1472.
154. Mortensen, J.J., Morikawa, Y., Hammer, B., and Noerskov, J.K. (1997) *J. Catal.*, **169**, 85–92.
155. Macyk, W., Stochel, G., and Szaciłowski, K. (2007) *Chem. Eur. J.*, **13**, 5676–5687.
156. Gosh, H.N., Ashbury, J.B., Weng, Y., and Lian, T. (1998) *J. Phys. Chem. B*, **102**, 10208–10215.
157. De Angelis, F., Tilocca, A., and Selloni, A. (2004) *J. Am. Chem. Soc.*, **126**, 15024–15025.
158. Macyk, W., Szaciłowski, K., Stochel, G., Buchalska, M., Kuncewicz, J., and Łabuz, P. (2010) *Coord. Chem. Rev.*, doi: 10.1016/j.ccr.2009.12.037
159. Regazzoni, A.E., Mandelbaum, P., Matsuyoshi, M., Schiller, S., Bilmes, S.A., and Blesa, M.A. (1998) *Langmuir*, **14**, 868–874.
160. Rodriguez, R., Blesa, M.A., and Regazzoni, A.E. (1996) *J. Colloid Interface Sci.*, **177**, 122–131.
161. Moser, J., Punchihewa, S., Infelta, P.P., and Grätzel, M. (1991) *Langmuir*, **7**, 3012–3018.
162. Araujo, P.Z., Mendive, C.B., Garcia Rodenas, L.A., Morando, P.J., Regazzoni, A.E., Blesa, M.A., and Bahnemann, D. (2005) *Colloids Surf. A: Physicochem. Eng. Aspects*, **265**, 73–80.
163. Liu, L.-M., Li, S.-C., Cheng, H., Diebold, U., and Selloni, A. (2011) *J. Am. Chem. Soc.*, **133**, 7816–7823.
164. Li, S.-C., Losovyj, Y., and Diebold, U. (2011) *Langmuir*, doi: 10.1021/la201553k
165. Abramavicius, D., Palmieri, B., Voronine, D.V., Šanda, F., and Mukamel, S. (2009) *Chem. Rev.*, **109**, 2350–2408.
166. Oszajca, M., Kwolek, P., Mech, J., and Szaciłowski, K. (2011) *Curr. Phys. Chem.*, **1**, 242–260.
167. Hug, S.J. and Bahnemann, D. (2006) *J. Electron. Spectrosc.*, **150**, 208–219.
168. Weisz, A.D., Rodenas, L.G., Morando, P.J., Regazzoni, A.E., and Blesa, M.A. (2002) *Catal. Today*, **76**, 103–112.
169. Li, S.-X., Zheng, F.-Y., Cai, W.-L., Han, A.-Q., and Xie, Y.-K. (2006) *J. Hazard. Mater. B*, **135**, 431–436.
170. Gaweda, S., Stochel, G., and Szaciłowski, K. (2008) *J. Phys. Chem. C*, **112**, 19131–19141.

171. Hebda, M., Stochel, G., Szaciłowski, K., and Macyk, W. (2006) *J. Phys. Chem. B*, **110**, 15275–15283.
172. Szaciłowski, K., Macyk, W., Hebda, M., and Stochel, G. (2006) *ChemPhysChem*, **7**, 2384–2391.
173. Szaciłowski, K. and Macyk, W. (2006) *C. R. Chim.*, **9**, 315–324.
174. Szaciłowski, K. and Macyk, W. (2006) *Solid State Electron.*, **50**, 1649–1655.
175. Campbell, W.M., Burrell, A.K., Officer, D.L., and Jolley, K.W. (2004) *Coord. Chem. Rev*, **248**, 817–833.
176. Ashkenasy, G., Cahen, D., Cohen, R., Shanzer, A., and Vilan, A. (2002) *Acc. Chem. Res.*, **35**, 121–128.
177. Hoffman, R. (1988) *Solids and Surfaces: A Chemist's View of Bonding in Extended Structures*, VCH Publication, Weinheim.
178. Bent, S.F. (2002) *J. Phys. Chem. B*, **106**, 2830–2842.
179. Ulman, A. (1996) *Chem. Rev.*, **96**, 1533–1554.
180. Hamers, R.J., Coulter, S.K., Ellison, M.D., Hovis, E.S., Padowitz, D.F., and Schwartz, M.P. (2000) *Acc. Chem. Res.*, **33**, 617–624.
181. Cohen, R., Kronik, L., Shanzer, A., Cahen, D., Liu, A., Rosenwaks, Y., Lorenz, J.K., and Ellis, A.B. (1999) *J. Am. Chem. Soc.*, **121**, 10545–10553.
182. Gawęda, S., Stochel, G., and Szaciłowski, K. (2007) *Chem. Asian J.*, **2**, 580–590.
183. Di Iorio, Y., Rodríguez, H.B., San Román, E., and Grela, M.A. (2010) *J. Phys. Chem. C*, **114**, 11515–11521.
184. Podborska, A. and Szaciłowski, K. (2010) *Aust. J. Chem.*, **63**, 165–168.
185. Szaciłowski, K. and Macyk, W. (2007) *Chimia*, **61**, 831–834.
186. Szaciłowski, K., Macyk, W., and Stochel, G. (2006) *J. Am. Chem. Soc.*, **128**, 4550–4551.
187. Mech, J., Kowalik, R., Podborska, A., Kwolek, P., and Szaciłowski, K. (2010) *Aust. J. Chem.*, **63**, 1330–1333.
188. Toffoli, T. (1998) *Phys. D*, **120**, 1–11.

8
Toward Arithmetic Circuits in Subexcitable Chemical Media
Andrew Adamatzky, Ben De Lacy Costello, and Julian Holley

8.1
Awakening Gates in Chemical Media

The design of logical gates in chemical systems can be traced back to the early 1990s when Hjemfelft *et al.* suggested a theoretical coupled mass-flow system for implementing logic gates and finite-state machines [1–5] and Lebender and Schneider proposed logical gates utilizing a series of flow-rate-coupled continuous stirred tank reactors and a bistable chemical reaction [6]. No experimental prototypes were implemented at that time. Mass-kinetic based computing is appealing theoretically but laboratory experiments are cumbersome to undertake. The implementations of mass-kinetic networks are inefficient, as most designs require the use of programmable pumping devices.

In 1994, the Showalter Laboratory presented the first ever experimental implementation of logical gates in the Belousov – Zhabotinsky (BZ) system [7, 8]. The logical gates were based on the geometrical configuration of the channels in which excitation waves propagate. The ratio between the channel diameter and the critical nucleation radii of the excitable media allowed various logical schemes to be realized. These original findings led to several innovative designs of computational devices, based on geometrically constrained excitable substrates. Designs incorporating assemblies of channels for excitation wave propagation were used to implement logical gates for Boolean and multiple-valued logic [9–12], many-input logical gates [13, 14], counters [15], coincidence detectors [16], and detectors of direction and distance [17, 18] All these chemical computing devices were realized in geometrically constrained media where excitation waves propagate along defined catalyst-loaded channels or tubes filled with the BZ reagents. The waves perform computation by interacting at the junctions between the channels. Despite its apparent novelty, the approach is just an implementation of conventional computing architectures in novel materials, namely, excitable chemical systems. There is, however, another way to undertake computation – by employing the principles of collision-based computing [19].

We provide an overview of our recent results on the implementation of collision-based circuits in the subexcitable BZ medium. In Section 8.2, we introduce a paradigm of collision-based computing. We discuss traveling localizations in the subexcitable BZ medium in Section 8.3. In Section 8.4, we show how to reach a compromise between geometrically constrained and free-space (collision-based) computing by encapsulating the BZ system in vesicles. The various types of interactions between the wave fragments in BZ vesicles are analyzed in Section 8.5. In Section 8.6, we demonstrate that each BZ vesicle is a universal collision-based computing device. Finally, we assemble the BZ vesicles to construct a binary adder in Section 8.7.

8.2
Collision-Based Computing

The paradigm of collision-based computing originates from the computational universality of the Game of Life [20], conservative logic, and the billiard-ball model [21] with its cellular automaton implementation [22].

A collision-based computer employs mobile, self-localized excitations to represent quanta of information in active nonlinear media. Information values, for example, truth values of logical variables, are given by either the absence or presence of the localizations or by other parameters such as direction or velocity. The localizations travel in space and collide with each other. The results of the collisions are interpreted as computation. There are no predetermined stationary wires; a trajectory of the traveling localization is a momentary wire. Almost any part of the reactor space can be used as a wire. Localizations can collide anywhere within this space. The localizations undergo transformations, form bound states, annihilate, or fuse. Information values of localizations are transformed as a result of these collisions [19].

8.3
Localizations in Subexcitable BZ Medium

To implement a collision-based scheme in a spatially extended chemical medium, we must employ traveling localized excitations. Such localizations, or wave fragments, emerge in a light-sensitive BZ medium when it is in a subexcitable state [23]. The ruthenium-catalyzed BZ medium shows a high degree of light sensitivity. At some levels of illumination, the medium behaves as a classical excitable medium in which a perturbation leads to the formation of omnidirectional propagating waves of the excitation. When the level of illumination exceeds a critical threshold, no excitation persists. There is a narrow range of illumination parameters where the BZ medium is in a subexcitable (weakly excitable) state between the nonexcitable and excitable states.

A perturbation of the subexcitable medium leads to the formation of localized traveling excitations, or wave fragments. The Wave fragments travel along their predetermined trajectories and preserve their shapes and velocity vectors for some time. The fate of each fragment is determined by the exact level of illumination and the size of the fragment. The smaller fragments usually collapse, whereas the large ones usually expand. If the illumination level is at a critical level, then appropriate sized fragments entering a subexcitable medium will preserve their size for appreciable distance/time intervals.

These excitation wave fragments behave like quasiparticles. They exhibit rich dynamics of collisions, including quasireflection, fission, fusion, and annihilation [24, 25]. Snapshots of collisions obtained in chemical laboratory experiments are shown in Figures 8.1 and 8.2. Let us discuss how such types of collisions can be interpreted in terms of logical functions.

Take a look at Figure 8.1. There are two trains of wave fragments – one traveling east and one traveling west. They represent input trajectories x and y. The presence of a fragment signifies the logical truth ("1"), and the absence the logical false ("0"). If just one fragment, for example, x, enters the space, it continues along its original trajectory and appears along input y. If both fragments enter the space ($x = 1$ and $y = 1$), they collide and annihilate. Such a collision setup represents a two-input two-output logical gate

$$\langle x, y \rangle \rightarrow \langle \bar{x}y, x\bar{y} \rangle$$

A two-input three-output logical gate

$$\langle x, y \rangle \rightarrow \langle \bar{x}y, xy, x\bar{y} \rangle$$

is realized by the collision of wave fragments shown in Figure 8.2. The wave fragment traveling west represents the logical variable x, and the fragment traveling north represents y. If the fragments collide, they merge into a single wave fragment which travels northwest (Figure 8.2a). This new fragment represents conjunction xy. A fragment travels along its original trajectory in the absence of the other fragment, and thus values $\bar{x}y$ and $x\bar{y}$ are produced (Figure 8.2b).

Figure 8.1 Snapshots of head-on collision between fragments traveling west and those traveling east. Note that the collision is slightly offset (not directly head-on), resulting in dominant daughter fragments traveling in northwards (will maintain stability) and the weaker daughter fragment traveling southward (collapses in the collision zone). From Ref. [24]. Images were obtained from laboratory experiments with the light-sensitive BZ reaction in a subexcitable state.

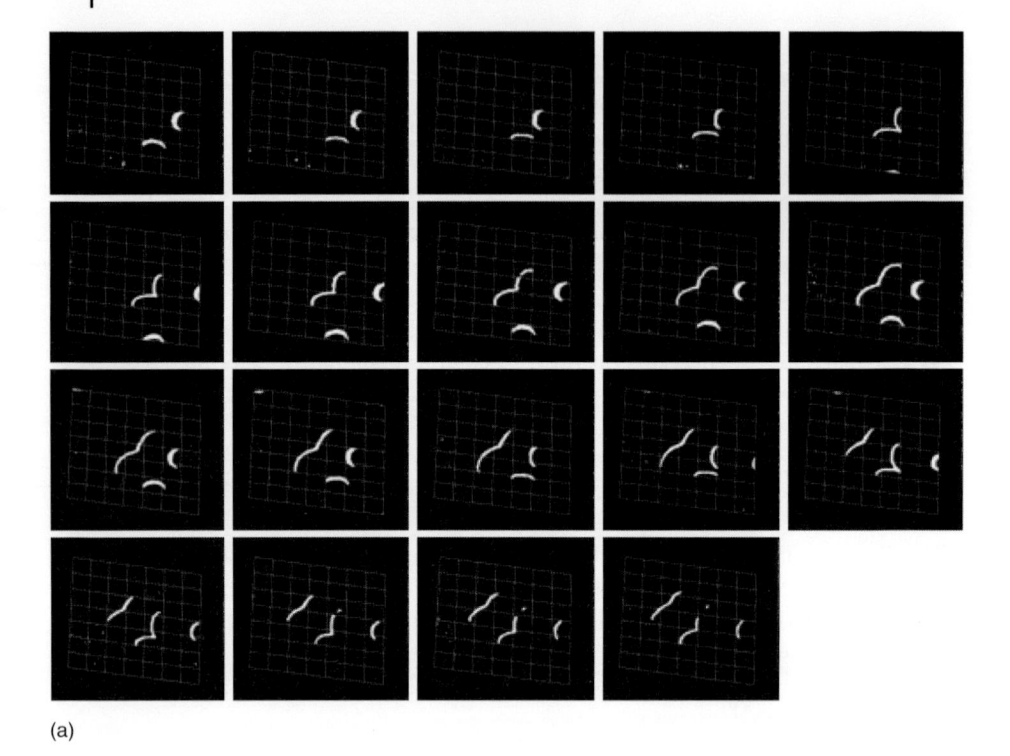

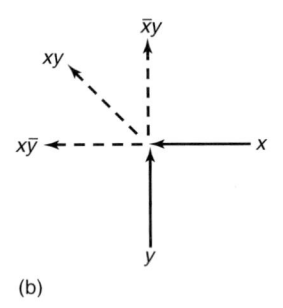

Figure 8.2 Snapshots of merging of two wave fragments. From Ref. [24]. Images were obtained from laboratory experiments with the light-sensitive BZ reaction in a subexcitable state.

Throughout this chapter, we use the two-variable Oregonator equation [26] adapted to a light-sensitive analog of the BZ reaction with the applied illumination [27]:

$$\frac{\partial u}{\partial t} = \frac{1}{\epsilon}\left(u - u^2 - (fv + \phi)\frac{u-q}{u+q}\right) + D_u \nabla^2 u$$

(a) $\phi = 0.07$ (b) $\phi = 0.077$ (c) $\phi = 0.07873$

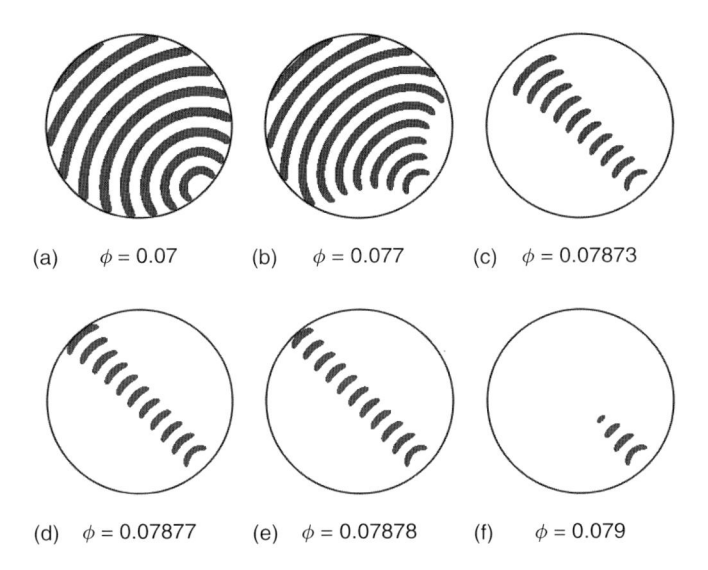

(d) $\phi = 0.07877$ (e) $\phi = 0.07878$ (f) $\phi = 0.079$

Figure 8.3 Demonstration that the development of the initial excitation is sensitively dependent on the level of illumination ϕ. Excitation is initiated at the southwest edge of the vesicle.

$$\frac{\partial v}{\partial t} = u - v$$

The variables u and v represent the local concentrations of the activator (or excitatory component) and the inhibitor, (or refractory component), respectively. The parameter ϵ sets up a ratio of the timescales for the variables u and v, q is a scaling parameter dependent on the rates of activation/propagation and inhibition, and f is a stoichiometric factor. The constant ϕ is the rate of inhibitor production. In the light-sensitive BZ, ϕ represents the rate of inhibitor production which is proportional to the intensity of illumination. We integrate the system using the Euler method with five-node Laplace operator, time step $\Delta t = 0.005$, and grid point spacing $\Delta x = 0.25$, $\epsilon = 0.022$, $f = 1.4$, and $q = 0.002$. The equations effectively map the space – time dynamics of excitation in the BZ medium and have proved to be an invaluable tool for studying the dynamics of collisions between traveling localized excitations in our previous work [24, 25, 28, 29].

The parameter ϕ characterizes the excitability of the simulated medium. The medium is excitable and exhibits "classical" target waves, for example, when $\phi = 0.07$ (Figure 3ab); and the medium is subexcitable with propagating localizations, or wave fragments, when ϕ is between 0.07873 and 0.07878 (Figure 8.3c–e). The medium becomes nonexcitable for $\phi \geq 0.79$, and after this point the wave fragments collapse after relatively short timescales (Figure 8.3f).

No advanced arithmetical circuit is made so far in reaction-diffusion chemical processors. However, all preparatory steps are completed:

1) Basic logical gates are experimentally implemented and principles of their cascading are established [24, 25, 28];
2) Generators of mobile localizations (BZ analogs of glider guns in Conway's Game of Life) necessary to implement the negation operation are discovered in light-sensitive BZ medium with a checkerboard pattern of high- and low-intensity illumination projected onto the BZ medium [29];
3) Basic arithmetical circuits – binary counters, adders, and multipliers – are verified in cellular automaton models of excitable media [30, 31] and in the two-variable Oregonator model [32];
4) Algorithms of evolving computing schemes in chemical media are developed and tested in laboratory experiments [33, 34], so most paradigms and implementation of evolutionary hardware can be applied in full to laboratory designs of chemical wetware.

In this chapter, we develop the theoretical constructions sufficient for future experimental implementations of BZ-based logical circuits.

8.4
BZ Vesicles

When the BZ reaction is in a subexcitable mode, asymmetric perturbations lead to the formation of propagating localized excitations, or excitation wave fragments. Wave fragments of this type may travel in a predetermined direction for a finite period. If the wave fragments kept their shape indefinitely, we would be able to build a collision-based computing circuit of any size. In reality, the wave fragments are inherently unstable: after traveling some period of conserved shape/distance, a wave fragment either collapses or expands.

A way to overcome the problem of the wave-fragment instability would be via the subdivision of the computing substrate into interconnected compartments, the so-called BZ vesicles, and allowing the waves to collide only inside the compartments [35, 36].

Each BZ vesicle has a membrane that is impassable for excitation (Gorecki, J.J. Private communication (2010) Private communication.), [35]. A pore, or a channel, between two vesicles is formed when two vesicles come into direct contact. The pore is small, so that, when a wave passes through the pore, there is insufficient time for the wave to expand or collapse before interacting with other waves entering through adjacent pores, or sites of contact.

A spherical compartment – BZ vesicle – is the best natural choice as it allows effortless arrangement of the vesicles into a regular lattice, has an almost unlimited number of input/output states, and also loosely conforms to a structure likely to be achieved in experiments involving the encapsulation of excitable chemical media in a lipid membrane (Gorecki, J.J. Private communication (2010) Private communication.), [35]. Scoping experiments with droplets of BZ mixture covered with phospholipid L-α-phosphatidylcholine provided evidence of excitation propagation between BZ vesicles in direct physical contact [37].

We simulate a vesicle filled with BZ solution as a disc with radius R centered in (x_0, y_0). Sites inside the disc are excitable; sites outside the disc are not excitable. We imitate a wave fragment entering the vesicle by exciting (assigning values $u = 1$) grid nodes inside the small disc with radius r, centered in $(x_0 + (R - s)\cos(\theta), y_0 + (R - s)\sin(\theta))$. The following parameters are used in the illustrations: $R = 100$, $r = 5$, $s = 5$, and $\theta \in [0, 2\pi]$. The time-lapse snapshots provided were recorded at every 150 time steps, and the grid sites with excitation level $u > 0.04$ were displayed.

8.5
Interaction Between Wave Fragments

All constructs presented in this chapter assume that the system of BZ vesicles is fully synchronized. Waves enter any single vesicle simultaneously. This is a very strong and somewhat unrealistic assumption. However, we use it as a starting point in order to construct the first reliable models of BZ-vesicle computers.

Proposition 1 *Let several wave fragments enter a vesicle. If at least two wave fragments have opposite velocity vectors, all wave fragments annihilate; otherwise the wave fragments merge and the velocity vector of the newly formed wave fragment is the sum of the velocity vectors of the incoming wave fragments [36].*

The constructive proof is illustrated by a representative scenario of wave interactions inside a single vesicle (vesicle x in Figure 8.4) shown in Figure 8.5. If just one neighbor of vesicle x is excited, the vesicle x acts as a conductor and a signal amplifier: the wave simply passes through the vesicle x slightly increasing in size and exits through the pore opposite to the wave's entry pore. Thus, in Figure 8.5a, northwest neighbors (vesicle x_0 in Figure 8.4) are activated. The excitation enters vesicle x, and a wave fragment traveling southeast is formed (Figure 8.5a). The wave fragment is transmitted to the southeast neighbor (vesicle x_3 in Figure 8.4) of vesicle x.

There are three possible scenarios when two neighbors of vesicle x are excited. In a situation where the excited neighbors of x are also the closest neighbors of each other (e.g., vesicles x_0 and x_1 in Figure 8.4), the wave fragments generated by them merge inside vesicle x (Figure 8.5b). The velocity vector of the newly formed wave fragment is the sum of the vectors of the two original wave fragments. Vectors of the original wave fronts orient toward the exit pores opposite to the excitation entry pores. The vector of a newly formed wave fragment aims between the pores, and hence no excitation leaves vesicle x. For example, in Figure 8.5b, the northwest

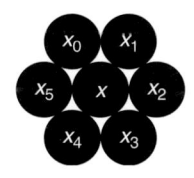

Figure 8.4 Structure of neighborhood of the disc compartment of packed BZ vesciles. From Ref. [36].

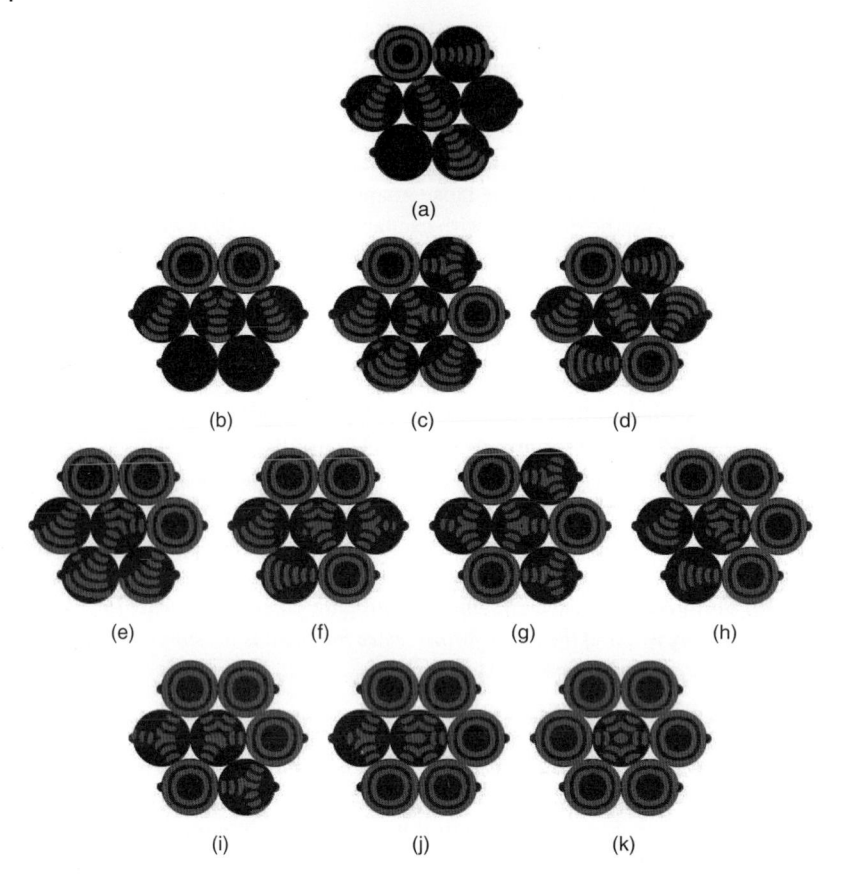

Figure 8.5 Time-lapse trajectories of wave interactions in BZ vesicles. Representative scenarios are given for (a) one input wave, (b)–(d) two input waves, (e)–(g) three input waves, (h)–(i) four input waves, (j) five input waves, and (k) six input waves. From Ref. [36].

and northeast neighbors of vesicle x are excited. Two wave fragments enter vesicle x: one fragment travels southeast, and the other southwest. The wave fragments merge and form a new wave fragment which travels south. This fragment collides with the part of vesicle x's wall lying between the southwest and southeast pores. The fragment is annihilated as a result of the collision.

If the excited neighbors of vesicle x are separated by another neighbor of x (e.g., vesicles x_0 and x_2 in Figure 8.4), then the vector of the newly formed wave fragment (the result of merging two input wave fragments) points exactly to one exit pore (e.g., pore connecting vesicle x and vesicle x_4 in Figure 8.4). In the example in Figure 8.5c, the northwest and east neighbors of vesicle x are excited. Two wave fragments enter vesicle x: one travels southeast and another travels west. They merge and form a wave fragment traveling southwest. This fragment leaves vesicle x for its southwest neighbor (vesicle x_4).

If the wave fragments travel toward each other (Figure 8.5d), they collide and annihilate as a result of the collision. Thus no excitation leaves vesicle x. Scenarios of collision between three wave fragments are shown in Figure 8.5e–g).

In the scenarios illustrated in Figure 8.5h–k, at least two wave fragments undergo head-on collision and all input wave fragments annihilate. The situation (Figure 8.5h) is similar to that in Figure 8.5c, with the only difference that three not two waves collide; however, the resulting wave fragment is the same: it travels southwest and excites the southwest neighbor (vesicle x_4) of vesicle x.

The configuration of the excited neighbors shown in Figure 8.5f results in the annihilation of all three incoming wave fragments. Three wave fragments entering vesicle x travel southeast, southwest, and northwest. The wave fragment traveling southeast collides head on with the wave fragment traveling northwest. Both wave fragments annihilate as a result of the collision. At the same time, the wave fragment traveling southwest collides with both the southeast and northwest fragments. The fragment traveling southwest is also annihilated (Figure 8.5f).

The situation shown in Figure 8.5g corresponds to the direct head-on collision between the wave fragments. The three incoming wave fragments travel southeast, west, and northeast. The sum of the velocity vectors of these wave fragments is nil, and therefore all fragments annihilate and no wave fragments leave vesicle x.

In the remaining scenarios (Figure 8.5h–k), at least two incoming wave fragments experience head-on collisions and all wave fragments annihilate.

8.6
Universality and Polymorphism

Theorem 1 *A BZ vesicle is a logically universal device: excitation wave fragments traveling and colliding within a BZ vesicle implement a polymorphic logical gate switchable between functional states* XNOR *and* NOR *by changing the degree of illumination* [38].

We prove this theorem via the constructive proofs of two propositions.

Proposition 2 *Let the Boolean values of x and y be represented by wave fragments; then a BZ vesicle implements a two-input three-output switchable logical gate $\langle x, y, \phi \rangle \rightarrow \langle xy, \chi(\phi)xy, \overline{x}y \rangle$, where $\chi(\phi) = 1$ (TRUE) if $\psi = \psi_{low}$, and 0 (FALSE) otherwise* [38].

Let there be at most two wave fragments entering a BZ vesicle. They enter the vesicle along trajectories x and y (Figure 8.6a). We assume that the presence of a wave fragment at the entry point x represents the logical value TRUE, and the absence the logical value FALSE. If there is a wave fragment entering a BZ vesicle along trajectory y, we assume $y =$ TRUE, otherwise $y =$ FALSE. When just one of the input values is TRUE, then the solitary wave fragment passes through the vesicle without significant modification and exits the vesicle at the site opposite to its entry point (Figure 8.7a–d, $x = 1, y = 0$ and $x = 0, y = 1$). When two wave fragments

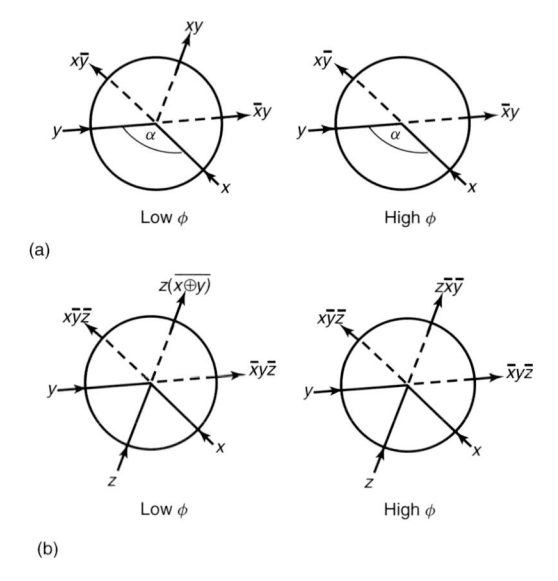

Figure 8.6 Two types of logical gates controllable by the illumination level ϕ. From Ref. [38].

enter the vesicle, they interact and do not follow their original trajectories. Thus the output trajectories along which the undisturbed wave fragments x and y move represent functions $x\bar{y}$ and $\bar{x}y$, respectively (Figure 8.6a).

The interaction of wave fragments is determined by the level of illumination. When the illumination is low enough, say ϕ_{low}, the colliding wave fragments merge into a new (i.e., traveling along a new trajectory) wave fragment (Figure 8.7a and c, $x = 1, y = 1$). The new wave fragment exiting the BZ vesicle represents the operation xy (Figure 8.6a, left). For a higher level of illumination, say ϕ_{high}, the colliding wave fragments annihilate each other (Figure 8.7b and d, $x = 1, y = 1$); no additional operation is realized.

Proposition 3 *Let the Boolean values of x, y, and z be represented by wave fragments; then a BZ vesicle implements a three-input three-output switchable logical gate $\langle x, z, y, \phi \rangle \rightarrow \langle x\bar{y}\,\bar{z}, \chi(\phi)z(x \oplus y) + \overline{\chi(\phi)}\bar{x}\,\bar{y}z, \bar{x}y\bar{z}\rangle$, where $\chi(\phi) = 1$ (TRUE) if $\phi = \phi_{low}$, and 0 (FALSE) otherwise [38].*

The outputs presented by trajectories of undisturbed signals $x(y) - x\bar{y}\,\bar{z}$ ($\bar{x}y\bar{z}$) – are determined as follows. Wave fragment x (y) continues traveling along its original trajectory only if neither wave fragment y (x) nor wave fragment z enters the vesicle (Figures 8.6 and 8.8a and b, $x = 1, y = 0, z = 0$ and $x = 0, y = 1, z = 0$).

The following scenarios take place for both low ϕ_{low} and high ϕ_{high} levels of illumination. If only wave fragment z is present, it travels through the vesicle undisturbed (Figure 8.8, $x = 0, y = 0, z = 1$). When wave fragment z is present and also either wave fragment x or y, the wave fragments collide and form a wave

| (a) | $x=1, y=0$ | $x=0, y=1$ $\alpha = \frac{7\pi}{18}, \phi = 0.07871$ | $x=1, y=1$ |

| (b) | $x=1, y=0$ | $x=0, y=1$ $\alpha = \frac{7\pi}{18}, \phi = 0.07873$ | $x=1, y=1$ |

| (c) | $x=1, y=0$ | $x=0, y=1$ $\alpha = \frac{2\pi}{3}, \phi = 0.07874$ | $x=1, y=1$ |

| (d) | $x=1, y=0$ | $x=0, y=1$ $\alpha = \frac{2\pi}{3}, \phi = 0.07875$ | $x=1, y=1$ |

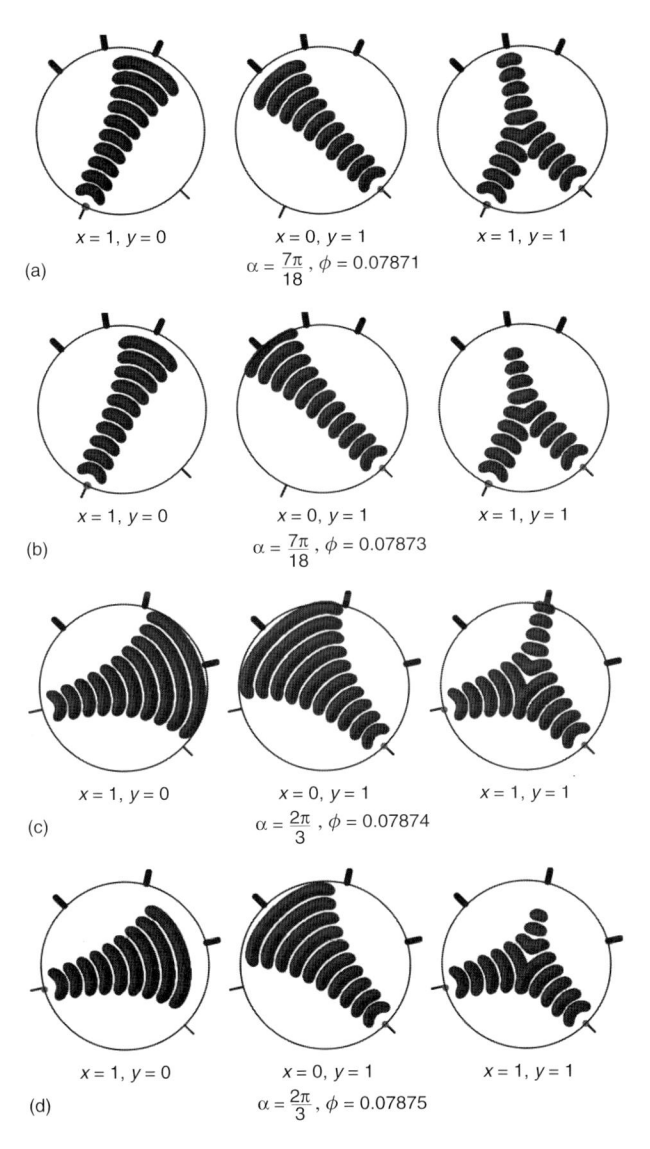

Figure 8.7 Implementation of a polymorphic gate $\langle x, y, \phi \rangle \rightarrow \langle x\bar{y}, \chi(\phi)xy, \bar{x}y \rangle$. The scheme of the gate is shown in Figure 8.6a. Time-lapse snapshots of the subexcitable media are shown for various illumination levels ϕ and collision angles α. Wave fragments represent the logical values of inputs x and y. Inputs (entry points, pores) are marked by thin lines, and outputs (exit points, pores) are marked by thick lines. From Ref. [38].

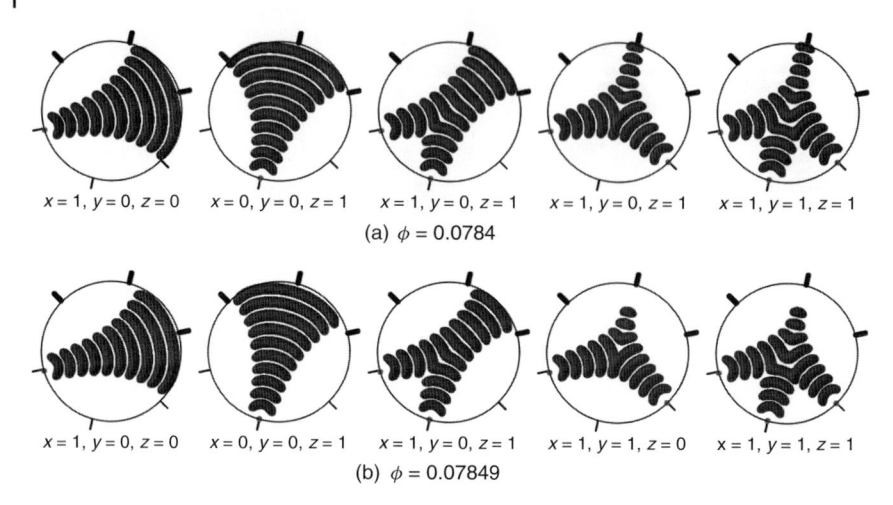

Figure 8.8 Implementation of the polymorphic gate $\langle x, z, y, \phi \rangle \to \langle x\bar{y}\bar{z}, \chi(\phi)z\overline{(x \oplus y)} + \overline{\chi(\phi)}\bar{x}\,\bar{y}z, \bar{x}y\bar{z}\rangle$. The scheme of the gate is shown in Figure 8.6b. Time-lapse snapshots of the subexcitable medium are shown for various illumination levels ϕ and collision angles α. Waves represent logical values of inputs x and y. Wave fragments x and z, and z and y collide at an angle $\frac{\pi}{6}$; wave fragments x and y collide at an angle $\frac{\pi}{3}$. Inputs (entry points, pores) are marked by thin lines, and outputs (exit points, pores) are marked by thick lines. From Ref. [38].

fragment whose velocity vector is the average of the velocity vectors of the colliding wave fragments. The newly formed wave fragment collides with the vesicle's wall just between the output channels and misses both potential exit points. Thus no output is generated (Figure 8.8, $x = 1, y = 0, z = 1$ and $x = 0, y = 1, z = 1$).

For two combinations of inputs – $x = 1, y = 1, z = 0$ and $x = 1, y = 1, z = 1$ – the outcomes depend on the level of illumination. If only wave fragments x and y or all three wave fragments enter the vesicle, they collide and annihilate when the level of illumination is high ϕ_{high} (Figure 8.8a). The fragments merge and form a new wave fragment which hits the output channel, thus generating the output value TRUE, when the level of illumination is low ϕ_{high} (Figure 8.8b). Thus the output channel opposite to the input channel z generates $z\overline{(x \oplus y)}$ when the level of illumination is low, and it generates $\bar{x}\,\bar{y}z$ when the level of illumination is high.

By assigning a constant TRUE to input z, we realize a two-input one-output gate $\langle x, y, \phi \rangle \to \langle \chi(\phi)z\overline{(x \oplus y)} + \overline{\chi(\phi)}\bar{x}\,\bar{y}z\rangle$.

8.7
Binary Adder

We simulate the interactions of wave fragments in BZ vesicles using a two-dimensional cellular automaton. The cellular automaton is a hexagonal array **L** of cells. A cell is a finite-state machine which updates its states in discrete time depending on the states of its six closest neighbors. All cells update their states simultaneously.

8.7 Binary Adder

A cell takes eight states from the set $\mathbf{Q} = \{\circ, -, \searrow, \swarrow, \leftarrow, \nwarrow, \nearrow, \rightarrow\}$. They are the resting state ($\circ$), when the cell is ready to be excited; the refractory state ($-$), when the cell does not react to the states of its neighbors; and six excited states representing wave fragments. The states representing wave fragments are coded as $\searrow$ (southeast traveling wave), $\swarrow$ (southwest traveling wave), $\leftarrow$ (westward traveling wave), $\nwarrow$ (northwest traveling wave), $\nearrow$ (northeast traveling wave), and $\rightarrow$ (eastward traveling wave). The excited state of a cell indicates what type of excitation wave fragment is leaving the cell.

Let $s : \mathbf{L} \times \mathbf{Q} \rightarrow \{0, 1\}$ be defined as follows. For every x with neighborhood $u(x) = \{x_0, x_1, \ldots, x_5\}$ (Figure 8.4): $s(x_0)^t = 1$ if $x_0^t = \searrow$ and $s(x_0)^t = 0$ otherwise; $s(x_1)^t = 1$ if $x_1^t = \swarrow$ and $s(x_1)^t = 0$ otherwise; $\ldots$; $s(x_5)^t = 1$ if $x_5^t = \rightarrow$ and $s(x_5)^t = 0$ otherwise. The cell-state transition rule is defined as follows:

$$x^{t+1} = \begin{cases} \searrow, & \text{if } x^t = \circ \text{ and } s(u(x)^t) \in \{(100000), (010001), (110001)\} \\ \swarrow, & \text{if } x^t = \circ \text{ and } s(u(x)^t) \in \{(010000), (101000), (111000)\} \\ \leftarrow, & \text{if } x^t = \circ \text{ and } s(u(x)^t) \in \{(001000), (010100), (011100)\} \\ \nwarrow, & \text{if } x^t = \circ \text{ and } s(u(x)^t) \in \{(000100), (001010), (001110)\} \\ \nearrow, & \text{if } x^t = \circ \text{ and } s(u(x)^t) \in \{(000010), (000101), (000111)\} \\ \rightarrow, & \text{if } x^t = \circ \text{ and } s(u(x)^t) \in \{(000001), (100010), (100011)\} \\ -, & \text{if } x^t \in \{\searrow, \swarrow, \leftarrow, \nwarrow, \nearrow, \rightarrow\} \\ \circ, & \text{otherwise} \end{cases} \quad (8.1)$$

An example of the cellular automaton evolution from an initial random configuration is shown in Figure 8.9. In case of absorbing boundary conditions, any random initial configurations lead to an "empty" global configuration where all cells are in the resting state. This is because wave fragments either annihilate each other (in most local configurations) or produce fewer new wave fragments than those involved in a collision.

Examples of the collision scenarios of Figure 8.5 represented in cellular automaton configurations are shown in Figure 8.10. The cellular automaton can be used as a "rapid prototyping" tool to design logical circuits in BZ vesicle arrays.

A binary adder based on BZ vesicles uses collisions between propagating wave fragments to perform the addition of three 1-bit binary numbers x, y, and z. The signal z depicts C_{in}. There are many versions of particular implementations of a full 1-bit adder. We adopt the most common one, where the adder outputs the sum of the signals $S = (x \oplus y) \oplus z$ (Figures 8.11 and 8.13a) and carry out value $C_{\text{out}} = xy + z(x \oplus y)$ (Figures 8.12 and 8.13b).

The positions of inputs in Figures 8.11 and 8.12 are shown by thick solid circles. If an input equals FALSE, the circle contains only a central dot; if the input equals TRUE, the circle contains a vector indicating the sites state. Results of the sum circuit is represented by a wave fragment traveling southeast, while the output of the carry out circuit is represented by a wave fragment traveling east. The circuit calculating S is packed in an array of 6×10 BZ vesicles, and the circuit calculating C_{out} occupies a subarray of 7×13 BZ vesicles. Let us discuss the functioning of circuit S (Figures 8.11) in detail.

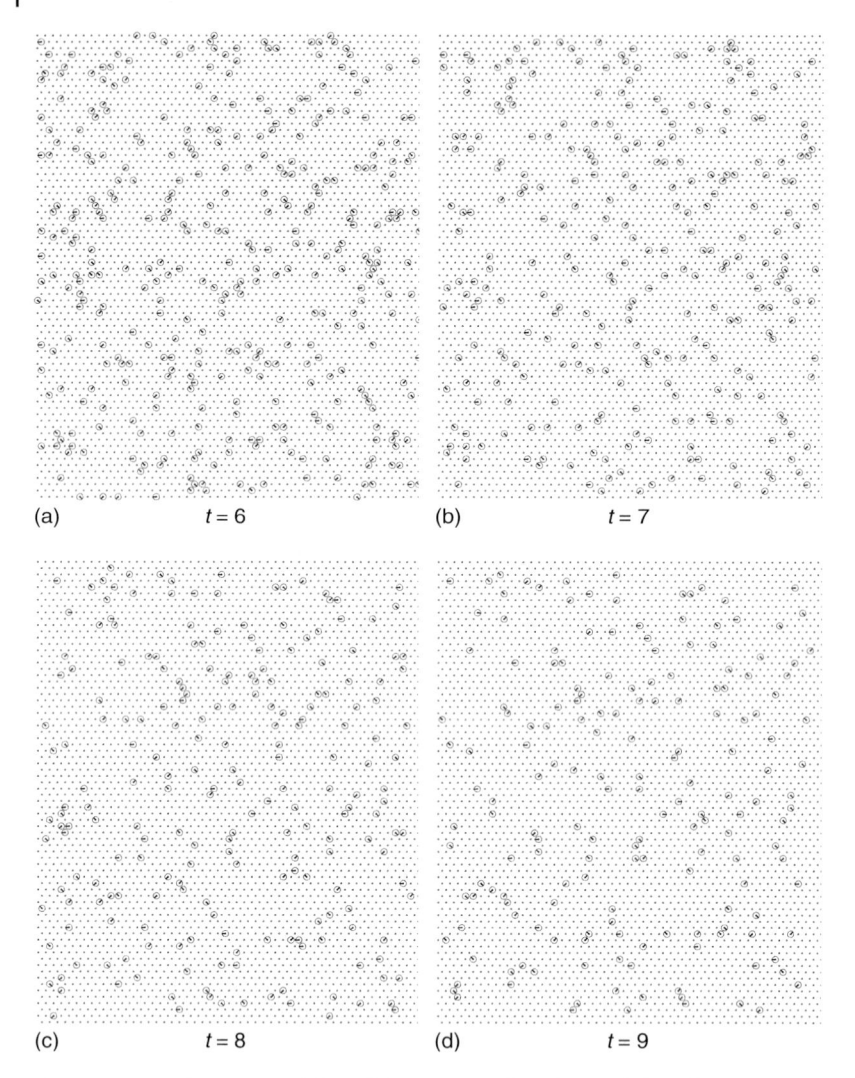

Figure 8.9 Configurations of the cellular automaton from an initially random configuration: a cell gets six wave-fragment states with probability 0.1 and gets a resting state otherwise. Boundaries are absorbing. Resting cell states are shown by small circles, and the refractory states are not shown. Cell states representing wave fragments are shown by large circles with vectors pointed southeast, southwest, west, northwest, northeast, and east. From Ref. [36].

8.7.1
Sum

If all three inputs are FALSE, no waves are initiated in sites w_3, w_5, and w_7 (Figures 8.11a and 8.13a). The wave fragments representing a constant TRUE are placed in sites w_1 and w_2 (southeast traveling wave fragments) and w_4 and w_6 (wave

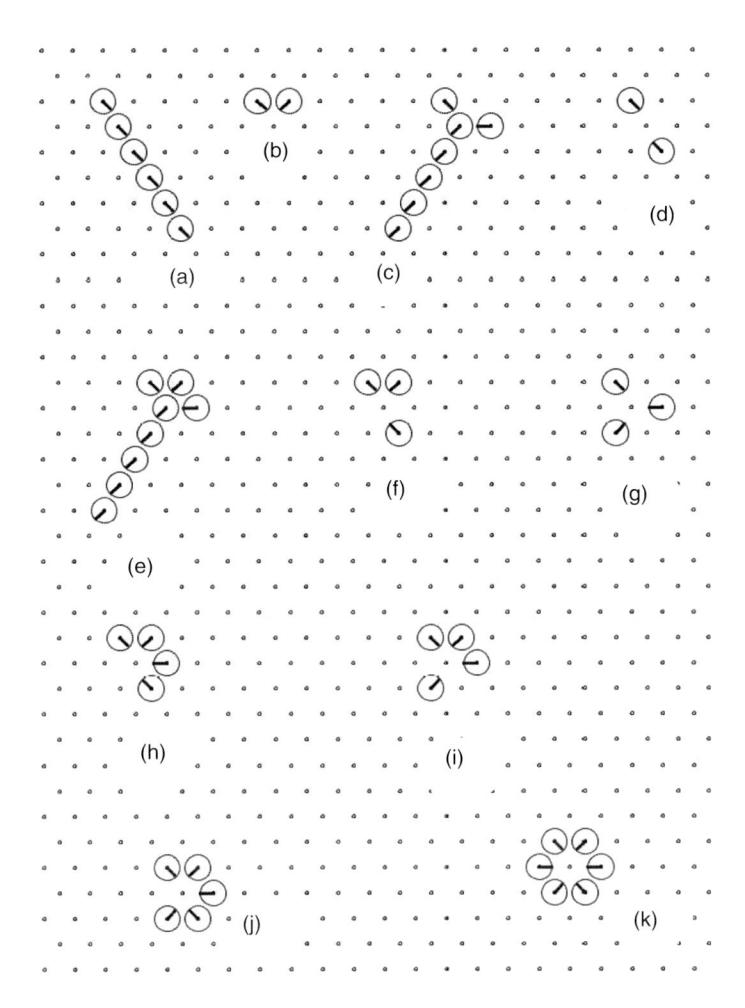

Figure 8.10 Traces of the 7 steps of cellular automaton development with 11 initial subconfigurations corresponding to the types of wave collisions in Figure 8.5. Cell states representing wave fragments are shown by large circles with vectors pointed southeast, southwest, west, northwest, northeast, and east. To show the trace of the cellular automaton development, we do not execute transition to the refractory state but leave a cell in its wave-fragment state instead. Such an approach is appropriate when the signals trajectories do not cross each other [36].

fragments traveling east). The wave fragment w_2 collides with wave fragment w_4, and both wave fragments annihilate (step three of cellular automaton simulation, Figure 8.11a). Wave fragments w_1 and w_6 travel a bit further but collide with each other and annihilate during the fifth step of the automaton simulation. All outputs are therefore FALSE.

If the carry in value z is TRUE and other input values are FALSE (Figures 8.11a and 8.13b), then northeast traveling wave w_7 annihilates the east traveling wave w_6.

(a) $x = 0, y = 0, z = 0$ (b) $x = 0, y = 0, z = 1$ (c) $x = 0, y = 1, z = 0$

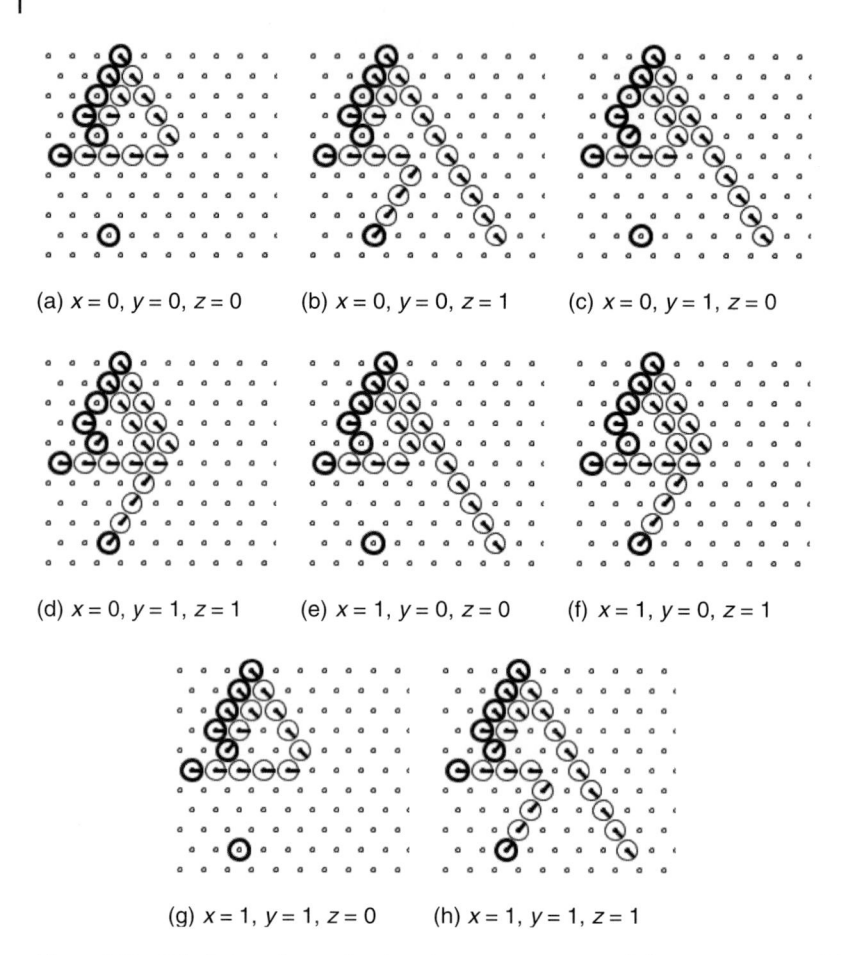

(d) $x = 0, y = 1, z = 1$ (e) $x = 1, y = 0, z = 0$ (f) $x = 1, y = 0, z = 1$

(g) $x = 1, y = 1, z = 0$ (h) $x = 1, y = 1, z = 1$

Figure 8.11 Calculation of sum $S = (x \oplus y) \oplus z$, $z = c_{in}$, in a cellular automaton model of BZ vesicle hexagonal array. (a)–(h) Traces of cell states, representing wave fragments, for all possible combinations of input values (Figure 8.13). Input cells are shown by circles with thick lines. From Ref. [38].

Therefore, wave w_1 continues traveling southeast, thus representing the TRUE value of $(x \oplus y) \oplus z$.

In situation $(x, y, z) = $ (FALSE, TRUE, FALSE), wave w_5 colliding with wave w_4 (Figures 8.11c and 8.13a), and both wave fragments annihilate. Therefore, the wave fragment w_2 continues undisturbed on its trajectory to the southeast, where it collides with wave fragment w_6. With both waves representing east traveling constant TRUE cancelled wave fragment w_1 reaches an output. Similarly, for input $(x, y, z) = $ (TRUE, FALSE, FALSE), wave fragments w_3 and w_4, and w_2 and w_6 annihilate in collision with each other (Figures 8.11e and 8.13a). Therefore, the wave fragment w_1 travels undisturbed. Other combinations of inputs can be considered in a similar manner.

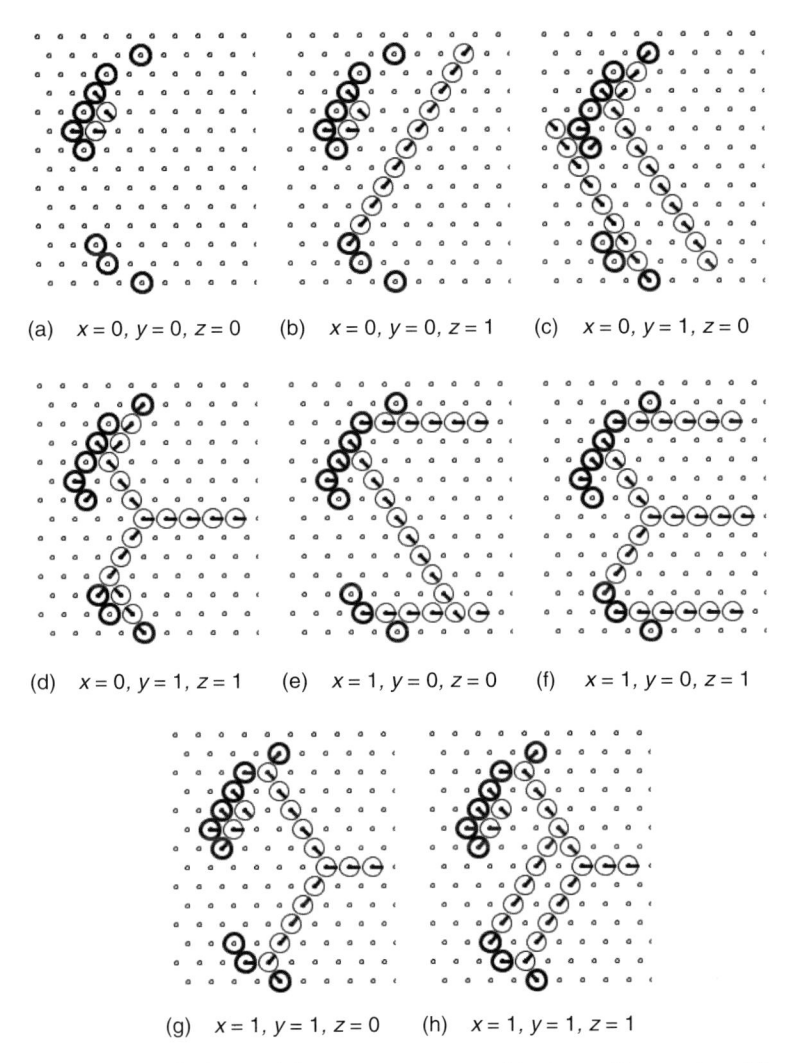

(a) $x = 0, y = 0, z = 0$
(b) $x = 0, y = 0, z = 1$
(c) $x = 0, y = 1, z = 0$
(d) $x = 0, y = 1, z = 1$
(e) $x = 1, y = 0, z = 0$
(f) $x = 1, y = 0, z = 1$
(g) $x = 1, y = 1, z = 0$
(h) $x = 1, y = 1, z = 1$

Figure 8.12 Calculation of the carry out value $C_{out} = xy + z(x \oplus y)$, $z = c_{in}$ in cellular automaton model of B7-vesicle hexagonal array. (a)–(h) Traces of cell states, representing wave fragments, for all possible combinations of input values (Figure 8.13). Input cells are shown by circles with thick lines. Eastward-traveling wave fragments in 3rd and 13th rows in (e,f) are unused by-products of the circuit. From Ref. [38].

8.7.2
Carry Out

Computation starts in a group of sites marked $w_1, \ldots w_4$ in Figure 8.13b: site w_2 represents x, site w_4 represents y, and sites w_1 and w_3 represent constant TRUE. The subcircuits' output is $x \oplus y$. Wave fragments w_1, traveling southeast, and w_3, traveling east, are always present in the system. If only one of the inputs x or y has

192 | *8 Toward Arithmetic Circuits in Subexcitable Chemical Media*

(a) Sum (S)

(b) Carry out (C_{out})

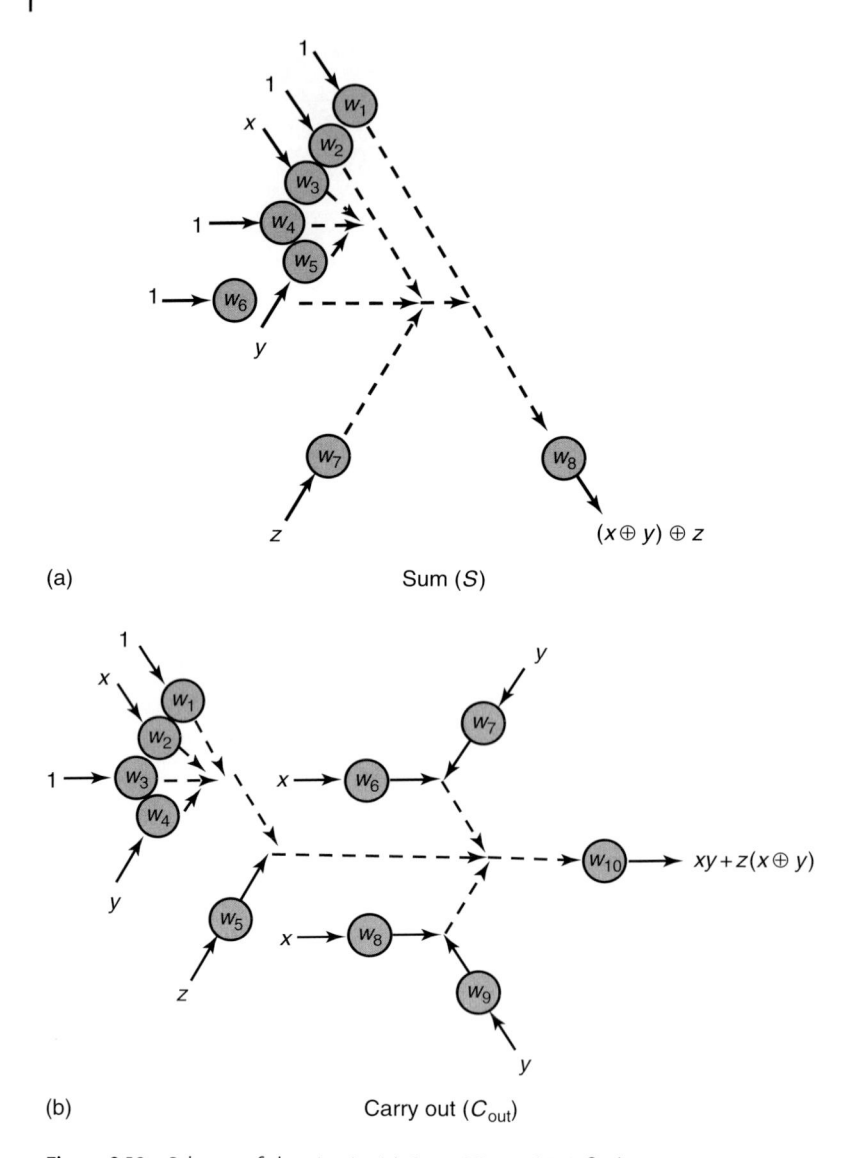

Figure 8.13 Scheme of the circuit. (a) Sum (Figure 8.11) & (b) Carry out (Figure 8.12) [36].

TRUE value, for example, the southeast traveling wavefront w_2 in Figure 8.12e and f, then the wave fragment representing this input collides with the wave fragment w_3 (constant TRUE), and both wave fragments annihilate. If both inputs x and y are TRUE (Figure 8.12g and h), then waves w_2 and w_4 collide with each other and the wave w_3, merge together, and produce a new wave traveling east. This new wave (seen in node (6, 5) in Figure 8.12g and h) collides with wave w_1, and both wave fragments annihilate. Thus, the subcircuit ($w_1, \ldots, w_4$) computes $x \oplus y$.

The output wave fragment of subcircuit $x \oplus y$ collides with the wave fragment w_5, which represents the carry in value z. These wave fragments collide at an angle 120°, therefore they merge and produce a new wave fragment (traveling east) when they collide. See examples in Figure 8.12d and f. Thus an intermediate result $z(x \oplus y)$ is calculated.

Two small subcircuits – w_6 and w_7, and w_8 and w_9 – are arranged symmetrically north and south of the trajectory of the wave fragment which represents $z(x \oplus y)$. Each of the subcircuits (w_6, w_7) and (w_8, w_9) produces xy. The final wave fragment w_{10} is produced only when either only the wave fragment $z(x \oplus y)$ travels east (Figure 8.12d and f), or all three wave fragments – wave fragment xy (output of subcircuits (w_6, w_7)) traveling southeast, wave fragment xy (output of subcircuits (w_8, w_9)) traveling northeast, and wave fragment $z(x \oplus y)$ – collide (Figure 8.12h).

8.8
Regular and Irregular BZ Disc Networks

In a parallel section of the work, we have also investigated two other related configurations of BZ discs as simple analogs of two-dimensional BZ vesicles: first, uniform-sized orthogonal disc networks connected with one of two different pore sizes; second, BZ disc networks with irregular diameters, connection angles, and pore efficiencies. In the first configuration, we could create elementary logic gates and then combine those gates to create more complex circuits analogous to their electronic counterparts. In the second configuration, more flexible designs with variable disc and pore sizes and connection angles lead to more compact devices. We illustrate these findings by creating a 1-bit half-adder circuit (HA) first with combinations of logic gates in an orthogonal network and then in an irregular network arrangement [39].

As described in Section 8.3, we have simulated the BZ reaction with a two-variable version of the Oregonator adapted for photosensitive modulation of the Ru-catalyzed reaction. Variables u and v are the local, instantaneous dimensionless concentrations of the bromous acid autocatalyst activator $HBrO_2$ and the oxidized form of the catalyst inhibitor $Ru(bpy)_3^{3+}$, respectively. Φ symbolizes the rate of bromide production which is proportional to the applied light intensity. Bromide Br^- is an inhibitor of the Ru-catalyzed reaction, therefore excitation can be modulated by light intensity; high-intensity light inhibits the reaction. Dependent on the rate constant and reagent concentration, ϵ represents the ratio of the timescales of the two variables u and v, and q is a scaling factor dependent on the reaction rates alone. The diffusion coefficients D_u and D_v of u and v were set to unity and zero, respectively. The coefficient D_v is set to zero because it is assumed that the diffusion of the catalyst is limited.

The parameters of the Oregonator model were as follows: ratio of the timescale for variables u and v, $\epsilon = 0.022$; propagation scaling factor $q = 0.0002$; stoichiometric coefficient $f = 1.4$; spatial step $\Delta x = 0.25$; time step $\Delta t = 0.001$; $D_u = 0$ and $D_v = 1.0$ the activator diffusion coefficient and inhibitor diffusion Coefficient, respectively. Φ varies between two levels, subexcited ($L1$) and inhibited ($L2$), $\phi_{L1} = 0.076$, $\phi_{L2} = 0.209$. Values of u & v represent the activator concentration of $HBrO_2$ and inhibitor $Ru(bpy)_3^{3+}$, respectively.

Networks of discs were created by mapping two different ϕ values (proportional to the light intensity) onto a rectangle of a homogeneous simulation substrate. The excitation levels $L1 \rightarrow L2$ relate to the partially active disc interiors and the non-active substrate.

Discs have a radius of 28 simulation points (SPs) and are always separated by a single SP-wide boundary layer. Connection apertures between discs are created by superimposing another small *link* disc at the point of connection, typically of 2–6 SP radius; SPs have a 1 : 1 mapping with on screen pixels. The reagent concentrations are represented by red and blue color mapping; the activator u is proportional to the red level, and the inhibitor v proportional to blue. The color gradation is automatically calibrated to minimum and maximum levels of concentration over the simulation matrix. The background illumination is monochromatically calibrated in the same manner, proportional to ϕ, with white areas inhibitory and dark areas excited. Wave-fragment flow is represented by a series of superimposed time-lapse images (unless stated otherwise), and the time lapse is 50 simulation steps. To improve clarity, only the activator (u) wavefront progression is recorded.

8.8.1
Elementary Logic Gates

Electronic logical gates form the building blocks of more complex digital circuitry forming the foundations of complex, high-level components such as microprocessors. Although we do not envisage creating traditional von Neumann architecture microprocessors in BZ vesicles, the ability to create simple logic gates with BZ vesicles demonstrates that (like electronics) the medium and architecture are *capable* of such processing. Logic gates and composite circuits of logic gates have been created several times before using the BZ substrate, see for example Refs [40–43]. Here, we illustrate a selection of key gates that can be created using only interconnected BZ discs. Figure 8.14 illustrates the operation of the most elementary gate, the inverter (NOT gate). The circuit operation starts with the simultaneous application of the circuit input (left-most disc) in conjunction with the source (permanent logical "1") input (top-most disc). The circuit operation terminates by observing the output disc (lower most disc) at a time when either result state would be present. If the progression of a wave fragment through a disc is considered as one step, then the output disc will hold a valid result after two steps from the application of the source input. Incorporating a parallel unmodulated source signal that travels from the output to the input could also be used to indicate the point at which the output discs holds a valid output. In this case, this would simply consist of three serial discs.

The operation of an AND gate and the inversion, the NAND gate, are shown in Figures 8.15 and 8.16, respectively. The result of a wave collision in the NOT

(a) 0→1   (b) 1→0

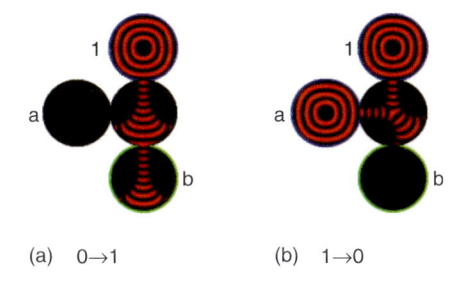

Figure 8.14 Inverter gate ($a = \bar{b}$) where the input (*a*) is center left (blue ring), bottom disc (green ring) is the output (*b*), and a supply, or source logical "1", is the top-most disc (blue ring). (a) $a = 0$. The gate initiates with the source pulse in the top disc. In this case, no signal is present at the input disc and the source pulse travels to the output disc (bottom), resulting in a logical 1 output (1→ 0). (b) $a = 1$. Again, the source pulse travels from top to bottom, but in this case a collision with a signal present on the input disc produces a logical 0 output. (0 → 1).

(a) 01→0   (b) 10→0   (c) 11→1

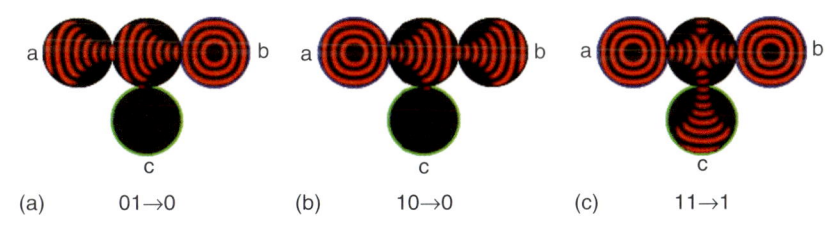

Figure 8.15 Two-input AND gate ($c = a \bullet b$) where inputs *a*, *b* are top-left and right discs (blue rings) and output *c* is the bottom central disc (green ring). (a) $(a, b)(0, 1)$. A wave from input *b* propagates uninterrupted and terminates in the opposing input disc *a*. (b) $(a, b)(1, 0)$. Likewise, a wave from input *b* propagates uninterrupted and terminates in the input disc *b*. (c) $(a, b)(1, 1)$. Waves from both input discs *a* and *b* collide in the central disc and eject two perpendicular waves, one of which propagates into the output disc (*c*).

gate was exploited to deflect and extinguish the source wave into the disc edge, whereas in the AND gate the collision between the two inputs results in two perpendicular fragments, one of which develops in the output cell to produce the result.

A NAND gate can be created by combining the NOT gate and the AND gate (Figure 8.16). The NAND gates are known as *universal* gates since all other gates can be created from arrangements of NAND gates alone. The NOT gate (Figure 8.14) is integrated below the AND gate in the lower row (Figure 8.15), where the activity of a horizontal source signal inverts the vertical output.

The OR gate is used to detect the presence of one or more signals. A logical "1" on any input results in an output (Figure 8.17). Common to all these gates, the output value of a logical "1" or "0" as indicated by the presence or absence of a wave is valid only at a specific point in the development and, in these instances approximated to be proportional to time. For example, the OR gate output is sampled after a wave

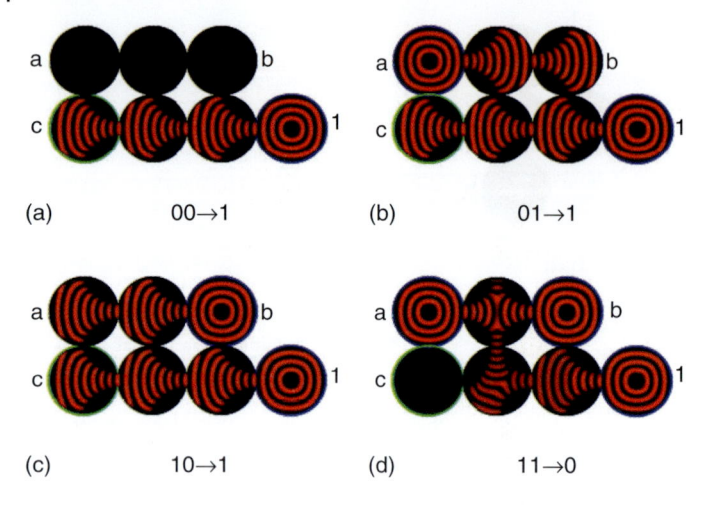

Figure 8.16 Two-input NAND gate ($c = \overline{a \cdot b}$) where inputs a, b are the top left and right discs (blue rings) and output c is the bottom left disc (green ring). The source input is located on the bottom right (blue ring). Its operation is identical to the AND gate (Figure 8.15) but with an inverter (Figure 8.14) integrated along the bottom disc row. (a)–(c) The source input provides a logical "1" output for all input combinations other than $(a, b)(1, 1)$. (d) $(a, b)(1, 1)$. Output from the AND gate portion of the gate collides with the source input creating a logical "0" output.

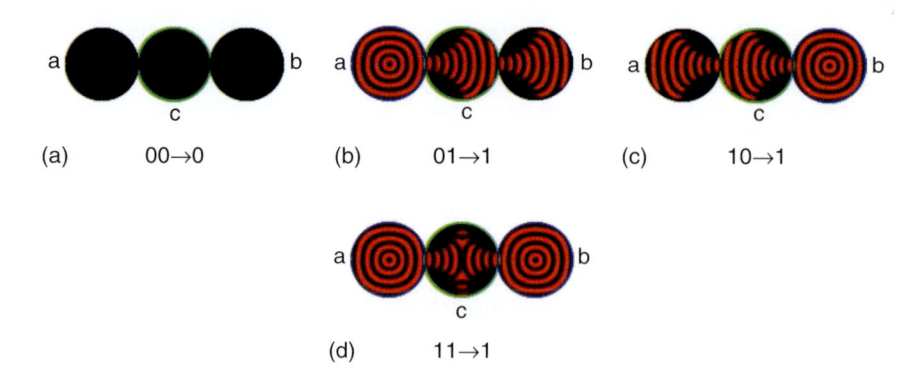

Figure 8.17 Two-input OR gate ($c = a + b$) where inputs a, b are the left and right discs (blue rings) and the central disc is the output c (green ring).

fragment has traveled by one disc unit ($t_d = 1$). Therefore, the annihilation of the $(a, b)(1, 1)$ case and the continuation of the waves into opposing input cells for cases $(a, b)(0, 1)$ and $(a, b)(1, 0)$ do not affect the outcome.

The XOR gate is used to signal a difference between signals, producing an output when the inputs alternate regardless of the composition of the difference. Figure 8.19 illustrates the BZ disc implementation along with the inversion NXOR in Figure 8.18.

(a) 00→1 (b) 01→0 (c) 10→0

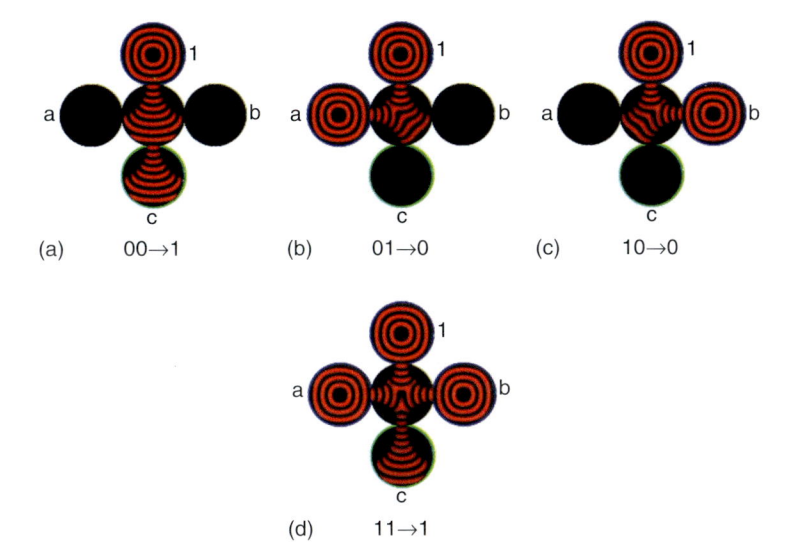

(d) 11→1

Figure 8.18 Two-input NXOR gate ($c = \overline{a \oplus b}$) where inputs are middle row left and right (blue rings), the output disc is center bottom (green ring), and the source input is at center top (blue ring). The OR structure (Figure 8.17) is repeated in the central row, the output of which deflects the source input from the center top disc, and the result is an NXOR gate. The output of the NXOR can then be inverted to create a XOR gate (Figure 8.19).

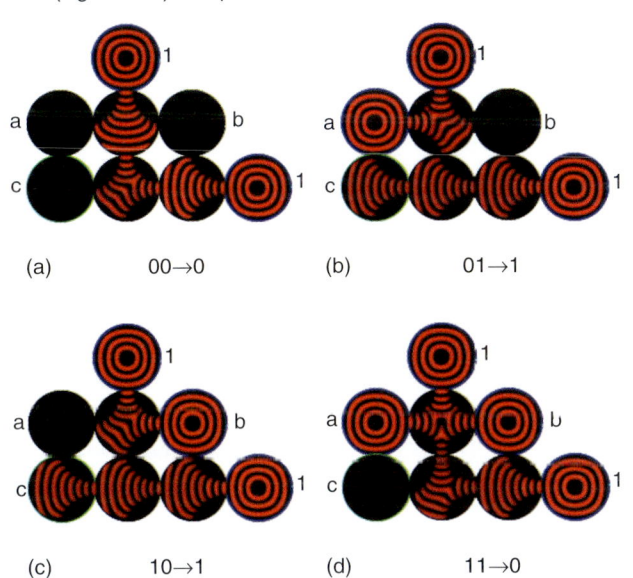

(a) 00→0 (b) 01→1

(c) 10→1 (d) 11→0

Figure 8.19 Two-input XOR gate ($c = a \oplus b$) where inputs are middle row left and right (blue rings), the output disc is bottom left (green ring) and there are two source inputs, center top and bottom right (blue rings). The gate is an extension of inverting the NXOR gate (Figure 8.18).

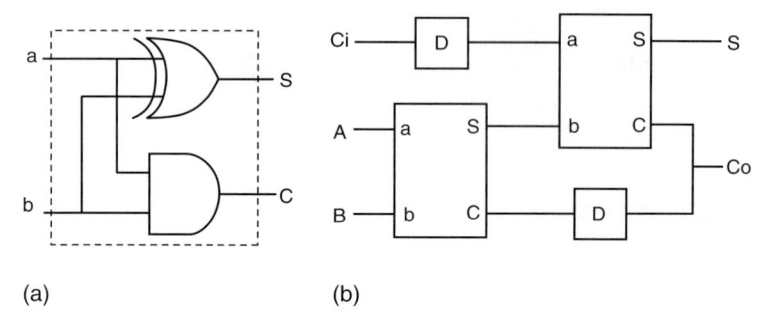

Figure 8.20 (a) One-bit half adder. The circuit comprises two Outputs: the sum (S) derived from the XOR gate ($S = a \oplus b$), and carry (C) derived from the AND gate ($C = a \bullet b$). (b) One-bit full adder. Two half adders can be cascaded together to create a full 1-bit adder ($S = C_i \oplus (a \oplus b)$, $C_o = a \bullet b + (a \oplus b)$). In turn, full 1-bit adders can be cascaded to create an n-bit adder. The D blocks represent signal delays required to synchronize signal pulses from different sources.

8.8.2
Half Adder

The half adder is a subsystem used in binary addition circuits. The half adder adds two binary digits and, when connected with another half adder, creates a full 1-bit adder. The 1-bit adders can in turn be connected together to make n-bit adders (Figure 8.20). A half adder can be constructed from a combination of two logic gates, the XOR and the AND gate. There are two inputs (a and b) and two outputs (S and C); the binary sum (S) of a and b is achieved by the XOR gate ($S = a \oplus b$) and inability of the configuration (overflow) to present the $1 + 1$ input is achieved with a carry (C) output ($C = a \bullet b$).

A 1-bit half adder created from BZ discs can also be constructed from connecting a BZ disc AND gate and XOR gate (Section 8.1). Figure 8.21 shows the BZ disc conjunction for the half-adder circuit. The input a needs to be repeated on the other side of input b for this circuit to work. This is necessary to overcome the *signal passing problem*, a universal problem for systems where signals propagate along specific planular channels. There are two ways to overcome this problem: either add identity to the signals in such a way that signals can share the medium, or share the medium at different times. How two or more waves could be identified and share the same space in this BZ system remains unclear because of the diffusive nature of the reaction. However, sharing a channel medium in time[1]. is possible if the time difference between the signals is large enough to prevent one refractory tail from extinguishing the other. Figure 8.22 illustrates one such temporal separation strategy, where signal a passes over signal b but becomes shifted in time. The circuit operates with two types of apertures, one that creates a narrow beam (type J1) wav, and the other that creates a broad beam (type J2) wave.

---

1) In communication systems, this is known as time division multiplexing (TDM).

(a) $ab(0, 0) \rightarrow S = 0, C = 0$

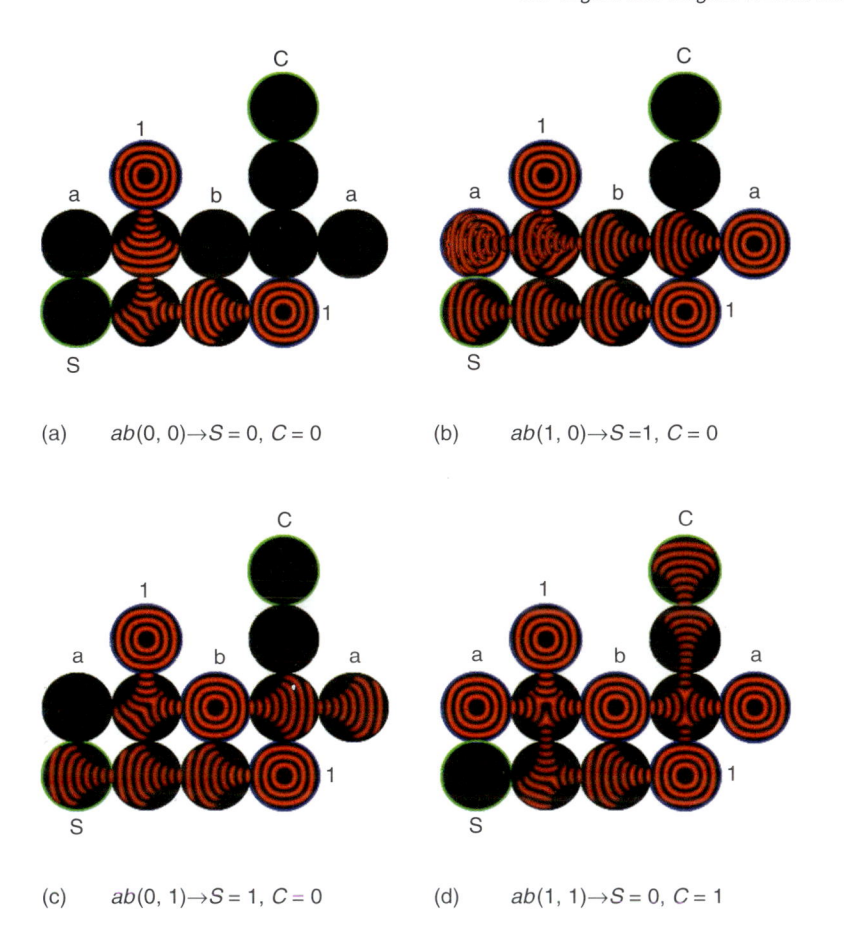

(b) $ab(1, 0) \rightarrow S = 1, C = 0$

(c) $ab(0, 1) \rightarrow S = 1, C = 0$

(d) $ab(1, 1) \rightarrow S = 0, C = 1$

Figure 8.21 Half-adder circuit ($S = a \oplus b$, $C = a \bullet b$) where the inputs are located along the central row, a far left, b central and then a repeated far right (blue rings), two source inputs located top left and bottom right (blue ring). The output cell is at bottom left (green ring). The circuit is constructed by combining an XOR gate (Figure 8.19) (bottom left) and the AND gate (Figure 8.15) (top right). The issue of the signal passing problem is obviated by replicating one of the inputs "a" (see *signal passing problem* below).

Signals a and b travel from bottom to top, with a on the left and b on the right. The signal a is split at the junction to the first disc, and a fragment a' travels horizontally toward b. Meanwhile, b is already traversing the first disc and has progressed into the final disc before a' crosses the b path, allowing a' to cross b. A time shift t_d now exists between a, b and a', b so any further processing between a' and b must delay b by t_d. This strategy relies on allowing sufficient time for the refractory tail of signal b to have a negligible effect on a. If the signals are not sufficiently separated, then b will extinguish a', which can, in another context, be used as another logical construction (Figure 8.23).

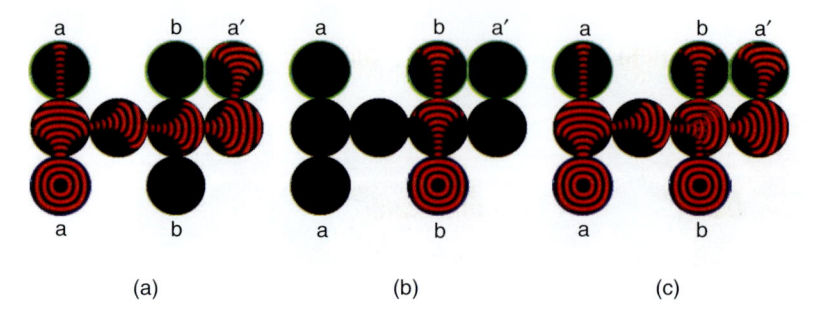

Figure 8.22 Signal passing problem resolved by sharing the same space (a *cross road*) but at different times. Signals travel from bottom to top. (a) The independent path of signal a. The signal is split by using a broad-band aperture at the junction between the first and second disc to create a'. (b) The independent path of signal b. (c) a' crossing b.

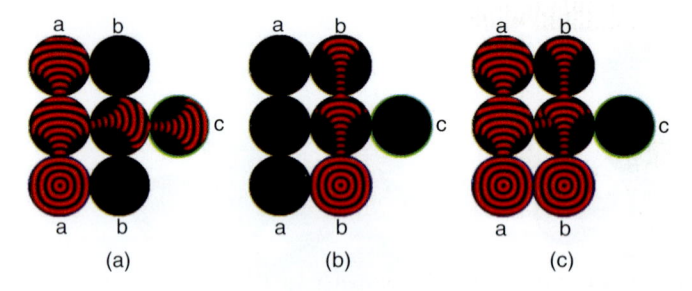

Figure 8.23 Signal passing gate. When the time difference between two signals trying to cross is very small, the refractory tail of one signal will cause extinction in the other. This feature can be exploited to create another logic gate. The output in this gate produces $c = a \cdot \bar{b}$. (a) The independent path of signal a. Signal a is split by using a broad-band aperture at the junction between the first and second disc to create a'. (b) The independent path of signal b. (c) The refractory tail of b extinguishing signal a'.

Venturing into three dimensions (3D) resolves the signal passing problem altogether, allowing signals to be routed vertically. At this stage, only two-dimensional (2D) structures of discs have been explored, but these are approximations of our target computation node, which is a 3D BZ vesicle. In this current 2D perspective, overcoming the signal passing problem via interconnecting linking layers above and below planular 2D functions seems the next logical step analogous to a methodology used in two-layer and multilayer electronic circuit boards.

Another specific solution for the half-adder circuit that removes the need to repeat one of the inputs is possible if all the signal modulation techniques are exploited: disc connection geometry, disc size, and aperture efficacy. Figure 8.24 demonstrates a half-adder design where most of the processing occurs in one central reactor disc. The central disc achieves the AND function (Figure 8.24c) and the XOR function (Figure 8.24b and c). By considering the central disc principally in terms of an AND gate, the XOR function can be considered as being derived from

(a) $ab\,(1, 0) \to S = 1, C = 0$ (b) $ab\,(0, 1) \to S = 1, C = 0$

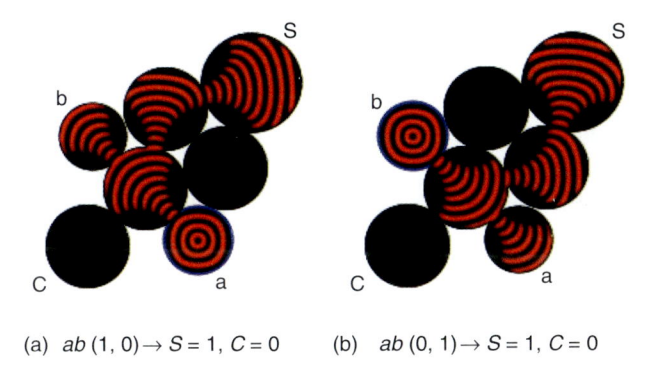

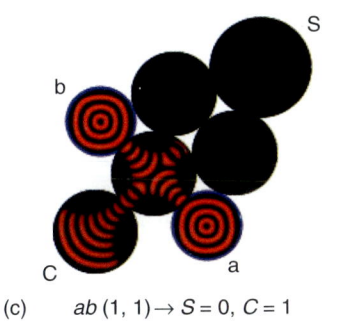

(c) $ab\,(1, 1) \to S = 0, C = 1$

Figure 8.24 Composite half-adder circuit ($S = a \oplus b$, $C = a \bullet b$) where inputs and outputs are all connected to a central reactor disc which can achieve both the AND and XOR function. The two *half* outputs from the XOR operation are recombined with an OR operation with additional discs in the top right. The circuit uses all three methods of modulation: connection angle, disc size, and aperture efficacy.

the AND gate response to input sets $(a, b)(0, 1)$ and $(a, b)(1, 0)$, the outputs of which are curved around into an OR gate in the S output disc creating the XOR function.

8.9
Memory Cells with BZ Discs

Memory is an essential facet of the adaptive behavior in both Nature and synthetic computation. It permits animals and machines to build an internal state independent of the current external world state. In this section we present a simple 1-bit volatile read/write memory cell constructed entirely with BZ discs. The cell design is independent of but similar to the previous designs [44, 45] insofar as the existence or absence of a rotating wave represents the setting or resetting of one bit of information.

When two BZ waves progress in opposite directions around an enclosed channel, loop, or ring of connected discs, at some point the two opposing wavefronts will meet and are always mutually annihilated (Figure 8.26a). Nevertheless, if a

(a) Pass a→b   (b) Block b→a

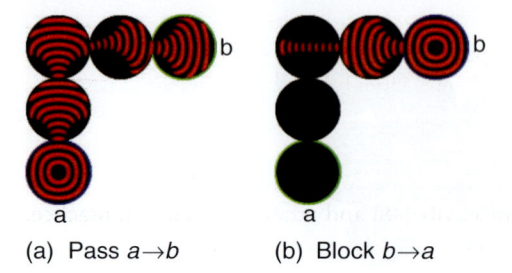

Figure 8.25 BZ disc diode. (a) Signal propagates from the bottom to the top ($a \to b$). A broad-band (type J2) aperture at the second (right angle) junction connection permits the signal to expand horizontally toward b. (b) Signal propagates from right to left ($b \to a$). Conversely, the narrow-band junction at the second (right angle) junction prohibits propagation toward a.

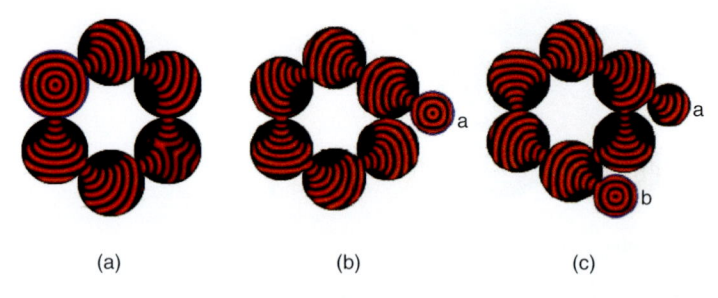

(a)   (b)   (c)

Figure 8.26 Memory cell development. (a) Hexagonal connected loop of discs perturbing the medium in any of the cells always leads to complete wave extinction as counterclockwise and clockwise wave fronts meet at some point during the path. (b) Insertion of an angled *diode* junction inserts a unidirectional (counterclockwise) wave that rotates indefinitely (in simulation). (b) Addition of another opposing diode junction provides for the insertion of a unidirectional (clockwise) wave. Insertion of a wave from either input disc (a/b) can be cancelled by inserting another asynchronous unidirectional wave from the other opposing disc. Inserting more than one wave is not sustainable and always results in a reduction to one wavefront.

unidirectional wave can be inserted into the loop, then that wavefront will rotate around the loop indefinitely[2]. (Figure 8.26). Furthermore, the rotating wave can be terminated by the injection of another asynchronous wave rotating in the opposite direction (Figure 8.26c). Opposing inputs into a loop are analogous to a memory *set* or *reset*. Reading the state of the cell without changing the state can be achieved by connecting another output node where a stream of pulses can be directed to modulate other circuits [46].

The loop and a unidirectional gate (diode) are the two key constructions of this type of memory cell. Unidirectional gates in BZ media have previously been created by exploiting asymmetric geometries or chemistry on either side of a barrier [47]. An alternative design is possible, however, using discs connected with different

2) For as long as the chemical reagents can sustain the reaction.

apertures. Figure 8.25 illustrates a diode constructed from a right-angle junction connected by a broad-band (type J2) aperture to a vertical column and by a narrow beam (type J1) aperture to a horizontal row. Signal flow is possible only from the bottom to the top ($a \rightarrow b$) because of the asymmetric apertures in the right-angle connecting the disc. The operation relies on the relationship between the wave expansion and the angle of the connection. Fine control of the wave beam would, in theory, allow other angles of connectivity [48] and other functions. In practice, fine control of wave diffusion is, however, difficult to achieve and hence we have restricted our choice between just two types.

As the rotating wave progresses around the loop in the memory cell illustrated in Figure 8.26, the opposing input cell also inadvertently becomes an output cell. This may be undesirable in some designs but can be easily resolved by adding another pair of diode junctions to the circuit. Figure 8.27 shows such a design, where the opposing inputs are not affected by the opposing input.

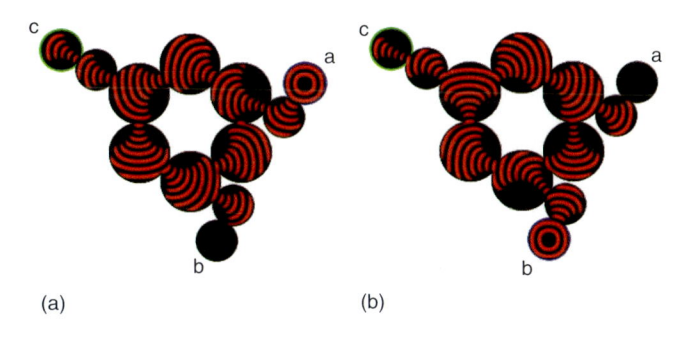

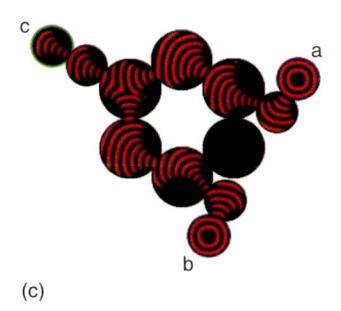

Figure 8.27 Memory cell with additional diodes on the cell inputs. Two additional angled diode junctions are added to each of the input discs (*a* & *b*). This prevents a reverse wave flowback down either of the inputs. An example output disc is also connected (top left). (a) Wave insertion at (top right) *a* input node results in a persistent counterclockwise wave. Reverse wave flow down the opposing (bottom) input is blocked by an angled diode junction. (b) Wave insertion at (bottom) *b* input node results in a persistent clockwise wave. Reverse wave flow down the opposing (top right) *a* input is likewise blocked by an angled diode junction. (c) Simultaneous *a* & *b* inputs produce one output pulse (*c*) and annihilate wave rotation.

8.10
Conclusion

In the BZ medium in a subexcitable state, localized traveling excitation waves are formed. We interpret these localizations as quanta of information, which are values of logical variables. When two or more localizations collide, they annihilate or form new localizations. We interpret post-collision trajectories of the localizations as the results of a computation. We demonstrated that, by colliding wave fragments in an encapsulated, excitable chemical medium, we can realize a number of logical gates. We showed that by changing the illumination of the chemical medium we could switch between different outcomes of the computation. Thus, we were able to realize a polymorphic logical gate which could execute either function XNOR or NOR depending on the level of illumination. The NOR gate is a universal gate, and, thus, as a by-product we demonstrated the computational universality of the BZ medium when in a subexcitable state.

Implementation of universal logical gates is a noble task; however, it is just a first step toward designing general-purpose computers based on assemblies of BZ vesicles. Assembling gates into a logical or arithmetical circuit is a paramount task, particularly when signals are transmitted via such sensitive and intrinsically unstable dissipative localizations. We categorized all interactions between wave fragments in the idealized BZ vesicles and constructed a cellular automaton, whose cell-state transition rules represent the dynamics of the interaction between excitation wave fragments. Using this cellular automaton, we designed a full 1-bit binary adder.

The next step of our studies would be the implementation of these theoretical designs in chemical laboratory experiments. Some experimental work has been done. For example, Kitahata and colleagues provided experimental evidence of excitation waves traveling inside a tiny sphere and even movement of a BZ vesicle partly controlled by wave propagation [49, 50]. Also, the BZ reaction mixture can be encapsulated via dispersion of the mixture in a water-in-oil microemulsion with a surfactant: BZ reagents become enclosed in a monolayer of anionic surfactant. "Unlimited" energy supply from the solution surrounding BZ vesicles enables the reactivity of BZ droplets to be sustained for extended periods [51]. And, finally, Gorecki's team demonstrated that the BZ mixture can be successfully encapsulated in a lipid membrane, and BZ vesicles can be formed. Moreover, there is transition of excitation between BZ vesicles in contact with each other [37].

Acknowledgments

The work is part of the European project 248992 funded under 7th FWP (Seventh Framework Programme) FET Proactive 3: Bio-Chemistry-Based Information Technology CHEM-IT (ICT-2009.8.3). The authors wish to thank the project coordinator Peter Dittrich and the project partners Jerzy Gorecki and Klaus-Peter Zauner for their encouragement and useful discussions.

References

1. Hjelmfelt, A., Weinberger, E.D., and Ross, J. (1991) Chemical implementation of neural networks and Turing machines. *Proc. Natl. Acad. Sci. U.S.A.*, **88**, 10983–10987.
2. Hjelmfelt, A., Weinberger, E.D., and Ross, J. (1992) Chemical implementation of finite-state machines. *Proc. Natl. Acad. Sci. U.S.A.*, **89**, 383–387.
3. Hjelmfelt, A. and Ross, J. (1993) Mass-coupled chemical systems with computational properties. *J. Phys. Chem.*, **97**, 7988–7992.
4. Hjelmfelt, A., Schneider, F.W., and Ross, J. (1993) Pattern recognition in coupled chemical kinetic systems. *Science*, **260**, 335–337.
5. Hjelmfelt, A. and Ross, J. (1995) Implementation of logic functions and computations by chemical kinetics. *Physica D*, **84**, 180–193.
6. Lebender, D. and Schneider, F.W. (1994) Logical gates using a nonlinear chemical reaction. *J. Phys. Chem.*, **98**, 7533–7537.
7. Tóth, A., Gáspár, V., and Showalter, K. (1994) Propagation of chemical waves through capillary tubes. *J. Phys. Chem.*, **98**, 522–531.
8. Tóth, A. and Showalter, K. (1995) Logic gates in excitable media. *J. Chem. Phys.*, **103**, 2058–2066.
9. Sielewiesiuk, J. and Górecki, J. (2001) Logical functions of a cross junction of excitable chemical media. *J. Phys. Chem.*, **A105**, 8189.
10. Motoike, I.N. and Yoshikawa, K. (2003) Information operations with multiple pulses on an excitable field. *Chaos Solitons Fractals*, **17**, 455–461.
11. Górecki, J., Górecka, J.N., and Igarashi, Y. (2009) Information processing with structured excitable medium. *Nat. Comput.*, **8**, 473–492.
12. Yoshikawa, K., Motoike, I.M., Ichino, T., Yamaguchi, T., Igarashi, Y., Gorecki, J., and Gorecka, J.N. (2009) Basic information processing operations with pulses of excitation in a reaction-diffusion system. *Int. J. Unconv. Comput.*, **5**, 3–37.
13. Górecki, J. and Górecka, J.N. (2006) Multi-argument logical operations performed with excitable chemical medium. *J. Chem. Phys.*, **124**, 084101.
14. Górecki, J. and Górecka, J.N. (2006) Information processing with chemical excitations – from instant machines to an artificial chemical brain. *Int. J. Unconv. Comput.*, **2**, 321–336.
15. Górecki, J., Yoshikawa, K., and Igarashi, Y. (2003) On chemical reactors that can count. *J. Phys. Chem. A*, **107**, 1664–1669.
16. Górecka, J.N. and Górecki, J. (2003) T-shaped coincidence detector as a band filter of chemical signal frequency. *Phys. Rev. E*, **67**, 067203.
17. Górecki, J., Górecka, J.N., Yoshikawa, K., Igarashi, Y., and Nagahara, H. (2005) Sensing the distance to a source of periodic oscillations in a nonlinear chemical medium with the output information coded in frequency of excitation pulses. *Phys. Rev. E*, **72**, 046201.
18. Yoshikawa, K., Nagahara, H., Ichino, T., Gorecki, J., Gorecka, J.N., and Igarashi, Y. (2009) On chemical methods of direction and distance sensing. *Int. J. Unconv. Comput.*, **5**, 53–65.
19. Adamatzky, A. (ed.) (2003) *Collision-Based Computing*, Springer.
20. Berlekamp, E.R., Conway, J.H., and Guy, R.L. (1982 *Winning Ways for your Mathematical Plays*, vol. 2, Academic Press.
21. Fredkin, F. and Toffoli, T. (1982) Conservative logic. *Int. J. Theor. Phys.*, **21**, 219–253.
22. Margolus, N. (1984 Physics-like models of computation. *Physica D*, **10**, 81–95.
23. Sendińa-Nadal, I., Mihaliuk, E., Wang, J., Pérez-Muńuzuri, V., and Showalter, K. (2001) Wave propagation in subexcitable media with periodically modulated excitability. *Phys. Rev. Lett.*, **86**, 1646–1649.
24. Toth, R., Stone, C., Adamatzky, A., De Lacy Costello, B., and Bull, L. (2009) Experimental validation of binary collisions between wave-fragments in the photosensitive Belousov-Zhabotinsky

reaction. *Chaos Solitons Fractals*, **41**, 1605–1615.

25. Adamatzky, A. and De Lacy Costello, B. (2007) Binary collisions between wave-fragments in a sub-excitable Belousov – Zhabotinsky medium. *Chaos Solitons Fractals*, **34**, 307–315.

26. Field, R.J. and Noyes, R.M. (1974) Oscillations in chemical systems. IV. Limit cycle behavior in a model of a real chemical reaction. *J. Chem. Phys.*, **60**, 1877–1884.

27. Beato, V. and Engel, H. (2003) Pulse propagation in a model for the photosensitive Belousov – Zhabotinsky reaction with external noise, in *Noise in Complex Systems and Stochastic Dynamics*, Proceedings of SPIE 5114 (eds L. Schimansky-Geier, D. Abbott, A. Neiman, and C. VandenBroeck), SPIE, pp. 353–362.

28. Adamatzky, A. (2004) Collision-based computing in Belousov – Zhabotinsky medium. *Chaos Solitons Fractals*, **21**, 1259–1264.

29. De Lacy Costello, B., Toth, R., Stone, C., Adamatzky, A., and Bull, L. (2009) Implementation of glider guns in the light-sensitive Belousov – Zhabotinsky medium. *Phys. Rev. E*, **79**, 026114.

30. Zhang, L. and Adamatzky, A. (2009) Collision-based implementation of a two-bit adder in excitable cellular automaton. *Chaos Solitons Fractals*, **41**, 1191–1200.

31. Zhang, L. and Adamatzky, A. (2009) Towards arithmetical chips in sub-excitable media: cellular automaton models. *Int. J. Nanotechnol. Mol. Comput.*, **1**, 73–81.

32. Adamatzky, A. (2010) Slime mould logical gates: exploring ballistic approach arXiv:1005.2301v1 [nlin.PS] (2010), http://arxiv.org/abs/1005.2301

33. Toth, R., Stone, C., De Lacy Costello, B., Adamatzky, A., and Bull, L. (2008) Dynamic control and information processing in the Belousov – Zhabotinsky reaction using a co-evolutionary algorithm. *J. Chem. Phys.*, **129**, 184708.

34. Toth, R., Stone, C., De Lacy Costello, B., Adamatzky, A., and Bull, L. (2009 Simple collision-based chemical logic gates with adaptive computing. *Int. J. Nanotechnol. Mol. Comput.*, **1**, 1–16.

35. NEUNEU. (2010) Artificial Wet Neuronal Networks from Compartmentalised Excitable Chemical Media, http://www.neu-n.eu/

36. Adamatzky, A., Holley, J., Bull, L., and De Lacy Costello, B. (2010) On computing in fine-grained compartmentalised Belousov – Zhabotinsky medium arXiv:1006.1900v1 [nlin.PS], http://arxiv.org/abs/1006.1900

37. Szymanski, J., Igarashi, Y., Gorecki, J., Gorecka, J.N., Zauner, K.-P., and de Planque, M. (2010) Belousov – Zhabotinsky reaction in lipid covered droplets, submitted.

38. Adamatzky, A., De Lacy Costello, B., and Bull, L. (2011) On polymorphic logical gates in sub-excitable chemical medium. *Int. J. Bifurcat. Chaos*, **21**(7), pp. 1977–1986.

39. Holley, J., Adamatzky, A., Bull, L., De Lacy Costello, B., and Jahan, I. (2011 Computational modalities of Belousov – Zhabotinsky encapsulated vesicles. *Nano Commun. Netw.*, **2**(1),), 50–61.

40. Toth, A. and Showalter, K. (1995 Logic gates in excitable media. *J. Chem. Phys.*, **103**, 2058–2066.

41. Steinbock, O., Kettunen, P., and Showalter, K. (1996 Chemical wave logic gates. *J. Chem. Phys.*, **100**, 18970–18975.

42. Górecki, J. and Górecka, J.N. (2009) Computing in geometrical constrained excitable chemical systems, in *Encyclopedia of Complexity and Systems Science* (ed. R.A. Meyers), Springer-Verlag.

43. Motoike, I. and Yoshikawa, K. (1999) Information operations with an excitable field. *Phys. Rev. E*, **59**(5),), 5354–5360.

44. Motoike, I.N., Yoshikawa, K., Iguchi, Y., and Nakata, S. (2001) Real-time memory on an excitable field. *Phys. Rev. E*, **63**(3), 036220.

45. Gorecki, J., Yoshikawa, K., and Igarashi, Y. (2003) On chemical reactors that can count. *J. Phys. Chem.*, **107**(10),), 1664–1669.

46. Agladze, K., Aliev, R.R., Yamaguchi, T., and Yoshikawa, K. (1996) Chemical diode. *J. Phys. Chem.*, **100**(33), 13895–13897.

47. Adamatzky, A., De Lacy Costello, B., and Bull, L. (2010a) On polymorphic logical

gates in sub-excitable chemical medium, http://arxiv.org/abs/1007.0034.

48. Kitahata, H., Aihara, R., Magome, N., and Yoshikawa, K. (2002) Convective and periodic motion driven by a chemical wave. *J. Chem. Phys.*, **116**, 5666.

49. Kitahata, H. (2006) Spontaneous motion of a droplet coupled with a chemical reaction. *Prog. Theor. Phys. Suppl.*, **161**, 220–223.

50. Epstein, I.R. and Vanag, V.K. (2005) Complex patterns in reactive microemulsions: self-organized nanostructures? *Chaos*, **15**, 047510.

51. Górecka, J.N., Górecki, J., and Igarashi, Y. (2009) On the simplest chemical signal diodes constructed with an excitable medium. *Int. J. Unconv. Comput.*, **5**, 129–143.

9
High-Concentration Chemical Computing Techniques for Solving Hard-To-Solve Problems, and their Relation to Numerical Optimization, Neural Computing, Reasoning under Uncertainty, and Freedom of Choice

Vladik Kreinovich and Olac Fuentes

9.1
What are Hard-To-Solve Problems and Why Solving Even One of Them is Important

9.1.1
What is so Good About Being Able to Solve Hard-To-Solve Problems from Some Exotic Class?

In this chapter, we will talk about applying chemical computing to a specific class of hard-to-solve (NP-complete) problems.

To a person who is not very familiar with the notions of NP-completeness, this may sound like a very exotic (and thus not very interesting) topic. For example, this person may ask: OK, we spend all these efforts and solve problems from this exotic class, but how will this help me solve my own hard-to-solve problems, problems which are formulated in completely different terms?

1) A *short answer* to this equation is: once we learn how to solve problems from *one* class, then we will be able to solve *all* hard-to-solve problems.
2) A *detailed answer* to this question – with appropriate explanations – is given in this section.

Since this volume is devoted to chemical computing, we expect most readers to be familiar with at least the basics of chemistry. Because of this, in our description of NP-completeness, we will try (whenever possible) to use examples from (computational) chemistry. Since some potential readers are computational scientists, who may not be very familiar with the details of computational chemistry problems, we will try to explain the related chemistry problems as much as possible.

Comment. Readers who are already very familiar with the notions of P, NP, and NP-completeness are welcome to skip this section. Readers interested in more details can read, for example, [1, 2].

Molecular and Supramolecular Information Processing: From Molecular Switches to Logic Systems,
First Edition. Edited by Evgeny Katz.
© 2012 Wiley-VCH Verlag GmbH & Co. KGaA. Published 2012 by Wiley-VCH Verlag GmbH & Co. KGaA.

9.1.2
In Many Applications Areas – In Particular in Chemistry – There are Many Well-Defined Complex Problems

In many application areas, we face well-defined problems. Many such problems are known in chemistry.

For example, it is known that, from the physical viewpoint, chemical reactions are interactions between electrons of different atoms. Thus, to get a good understanding of the chemical reactions, it is desirable to describe possible electronic states and their energies. There exist known fundamental equations – partial differential equations originally proposed by Schrödinger – that exactly describe these states.

On a larger scale, changes in concentrations that occur during a complex chemical reaction are described by a system of ordinary differential equations – equations of chemical kinetics. If we also want to take into account spatial inhomogeneities, we need to use the corresponding partial differential equations, and so on.

In some applications, we need to solve optimization problems. For example, in bioinformatics applications, we know how to describe, for each possible folding of a protein, the resulting potential energy. Based on this description, we need to find the folding for which this potential energy is the smallest possible – because this is the shape into which proteins fold within a cell.

9.1.3
In Principle, There Exist Algorithms for Solving These Problems

In computational mathematics, there exist algorithms for solving the corresponding problems: that is, algorithms for solving systems of ordinary or partial differential equations, algorithms for finding where a complex function attains its minimum, and so on.

9.1.4
These Algorithms may Take Too Much Time to be Practical

The problem is that, often, these algorithms, when applied to practical chemistry-related (and other) problems, require too much time to be practically useful.

9.1.5
Feasible and Unfeasible Algorithms: General Idea

When the algorithm takes too much time, the big question is: how much? For some problems, the required computation time is, for example, 10 or 100 times larger that what are accessible now. In this case, there is a good chance that this problem will be soon solved:

- it can either be solved right away, by running the algorithm on a high-performance supercomputer – which is usually several orders of magnitude faster than usual university computers,
- or it can be solved in a few years, since the computer speed approximately doubles every few years or so (this empirical fact is known in computer science as *Moore's law*).

Informally, we can say that such algorithms may not be practical on a typical computer, and they may not be practical on all existing computers – but they are *feasible* in the sense that in a reasonable amount of time, and with appropriate resources, these algorithms can be implemented.

On the other hand, there are other algorithms for which the required computation time may be 10^{20} times larger than what we have available; an example will be given soon. For such algorithms, we can use the fastest supercomputers, we can wait 10 years – none of this will overcome this enormous gap between the desired computation speed and the available speed of computations. From the practical viewpoint, such algorithms are *unfeasible*.

Let us give examples of feasible and unfeasible algorithms.

9.1.6
Solving Equations of Chemical Kinetics: An Example of a Feasible Algorithm

Let us consider the most realistic case of equations that take into account the spatial inhomogeneity. These partial differential equations describe the dependence $c_i(x, y, z, t)$ of the concentration of each substance i at different spatial points (described by spatial coordinates x, y, and z) at different moments of time.

Most computational techniques for solving partial differential equations are based on the following straightforward idea:

- instead of considering all infinitely many moments of time, we consider only moments on a grid, for example, moments t_0, $t_1 = t_0 + \Delta t$, $t_2 = t_1 + \Delta t = t_0 + 2\Delta t, \ldots, t_k = t_0 + k \cdot \Delta t, \ldots$, for some small step Δt;
- similarly, instead of considering all infinitely many values (x, y, z), we consider finitely many points on a grid – for example, values with $x = x_0 + k_x \cdot \Delta x$, $y = y_0 + k_y \cdot \Delta y$, and $z = z_0 + k_z \cdot \Delta z$.

Of course, this is an oversimplified description. In more sophisticated algorithms, such as finite element methods, instead of using a predetermined grid, we select points as we go – more points in areas of drastic change and fewer points in areas where there is practically no spatial dependence.

In this discrete approximation, the original partial differential equation becomes a difference equation, and we can solve it by consequently computing the values at moment t_0, at moment t_1, and so on. If N is a total number of grid points in each of the four directions x, y, z, and t, then we have N^4 possible combinations of these values, that is, N^4 nodes on a four-dimensional (4-D) grid at each of which we need to perform appropriate computations.

If we have n substances i, and we only consider reactions with two inputs (i.e., of the type $i + j \to \ldots$), the right-hand side of the corresponding equations of chemical kinetics contains terms such as $c_i \cdot c_j$. Since we have n different substances, there are no more than n^2 such terms, and thus, the total computation time grows with number of substances as n^2.

In this case, if we double the number of substances – for example, take into consideration important short-term intermediate substances whose study is very important for studying catalysis – the whole computation time increases only by a factor of four. We can usually afford such an increase – either by waiting four times longer for the computations to finish, or by going to a four times faster computer, or by waiting $2 + 2 = 4$ years during which, according to Moore's law, computers will twice double in speed and thus, become $2 \times 2 = 4$ times faster.

If we take into account reactions with three inputs, then the computation time starts growing as n^3. If we double n, the total computation time increases by a factor of eight – still feasible.

So, straightforward algorithms for solving equations of chemical kinetics are feasible.

9.1.7
Straightforward Solution of Schrödinger Equation: An Example of an Unfeasible Algorithm

Schrödinger's equation is the main equation of quantum physics. To describe an atom with n electrons in quantum physics, we need to describe a complex-valued function $\Psi(t, x_1, y_1, z_1, \ldots, x_n, y_n, z_n)$ called *wave function*. Here, x_i, y_i, and z_i are spatial coordinates of the ith particle. A straightforward way to solve this partial differential equation is the same as for chemical kinetics: select a grid and consider only points from a grid. The difference is that, when we select N options for x_1, N options for y_1, N options for $z_1, \ldots$, and N options for z_n, then we get N^{3n+1} possible combinations (grid points) $(t, x_1, y_1, z_1, \ldots, x_n, y_n, z_n)$. Processing each grid point requires at least one computational step, so the overall number of computational steps – and thus the overall computation time – grows exponentially with n, as c^n for some constant c.

Such computations are realistically possible for hydrogen (H), for which there is only one electron ($n = 1$), and for helium (He), for which $n = 2$. But, for iron (Fe), for example, with $n = 26$, even for the simplest case when we take only two points $N = 2$ in each direction, we need $2^{3n+1} = 2^{79} \approx 3 \times 10^{24}$ steps. The fastest supercomputer performs 10^{12} operations per second, so in a year it can perform 3×10^7 s/year $\times 10^{12}$ oper/s $= 3 \times 10^{19}$. Thus, we need 30 000 years to finish these computations – and we only considered a rather useless approximation with only two values per spatial dimension. For a somewhat better (but still lousy) approximation, we can take $N = 10$ points per dimension, and in this case we need 10^{79} steps: much more than the fastest computer can perform during the lifetime of the universe.

This algorithm is clearly unfeasible.

9.1.8
Straightforward Approach to Protein Folding: Another Example of an Unfeasible Algorithm

A guaranteed way to find a global minimum of a function of n variables is to compute its values on all the points of a grid and to find the smallest of these values. This method requires N^n steps and is thus feasible, for example, for $n = 1$, $n = 2$, or even $n = 3$ variables. However, in protein folding, we need to find spatial locations of several thousand atoms forming the protein, so $n \approx 10^3$, and the value N^n is not even astronomical: it is much larger that the lifetime of the universe!

9.1.9
Feasible and Unfeasible Algorithms: Toward a Formal Description

In the above examples, for some algorithms the computation time grows polynomially with the number n of inputs, as $C \cdot n^k$ for some k; these algorithms are feasible. For some algorithms, the computation time grows exponentially with n, as c^n; these algorithms are unfeasible.

This distinction underlies the current formal definition of a feasible algorithm: an algorithm is called *feasible* if there exists a polynomial P such that on every input of size n, this algorithm finishes computations in time $\leq P(n)$. All other algorithms are considered *unfeasible*.

Comment. It is well known that this definition does not always properly capture the intuitive idea of feasibility. For example, an algorithm that requires computation time $10^{40} \cdot n$ is not practically feasible but it is feasible in the sense of the above definition. On the other hand, an algorithm that requires time $\exp(10^{-9} \cdot n)$ is practically feasible but not feasible in the sense of the above definition – since the exponential function cannot be bounded from above by any polynomial. However, this is the best definition we have.

9.1.10
Maybe the Problem Itself is Hard to Solve?

When an algorithm for solving a problem is not feasible, a natural idea is to look for a faster algorithm. But maybe it is not the algorithm's fault; maybe the problem itself is hard to solve, so that no feasible algorithm is possible that would solve all particular cases of this problem.

To be able to decide whether a problem is hard to solve, we need to first provide a precise definition of what is a problem and what it means for a problem to be hard to solve. Let us start with describing what a problem is.

9.1.11
What Is a Problem in the First Place?

In the previous discussion, we tried our best to relate to chemistry. However, when we analyze what a problem is, we want our answer to be as general as possible – to make sure that we do not miss important real-life problems. Thus, let us now consider the activity of other disciplines as well.

9.1.12
What is a Problem: Mathematics

For example, the main activity of a mathematician is proving theorems. We are given a mathematical statement x, and we need to find a proof y:

- either a proof that x is true
- or, if x is not true, a proof that the original statement x is false.

Mathematicians are usually interested in proofs that can be checked by human researchers and are, thus, of reasonable size. This notion of "reasonable size" can be formalized in the same way as in the definition of a feasible algorithm: as the existence of a polynomial P_l for which the length $\text{len}(y)$ of the proof y does not exceed the result $P_l(\text{len}(x))$ of applying this polynomial to the length $\text{len}(x)$ of the input x.

In the usual formal systems of mathematics, the correctness of a formal proof can be checked in polynomial time. So, the main problem of mathematics can be formulated as follows:

9.1.13
A Description of a General Problem

- *A description of a general problem.* We are given
    - a feasible algorithm $C(x, y)$ that, given two strings x and y, returns "true" or "false"; and
    - a polynomial P_l.
- *A description of the particular case (instance) of the general problem.*
    - we are given a string x;
    - we must find a string y of length $\text{len}(y) \leq P_l(\text{len}(x))$ for which $C(x, y) =$ "true" – or produce the corresponding message if there is no such string.

Comment. The possibility that we have neither a proof of x nor a proof of its negation $\bar{x}$ is quite real: there are known statements x that are independent of the axioms.

9.1.14
What About Other Activity Areas?

The above description was derived from the analysis of mathematics but, as we will now show, a similar description applies to other activity areas as well.

9.1.15
What is a Problem: Theoretical Physics

In theoretical physics, one of the main challenges is to find a formula y that describes the observed data x. The size of such a formula cannot exceed the amount of data: otherwise, we could simply enumerate all the observations and call it a formula. So, here, $\text{len}(y) \leq \text{len}(x)$, that is, we have the above inequality for $P_l(n) = n$. Once a formula y is proposed, it is easy to check whether it is consistent with all the observations x: this can be done observation by observation, so this checking can be performed in linear time. If we denote by $C(x, y)$ the statement "the formula y is consistent with observations x," then we get exactly the above formulation.

9.1.16
What is a Problem: Engineering

In engineering, one of the main challenges is to find a design y that satisfies given specifications x. For example, a design for a bridge must be able to withstand winds up to a certain speed and loads up to a certain amount, and its building cost should not increase the amount allocated in the budget.

This design has to be practical, so its description cannot be too long; thus, a condition of the type $\text{len}(y) \leq P_l(\text{len}(x))$ sounds quite reasonable. Once a design y is proposed, we can use known engineering software tools to efficiently check whether the design y satisfies the specifications x; so, we have a feasible checking algorithm $C(x, y)$. Thus, we also get exactly the above formulation.

9.1.17
Class NP

In all these general problems, once we guess a solution candidate y, we can check, in polynomial time, whether this guess y is indeed a solution. In theoretical computer science, computations with guessing steps are called *nondeterministic*. Thus, this class is called *NP*.

9.1.18
Class P and the $P\overset{?}{=}NP$ Problem

For some of the problems from the class NP, there exist algorithms that solve these problems in polynomial time (i.e., feasibly). The class of all such problem is denoted by P.

By definition, the class P is a subset of the class NP: $P \subseteq NP$. A natural question is: is P a proper subclass of NP? In other words, do there exist problems from the class NP that cannot be solved in polynomial time, or, vice versa, can every problem from the class NP be feasibly solved and, thus, P=NP? The answer to this question is unknown. Checking whether P is equal to NP has been an open problem for

40 years already. Most computer scientists believe that these classes are different, but no one knows for sure.

9.1.19
Exhaustive Search: Why it is Possible and Why it is Not Feasible

In principle, since the length len(y) of a possible solution is *a priori* restricted (by the value $P_l(\text{len}(x))$), we can simply try all the words y of length $\leq P_l(\text{len}(x))$ until we find a string y that satisfies the desired condition $C(x, y)$. There are finitely many words of a given length, so this procedure always produces the desired result.

This "exhaustive search" algorithm works for small lengths, but, in general, this algorithm is not feasible. Indeed, even for the binary alphabet, we need to try $2^{P_l(\text{len}(x))}$ possible words y, and we have already shown that even for reasonable values m it is not feasible to perform 2^m computational steps.

9.1.20
Notion of NP-Complete Problems

The fact that no one knows whether P is equal to NP does not mean that we have no information about the relative complexity of different problems from the class NP. There are known *reducibility* relations between different problems A and A': sometimes, every instance of the problem A can be feasibly reduced to an instance of the problem A'. In this case, the problem A' is harder than – of the same hardness as – the problem A, in the sense that, if we can efficiently solve every instance of the problem A', then we can also solve every instance of the problem A.

For example, if we know how to solve systems with three unknowns, then we can solve every system with two unknown – by introducing a dummy third variable and applying the algorithm for solving systems with three unknowns. Thus, solving systems with three unknowns is harder than (or of the same hardness as) solving systems of two unknowns.

Similarly, if we know how to solve a system of linear *inequalities*, then we can also solve systems of linear *equalities* – since each equality $f = 0$ is equivalent to two inequalities $f \geq 0$ and $f \leq 0$. Thus, solving systems of linear inequalities is harder than (or of the same hardness as) solving systems of linear equalities.

An important discovery made in the early 1970s – a discovery that started the whole area of research about P, NP, and NP-completeness – that in the class NP, there exist problems to which every other problem from the class NP can be reduced. Thus, each of these problems is harder than (or of the same hardness as) the complete class NP. Such hard-to-solve problems are called *NP-complete*.

9.1.21
Why Solving Even One NP-Complete (Hard-To-Solve) Problem is Very Important

Because of the reduction-related definition of NP-completeness, once we know how to efficiently solve *one* NP-complete problem, we will then be able to efficiently solve *all* problems from the class NP. Similarly, once we have an algorithm that efficiently solves *many* instances of one NP-complete problem, we can use it to solve many instances of other problems from the class NP. Thus, any progress in solving *one* of the NP-complete problems automatically leads to a progress in *all* of them. As a result, solving even one NP-complete problem – no matter how exotic it looks, no matter how unrelated it seems to the problems in which we are actually interested – is very important because it will help solving other problems.

9.1.22
Propositional Satisfiability: Historically the First NP-Complete Problem

At present, thousands of different NP-complete problems are known. Historically, the first problem for which NP-completeness was proved was the *propositional satisfiability* problem. This problem is still being actively studied as an example of an NP-complete problem, so let us describe this problem.

For convenience, instead of describing the original satisfiability problem, we will describe an easy-to-describe class 3-SAT which is also known to be NP-hard. We start with n propositional variables $v_1, \ldots, v_n$, that is, variables each of which can take only two values: "true" (usually represented, in the computers, by 1) and "false" (usually represented, in the computers, by 0). By a *literal a*, we mean a variable v_i or its negation $\bar{v}_i$. By a *clause C*, we mean an expression of one of the following types: $a \vee b$ or $a \vee b \vee c$, where a, b, and c are literals. Finally, by a *formula F*, we mean an expression of the type $C_1 \& C_2 \& \ldots \& C_m$, where $C_1, \ldots, C_m$ are clauses.

To illustrate this concept, let us give a simple example of the formula

$$(v_1 \vee \bar{v}_2) \& (v_1 \vee v_2 \vee v_3).$$

This formula had $m = 2$ clauses: $v_1 \vee \bar{v}_2$ and $v_1 \vee v_2 \vee v_3$.

The problem is: *given* a formula F, *find* the values of the variables $v_1, \ldots, v_n$ that make this formula true (i.e., for which the formula F is *satisfied*) – or return a message that such values do not exist. Once we have a sequence of values $v_1, \ldots, v_n$, we can plug these values into the formula F and easily check whether the formula is true. Thus, this problem belongs to the class NP. (The proof that this problem is NP-complete is beyond the scope of this chapter.)

9.1.23
What We Do

In this chapter, we study the above-described propositional satisfiability problem: namely, we show how chemical computing can solve this problem.

9.2 How Chemical Computing Can Solve a Hard-To-Solve Problem of Propositional Satisfiability

9.2.1 Chemical Computing: Main Idea

When a person needs to perform a complex task – for example, build a house, dig a ditch – and realizes that it would take too much time for him to do it alone, he gets himself a helper. When they work simultaneously, in parallel, they finish the task faster. To perform this task even faster, he can get many helpers; the more helpers (up to a certain limit), the better.

Similarly, when a computational problem requires too much computation time, a natural way to finish computations faster is to have many computers working in parallel. From this viewpoint, what can be faster than having all $\approx 10^{23}$ molecules work in parallel to perform the desired computations? In other works, ideally, we should make chemical reactions – on the level of individual molecules – perform the desired computations.

This is the main idea behind chemical computing.

9.2.2 Why Propositional Satisfiability was Historically the First Problem for Which a Chemical Computing Scheme was Proposed

This may be not a widely known fact, but the main idea of chemical computing was first proposed by Yuri Matiyasevich exactly for the purpose of solving a hard-to-solve problem of propositional satisfiability. This idea was first presented at a meeting; it was first published in [3].

It makes sense to have selected a hard-to-solve problem: these problems require a lot of computation time, and, thus, for them the need to reduce this time is the most urgent. But why propositional satisfiability and not any other hard-to-solve problem? The answer is that, surprisingly, the propositional satisfiability problem can be naturally represented in terms that are very similar to chemistry.

To explain this representation, let us recall the meaning of each clause. A clause $a \vee b$ means that either a is true or b is true. Thus, if a is false, then b is true; similarly, if b is false, then a should be true. In other words, this clause can be represented by two implications

$$\overline{a} \to b; \overline{b} \to a.$$

Vice versa, if both these implications are true, this means that the clause $a \vee b$ is true. Indeed, in general, either a is true or a is false.

- If a is true, then the clause $a \vee b$ is also true.
- If a is false, then, due to the implication $\overline{a} \to b$, the literal b is true.

In both cases, the clause is true. (Notice that we used only one implication.)

Similarly, a clause $a \vee b \vee c$ means that one of the three literals a, b, or c must be true. Thus, if both a and b are false, then c must be true; if a and c are both false, then b must be true; and if b and c are both false, then a must be true. In other words, this clause can be represented by three implications:

$$\overline{a}, \overline{b} \rightarrow c; \overline{a}, \overline{c} \rightarrow b; \overline{b}, \overline{c} \rightarrow a.$$

Vice versa, one can check that if these implications are true, then the original clause is true is well. (Actually, it is sufficient to require that one of these implications is true.)

For example, the above formula $(v_1 \vee \overline{v}_2) \,\&\, (v_1 \vee v_2 \vee v_3)$ can be represented by the following five implications:

$$\overline{v}_1 \rightarrow \overline{v}_2; v_2 \rightarrow v_1; \overline{v}_1, \overline{v}_2 \rightarrow v_3; \overline{v}_1, \overline{v}_3 \rightarrow v_2; \overline{v}_2, \overline{v}_3 \rightarrow v_1.$$

9.2.3
How to Apply Chemical Computing to Propositional Satisfiability: Matiyasevich's Original Idea

Matiyasevich noticed that these implications look exactly like chemical reactions involving substances v_i and $\overline{v}_i$. Thus, he proposed to solve the original propositional satisfiability problem with variables $v_1, \ldots, v_n$ by finding $2n$ substances which have exactly these implications $a, b \rightarrow c$ as chemical reactions $a + b \rightarrow c$. For each variable v_i:

- the larger concentration of the substance v_i in comparison with the concentration of the "opposite" substance $\overline{v}_i$ indicates that this variable v_i is true, while
- the larger concentration of substance $\overline{v}_i$ in comparison with the concentration of v_i indicates that v_i is false.

These reactions work in such a way as to make all the implications true, and thus, the whole formula true. For example, the reaction $\overline{a} \rightarrow b$ means that, if we have a prevalence of the substance $\overline{a}$ (i.e., if, in our interpretation, a is false), then this reaction would create a prevalence of the substance b – that is, b will become true as well. Once all the implications are true, this means that all the clauses are true, and thus, the original formula is satisfied.

Of course, the original propositional formula may be always false. In this case, no matter what truth values we plug in, we will always get false. Therefore, once we get the values "true" and "false" from chemical computations, we must check whether the authors of [1] make the formula true. If they do, we return these values; if they do not, we return the message that the original formula was not satisfiable.

9.2.4
A Precise Description of Matiyasevich's Chemical Computer: First Example

To analyze the behavior of Matiyasevich's chemical computer, they wrote down – and analyzed – the corresponding system of chemical kinetic equations.

For simplicity, they assumed that the chemical reactions corresponding to each implication has exactly the same intensity.

Before we describe a general formula, let us describe these chemical kinetic equations on the example of the above simple propositional formula. In these equations, we will denote concentrations of each substance v_i by c_i, and concentration of the "opposite" substance $\bar{v}_i$ by c_{-i}.

None of the above five chemical reactions consumes v_1 and two reactions produce v_1: the reactions $v_2 \to v_1$ and $\bar{v}_2 + \bar{v}_3 \to v_1$. According to chemical kinetics, the rate of the first reaction is proportional to c_2 and the rate of the second reaction is proportional to the product $c_{-2} \cdot c_{-3}$. Thus, the differential equation describing the changes in the concentration c_1 of the substance v_1 has the form

$$\dot{c}_1 = c_2 + c_{-2} \cdot c_{-3}.$$

For the substance $\bar{v}_1$, the opposite is true: none of the reactions produces this substance, but we have three reactions that consume it: $\bar{v}_1 \to v_2$, $\bar{v}_1, \bar{v}_2 \to v_3$, and $\bar{v}_1, \bar{v}_3 \to v_2$. The rate of the first reaction is c_{-1}, the rate of the second reaction is $c_{-1} \cdot c_{-2}$, and the rate of the third reaction is $c_{-1} \cdot c_{-3}$. Thus,

$$\dot{c}_{-1} = -c_{-1} - c_{-1} \cdot c_{-2} - c_{-1} \cdot c_{-2}.$$

For the substance v_2, we have one reaction that produces it: the reaction $\bar{v}_1, \bar{v}_3 \to v_2$, and one reaction that consumes it: the reaction $v_2 \to v_1$. Thus, we get

$$\dot{c}_2 = c_{-1} \cdot c_{-3} - c_2.$$

Similarly, we have

$$\dot{c}_{-2} = c_{-1} - c_{-1} \cdot c_{-2} - c_{-1} \cdot c_{-3};$$
$$\dot{c}_3 = c_{-1} \cdot c_{-2};$$
$$\dot{c}_{-3} = -c_{-1} \cdot c_{-3} - c_{-2} \cdot c_{-3}.$$

From this system of equations, it is easy to see why the chemical reactions will lead to values v_i that satisfy the original formula. Indeed, in these reactions, the substance v_1 is only produced and never consumed, and the substance $\bar{v}_1$ is always consumed and never produced. Thus, after a sufficiently long time, the concentration of the substance v_1 will becomes larger than the concentration of the substance $\bar{v}_1$. According to our interpretation, this means that we will select v_1 to be true.

Similarly, the substance v_3 is only produced, and the substance $\neg v_3$ is only consumed, which means that the substance v_3 will prevail – that is, that we will select x_3 to be true as well.

We cannot make a similar conclusion about v_2 without performing detailed computations, but we do not actually need to perform these computations: if we select v_1 and v_3 to be true, then, no matter what value we select for v_2, both clauses are satisfied, and thus the original formula is satisfied.

9.2.5
A Precise Description of Matiyasevich's Chemical Computer: Second Example

The conclusion is not always as simple and as straightforward as in the above example. For example, for a formula $(v_1 \vee v_2) \,\&\, (\bar{v}_1 \vee \bar{v}_2)$, by trying all four possible combinations, we can see that it has two possible solutions:

- $v_1 =$ "true" and $v_2 =$ "false"; and
- $v_1 =$ "false" and $v_2 =$ "true."

The corresponding equations of chemical kinetics take the form

$$\dot{c}_1 = c_{-2} - c_1;\ \dot{c}_{-1} = c_2 - c_{-1};\ \dot{c}_2 = c_{-1} - c_2;\ \dot{c}_{-2} = c_2 - c_{-2}.$$

According to our interpretation, what we are really interested in whether $c_1 > c_{-1}$ and whether $c_2 > c_{-2}$. From this viewpoint, it makes sense to consider the differences $\Delta c_1 \stackrel{\text{def}}{=} c_1 - c_{-1}$ and $\Delta c_2 \stackrel{\text{def}}{=} c_2 - c_{-2}$: for each i, we select v_i to be true if $\Delta c_i > 0$ and to be false if $\Delta c_i < 0$.

By subtracting the above expressions for the rate changes of the concentrations c_i and c_{-i}, we can get the expressions for the rate changes of the differences Δc_i:

$$\Delta \dot{c}_1 = -(c_2 - c_{-2}) - (c_1 - c_{-1});\ \Delta \dot{c}_2 = -(c_1 - c_{-1}) - (c_2 - c_{-2}),$$

that is,

$$\Delta \dot{c}_1 = -\Delta c_1 - \Delta c_2;\ \Delta \dot{c}_2 = -\Delta c_1 - \Delta c_2.$$

By adding these two equations, we conclude that, for $\Delta \stackrel{\text{def}}{=} \Delta c_1 + \Delta c_2$, we get $\dot{\Delta} = -2\Delta$, and hence $\Delta(t) = \Delta(0) \cdot \exp(-2t)$. When $t \to \infty$, we get $\Delta(t) \to 0$; thus, for large t, we have $\Delta(t) \approx 0$. By definition of Δ, this means that $\Delta c_1 + \Delta c_2 \approx 0$, that is, that $\Delta c_2 \approx -\Delta c_1$. Thus,

- If $\Delta c_1 > 0$, that is, if v_1 is true, then we should have $\Delta c_2 < 0$, that is, v_2 should be false.
- Vice versa, if $\Delta c_1 < 0$, that is, if v_1 is false, then we should have $\Delta c_2 > 0$, that is, v_2 should be true.

So, for this formula, chemical kinetic equations lead to both solutions; which one we get depends on the initial conditions.

9.2.6
A Precise Description of Matiyasevich's Chemical Computer: General Formula

In general, similar ideas result in the following formula for the rate at which the concentration c_a of each literal changes:

$$\dot{c}_a = \sum_{C: a \in C} \left(\prod_{b \in C, b \neq a} c_{\bar{b}} \right) - c_a \cdot \sum_{C: \bar{a} \in C, |C|=3} \left(\sum_{b \in C \,\&\, b \neq \bar{a}} c_{\bar{b}} \right) -$$

$$c_a \cdot \#\{C : \bar{a} \in C \,\&\, |C| = 2\}.$$

Here, C goes over all the clauses, $|C|$ is the number of literals in the clause, and $\#S$ is the number of elements in the set S.

Indeed, a substance corresponding to the literal a is produced if a belongs to a clause. Each such clause $a \vee b \vee c$ leads to the chemical reaction $\overline{b} + \overline{c} \to a$ and, thus, to the term $c_{\overline{b}} \cdot c_{\overline{c}}$ in the expression for $\dot{a}$.

Similarly, a substance corresponding to the literal a is consumed if the negation $\overline{a}$ belongs to a clause. Each such clause $\overline{a} \vee b \vee c$ leads to the chemical reaction $a + \overline{b} \to c$ and, thus, to the term $-c_a \cdot c_{\overline{b}}$ in the expression for $\dot{a}$. If the negation $\overline{a}$ belongs to a clause $\overline{a} \vee b$, then the consuming chemical reaction is $a \to b$, which leads to the term $-c_a$ in the expression for $\dot{c}$.

9.2.7
A Simplified Version (Corresponding to Catalysis)

In the above system of chemical reactions, each substance is both produced and consumed. To make the analysis of the resulting system of equations simpler, is may be desirable to avoid consumption and consider only production. We can do this if we introduce a new universal substance U and, to each implication $a, b \to c$, assign a modified chemical reaction $U + a + b \to a + b + c$. In this reaction, the input substances a and b are not consumed: in chemical terms, they play the role of *catalysts* that enhance the transformation of the universal substance into the generated substance c.

In principle, in this case, we should also take into account the changes in the concentration of substance U. To maximally simplify the situation, we assume that we have a large (practically unlimited) supply of the substance U, so that the consumption of U during our reactions is negligible in comparison with its original concentration. In this case, we only need to take into account production of each substance, and the resulting differential equations take a simplified form:

$$\dot{c}_a = \sum_{C : a \in C} \left(\prod_{b \in C, b \neq a} c_{\overline{b}} \right).$$

This simplification makes perfect sense from the logical viewpoint:

- In chemical kinetics, a reaction $a + b \to c$ means not only that c is produced but also that a and b are consumed.
- In contrast, in logic, an implication $a, b \to c$ means that if we have some reasons to believe in a and b are true, this increases our belief in c, but it *does not* mean that we somehow decrease our beliefs in a and b.

The above modification of the original system of chemical kinetics equations allows us to avoid this discrepancy.

9.2.8
Simplified Equations: Example

Let us give an example of such simplified equations. For the above propositional formula $(v_1 \vee \bar{v}_2) \& (v_1 \vee v_2 \vee v_3)$, the corresponding equations of chemical kinetics take the following simplified form:

$$\dot{c}_1 = c_2 + c_{-2} \cdot c_{-3}; \; \dot{c}_{-1} = 0; \; \dot{c}_2 = c_{-1} \cdot c_{-3}; \; \dot{c}_{-2} = c_{-1}; \; \dot{c}_3 = c_{-1} \cdot c_{-2}; \; \dot{c}_{-3} = 0.$$

9.2.9
Chemical Computations Implementing Matiyasevich's Idea Are Too Slow

Matiyasevich is a star of the mathematical world; he has the distinction of having solved 1 of the 23 famous Hilbert's problems – 23 important problems that, at the 1900 World Congress of Mathematics, the nineteenth century mathematics proposed as a challenge for the twentieth century. Whatever Matiyasevich writes is therefore taken seriously by mathematicians and computer scientists. Immediately, Gurevich, one of the world leaders in theoretical computer science, engaged his colleagues in the analysis of Matiyasevich's idea: how efficient is it?

Alas, the results of this analysis, published in [4], were not very promising: even for simple propositional formulas, for which simple algorithms produce satisfactory propositional values $v_1, \ldots, v_n$, Matiyasevich's system requires exponential time to converge to a correct solution – that is, to concentrations c_i and c_i for which the vector $v_1, \ldots, v_n$ for which v_i is true if and only if $c_i > c_{-i}$ is indeed satisfactory.

9.2.10
Natural Idea: Let us Use High-Concentration Chemical Reactions Instead

Since the original chemical reactions are too slow, we need to speed them up. The reaction rate is proportional to the product of the concentrations. Thus, to drastically speed up the reaction, we need to drastically speed up the concentrations c_i and c_{-i}.

The interesting thing is that, when the concentrations become very high, the formulas for the rate of chemical reaction change. Indeed, the usual formulas of chemical kinetics are based on the natural idea that, when concentrations are small, then, for the reaction to take place, all the molecules have to physically meet. For example, for a reaction $a + b \to c$ to take place, we need the molecules of a and b to meet.

The total number of molecules of a is proportional to the concentration c_a of the substance a. For each molecule of a, the probability of meeting a molecule of b is proportional to the concentration c_b of the molecules b. Thus, the total number of reactions per unit time is proportional to the product $c_a \cdot c_b$ of these concentrations.

In the case of very high concentrations, the molecules are there already, so the reaction always takes place. The rate of this reaction is thus proportional to the total number of pairs (a, b).

- If the concentration c_a of the substance a is higher, then the rate is determined by a concentration c_b of the substance b.
- If the concentration c_b of the substance b is higher, then the rate is determined by a concentration c_a of the substance a.

We can describe both cases by saying that the reaction rate is proportional to the minimum $\min(c_a, c_b)$ of the two concentrations.

This argument may be not absolutely clear when presented on the example of chemical kinetics where we do not have much of an intuition, but it can be made clearer if we use an example of the similar predator–prey equations. When the concentrations of rabbits and wolves are small, the rate with which wolves consume rabbits is proportional to the product $c_w \cdot c_r$ of the concentration of wolves c_w and the concentration of rabbits c_r. Indeed, in this case, a wolf has to run around the forest to find his rabbit meal.

On the other hand, if we place all the wolves and all the rabbits together – in a small area where rabbits cannot run and cannot hide – then each wolf will immediately start consuming a rabbit – provided, of course, that there are enough rabbits for all the wolves. So, if the number of rabbits is larger than the number of wolves, the reaction speed will be determined by the number of wolves – hence, by the concentration of wolves c_w: each wolf eats a rabbit. In the opposite situation $c_w > c_r$, when there are more wolves than rabbits, this rate will be proportional to the concentration of rabbits: each rabbit is being eaten by a wolf. In both cases, the reaction rate is proportional to $\min(c_w, c_r)$.

9.2.11
Resulting Equations

The main difference between the usual chemical kinetics equations and equations corresponding to high concentrations is that these high-concentration equations have minimum instead of the product. Thus, by applying this high-speed, high-concentration kinetics to the (simplified) chemical reactions emerging from a propositional formula, we get the following system of differential equations:

$$\dot{c}_a = \sum_{C: a \in C} \left(\min_{b \in C, b \neq a} c_{\bar{b}} \right).$$

For our example of a propositional formula $(v_1 \vee \bar{v}_2) \& (v_1 \vee v_2 \vee v_3)$, we thus get the following equations:

$$\dot{c}_1 = c_2 + \min(c_{-2}, c_{-3}); \ \dot{c}_{-1} = 0; \ \dot{c}_2 = \min(c_{-1}, c_{-3}); \ \dot{c}_{-2} = c_{-1};$$

$$\dot{c}_3 = \min(c_{-1}, c_{-2}); \ \dot{c}_{-3} = 0.$$

9.2.12
Discrete-Time Version of These Equations Have Already Been Shown to be Successful in Solving the Propositional Satisfiability Problem

How good is the new system of differential equations? Is it indeed faster than the original one?

The ideal answer to this question would come if we could actually find the substances that have these chemical reactions. Alas, finding such substances is difficult, so we have to restrict ourselves to simulating this system of equations on a computer.

In order to simulate a system of differential equations $\dot{x}_i = f_i(x_1, \ldots, x_n)$, we can use the fact that the derivative $\dot{x}_i$ is defined as a limit of the ratios $\dfrac{x_i(t + \Delta t) - x_i(t)}{\Delta t}$. By definition of the limit, this means that, when Δt is small, the ratio is approximately equal to the derivative:

$$\frac{x_i(t + \Delta t) - x_i(t)}{\Delta t} \approx f_i(x_1, \ldots, x_n),$$

hence $x_i(t + \Delta t) = x_i(t) + \Delta t \cdot f_i(x_1(t), \ldots, x_n(t))$.

Thus, if we know the values $x_i(t)$ for some moment of time, we can use this formula to compute the values of all the variables x_i in the next moment of time $t + \Delta t$. Based on the values $x_i(t + \Delta t)$, we compute the values $x_i(t + 2\Delta t)$, and so on. If we start at a moment t_0 and we are interested in the values of x_i at the moment t_f, then we need $k = \dfrac{t_f - t_0}{\Delta t}$ iterations of this procedure.

For our system of equations, this means that, once we know the values of the concentrations at each moment of time, we can compute the new values of the concentrations as

$$c'_a = c_a + \Delta t \cdot \sum_{C : a \in C} \left(\min_{b \in C, b \neq a} c_{\overline{b}} \right).$$

We repeat this iterative procedure many times, and then select each variable v_i to be true if and only if $c_i > c_{-i}$.

Interestingly, we get exactly the same formulas that were proposed by Maslov in 1980 [5]; see a detailed description and analysis in [6, 7]. In particular, in [6, 7], it was shown that (in contrast to the original Matiyasevich's equations) Maslov's method performs very well on many classes of propositional formulas. For example, for many classes of propositional formulas for which efficient algorithms are known, Maslov's method also comes up with a solution in feasible time.

Thus, we arrive at the following conclusion:

9.2.13
Conclusion

The use of high-concentration chemical computations is indeed an efficient approach to hard-to-solve problems.

Historical comment. Maslov's method was originally proposed on a purely heuristic basis, without mentioning chemical computing. The high-concentration interpretation of this method – providing an explanation of why these formulas are used, and a physical justification of why this method should be faster than, for example, Matiyasevich's approach – is described in [8–12].

Pragmatic comment. Since Maslov's method was known earlier, what do we gain by finding out that it coincides with the result of fast chemical computing?

- First, we gain a new justification: Maslov's method is heuristic, and to be able to explain its formulas and explain why they work fast is an advantage.
- We also gain the possibility to naturally modify the original method – for example, by applying it to the original system of chemical reactions instead of the reactions with a universal substance U – and maybe to find a modification which will work even faster.
- Third, we gain an understanding of how to optimally select a parameter of the Maslov's method: since this is interpreted as an integration step of the system of differential equations, we can use known techniques to optimally select this step (see below).
- Fourth, we gain an ability to extend Maslov's technique to problems beyond propositional satisfiability – as long as these problems can be naturally interpreted in terms of chemical processes. For example, in [8–10, 12], a similar approach was used to find so-called stable models of logic programs. The main difference between a propositional formula and a logic program is that, in a formula, an implication $a, b \to c$ automatically leads to $a, \neg c \to \neg b$ – this is why we used three implications and three chemical reactions for each clause; in a logic program, this is not automatically true: rules involving negations have to be explicitly formulated. This difference is easy to describe in chemical computing terms: just add only the rules of the original logic program as chemical reactions (and not the extra rules). As a result, we get an efficient way of computing stable models of logic programs.
- Finally, last but not least, if we find actual substances that have these chemical reactions, then, by performing these reactions we can actually solve hard-to-solve problems.

9.2.14
Auxiliary Result: How to Select the Parameter Δt

In our chemical computing model, we start with some concentrations c_i and c_{-i} of the substances corresponding to v_i and $\bar{v}_i$. For these arbitrary concentrations, selecting each propositional variable v_i to be true when $c_i > c_{-i}$ will not, in general, lead to the values that satisfy the original propositional formula F. In the process of chemical reactions, the original inequalities $c_i > c_{-i}$ and $c_j < c_{-j}$ change and, eventually, the process (hopefully) *stabilizes* in the sense that the differences $\Delta c_i = c_i - c_{-i}$ no longer change sign. Once the process stabilizes, there is no need to perform further simulations, and we can already find the appropriate values of the propositional variables v_i.

9.2 How Chemical Computing Can Solve a Hard-To-Solve Problem of Propositional Satisfiability

Let us denote by T the time that it takes for a process to stabilize. In terms of T, if we select a value Δt, then we will need $T/\Delta t$ iteration steps to find the desired solution to the original propositional satisfiability problem. The smaller the value of Δt, the more iterations we need; from this viewpoint, if we want to find the solution faster, we must choose the largest possible value Δt. However, we cannot take Δt too large: otherwise, a linear approximation $x_i(t) + \Delta t \cdot \dot{x}_i$ will not be a good approximation for $x_i(t + \Delta t)$. So, we need to select the largest Δt for which the error of this linear approximation is not too large.

The above linear approximation can be viewed as the sum of the first two terms of the Taylor expansion

$$x_i(t + \Delta t) = x_i(t) + \Delta t \cdot \dot{x}_i(t) + \frac{1}{2} \times (\Delta t)^2 \cdot \ddot{x}_i(t) + \cdots.$$

The approximation error of the linear approximation is equal to the sum of all the terms that we ignored in this linear approximation, that is,

$$x_i(t + \Delta t) - (x_i(t) + \Delta t \cdot \dot{x}_i(t)) = \frac{1}{2} \times (\Delta t)^2 \cdot \ddot{x}_i(t) + \cdots.$$

In this expansion, each term is (for sufficiently small Δt) much smaller than the next one. Thus, the first (quadratic) term on the right-hand side provides a good approximation for the size of the approximation error. So, to make sure that this approximation error is small, we should require that it does not exceed a certain given portion $\delta > 0$ of the linear approximation, for example, that

$$\left\| \frac{1}{2} \times (\Delta t)^2 \cdot \ddot{x}_i \right\| \leq \delta \cdot \|\Delta t \cdot \dot{x}_i\|,$$

where $\|a_i\| = \sqrt{a_1^2 + \cdots + a_N^2}$ denotes the length of a vector $a = (a_1, \ldots, a_N)$. From this inequality, we can find the largest value of Δ for which this inequality is still satisfied, that is, the largest value of Δt for which linear approximation still works well, as

$$\Delta t = 2 \times \delta \cdot \frac{\|\dot{x}_i(t)\|}{\|\ddot{x}_i(t)\|}.$$

In our case, the variables x_i are concentrations c_a corresponding to different literals a. We already have the formula for the first derivatives of these variables:

$$\dot{c}_a = \sum_{C: a \in C} \left(\min_{b \in C, b \neq a} c_{\bar{b}} \right).$$

To find the formula for the second derivatives $\ddot{c}_a$, we need to differentiate the expression for $\dot{c}_a$. A (minor) problem here is that this expression contains a minimum. For each t, the minimum of several terms coincides with the smallest of these terms. Thus, the derivative of the minimum is simply equal to the derivative of the smallest term. This leads to the following formula

$$\ddot{c}_a = \sum_{C: a \in C} \dot{c}_{\bar{b}_{C,a}},$$

where $b(C, a)$ denotes the literal for which the value $c_{\bar{b}}$ is the smallest value among all $b \in C$ that are different from a. Once we know the values $b(C, a)$, we can compute the value $\dot{c}_{\bar{b}_{C,a}}$ by using the above general formula for the first derivative $\dot{c}_a$.

As a result, we select the value

$$\Delta t = 2 \times \delta \cdot \frac{\|\dot{c}\|}{\|\ddot{c}\|},$$

where

$$\|\dot{c}\| \stackrel{\text{def}}{=} \sqrt{\sum_a (\dot{c}_a)^2} \text{ and } \|\ddot{c}\| \stackrel{\text{def}}{=} \sqrt{\sum_a (\ddot{c}_a)^2}.$$

9.3
The Resulting Method for Solving Hard Problems is Related to Numerical Optimization, Neural Computing, Reasoning under Uncertainty, and Freedom of Choice

9.3.1
Relation to Optimization: Why it is Important

The fact that Maslov's method turned out to be equivalent to fast chemical computations is nice, but this fact only shows that this method is faster than other *chemistry-motivated* methods of solving the propositional satisfiability problem. Chemical computing has a clear advantage when implemented *in vitro* – we drastically parallelize computations. However, if we simply simulate the corresponding chemical reactions on a computer, then there is no convincing reason to restrict ourselves to algorithms that come from such simulations. Instead, we should search for the methods that are the fastest among *all* algorithms, chemistry-motivated or not.

In such a search, we can use the experience of computational mathematics. We cannot *directly* use this experience, because propositional satisfiability – and probably any other NP-complete problem – is *not* something we normally solve in numerical methods. This absence of hard problems from the numerical methods experience makes perfect sense:

- Numerical methods are designed to solve *feasible* problems such as optimization (when it is feasible), solving systems large systems of equations, or solving a system of ordinary differential equations, problems in which, in principle, an algorithm is known, but because of the large size of the problem, we need to find a faster modification. For these problems, it is possible to find general modifications that allow us to solve the original problems much faster.
- In contrast, for hard-to-solve (NP-complete) problems, there is no general feasible algorithm, so solving each of these problems requires creative thinking. Thus, there is no hope (unless P=NP) that we can find a general feasible modification for solving these problems faster.

Since we cannot use a *direct* experience of solving the original propositional satisfiability problem, we must use – *indirectly* – the experience of solving more traditional numerical problems. We have already mentioned that we can use the experience of solving systems of differential equations. Let us now show that we can also use an experience of solving optimization problems.

9.3.2
Relation to Optimization: Main Idea

In the propositional satisfiability problem, we need to find truth values of all the literals a that make the formula $C_1 \& C_2 \& \ldots \& C_m$ true, that is, that makes all the clauses $C_1, \ldots, C_m$ true [11, 13].

In the computer, everything is represented as 0s and 1s. In particular, a truth value is represented as 0 or 1: "true" corresponds to 1, and "false" corresponds to 0. Let us denote, by c_a, the truth value of a literal a. If the literal a is true, then its negation $\bar{a}$ is false, and vice versa; in both cases, we have $c_a + c_{\bar{a}} = 1$.

A clause $a \vee b \vee c$ is true if and only if at least one of the three literals a, b, or c is true. (Similarly, a clause $a \vee b$ is true if at least one of the two literals a or b is true.) To make it easier to compare with the chemical approach, in which each clause leads to equations $\bar{a} + \bar{b} + U \to \bar{b} + \bar{c} + a$ that mostly contain negations, let us reformulate the above condition in terms of negations: a clause $a \vee b \vee c$ is true if and only if at least one of the literals $\bar{a}$, $\bar{b}$, or $\bar{c}$ is false. In terms of truth values $c_{\bar{a}}$, $c_{\bar{b}}$, and $c_{\bar{c}}$, this means that a clause is true if at least one of the nonnegative values $c_{\bar{a}}$, $c_{\bar{b}}$, or $c_{\bar{c}}$ is equal to 0. This, in turn, is equivalent to requiring that the minimum $\min_{a \in C} c_{\bar{a}}$ of these values is equal to 0.

In general, we want the (nonnegative) expressions $\min_{a \in C} c_{\bar{a}}$ corresponding to all the clauses C to be equal to 0. This is equivalent to requiring that the sum J of all these expressions is equal to 0, where we denoted

$$J \stackrel{\text{def}}{=} \sum_C \left(\min_{a \in C} c_{\bar{a}} \right).$$

It is possible that the original formula does not have any satisfactory propositional values $v_1, \ldots, v_n$. In this case, the value J will never become equal to 0. Thus, we can reformulate the original problem as follows: find the values $c_a \in \{0, 1\}$, with $c_a + c_{\bar{a}} = 1$ for all a, for which the expression J attains its minimum. If this minimum is 0, then we get satisfactory values v_i. If this minimum is not zero, this means that the original propositional formula cannot be satisfied.

We have reduced the original propositional satisfiability problem to a *discrete* optimization problem, in which the set of possible values of each variable c_a is discrete: it actually consists of two values 0 and 1. This is not exactly what we wanted:

- our goal was to use the experience of numerical methods;

- however, in general, discrete optimization problems are at least as hard as NP-complete problems (see, e.g., [1, 2]); thus, they are not usually solved by numerical methods.

So, to use the desired experience, we must reduce the above *discrete* optimization problem to a *continuous* one. In the above formulation, this can be easily done: just replace each discrete range $c_a \in \{0, 1\}$ by a continuous range $c_a \in [0, 1]$. Thus, we arrive at the following problem: find the values $c_a \in [0, 1]$, with $c_a + c_{\bar{a}} = 1$ for all a, for which the expression J attains its minimum.

Now, we can use the experience of numerical optimization. There exist many techniques for minimizing a function $f(x_1, \ldots, x_n)$. Most of these techniques use derivatives of the minimized function f. Among the techniques that use only the first derivatives of f, the fastest is the *gradient descent* method, in which, at each iteration, we replace the original values x_i with new values

$$x'_i = x_i - \lambda \cdot \frac{\partial f}{\partial x_i},$$

for an appropriate value λ.

In principle, we could directly use this formula to the above function J, by computing

$$\frac{\partial J}{\partial c_a} = \lim_{\Delta \to 0} \frac{J(\ldots, c_b, c_a + \Delta, c_c, \ldots) - J(\ldots, c_b, c_a, c_c, \ldots)}{\Delta}.$$

However, this would mean, in general, that we use the values $c_a + \Delta$, $c_a \in \{0, 1\}$ that were artificially added to the original values $c_a \in \{0, 1\}$. To make these computations more adequate for the original problem, it may be better to only consider values from the original set $\{0, 1\}$, that is, to use the following discrete approximation $\frac{DJ}{Dc_a}$ to the partial derivative $\frac{\partial J}{\partial c_a}$:

$$\frac{DJ}{Dc_a} \stackrel{\text{def}}{=} \frac{J(\ldots, c_b, 1, c_c, \ldots) - J(\ldots, c_b, 0, c_c, \ldots)}{1 - 0} =$$

$$J(\ldots, c_b, 1, c_c, \ldots) - J(\ldots, c_b, 0, c_c, \ldots).$$

Then, we can take $c'_a = c_a - \lambda \cdot \frac{DJ}{Dc_a}$.

The minimized function J is the sum of several terms t_C corresponding to different clauses C. One can easily check that the discrete derivative of the sum of several terms is equal to the sum of discrete derivatives of each term. For each term t_C of the type $\min(c_a, c_b, c_c)$, we have $\frac{Dt_C}{dc_a} = \min(1, c_b, c_c) - \min(0, c_b, c_c)$. Here, all the values c_b and c_c are in the interval $[0, 1]$; thus,

$$\min(1, c_b, c_c) = \min(c_b, c_c), \min(0, c_b, c_c) = 0,$$

and, therefore, $\frac{Dt_C}{dc_a} = \min(c_b, c_c)$. So, we conclude that

$$c'_{\bar{a}} = c_{\bar{a}} - \lambda \cdot \sum_{C : a \in C} \left(\min_{b \in C, b \neq a} c_{\bar{b}} \right).$$

Since $c_a + c_{\bar{a}} = 1$, any term subtracted from $c_{\bar{a}}$ means that an equal term is added to c_a; so we have

$$c'_a = c_a + \lambda \cdot \sum_{C: a \in C} \left(\min_{b \in C, b \neq a} c_{\bar{b}} \right).$$

This is exactly the chemical kinetics formulas, with λ instead of Δt.

9.3.3
Relation to Numerical Optimization: Conclusion

We can conclude that *the chemistry-motivated formulas for solving hard-to-solve problems can also be justified by the experience of numerical optimization.*

9.3.4
Relation to Numerical Optimization: What Do We Gain from It?

We can ask the same pragmatic question that we asked before: what did we gain by this optimization justification? Well, first, we gained a new justification, but – similar to the previous section – there are more pragmatic gains as well:

- First, we now use the experience of numerical optimization to come up with a new method for selecting $\Delta t = \lambda$ (and for checking whether the selected parameter Δt is adequate). Namely, we can estimate the quality of each iteration c_a if we normalize the corresponding values to the condition $c_a + c_{\bar{a}} = 1$, by taking $\widetilde{c}_a = \dfrac{c_a}{c_a + c_{\bar{a}}}$ and computing the value $J(\{\widetilde{c}_a\})$ of the minimized function. If the value of J on the next iteration is larger than the value of the previous iteration, this means that we moved too fast, and we should decrease the value λ; numerical optimization techniques recommend halving λ. Vice versa, if the value J on the next iteration is smaller, this means that maybe we can move faster, so we can try doubling λ and see what happens.
- Another idea is that, instead of gradient methods that only use the first derivatives, we can use faster second-order methods that use second derivatives as well; see, for example, [13–15].

9.3.5
Relation to Neural Computing

Neural computing is a way to perform computations by simulating how such computations are performed in the human brain. In the human brain, the state of each neuron is usually well represented by a real number – the frequency with which this neuron generates pulses. When the neuron is active, it generates a large number of pulses; when the neuron is inactive, it generates only a few pulses. Neurons send these pulses to other neurons, and the received signal changes the state of the receiving neurons.

In the first approximation, we can say that a neuron can be in two states: active and inactive. For example, a neuron receiving optical signals from the eye is active if there is light coming to the corresponding portion of the eye, and inactive if there is no light coming to this portion. Similarly, when we think about an object or a person, certain neurons are activated. Thus, when we think about how to solve a propositional satisfiability problem with propositional variables $v_1, \ldots, v_n$, it makes sense to assume that to each variable there is an appropriate neuron that becomes active when we have reasons to believe that this variable is true, and inactive if there are no such reasons.

In the brain, frequently, two different neurons (or groups of neurons) correspond to each property. For example, when we are asleep, neurons that are normally very active are deactivated; on the other hand, other neurons – which are normally not active at all – become very active. So, it makes sense to assign neurons also to negations $\bar{v}_i$: such neurons becomes active when $\bar{v}_i$ is true (i.e., when v_i is false). Thus, we assign a neuron to each literal a.

Each rule $\bar{b}, \bar{c} \to a$ means that, if we have reasons to believe in $\bar{b}$ and in $\bar{c}$, then this gives us extra reasons to believe in a. In neural terms, this means that, if the neurons $\bar{b}$ and $\bar{c}$ are active, then the neuron a also becomes more active, that is, we add a term to the original activation level c_a. This term is added only when both neurons are activated, that is, when $c_{\bar{b}} > 0$ and $c_{\bar{c}} > 0$. Similar to our analysis of the optimization relation, we can show that this combined condition is equivalent to $\min(c_{\bar{b}}, c_{\bar{c}}) > 0$. Thus, it makes sense to add to the original activation level c_a, for each implication of the type $\bar{b}, \bar{c} \to a$, a term proportional to this minimum.

As a result, when we take into account all the implications corresponding to all the clauses, we get the same Maslov's formula as in the case of chemical computing. Thus, this formula can also be interpreted in neural terms.

Comment. Minimum $\min(a, b)$ is, of course, not the only function with the property that it is positive only if both a and b are positive; we can therefore try other such functions as well. In particular, Maslov himself proposed the use of the functions $f_r(a, b) = (a^r + b^r)^{-1/r}$ for $r > 0$. When $r \to \infty$, these functions tends to $\min(a, b)$. When using these functions instead of minimum in the iterative method of solving the propositional satisfiability problems, he also got very good results; the justification of using this family of functions is given in [11].

Historical comment. Maslov himself presented this heuristic neural derivation of his iterative method in numerous talks, but he never published it. The details of Maslov's derivation were first published in [16].

9.3.6
Relation to Reasoning Under Uncertainty

In traditional mathematical reasoning, each statement is either true of false. In reasoning under uncertainty – for example, in reasoning about expert knowledge – it is important to take into account that we may have different degrees of confidence in different statements. For example, a medical expert can say that a bleeding, large-size irregularly shaped skin tumor is probably cancerous, but this expert understands well that this statement is sometimes false.

A natural idea is to represent this degree of certainty by a number from the interval [0, 1], so that:

- a complete certainty – meaning that the statement is true – corresponds to 1,
- the absence of any argument in favor of this statement (meaning probably that this statement is false) corresponds to 0, and
- intermediate degrees of certainty are represented by numbers in the interval (0, 1).

We may have different arguments in favor of a statement a and in favor of its negation $\bar{a}$; in this case, when we need to make a definite decision:

- we select a if our confidence c_a in the statement a is larger than our confidence $c_{\bar{a}}$ in its negation $\bar{a}$, that is, if $c_a > c_{\bar{a}}$;
- we select $\bar{a}$ if our confidence in the negation $\bar{a}$ is larger, that is, if $c_{\bar{a}} > c_a$.

In these terms, an implication $a, b \to c$ means that, if we believe in both a and b, then we have additional reason to believe in c, that is, that our degree of certainty in c increases. Arguments similar to the neural case show that it is reasonable to add, to the degree of certainty of a, a term proportional to $\min(c_b, c_c)$ (or to $f(c_b, c_c)$ for some other combination function). Thus, we also arrive at Maslov's iterative formulas.

9.3.7
Relation to Freedom of Choice

Freedom of choice was the original motivation of Maslov's iterative method – it is explicitly mentioned in the title of his first paper [5] describing this method; see [7, 17] for details.

This idea is easy to explain: Initially, we have a large search space, whose size grows exponentially with the length of the input. For example, for propositional satisfiability with n Boolean variables $v_1, \ldots, v_n$, this search space includes 2^n possible combinations of "true" and "false" values. Because of the huge size of this space, we cannot test all its elements. Instead, we must test only a few "most possible" candidates for a solution. For example, for propositional formulas, we can cut the size of the search space into half if we fix a value of one of the Boolean variables v_i to a certain value ε_i ("true" or "false").

Since we are not testing all the elements of the search space, we may miss a solution. So, we must select a subclass with the smallest "probability" of losing a solution. In particular, for propositional satisfiability, we must select a variable v_i and a value ε_i for which the probability of losing the solution is the smallest possible. After each choice (v_i, ε_i), there may be several solutions.

If we knew exactly the number of solutions $N(v_i, \varepsilon_i)$ left after each choice, then we could simply take a solution for which $N(v_i, \varepsilon_i) > 0$. In reality, however, we do not know these values $N(v_i, \varepsilon_i)$. At best, we know the *estimates* $\widetilde{N}(v_i, \varepsilon_i)$ for these numbers.

Usually, we have no information about the errors $\widetilde{N}(v_i, \varepsilon_i) - N(v_i, \varepsilon_i)$ of these estimates. Therefore, it is natural to assume that larger values of error are less probable than smaller ones. Hence, the larger the estimate $\widetilde{N}(v_i, \varepsilon_i)$, the larger the probability that for this choice (v_i, ε_i) the actual number of solutions will be positive, and therefore, that we will not miss a solution.

As a result, a reasonable method is to look for a choice (v_i, ε_i) after which the estimated number of solutions $\widetilde{N}(v_i, \varepsilon_i)$ is the largest possible. In other words, we must make a choice after which *the remaining freedom of choice is the largest possible*. Maslov called this idea "the strategy of increasing the freedom of choice."

Let us denote the estimate $\widetilde{N}(v_i, \varepsilon_i)$ by c_i if $\varepsilon_i =$ "true" and by c_{-i} when $\varepsilon_i =$ "false".

Each clause $a \vee b \vee c$ can be reformulated in the form $\neg a \& \neg b \to c$. From the viewpoint of the *freedom of choice* strategy, this means that if, according to our estimate, there are many solutions for which $\neg a$ and $\neg b$ are true, then the estimate for the number of solutions for which c is true must also increase. By arguing as in the neural case, we conclude that for the corresponding estimates c_a, we get exactly Maslov's iterations.

Thus, Maslov's iterative formulas can be justified on the basis of the freedom of choice as well.

Comment. While Maslov's method prompted by this freedom-of-choice principle is new, the principle itself have been formulated, in various forms, by different researchers. For example, David Marr, a well-known researcher in computer vision, described a similar principle as the *Principle of Least Commitment*.

Acknowledgments

This work was supported in part by the National Science Foundation Grants HRD-0734825 and DUE-0926721, and by Grant 1 T36 GM078000-01 from the National Institutes of Health. The authors are thankful to Evgeny Katz for encouragement.

References

1. Kreinovich, V., Lakeyev, A., Rohn, J., and Kahl, P. (1997) *Computational Complexity and Feasibility of Data Processing and Interval Computations*, Kluwer, Dordrecht.
2. Papadimitriou, C.H. (1994) *Computational Complexity*, Addison Wesley, Boston, MA.
3. Matiyasevich, Yu. (1987) Possible nontraditional methods of establishing unsatisfiability of propositional formulas, in *Problems of Cybernetics,* Moscow, vol. 131,; English translation in: (eds V. Kreinovich and G. Mints), Problems of Reducing the Exhaustive Search, American Mathematical Society, Providence, RI, 1966, pp. 75–77.
4. Blass, A. and Gurevich, Yu. (1989) On Matiyasevich's nontraditional approach to search problems. *Inform. Process. Lett.*, 32, 41–45.
5. Maslov, S.Yu. and Kurierov, Yu.N. (1980) Strategy of increasing the freedom of choice when recognizing

propositional satisfiability. Abstracts of the All-Union Conference "Methods of Mathematical Logic in Artificial Intelligence and System Programming, Vilnius, Lithuania, Part 1, pp. 130–131 (in Russian).
6. Kreinovich, V. and Mints, G. (1996) *Problems of Reducing the Exhaustive Search*, American Mathematical Society, Providence, RI.
7. Maslov, S.Yu. (1987) *Theory of Deductive Systems and its Applications*, MIT Press, Cambridge, MA.
8. Fuentes, L.O. (1991) Applying uncertainty formalisms to well-defined problems. Master's thesis. Department of Computer Science, University of Texas at El Paso.
9. Fuentes, L.O. and Kreinovich, V.Ya. (1990) Simulation of chemical kinetics as a promising approach to expert systems. Abstracts of the Southwestern Conference on Theoretical Chemistry, El Paso, Texas, November.
10. Fuentes, L.O. and Kreinovich, V. (1991) *A touch of Mexican soul makes computers smarter*. Technical Report UTEP-CS-91-6. University of Texas at El Paso, Department of Computer Science.
11. Kreinovich, V. (1987) Semantics of S. Yu. Maslov's iterative method, in *Problems of Cybernetics*, Moscow, vol. 131, pp. 30–62; English translation in: (eds V. Kreinovich and G. Mints), Problems of Reducing the Exhaustive Search, American Mathematical Society, Providence, RI, 1996, pp. 23–51.
12. Kreinovich, V. and Fuentes, L.O. (1991) Simulation of chemical kinetics – a promising approach to inference engines, in *Proceedings of the World Congress on Expert Systems, Orlando, Florida, 1991*, vol. 3, (ed J. Liebowitz), Pergamon Press, New York, pp. 1510–1517.
13. Zakharevich, M.I. (1987) Ergodic properties of Maslov's iterative method, in *Problem of Cybernetics*, Moscow, Vol. 131, (in Russian); English translation in [6], pp. 53–64.
14. Freidzon, R.I., Zakharevich, M.I., Dantsin, E.Ya., and Kreinovich, V. (1988) Hard problems: formalizing creative intelligent activity (new directions), *Proceedings of the Conference on Semiotic aspects of Formalizing Intelligent Activity, Borzhomi, Republic of Georgia, 1988, Moscow*, Nauka, pp. 407–408 (in Russian).
15. Zakharevich, M.I. (1988) Ergodic properties of Maslov's iterative method. Proceedings of the Conference on Semiotic aspects of Formalizing Intelligent Activity, Borzhomi, Republic of Georgia, 1988, Moscow, 1988, pp. 141–145 (in Russian).
16. Sirisaengtaksin, O., Fuentes, L.O., and Kreinovich, V. (1995) Non-traditional neural networks that solve one more intractable problem: propositional satisfiability. Proceedings of the 1st International Conference on Neural, Parallel, and Scientific Computations, Atlanta, Georgia, May 28–31, 1995, vol. 1, pp. 427–430.
17. Kreinovich, V. (1996) S. Maslov's Iterative Method: 15 Years Later (Freedom of Choice, Neural Networks, Numerical Optimization, Uncertainty Reasoning, and Chemical Computing), in *Problems of Reducing the Exhaustive Search* (eds V. Kreinovich and G. Mints), American Mathematical Society, Providence, Rhode Island, pp. 175–189.

10
All Kinds of Behavior are Possible in Chemical Kinetics: A Theorem and its Potential Applications to Chemical Computing
Vladik Kreinovich

10.1
Introduction

10.1.1
Chemical Computing: A Brief Reminder

No matter how fast our computers become, there are problems – such as weather prediction – that still require a large amount of computation time. A natural way to speed up computations is to use many processors working in parallel: the more processors we use, the faster we come up with the answer. Parallelization is the main reason why we humans solve many problems (such as face recognition) faster than modern computers:

- In comparison with computers that can perform billions of operations per second, a neuron is a very slow computational device, performing only between 10 and 100 operations per second.
- However, because in the brain we have billions of neurons working in parallel, the resulting image processing occurs much faster.

Once we decide on the amount of space allocated for computing, the desire to have more processors working in parallel can be re-formulated as the need to make computational units smaller and smaller. In some modern computers, individual electronic units are already of the size of several hundred molecules. A natural next step is to reduce these units to the size of a single molecule. In this case, elementary computational operations consist of interactions between molecules. Such interactions are exactly what chemistry is about. Thus, the ideal case is when controlled chemical reactions perform computations for us. This is the main idea behind *chemical computing*.

This idea sounds very promising, because with $\approx 10^{23}$ molecules, we have a potential of 10^{23} processors working in parallel – many orders of magnitude more than what we can achieve today. This idea also sounds promising because this is, in effect, how we humans process data: in the neurons, all the processes are performed by appropriate chemical reactions.

To the best of our knowledge, the idea of chemical computing was first proposed by Matiyasevich, a mathematician famous for having solved one of Hilbert's problems (the tenth). This ideas was first published in Matiyasevich's paper [1]. This idea was noticed; for example, it was discovered that, while the general idea was interesting, its specific implementation suggested by Matiyasevich did not fully explore the natural parallelism; see, for example, [2]. After that, several alternative schemes were proposed that had theoretically better computation speed-up potential; see, for example, [3–9].

The situation change drastically when, in 1994, Adleman actually performed chemical computations "*in vitro*" [10]. Since then, chemical computing has become a thriving research area; see, for example, [11–19].

10.1.2
Chemical Computing: Remaining Theoretical Challenge

From the practical viewpoint, in chemical computing we have impressive results and even more impressive potential applications. However, from the theoretical viewpoint, there is still a challenge:

- Every time we need to implement a new computation-related process in chemical computing,
- it is an intellectual challenge; and When a creative idea makes this implementation possible, it is a great result.

By applying all these creative ideas, for many processes, researchers have shown that there processes can be indeed implemented by appropriate chemical reactions. However, a more general question remains open: can *any* possible process (i.e., process described by a general system of differential equations) be implemented by an appropriate system of chemical reactions? Or, are there processes (behaviors) that chemical computing cannot directly emulate?

10.1.3
What We Do

In this chapter, we prove that every possible behavior is also possible in chemical kinetics – and thus, in principle, can be implemented by an appropriate system of chemical reactions.

Thus, whatever computational device with however weird behavior one can invent, it is, in principle, possible to implement this device chemically.

10.2
Main Result

10.2.1
Chemical Kinetics Equations: A Brief Reminder

In order to formulate our result in precise terms, we need to recall the differential equations of chemical kinetics. Readers who are well familiar with the chemical kinetics equations can skip this subsection; we added it for the benefit of computer science readers who may be interested in chemical computing but not well acquainted with chemical equations.

When chemical reactions occur, concentrations of chemical substances change; see, for example, [20–22]. General chemical reactions have the form

$$k_1 A_1 + \cdots + k_p A_p \to l_1 B_1 + \cdots + l_q B_q,$$

where A_i and B_j are molecules, and k_i and l_j describe how many molecules participate in an individual reaction.

For example, the standard reaction of combining hydrogen and oxygen into water has the form

$$2H_2 + O_2 \to 2H_2O.$$

Here, we have $p = 2$ input substances $A_1 = H_2$ and $A_2 = O_2$, with $k_1 = 2$ and $k_2 = 1$, and $q = 1$ output substance $B_1 = H_2O$, with $l_1 = 2$.

The speed of each chemical reaction depends on the intensity i_r of this reaction r and on the concentrations of the substances that take part in this reaction. For a reaction to occur, the molecules of all the input substances have to meet. The probability of such an encounter is proportional to all the concentrations C_{A_i}, so the reaction rate v_r is proportional to the product of all the concentrations:

$$v_r = i_r \cdot (C_{A_1})^{k_1} \times \cdots \times (C_{A_p})^{k_p}.$$

Because of this reaction,

- the concentration of each input substance A_i decreases with a rate $k_i \cdot v_r$: $\dot{C}_{A_i} = -k_i \cdot v_r$, where, as usual, $\dot{C}$ means the time derivative $\dfrac{dC}{dt}$, and
- the concentration of each output substance B_j increases with a rate $l_j \cdot v_r$: $\dot{C}_{B_j} = l_j \cdot v_r$.

(It is worth mentioning that in some reactions, for example, in catalysis, a substance A_i can be both an input and an output. In this case, we have $\dot{C}_{A_i} = (-k_i + l_i) \cdot v_r$.)

For example, for the above reaction r, the reaction rate is equal to

$$v_r = i_r \cdot \left(C_{H_2}\right)^2 \cdot C_{O_2}.$$

Usually, several chemical reactions $r, r', \ldots$ are going on. In this case, to describe the rate with which the concentration C_A of each substance A changes, we simply add the rates of change corresponding to different reactions. For example, for water, there is also an inverse reaction r':

$$2H_2O \rightarrow 2H_2 + O_2,$$

whose reaction rate is equal to $V_{R'} = i_{r'} \cdot (C_{H_2O})^2$. Because of these two reactions, the concentrations of hydrogen, oxygen, and water change according to the following differential equations:

$$\dot{C}_{H_2} = -2 \times i_r \cdot (C_{H_2})^2 \cdot C_{O_2} + 2 \times i_{r'} \cdot (C_{H_2O})^2;$$

$$\dot{C}_{O_2} = i_r \cdot (C_{H_2})^2 \cdot C_{O_2} + i_{r'} \cdot (C_{H_2O})^2;$$

$$\dot{C}_{H_2O} = 2 \times i_r \cdot (C_{H_2})^2 \cdot C_{O_2} - 2 \times i_{r'} \cdot (C_{H_2O})^2.$$

10.2.2
Chemical Kinetics Until Late 1950s

It is known that *general* differential equations can exhibit all kinds of behavior. Newton's equations describe the periodic motion of celestial bodies, Lorentz equations describe chaotic behavior, and so on.

In comparison with the variety of behaviors that describe general differential equations, the behavior described by most *chemical* kinetics equations is simple: some concentrations decrease, while others increase. Until late 1950s, it was expected that all chemical systems behave in this simple manner.

10.2.3
Belousov – Zhabotinsky Reaction and Further Discoveries

Our understanding of the possible behavior of chemical systems changed drastically when it was discovered that, contrary to the original expectations, the equations of chemical kinetics could exhibit a periodic behavior [23–25].

Later on, it was discovered that other chemical systems show an even more complex behavior, a chaotic behavior or a behavior corresponding to some of the patterns described by catastrophe theory; see, for example, [21, 26, 27].

10.2.4
A Natural Hypothesis

Since many kinds of weird behaviors originally observed in general differential equations have been observed in actual chemical system, it is natural to conjecture that *all* kinds of general behaviors are possible in chemical systems as well. This is what we prove in this chapter.

10.2.5
Dynamical Systems

Here, we consider *dynamical systems*, that is, systems of differential equations of the type $\dot{x}_i = f_i(x_1(t), \ldots, x_n(t))$, where f_i are continuous functions. Such systems describe most physical phenomena.

We will prove that for each observed behavior of such a system, there exists a chemical system which has exactly the same behavior.

Since a chemical system describes concentrations, and concentrations are always nonnegative, we have to restrict ourselves to dynamical systems for which $x_i(t) \geq 0$ for all i and t.

Comment. In this chapter, we consider "stationary" dynamical systems in which the rate does not explicitly depend on time t. If needed, we can also allow an explicit dependence on time, that is, we can also allow systems of the type $\dot{x}_i = f_i(t, x_1(t), \ldots, x_n(t))$.

Indeed, as it is well known in dynamical systems theory, we can easily reduce this case to the stationary case if we introduce a new auxiliary variables x_0 (whose meaning is time) with the corresponding differential equation $\dot{x}_0 = 1$ and initial value $x_0(0) = 0$. Then, the system consisting of the equations $\dot{x}_0 = 1$ and $\dot{x}_i = f_i(x_0(t), x_1(t), \ldots, x_n(t))$ is stationary and describes exactly the same solutions.

10.2.6
W.l.o.g., We Start at Time $t = 0$

Without loss of generality, we can assume that our observations started at moment $t = 0$.

Indeed, if we started at some other moment of time t_0, then we can take this moment as a new starting point for measuring time. With this new starting point, what was originally time t becomes $t' = t - t_0$. Thus, in the new timescale, we start at the moment $t' = t_0 - t_0 = 0$.

10.2.7
Limited Time

At any moment, we only have observations corresponding to finitely many *trajectories* – that is, finitely many processes whose dynamics is described by the given system of differential equations.

For each trajectory, at any given moment of time, we have only finitely many observations. Thus, we have only finitely many moments of time at which one of these processes was observed – and we also have finitely many moments of time at which we want to predict these values.

Let T denote the largest of all these moments of time. At these terms, when comparing the chemical system with the original dynamical system, it is sufficient to consider values $t \in [0, T]$.

10.2.8
Limited Values of x_i

In each of the finitely many observed processes, we have some initial values $x_i(0)$. If we denote the largest of these values by X_0, then we can conclude that all the initial conditions satisfy the inequalities

$$0 \leq x_i(0) \leq X_0, i = 1, \ldots, n.$$

For each of these processes, each function $x_i(t)$ is differentiable – we have an explicit expression for its derivative – and, therefore, it is a continuous function of time $t \in [0, T]$. Each continuous function in a closed interval is bounded; thus, each of the components x_i is bounded for each of the observed trajectories. Let X denote the largest of these bounds. So, we are guaranteed that, for all moments of time $t \in [0, T]$, all the values $x_i(t)$ are bounded by X: $x_i(t) \leq X$.

10.2.9
Limited Accuracy

Observations are never absolutely accurate, and there is always some measurement uncertainty. Once we upper-bound $\varepsilon > 0$ on the corresponding inaccuracy, then

- the results $x'_i(t)$ of the chemical system are indistinguishable from the results $x_i(t)$ of the original dynamical systems;
- if for all moments $t \in [0, T]$, these results differ by no more than ε:

$$|x'_i(t) - x_i(t)| \leq \varepsilon.$$

10.2.10
Need to Consider Auxiliary Chemical Substances

In the above, we provided a somewhat simplified description of chemical kinetics. This simplified description corresponds to the ideal case in which the input substances are directly transformed into the output ones. In reality, in most real-life chemical reactions, there are intermediate stages in which some auxiliary chemical substances are formed.

For example, when hydrogen and oxygen combine to form water, we have intermediate reactions such as $H_2 \rightarrow H + H$ or reactions of the type $H_2 + Cat \rightarrow CatH_2$ for some catalyst Cat.

To adequately describe chemical kinetics, we thus need to consider not only concentrations of the original substances but also concentrations of these auxiliary substances as well. In other words, in order to describe how the concentrations of chemical substances change during chemical reactions, we need to also consider auxiliary variables.

Since auxiliary variables are needed even for a correct description of *chemical* dynamics, we will allow auxiliary variables in the *general* case as well.

Thus, we arrive at the following definition.

Definition 1. Let $T > 0$, $X_0 > 0$, and $X > 0$ be positive real numbers, and let n be a positive integer. By a (T, X_0, X)-*dynamical system* (or simply *dynamical system*, for short), we mean a tuple $f = (f_1, \ldots, f_n)$ consisting of n continuously differentiable functions $f_i : [0, X]^n \to \mathbb{R}$, $i = 1, \ldots, n$ with the following property: For all initial values $x_1(0), \ldots, x_n(0) \in [0, X_0]$, the solution $x_1(t), \ldots, x_n(t)$ of the corresponding system of differential equations

$$\frac{dx_i}{dt} = f_i(t, x_1(t), \ldots, x_n(t)), \, i = 1, \ldots, n,$$

satisfies the inequality $0 \leq x_i(t) \leq X$ for all $t \in [0, T]$ and for all $i = 1, \ldots, n$. The values $x_i(t)$ are called the *solution* to the dynamical system (corresponding to the given initial conditions).

Comment. The requirement that the functions $f_i(x_1, \ldots, x_n)$ be continuously differentiable is introduced to make sure that the trajectory is uniquely determined by the initial conditions. If this requirement is not satisfied, we may have non-uniqueness.

Let us give a simple example of this non-uniqueness: $n = 1$, and the function $f_1(x_1)$ is defined as follows:

- $f_1(x_1) = \sqrt{x_1}$ when $x_1 \geq 0$, and
- $f_1(x_1) = 0$ when $x_1 \leq 0$.

In this case, both $x_1(t) = 0$ and $x_1'(t) = \frac{1}{2} \cdot t^2$ are solutions of the equation $\dot{x}_1 = f_1(x_1)$ with the same initial condition $x_1(0) = x_1'(0) = 0$.

Let us now formally describe the notion of chemical equations. To the previous description, we must add the need to avoid "ex nihil" ("from nothing") reactions of the type $A \to A + B$, by requiring that a *conservation law* is satisfied: in each reaction, the total atomic mass of the input should be equal to the total atomic mass of the output.

Definition 2. Let $S = \{A_1, \ldots, A_N\}$ be a finite set. Its elements will be called *substances*. Let $m_1 > 0, \ldots, m_N > 0$ be integers called *atomic masses* of the corresponding substances.

- By a *state* of the system of substances, we mean a nonnegative vector

$$x = (x_1, \ldots, x_N).$$

For each i from 1 to N, the ith component x_i of the state is called the *concentration* of the substance A_i.

- By a *reaction*, we mean a triple $r = \langle k_r, l_r, i_r \rangle$ consisting of two nonnegative integer vectors $k_r = (k_{r,1}, \ldots, k_{r,N})$ and $l_r = (l_{r,1}, \ldots, l_{r,N})$ and a positive real number i_r for which $\sum_{i=1}^{N} k_{r,i} \cdot m_i = \sum_{i=1}^{N} l_{r,i} \cdot m_i$. A reaction will also be denoted as

$$k_{r,1} A_1 + \cdots + k_{r,N} A_N \xrightarrow{i_r} l_{r,1} A_1 + \cdots + l_{r,N} A_N.$$

- By a *system of chemical reactions*, we mean a finite set R of reactions.

- By a *reaction speed* v_r corresponding to the reaction r and concentrations $x = (x_1, \ldots, x_N)$, we mean a number $v_r(x) = i_r \cdot \prod_{i=1}^{N} (x_i)^{k_{r,i}}$.
- For each set R of reactions, once we fix N initial values

$$x_1(0) \geq 0, \ldots, x_N(0) \geq 0,$$

we can then find the solution $x_i(t)$ to the system of differential equations

$$\frac{dx_i}{dt} = \sum_{r \in R} (l_{r,i} - k_{r,i}) \cdot v_r(x)$$

with the given initial values $x_1(0), \ldots, x_N(0)$. For each real number $t \geq 0$ and for each integer $i = 1, \ldots, N$, the value $x_i(t)$ is called the *solution to the chemical system*.

Comment. In the description of a chemical reaction, terms corresponding to $k_{r,i} = 0$ or to $l_{r,j} = 0$ can be omitted. Also, the coefficients $k_{r,i} = 1$ and $l_{r,j} = 1$ can also be omitted.

For example, we will simply write $2H_2 + O_2 \xrightarrow{i_r} 2H_2O$, while, strictly speaking, Definition 2 requires us to write

$$2H_2 + 1O_2 + 0H_2O \xrightarrow{i_r} 0H_2 + 0O_2 + 2H_2O.$$

Theorem 1 *For all positive numbers $T > 0$, $X_0 > 0$, and $X > 0$, and for every (T, X_0, X)-dynamical system f, there exists an integer N, a system R of chemical reactions, and values $x_{n+1}(0), \ldots, x_N(0)$ such that for all initial conditions $x_1(0), \ldots, x_n(0) \in [0, X_0]$, the solution $x_i(t)$ of the dynamical system f and the solution $x'_i(t)$ of the chemical system are ε-close for all $i \leq n$ and for all $t \in [0, T]$:*

$$|x_i(t) - x'_i(t)| \leq \varepsilon.$$

Comment. This result was first announced in [28].

10.2.11
Discussion

In practice, it is important to take into account that the initial conditions can only be implemented with some accuracy.

For computations, it makes sense to start with a dynamical system performing computations that is "deterministic" – in the sense that its behavior is not affected by minor changes in the initial conditions. Theorem 1 states that, for each dynamical system, there is a chemical system that has (within a given accuracy) exactly the same behavior. When we start with a "deterministic" system, for which trajectories do not change much if we slightly change the initial conditions, the approximating chemical system has (approximately) the same trajectories – that is, its dependence on the initial conditions is equally small. In other words, if we start with a dynamical system that is appropriate for computations, we end up with a similarly "deterministic" – and thus, equally appropriate, chemical system for computations.

Not all dynamical systems are "deterministic" (appropriate for computations) in this sense. For some dynamical systems, small changes in initial conditions lead to huge changes in the resulting trajectory $x_i(t)$. Such dynamical systems are used, for example, to generate (pseudo-)random numbers. For such systems, because of the closeness of the trajectories $x_i(t)$ and $x'_i(t)$, the approximating chemical systems have exactly the same property – small changes in initial conditions lead to huge changes in the resulting trajectory $x'_i(t)$. In other words, if we start with a "chaotic," random-number-generator-appropriate dynamical system, we end up with a similarly "chaotic" – and thus, equally random-number-generator-appropriate – chemical system.

10.2.12
Effect of External Noise

In practice, in addition to the fact that initial conditions cannot be set exactly, we also need to take into account the fact that the very equations of chemical kinetics provide only an idealized description of the dynamics. In real life, there is always some external noise that slightly changes the dynamics. A natural question is: Will the result of chemical reactions be stable under such noise?

Definition 3. Let $\alpha > 0$ be a real number. We say that two dynamical systems $f = (f_1, \ldots, f_n)$ and $f_\approx = (f_{\approx, 1}, \ldots, f_{\approx, n})$ are α-*close* if

$$|f_i(x_1, \ldots, x_n) - f_{\approx, i}(x_1, \ldots, x_n)| \leq \alpha$$

for all possible values $x = (x_1, \ldots, x_n)$.

Theorem 2 *For all positive numbers $T > 0$, $X_0 > 0$, and $X > 0$, and for every (T, X_0, X)-dynamical system f, there exists an integer N, a real number $\alpha > 0$, a system R of chemical reactions, and values $x_{n+1}(0), \ldots, x_N(0)$ such that for all initial conditions $x_1(0), \ldots, x_n(0) \in [0, X_0]$, and for every dynamical system $f_\approx$ which is α-close to the chemical system, the solution $x_i(t)$ of the dynamical system f and the solution $x_{\approx, i}(t)$ of the dynamical system $f_\approx$ are ε-close for all $i \leq n$ and for all $t \in [0, T]$:*

$$|x_i(t) - x_{\approx, i}(t)| \leq \varepsilon.$$

Thus, if the noise is sufficiently small, the solution is stable.

10.3
Proof

In this, we will prove both Theorems 1 and 2.

1°. We want to approximate trajectories of a dynamical system by the trajectories of a corresponding chemical system. Let us show that, if we can approximate the original functions $f_i(x_1, \ldots, x_n)$ by sufficiently close functions $f'_i(x_1, \ldots, x_n)$, then the trajectories of the new system will be close to the trajectories of the original system corresponding to the same initial conditions $x'_i(0) = x_i(0)$.

For this proof, we will assume that the dynamical systems are α-close, that is, for every i, the functions f_i and f'_i differ by no more than some small number α:

$$|f_i(x_1, \ldots, x_n) - f'_i(x_1, \ldots, x_n)| \leq \alpha$$

for all possible values $x = (x_1, \ldots, x_n)$ from the given box $[0, X]^n$.

We will then find the bound on the absolute value of the difference

$$\Delta x_i(t) \stackrel{\text{def}}{=} x'_i(t) - x_i(t)$$

in terms of α. From this bound, it will be clear that, by choosing α to be sufficiently small, we can make the bound on $|\Delta x_i(t)|$ also as small as possible. In other words, we will prove what we intended to: that if the dynamical systems f_i and f'_i are sufficiently close, then we can guarantee that their trajectories will be close as well.

By definition of the trajectories, we have

$$\dot{x}_i = f_i(x_1, \ldots, x_n) \text{ and } \dot{x}'_i = f'_i(x'_1, \ldots, x'_n).$$

Thus, for the difference Δx_i, we have

$$\frac{d}{dt}(\Delta x_i) = f'_i(x'_1, \ldots, x'_n) - f_i(x_1, \ldots, x_n).$$

To use the fact that the functions f' and f are close, we represent the right-hand side as the sum of the two terms, one of which describes the difference between f' and f:

$$\frac{d}{dt}(\Delta x_i) = (f'_i(x'_1, \ldots, x'_n) - f_i(x'_1, \ldots, x'_n)) + (f_i(x'_1, \ldots, x'_n) - f_i(x_1, \ldots, x_n)).$$

The absolute value of the first term is bounded by α. To estimate the value of the second term – in which all n variables change – we will represent it, in turn, as the sum of several terms corresponding to a change in a single variable. First, we change x_1, then we change x_2, and so on:

$$f_i(x'_1, \ldots, x'_n) - f_i(x_1, \ldots, x_n) =$$

$$(f_i(x'_1, x_2, \ldots, x_n) - f_i(x_1, \ldots, x_n)) +$$

$$(f_i(x'_1, x'_2, x_3, \ldots, x'_n) - f_i(x'_1, x_2, x_3, \ldots, x_n)) + \cdots +$$

$$(f_i(x'_1, \ldots, x'_{n-1}, x_n) - f_i(x'_1, \ldots, x'_{n-1}, x'_n)).$$

To estimate each of these differences, we can use the fact that all the functions $f_i(x_1, \ldots, x_n)$ are continuously differentiable, that is, that each partial derivative (exists and) is continuous. Every continuous function on a bounded closed box is bounded. Let M denote the largest of the maxima of all these partial derivatives. Then, for all points x in this box and for all i and j, we have $\left|\dfrac{\partial f_i}{\partial x_j}\right| \leq M$.

For a function of one variable whose derivative is bounded by some number M, from

$$f(x') - f(x) = \int_x^{x'} \frac{df}{dx}\, dx,$$

we conclude that

$$|f(x') - f(x)| = \left|\int_x^{x'} \frac{df}{dx}\, dx\right| \leq M \cdot |x' - x|.$$

In the general case of a function of several variables, we can apply this argument to the dependence on x_1 and conclude that

$$|f_i(x_1', x_2, \ldots, x_n) - f_i(x_1, x_2, \ldots, x_n)| \leq M \cdot |\Delta x_1|.$$

By similarly considering a change in x_2, we conclude that

$$|f_i(x_1', x_2', \ldots, x_n) - f_i(x_1', x_2, \ldots, x_n)| \leq M \cdot |\Delta x_2|,$$

and so on, until we reach the last variable:

$$|f_i(x_1', \ldots, x_{n-1}', x_n) - f_i(x_1', \ldots, x_{n-1}', x_n')| \leq M \cdot |\Delta x_n|.$$

From the above equality,

$$f_i(x_1', \ldots, x_n') - f_i(x_1, \ldots, x_n) =$$

$$(f_i(x_1', x_2, \ldots, x_n) - f_i(x_1, \ldots, x_n)) +$$

$$(f_i(x_1', x_2', x_3, \ldots, x_n') - f_i(x_1', x_2, x_3, \ldots, x_n)) + \cdots +$$

$$(f_i(x_1', \ldots, x_{n-1}', x_n) - f_i(x_1', \ldots, x_{n-1}', x_n')),$$

we can now conclude that

$$|f_i(x_1', \ldots, x_n') - f_i(x_1, \ldots, x_n)| \leq$$

$$|f_i(x_1', x_2, \ldots, x_n) - f_i(x_1, \ldots, x_n)| +$$

$$|f_i(x_1', x_2', x_3, \ldots, x_n') - f_i(x_1', x_2, x_3, \ldots, x_n)| + \cdots +$$

$$|f_i(x_1', \ldots, x_{n-1}', x_n) - f_i(x_1', \ldots, x_{n-1}', x_n')|.$$

We already know the bounds for each terms in the right-hand side, so we conclude that

$$|f_i(x_1', \ldots, x_n') - f_i(x_1, \ldots, x_n)| \leq M \cdot |\Delta x_1| + M \cdot |\Delta x_2| + \cdots + M \cdot |\Delta x_n|,$$

that is,

$$|f_i(x'_1, \ldots, x'_n) - f_i(x_1, \ldots, x_n)| \leq M \cdot \sum_{i=1}^{n} |\Delta x_i|.$$

For every i, from the above formula,

$$\frac{d}{dt}(\Delta x_i) = (f'_i(x'_1, \ldots, x'_n) - f_i(x'_1, \ldots, x'_n)) + (f_i(x'_1, \ldots, x'_n) - f_i(x_1, \ldots, x_n)),$$

and we conclude that

$$\left|\frac{d}{dt}(\Delta x_i)\right| \leq |f'_i(x'_1, \ldots, x'_n) - f_i(x'_1, \ldots, x'_n)| + |f_i(x'_1, \ldots, x'_n) - f_i(x_1, \ldots, x_n)|.$$

We know that the first term is bounded by α and the second by $M \cdot \Delta$, where we denoted $\Delta \stackrel{def}{=} \sum_{i=1}^{n} |\Delta x_i|$. Thus, we have

$$\left|\frac{d}{dt}(\Delta x_i)\right| \leq \alpha + M \cdot \Delta.$$

Based on these inequalities, we want to deduce an inequality in terms of $\Delta = \sum_{i=1}^{n} |\Delta x_i|$ and its derivative. By the chain rule,

$$\frac{d}{dt}(|\Delta x_i|) = \text{sign}(\Delta x_i) \cdot \frac{d}{dt}(\Delta x_i),$$

where $\text{sign}(z) = \frac{d}{dz}(|z|)$ is equal to 1 for $z \geq 0$ and to -1 for $z \leq 0$. Since the absolute value of $\text{sign}(z)$ is always equal to 1, we get

$$\left|\frac{d}{dt}(|\Delta x_i|)\right| = \left|\frac{d}{dt}(\Delta x_i)\right|,$$

hence

$$\left|\frac{d}{dt}(|\Delta x_i|)\right| \leq \alpha + M \cdot \sum_{i=1}^{n} |\Delta x_i|.$$

By definition of $\Delta = \sum_{i=1}^{n} |\Delta x_i|$, we have

$$\frac{d}{dt}(\Delta) = \sum_{i=1}^{n} \frac{d}{dt}(|\Delta x_i|).$$

We have already shown that each term in the sum is bounded by $\alpha + M \cdot \Delta$, so we conclude that

$$\dot{\Delta} = \frac{d}{dt}(\Delta) \leq n \cdot \alpha + n \cdot M \cdot \Delta.$$

Therefore, to estimate the difference Δ, we must make conclusions based on this differential inequality.

The process of finding solutions of an inequality usually starts with solving the corresponding equality

$$\frac{d\Delta}{dt} = n \cdot \alpha + n \cdot M \cdot \Delta.$$

To simplify this expression, we can move all the terms containing Δ to the left-hand side – by dividing both sides by the original right-hand side $n \cdot \alpha + n \cdot M \cdot \Delta$. As a result, we get

$$\frac{1}{dt} \cdot \frac{d\Delta}{n \cdot \alpha + n \cdot M \cdot \Delta} = 1.$$

This expression can be simplified if we take into account that for the denominator

$$\widetilde{\Delta} \stackrel{\text{def}}{=} n \cdot \alpha + n \cdot M \cdot \Delta,$$

we have $d\widetilde{\Delta} = n \cdot M \cdot d\Delta$. Thus, multiplying both sides of the above inequality by $n \cdot M$, we get

$$\frac{1}{dt} \cdot \frac{d\widetilde{\Delta}}{\widetilde{\Delta}} = n \cdot M.$$

The left-hand side is equal to the derivative $\dfrac{d}{dt}(\ln(\widetilde{\Delta}))$ of the logarithm of $\widetilde{\Delta}$. The derivative is constant, so the logarithm is a linear function of time: $\ln(\widetilde{\Delta}(t)) = C + n \cdot M \cdot t$ for some constant C.

To find the integration constant, let is consider the initial moment of time $t = 0$. For $t = 0$, we have $\widetilde{\Delta}(0) = n \cdot \alpha + n \cdot M \cdot \Delta(0)$. Here,

$$\Delta(0) = \sum_{i=1}^{n} |\Delta x_i(0)| = \sum_{i=1}^{n} |x_i'(0) - x_i(0)|.$$

Since we start with the same initial conditions, we have $|x_i'(0) - x_i(0)| = 0$ and, thus, $\Delta(0) = 0$ and $\widetilde{\Delta}(0) = n \cdot \alpha$. Hence, $\ln(\widetilde{\Delta}(0)) = C = \ln(n \cdot \alpha)$ and, thus, $\ln(\widetilde{\Delta}(t)) = \ln(n \cdot \alpha) + n \cdot M \cdot t$. Therefore,

$$\widetilde{\Delta}(t) = \exp(\ln(\widetilde{\Delta}(t))) = n \cdot \alpha \cdot \exp(n \cdot M \cdot t).$$

So, for $\Delta(t) = \dfrac{\widetilde{\Delta}(t) - n \cdot \alpha}{n \cdot M}$, we get

$$\Delta(t) = \frac{n \cdot \alpha \cdot (\exp(n \cdot M \cdot t) - 1)}{n \cdot M} = \frac{\alpha}{M} \cdot (\exp(n \cdot M \cdot t) - 1).$$

Similarly, from the above inequality $\dfrac{d}{dt}(\Delta) \leq n \cdot \alpha + n \cdot M \cdot \Delta = \widetilde{\Delta}$, we conclude that

$$\frac{d\widetilde{\Delta}}{dt} = n \cdot M \cdot \frac{d\Delta}{dt} \leq n \cdot M \cdot \widetilde{\Delta};$$

hence for

$$\frac{d(\ln(\widetilde{\Delta}))}{dt} = \frac{1}{dt} \cdot \frac{d\widetilde{\Delta}}{\widetilde{\Delta}},$$

we get
$$\frac{d(\ln(\widetilde{\Delta}))}{dt} \leq n \cdot M.$$
Thus,
$$\ln(\widetilde{\Delta}(t)) = \ln(\widetilde{\Delta}(0)) + \int_0^t \frac{d(\ln(\widetilde{\Delta}))}{dt}\, dt \leq$$
$$\ln(\widetilde{\Delta}(0)) + n \cdot M \cdot t = \ln(n \cdot \alpha) + n \cdot M \cdot t.$$
Therefore,
$$\widetilde{\Delta}(t) = \exp(\ln(\widetilde{\Delta}(t))) \leq n \cdot \alpha \cdot \exp(n \cdot M \cdot t).$$
Thus, for $\Delta(t) = \dfrac{\widetilde{\Delta}(t) - n \cdot \alpha}{n \cdot M}$, we get
$$\Delta(t) \leq \frac{n \cdot \alpha \cdot (\exp(n \cdot M \cdot t) - 1)}{n \cdot M} = \frac{\alpha}{M} \cdot (\exp(n \cdot M \cdot t) - 1).$$
The right-hand side of this inequality is an increasing function of time t, so its largest value is attained for the largest possible value $t = T$; therefore,
$$\Delta(t) \leq \frac{\alpha}{M} \cdot (\exp(n \cdot M \cdot T) - 1).$$
Since $\Delta(t) = \sum_{i=1}^{n} |\Delta x_i(t)| = \sum_{i=1}^{n} |x'_i(t) - x_i(t)|$, this implies that, for every t and for every i, we have
$$|x'_i(t) - x_i(t)| \leq \frac{\alpha}{M} \cdot (\exp(n \cdot M \cdot T) - 1).$$
So, for any $\varepsilon > 0$, if we want to make sure that $|x'_i(t) - x_i(t)| \leq \varepsilon$ for all i and t, it is sufficient to choose α for which $\dfrac{\alpha}{M} \cdot (\exp(n \cdot M \cdot T) - 1) \leq \varepsilon$: that is, to choose
$$\alpha = \frac{M}{\exp(n \cdot M \cdot T) - 1} \cdot \varepsilon.$$
The statement is proven.

2°. We want to prove that an arbitrary dynamical system can be approximated by an appropriate chemical system. Our proof of this approximation result consists of two stages:

- First, we will prove this result for a certain class of dynamical systems, a class selected because for systems from this class the desired approximation is easier to construct;
- After that, we will prove that an arbitrary dynamical system can be approximated by a system of this simpler type, and how the chemical approximation of this simpler system can be modified into a chemical approximation to the original dynamical system.

3°. The definition of a special class of dynamical systems – with which we start the approximation result – is based on the fact that for every i, and for all points $x = (x_1, \ldots, x_n)$ for which $x_i > 0$, the ratio $g_i(x_1, \ldots, x_n) \stackrel{\text{def}}{=} \dfrac{f_i(x_1, \ldots, x_n)}{x_i}$ is continuous – as a ratio of two continuous functions.

As a special class, we will consider all dynamical systems for which each of these functions $g_i(x_1, \ldots, x_n)$ can be continuously extended to the values $x_i = 0$. In this case, $f_i(x_1, \ldots, x_n) = x_i \cdot g_i(x_1, \ldots, x_n)$ for some continuous function $g_i(x_1, \ldots, x_n)$. Then, to construct the desired α-approximation $f'_i(x_1, \ldots, x_n)$ to the given function $f_i(x_1, \ldots, x_n)$, it is sufficient to find, for an appropriate small $\beta > 0$, a β-approximation $g'_i(x_1, \ldots, x_n)$ to the ratio $g_i(x_1, \ldots, x_n)$. Once this approximation is found, for the function $f'_i(x_1, \ldots, x_n) = x_i \cdot g'_i(x_1, \ldots, x_n)$, the inequality

$$|g'_i(x_1, \ldots, x_n) - g_i(x_1, \ldots, x_n)| \leq \beta$$

implies that

$$|f'_i(x_1, \ldots, x_n) - f_i(x_1, \ldots, x_n)| = x_i \cdot |g'_i(x_1, \ldots, x_n) - g_i(x_1, \ldots, x_n)| \leq X \cdot \beta.$$

Thus, it is sufficient to take β for which $X \cdot \beta \leq \alpha$: for example, to take $\beta = \dfrac{\alpha}{X}$.

4°. It is known that an arbitrary, continuous function on a box can be approximated, with any given accuracy, by a polynomial.

We will use this result and approximate the original ratio $g_i(x_1, \ldots, x_n)$ by a polynomial $g'_i(x_1, \ldots, x_n)$. A polynomial is, in general, a linear combination of monomials:

$$g'_i(x_1, \ldots, x_n) = \sum_m a_m \cdot P_m(x_1, \ldots, x_n),$$

where each a_m is a constant, and

$$P_m(x_1, \ldots, x_n) = x_1^{d_{m,1}} \cdot \ldots \cdot x_n^{d_{m,n}},$$

with $d_{m,j} \geq 0$. In this case,

$$f'_i(x_1, \ldots, x_n) = x_i \cdot g'_i(x_1, \ldots, x_n) = \sum_m a_m \cdot (x_i \cdot P_m(x_1, \ldots, x_n)),$$

where each term $x_i \cdot P_m(x_1, \ldots, x_n)$ has the form

$$x_i \cdot P_m(x_1, \ldots, x_n) = x_1^{d_{m,1}} \cdot \ldots \cdot x_{i-1}^{d_{m,i-1}} \cdot x_i^{d_{m,i}+1} \cdot x_{i+1}^{d_{m,i+1}} \cdot \ldots \cdot x_n^{d_{m,n}}.$$

5°. For each monomial $P_m(x_1, \ldots, x_n)$, we will find a reaction r for which its intensity v_r is close to the corresponding term $a_m \cdot x_i \cdot P_m(x_1, \ldots, x_n)$.

To form these reaction, to each variable x_i we assign a substance A_i whose concentration will be described by x_i. In addition to the resulting n substances, we will use an auxiliary "universal" substance U – a substance that can be, in principle, transformed into any other substance. The concentration of the universal substance U will be denoted by u.

We assume that all $n + 1$ substances $A_1, \ldots, A_n, U$ have the same atomic weight $m_1 = \cdots = m_n = m_U$.

A specific rule corresponding to the monomial $P_m(x_1, \ldots, x_n)$ will be different depending on the whether the coefficient a_m is positive or negative.

6°. For monomials for which $a_m > 0$, we take the reaction

$$U + d_{m,1}A_1 + \cdots + d_{m,i-1}A_{i-1} + (d_{m,i} + 1)A_i + d_{m,i+1}A_{i+1} + \cdots + d_{m,n}A_n \to$$

$$d_{m,1}A_1 + \cdots + d_{m,i-1}A_{i-1} + (d_{m,i} + 2)A_i + d_{m,i+1}A_{i+1} + \cdots + d_{m,n}A_n.$$

In this reaction, for all the substances except for the ith substance A_i and the auxiliary universal substance U, the number of molecules entering the reaction is the same as the number of molecules leaving this reaction. The only difference is that a molecule of the universal substance is transformed into a molecule of A_i. In this transformation, all the other substances play the role of catalysts – in the sense that

- while the presence of these other substances is necessary for the reaction to occur, and
- while the speed of this reaction depends on the concentrations of these other substances,
- in the long run, each of this other substances is neither consumed nor produced – in this reaction their concentration does not change.

A precise description of the corresponding terms in chemical equations confirms this qualitative analysis. Indeed, according to the general formula for the chemical equations (as given in Definition 2), the speed v_r of this reaction is equal to

$$v_r = i_r \cdot u \cdot x_1^{d_{m,1}} \cdot \cdots \cdot x_{i-1}^{d_{m,i-1}} \cdot x_i^{d_{m,i}+1} \cdot x_{i+1}^{d_{m,i+1}} \cdot \cdots \cdot x_n^{d_{m,n}},$$

where i_r denotes the intensity of this reaction and u denotes the concentration of the universal substance U. In terms of the monomial $P_m(x_1, \ldots, x_n)$, this formula takes the form

$$v_r = i_r \cdot u \cdot x_i \cdot P_m(x_1, \ldots, x_n).$$

According to the same Definition 2, as a result of this reaction, there is no contribution to differential equations describing $\dfrac{dx_j}{dt}$ for $j \neq i$. The only contributions are to the terms $\dfrac{dx_i}{dt}$ and $\dfrac{du}{dt}$: namely, we get the terms

$$\frac{dx_i}{dt} = \cdots + i_r \cdot u \cdot x_i \cdot P_m(x_1, \ldots, x_n) + \cdots,$$

$$\frac{du}{dt} = \cdots - i_r \cdot u \cdot x_i \cdot P_m(x_1, \ldots, x_n) + \cdots$$

When $i_r \cdot u = a_m$, the term corresponding to $\dfrac{dx_i}{dt}$ is exactly the desired term $a_m \cdot x_i \cdot P_m(x_1, \ldots, x_n)$. However, this does not mean that we have solved our problem; indeed,

- we can select the initial concentration $u(0)$ of the universal quantity U to satisfy the equality $i_r \cdot u(0) = a_m$;

- however, due to the differential equation describing u, the amount u decreases with time, so at the next moments of time, this amount will be smaller, and the equality $i_r \cdot u(t) = a_m$ will no longer be satisfied.

Let us show that by selecting $u(0)$ to be large enough – and by correspondingly selecting $i_r = \dfrac{a_m}{u(0)}$ – we will be able to guarantee that the approximate equality $i_r \cdot u(t) \approx a_m$ holds for all moments $t \in [0, T]$ with a given accuracy.

Indeed, for $i_r = \dfrac{a_m}{u(0)}$, the product $i_r \cdot u(t)$ takes the form $\dfrac{u(0)}{u(t)} \cdot a_m$. Thus, if we find $u(0)$ for which the ratio $\dfrac{u(0)}{u(t)}$ is close to 1 for all t, we get the desired approximate equality $i_r \cdot u(t) \approx a_m$ for all moments $t \in [0, T]$.

In the ideal case when $i_r \cdot u \approx a_m$, the corresponding term in $\dfrac{du}{dt}$ takes the form

$$\frac{du}{dt} = \cdots - a_m \cdot x_i \cdot P_m(x_1, \ldots, x_n) + \cdots$$

Equations corresponding to different monomials lead to other such terms (and other terms comes from monomials with $a_m < 0$, see the following part of the proof). The good news for us is that these terms do not depend on the selected value $u(0)$, so we have

$$\frac{du}{dt} = d(t),$$

for some function $d(t)$ which does not depend on $u(0)$. Thus,

$$u(t) = u(0) + D(t),$$

where we denoted

$$D(t) \stackrel{\text{def}}{=} \int_0^t d(s) \, ds.$$

The function $D(t)$ is a (differentiable hence) continuous function of t, so its absolute value $|D(t)|$ has the largest possible value D. Thus, for all t, we have $|u(0) - u(t)| \le D$, and hence $u(0) - D \le u(t) \le u(0) + D$. So, the ratio $\dfrac{u(0)}{u(t)}$ satisfies the inequality

$$\frac{u(0)}{u(0) + D} \le \frac{u(0)}{u(t)} \le \frac{u(0)}{u(0) + D},$$

or, equivalently,

$$\frac{1}{1 + \dfrac{D}{u(0)}} \le \frac{u(0)}{u(t)} \le \frac{1}{1 - \dfrac{D}{u(0)}}.$$

When $u(0)$ increases, both the left-hand side and the right-hand side terms in this inequality tend to 1. Hence, whatever accuracy we want in approximating the original monomial term by the corresponding chemical term, we can indeed guarantee it by selecting an approximately large value $u(0)$.

7°. For monomials for which $a_m < 0$, we take the reaction

$$d_{m,1}A_1 + \cdots + d_{m,i-1}A_{i-1} + (d_{m,i} + 1)A_i + d_{m,i+1}A_{i+1} + \cdots + d_{m,n}A_n \to$$

$$U + d_{m,1}A_1 + \cdots + d_{m,i-1}A_{i-1} + d_{m,i}A_i + d_{m,i+1}A_{i+1} + \cdots + d_{m,n}A_n.$$

In this reaction, for all the substances except for the ith substance A_i and the auxiliary universal substance U, the number of molecules entering the reaction is the same as the number of molecules leaving this reaction. The only difference is that a molecule of the ith substance A_i is transformed into a molecule of the universal substance U. In this transformation, all the other substances play the role of catalysts – in the sense that

- while the presence of these other substances is necessary for the reaction to occur, and
- while the speed of this reaction depends on the concentrations of these other substances,
- in the long run, each of this other substances is neither consumed nor produced – in this reaction their concentration does not change.

A precise description of the corresponding terms in chemical equations also confirms this qualitative analysis. Indeed, according to the general formula for the chemical equations (as given in Definition 2), the speed v_r of this reaction is equal to

$$v_r = i_r \cdot x_1^{d_{m,1}} \cdot \ldots \cdot x_{i-1}^{d_{m,i-1}} \cdot x_i^{d_{m,i}+1} \cdot x_{i+1}^{d_{m,i+1}} \cdot \ldots \cdot x_n^{d_{m,n}},$$

where i_r denotes the intensity of this reaction. In terms of the monomial $P_m(x_1, \ldots, x_n)$, this formula takes the form

$$v_r = i_r \cdot x_i \cdot P_m(x_1, \ldots, x_n).$$

According to the same Definition 2, as a result of this reaction, there is no contribution to the differential equations describing $\dfrac{dx_j}{dt}$ for $j \neq i$. The only contributions are to the terms $\dfrac{dx_i}{dt}$ and $\dfrac{du}{dt}$: namely, we get the terms

$$\frac{dx_i}{dt} = \cdots - i_r \cdot x_i \cdot P_m(x_1, \ldots, x_n) + \cdots,$$

$$\frac{du}{dt} = \cdots + i_r \cdot x_i \cdot P_m(x_1, \ldots, x_n) + \cdots$$

When $i_r = |a_m|$, the term corresponding to $\dfrac{dx_i}{dt}$ is exactly the desired term $a_m \cdot x_i \cdot P_m(x_1, \ldots, x_n)$. The corresponding term for $\dfrac{du}{dt}$ takes the form

$$\frac{du}{dt} = \cdots + |a_m| \cdot x_i \cdot P_m(x_1, \ldots, x_n) + \cdots$$

8°. Thus, the desired approximation result is proven for the case when each ratio $g_i(x_1, \ldots, x_n) = \dfrac{f_i(x_1, \ldots, x_n)}{x_i}$ can be continuously extended to values $x = (x_1, \ldots, x_n)$ for which $x_i = 0$.

9°. Let us now show how a general dynamical system can be reduced to this special case. The main idea of this reduction is to avoid the zone $x_i \approx 0$ where the above ratio condition is not satisfied. For this purpose, we will add, to each substance, a small amount $\delta > 0$, and perform all the dynamics after that as before; the dynamics for the zone $x_i < \delta$ – which does not affect our processes, except for the short first period of time when the concentrations are increases – will then be defined in such a way that the above condition about the ratio $g_i(x_1, \ldots, x_n)$ is satisfied.

To implement this idea, we will introduce new functions $\widetilde{f}_i(x_1, \ldots, x_n)$, which, for $x_i \geq \delta$, have the form

$$\widetilde{f}_i(x_1, \ldots, x_n) = f_i(x_1 - \delta, \ldots, x_n - \delta).$$

The meaning of this definition is that, in describing the changes in all the variables x_i, we do not take into account the extra amount δ that we added to the concentrations x_i. Thus, trajectories of the new system have the form $\widetilde{x}_i(t) = \delta + x_i(t)$, where $x_i(t)$ is a trajectory of the original dynamical system. When δ is small, these trajectories are close.

For values $x_i < \delta$, the above formula does not work, since the original function $f_i(x_1, \ldots, x_n)$ is only defined for $x_i \geq 0$. Thus, we need to extend the above expression to cover such values. We want to make sure that there is a limit of the ratio $\widetilde{g}_i(x_1, \ldots, x_n) = \dfrac{\widetilde{f}_i(x_1, \ldots, x_n)}{x_i}$ when $x_i \to 0$. When such a limit exists, then $\widetilde{f}_i(x_1, \ldots, x_n) = x_i \cdot \widetilde{g}_i(x_1, \ldots, x_n)$ for some continuous function $\widetilde{g}_i(x_1, \ldots, x_n)$; in this case, for $x_i = 0$, we have

$$\widetilde{f}_i(x_1, \ldots, x_{i-1}, 0, x_{i+1}, \ldots, x_n) = 0.$$

$$\widetilde{f}_i(x_1, \ldots, x_{i-1}, 0, x_{i+1}, \ldots, x_n) = 0.$$

Vice versa, if $\widetilde{f}_i(x_1, \ldots, x_{i-1}, 0, x_{i+1}, \ldots, x_n) = 0$, then the limit of the desired ratio takes the following form:

$$\lim_{x_i \to 0} \frac{\widetilde{f}_i(x_1, \ldots, x_{i-1}, x_i, x_{i+1}, \ldots, x_n)}{x_i} =$$

$$\lim_{x_i \to 0} \frac{\widetilde{f}_i(x_1, \ldots, x_{i-1}, x_i, x_{i+1}, \ldots, x_n) - \widetilde{f}_i(x_1, \ldots, x_{i-1}, 0, x_{i+1}, \ldots, x_n)}{x_i}.$$

One can easily check that this is exactly the definition of the partial derivative $\dfrac{\partial \widetilde{f}_i}{\partial x_i}$ at the point where $x_i = 0$. So, when we extend the above expression $\widetilde{f}_i(x_1, \ldots, x_{i-1}, x_i, x_{i+1}, \ldots, x_n)$ to a continuously differentiable function that is defined for all $x_i \geq 0$, all we need to do to satisfy the above limit condition is to make sure that

$$\widetilde{f}_i(x_1, \ldots, x_{i-1}, 0, x_{i+1}, \ldots, x_n) = 0.$$

for all i. For each i, we thus extend the original smooth functions $\widetilde{f}_i(x_1, \ldots, x_n)$ defined for $x_i \geq \delta$ to the zone $x_i \geq 0$ in such a way that $\widetilde{f}_i(x_1, \ldots, x_n) = 0$ for $x_i = 0$. The existence of such extension is well known in mathematical analysis.

To move the behavior into the zone $x_i \geq \delta$, we add – very fast – the amount δ to the concentrations of each of n substances $A_1, \ldots, A_n$. This can be done, for example, if we introduce n new auxiliary substances $U_1, \ldots, U_n$ each of which has the initial concentration $u_i(0) = \delta$, and add n very fast reactions $U_i \to A_i$ for each i from 1 to n. As a result of these reactions, each substance A_i indeed acquires an additional concentration δ. When the time is short enough, this transition does not affect the remaining dynamics, so the new trajectories $x_i(t)$ will be close to the original ones.

$10°$. The reduction of a generic dynamical system to a dynamical system of a special type is explained. Since we have already proven that dynamical systems of the special type can be approximated by chemical systems, we can thus also an approximate generic dynamical system by a chemical system.

Theorem 1 is proven.

$11°$. Let us now prove Theorem 2. Because of Theorem 1, we can find a system of chemical reactions for which the trajectories $x_i(t)$ of the original system f are $(\varepsilon/2)$-close to the trajectories $x'_i(t)$ of the chemical system:

$$|x_i(t) - x'_i(t)| \leq \frac{\varepsilon}{2}.$$

On the other hand, owing to Part 1 of this proof, for sufficiently small $\alpha > 0$, the trajectories $x_{\approx,i}(t)$ of the realistic (approximately chemical) dynamical system $f\approx$ are $(\varepsilon/2)$-close to the trajectories $x'_i(t)$ of the chemical system

$$|x'_i(t) - x_{\approx,i}(t)| \leq \frac{\varepsilon}{2}.$$

Thus, combining these two inequalities, we conclude that

$$|x_i(t) - x_{\approx,i}(t)| \leq |x_i(t) - x'_i(t)| + |x'_i(t) - x_{\approx,i}(t)| \leq \frac{\varepsilon}{2} + \frac{\varepsilon}{2} = \varepsilon.$$

So, the trajectories $x_{\approx,i}(t)$ of the realistic dynamical system $f\approx$ are indeed ε-close to the trajectories $x_i(t)$ of the original dynamical system.

This conclusion proves Theorem 2.

Acknowledgments

This work was supported in part by the National Science Foundation Grants HRD-0734825 and DUE-0926721 and by Grant 1 T36 GM078000-01 from the National Institutes of Health. The author wishes to thank Larry Ellzey, Olac Fuentes, and Grigoriy Yablonsky for valuable discussions, and the anonymous referees for important advice.

References

1. Matiyasevich, Yu. (1987) Possible nontraditional methods of establishing unsatisfiability of propositional formulas, in *Problems of Cybernetics*, Moscow, vol. 131; English translation in: (eds V. Kreinovich and G. Mints), Problems of Reducing the Exhaustive Search, American Mathematical Society, Providence, RI, 1987, pp. 75–77.
2. Blass, A. and Gurevich, Yu. (1989) On Matiyasevich's nontraditional approach to search problems. *Inf. Process. Lett.*, **32**, 41–45.
3. Fuentes, L.O. (1991) Applying Uncertainty Formalisms to Well-Defined Problems. Master's thesis. Department of Computer Science, University of Texas at El Paso.
4. Fuentes, L.O. and Kreinovich, V.Ya. (1990) Simulation of chemical kinetics as a promising approach to expert systems. Abstracts of the Southwestern Conference on Theoretical Chemistry, El Paso, TX, November.
5. Fuentes, L.O. and Kreinovich, V. (1991) A touch of Mexican soul makes computers smarter. Technical Report UTEP-CS-91-6. Department of Computer Science, University of Texas at El Paso, El Paso, Texas.
6. Hjelmfelt, A., Weinberger, E.D., and Ross, J. (1991) Chemical implementation of neural networks and Turing machines. *Proc. Natl. Acad. Sci. U.S.A.*, **88**, 10983–10987.
7. Hjelmfelt, A., Weinberger, E.D., and Ross, J. (1992) Chemical implementation of finite-state machines. *Proc. Natl. Acad. Sci. U.S.A.*, **89**, 383–387.
8. Kreinovich, V. (1987) Semantics of S. Yu. Maslov's iterative method, in *Problems of Cybernetics*, Moscow, vol. 131, pp. 30–62; English translation in: (eds V. Kreinovich and G. Mints), Problems of Reducing the Exhaustive Search, American Mathematical Society, Providence, RI, 23–51.
9. Krcinovich, V. and Fuentes, L.O. (1991) Simulation of chemical kinetics – a promising approach to inference engines, in *Proceedings of the World Congress on Expert Systems, Orlando, Florida, 1991*, vol. 3, (ed. J. Liebowitz), Pergamon Press, New York, pp. 1510–1517.
10. Adleman, L.M. (1994) Molecular computation of solutions to combinatorial problems. *Science*, **266**, 11, 1021–1024.
11. Adamatzky, A., Holley, J., Bull, L., and De Lacy Costello, B. (2011) On computing in fine-grained compartmentalised Belousov-Zhabotinsky medium. *Chaos, Solitons & Fractals*, **44**, 10, 779–790.
12. Adamatzky, A., De Lacy Costello, B., and Bull, L. (2011) On polymorphic logical gates in sub-excitable chemical medium. *Int. J. Bifurcat. Chaos*, **21**, 7, 1977–1986.
13. Adamatzky, A., Martinez, G.J., Zhang, L., and Wuensche, A. (2010) Operating binary strings using gliders and eaters in reaction-diffusion cellular automaton. *Math. Comput. Model.*, **52**, 177–190.
14. De Lacy Costello, B., Adamatzky, A., Jahan, I., and Zhang, Z. (2011) Towards constructing one-bit binary adder in excitable chemical medium. *Chem. Phys.*, **381**, 88–99.
15. Holley, J., Adamatzky, A., Bull, L., De Lacy Costello, B., and Jahan, I. (2011) Computational modalities of Belousov-Zhabotinsky encapsulated vesicles. *Nano Commun. Netw.*, **2**, 50–61.
16. Katz, E. and Privman, V. (2010) Enzyme-based logic systems for information processing. *Chem. Soc. Rev.*, **39**, 1835–1857.
17. Martinez, G., Adamatzky, A., Morita, K., and Margenstern, M. (2010) Computation with competing patterns in Life-like automaton, in *Game of Life Cellular Automata* (ed. A. Adamatzky), Springer, pp. 547–572.
18. Pita, M., Strack, G., MacVittie, K., Zhou, J., and Katz, E. (2009) Set-reset flip-flop memory based on enzyme reactions: Towards memory systems controlled by biochemical pathways. *J. Phys. Chem. B*, **113**, 16071–16076.
19. Strack, G., Pita, M., Ornatska, M., and Katz, E. (2008) Boolean logic gates that use enzymes as input signals. *J. Am. Chem. Soc.*, **130**, 4234–4235.

20. Amundson, N.R. and Aris, R. (1973) *Mathematical Methods in Chemical Engineering*, Prentice Hall, Englewood Cliffs, NJ.
21. Aris, R. (1978) *Mathematical Modeling Techniques*, Pitman, London.
22. Aris, R. (2000) *Elementary Chemical Reactor Analysis*, Dover Publ., New York.
23. Belousov, B.P. (1959) A periodic reaction and its mechanism, in *Collection of Abstracts on Radiation Medicine in 1958*, Medicine Publ., Moscow, pp. 145–147 (in Russian); translated in: (eds R.J. Field and M. Burger), Oscillations and Traveling Waves in Chemical Systems, John Wiley & Sons, Inc, New York.
24. Field, R.J. and Burger, M. (1985) *Oscillations and Traveling Waves in Chemical Systems*, John Wiley & Sons, Inc, New York.
25. Zhabotinsky, A.M. (1964) Periodic processes of malonic acid oxidation in a liquid phase. *Biofizika*, **9**, 306–311 (in Russian).
26. Epstein, I.R. and Pojman, J.A. (1998) *An Introduction to Nonlinear Chemical Dynamics: Oscillations, Waves, Patterns, and Chaos*, Oxford University Press, New York.
27. Ndassa, I.M., Silvi, B., and Volatron, F. (2010) Understanding reaction mechanisms in organic chemistry from catastrophe theory: ozone addition on benzene. *J. Phys. Chem. A*, **114**, 49, 12900–12906.
28. Kreinovich, V. (1990) All Kinds of Behavior are Possible in Chemical Kinetics. Abstracts of the Southwestern Conference on Theoretical Chemistry, El Paso, Texas, November 1990, p. 21.

11
Kabbalistic–Leibnizian Automata for Simulating the Universe
Andrew Schumann

11.1
Introduction

Conventional computers are based on logic that was obtained more than a century ago. Now, the problem is whether unconventional computers (chemical, biological, etc.) are possible. For designing them, we should know what logic they need and whether this logic is conventional. Many (if not all) experts claim that the logic of unconventional computing (e.g., logic of circuits) should differ from that of the conventional one. Some new logic, such as glider logic, has been proposed. However, universal logical methods for unconventional computing have not been obtained still.

This chapter provides a sketch of the massively parallel proof theory (Section 11.3) that could be considered as a required logic for unconventional computing. The advantages of this logic are that it is an absolutely formal system in which we build proofs purely mechanically, and that it can formalize feedback and circular relations which are very important in unconventional computing.

In Section 11.4, a new logic to simulating the Belousov–Zhabotinsky reaction is applied. Hence, this means that massively parallel proof theory could be used in designing chemical computers. In Section 11.5, new logic is applied to simulating the dynamics of Plasmodium of *Physarum polycephalum*. As we will see, massively parallel proof theory may be used in designing biological computers too.

Section 11.2 contains some historical data concerning massive parallelism in logic and could be ignored by readers who are not interested in the philosophical background of unconventional computing.

11.2
Historical Background of Kabbalistic–Leibnizian Automata

In Judaism, there is a claim that only the Lord exists. He has created the world from nothing. He has control over anything. He stands above all. In a hymn of the

Jewish liturgy, there is the following statement about Him:

אדון עולם אשר מלך
בטרם כל יציר נברא
לעת נעשה בחפצו כל
אזי מלך שמו נקרא
ואחרי ככלות הכל
לבדו ימלוך נורא
והוא היה והוא הוה

Eternal master, who reigned supreme,
Before all of creation was drawn;
When it was finished according to his will,
Then "King" his name was proclaimed to be
When this our world shall be no more,
In majesty he still shall reign,
And he was, and he is,
And he will be in glory.

In Kabbalah, the esoteric teaching of Judaism, the illusiveness of the world is proved by the fact that the world does not exist in itself. The world is a combination of the characters of the sacred (Hebrew) language in which the Holy Scripture is written. According to the *Sefer Yetzirah*, the best known Kabbalistic source, the entire Universe was created from the set of 22 Hebrew letters, each of which is also a number. God's holy words have created all:

בשלשים ושתים נתיבות פליאות חכמה חקק י"ה יהו"ה לבאות
את עולמו בשלשה ספרים ספריס בספר וספור ועשר ספירות
בלי מה במספר עשר אלצעות חמש כנגד חמש וברית יחוד
מכוונת באמצע במלת לשון ובמלת המעור

In two and thirty most occult and wonderful paths of wisdom did JAH the Lord of Hosts engrave his name: God of the armies of Israel, ever-living God, merciful and gracious, sublime, dwelling on high, who inhabiteth eternity. He created this universe by the three Sepharim, Number, Writing, and Speech. Ten are the numbers, as are the Sephiroth, and twenty-two the letters, these are the Foundation of all things. Of these letters, three are mothers, seven are double, and twelve are simple [1].

Many Kabbalists had attempted to recreate some samples of this Divine speech. They supposed that the Divine words may be generated by a purely mechanical process and have no meaning in human language. In fact, as we see, the idea of automaton, a self-operating machine, flashed upon Kabbalists for the first time (for more details see [2, 3]).

11.2 Historical Background of Kabbalistic–Leibnizian Automata

As an example of Kabbalistic automata, let us consider one of the automata simulating the Divine speech that has been proposed by Eleazar ben Judah ben Kalonymus of Worms (Rokeach), the Talmudist and Kabbalist born probably at Mayence about 1176 and died at Worms in 1238. He was a descendant of the great Kalonymus family of Mayence.

In the following example of Kabbalistic automaton, we can see how the letters of word "Talmud" emerge from *'alef* (the first letter of the Hebrew alphabet):

Example Halting pattern: {*t, l, m, w, d*} (Talmud)
Input string: *'a l f*
Generation1: *lf md 'a*
Generation2: *md 'a m lt lf*
Generation3: *m l **t** lf **m** md **w** md *'a**

All five letters of the word "Talmud" appear after three generations. The letter *'alef* was first spelt out horizontally and then branches into a tree:

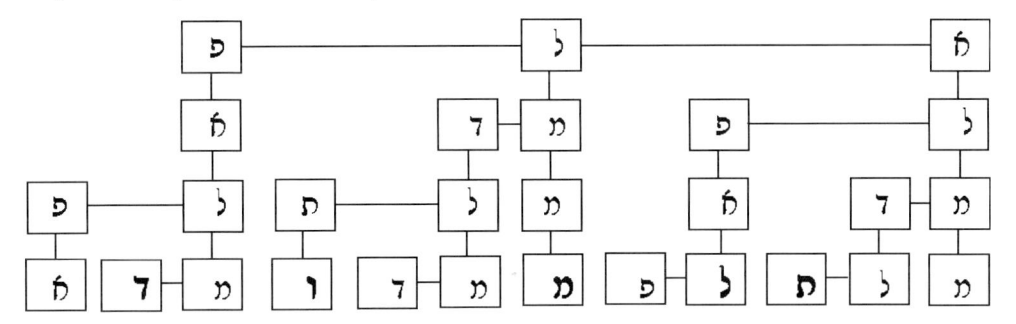

The idea of Kabbalistic automata, in which operations over characters of the Hebrew alphabet should simulate physical and biological processes, suggested to Leibniz the idea to work at a *characteristica universalis* or "*universal characteristic*" built on an alphabet of human thought in which each fundamental concept would be represented by a unique character:

> It is obvious that if we could find characters or signs suited for expressing all our thoughts as clearly and as exactly as arithmetic expresses numbers or geometry expresses lines, we could do in all matters insofar as they are subject to reasoning all that we can do in arithmetic and geometry. For all investigations which depend on reasoning would be carried out by transposing these characters and by a species of calculus [4].

In universal characteristic, complex thoughts would be represented by combining characters for atomic thoughts. Characters of the Hebrew alphabet, which in Kabbalists' intention are atoms of any information, were prototypes of characters in Leibniz' meaning. For Leibniz, such atoms should be presented by a system of coding of all elementary concepts by using prime numbers because of the uniqueness of prime factorization (also this idea was used in Gödel numbering).

Universal characteristic should have become an automaton simulating all physical processes, an automaton that knows the "Answer to the Ultimate Question of Life, the Universe, and Everything." This automaton should have become a universal checker that can check and verify any thought mechanically:

> The only way to rectify our reasonings is to make them as tangible as those of the Mathematicians, so that we can find our error at a glance, and when there are disputes among persons, we can simply say: Let us calculate [calculemus], without further ado, to see who is right [5].

Owing to such possibilities, universal characteristic may be regarded as sacral language: that is, a unique language in which it is possible to think unmistakably, a language that objectively reflects any physical and biological process, and a language which already coincides with reality. It is obvious that the sacred language in Kabbalistic meaning was a prototype of a universal language in Leibnizian meaning. The latter should have become a universal language for communication of all people:

> And although learnt men have long since thought of some kind of language or universal characteristic by which all concepts and things can be put into beautiful order, and with whose help different nations might communicate their thoughts and each read in his own language what another has written in his, yet no one has attempted a language or characteristic which includes at once both the arts of discovery and judgment, that is, one whose signs and characters serve the same purpose that arithmetical signs serve for numbers, and algebraic signs for quantities taken abstractly. Yet it does seem that since God has bestowed these two sciences on mankind, he has sought to notify us that a far greater secret lies hidden in our understanding, of which these are but the shadows [6].

Leibniz proposed some examples of how it is possible to design universal characteristic. In these examples, Leibniz presented "character" as a certain number, assigned to a subject or a predicate as an eigenvalue in the way that the derived number corresponds to a combination (superposition) of numbers of the initial terms. This combination can be presented, for example, by multiplication of numbers which are "characters" of the initial terms (atoms of universal characteristic). For instance, the expression "rational animal", consisting of two terms: "rational" with character a, and "animal" with character r, obtains a new character ar, that is, "man" ("man" = "rational animal"). In particular, if $a = 3$ and $r = 2$, then character of "man" $= 6$.

In categorical propositions, relations between characters are defined by means of the following rules:

1) The universal affirmative proposition is true if and only if the character of the subject is divided by the character of predicate without remainder of division: for example, in the proposition "man is a rational animal" we have: "man" (S)/"animal" $(P) = ar/r = a$.
2) The particular affirmative proposition is true if and only if the character of the subject is divided by the character of predicate without remainder of division, or the character of the predicate is divided by the character of subject without remainder of division.
3) The universal negative proposition is true if and only if the character of the subject cannot be divided by the character of the predicate without remainder

of division and also the character of the predicate cannot be divided by the character of the subject without remainder of division.
4) The particular negative proposition is true if and only if the character of the subject cannot be divided by the character of the predicate without remainder of division.

In case categorical propositions do not satisfy the above-mentioned rules, they are considered false. The advantage of a similar numerical interpretation of the atomic propositions consists in the fact that, by means of simple operations over numbers, it is possible to prove all syllogistic laws. At the same time, the evident assumption is as follows: characters $H = ABC$ and $J = BC$ for combinations of characters of simpler terms A, B, C are divided by the same number of terms X without remainder of division if $X < H$ and $X < J$. For example, let the number of "man" be H and the number of "animal" be A in the proposition "man is animal". Then fraction H/A is reduced to prime numbers r/v such that r/v is an integer, that is, $v = 1$. From this assumption it follows that

1) "Every H is A" iff $vH = rA$ and $v = 1$.
2) "Some A are H" (or "Some H are A") iff $rA = vH$ and ($r = 1$ or $v = 1$).
3) "No H is A" (or "No A is H") iff $vH = rA$ and ($r > 1$ and $v > 1$).
4) "Some A are not H" iff $rA = vH$ and $r > 1$.

The given interpretation satisfies the square of opposition. Indeed, in (2) and (3) (resp. in (1) and (4)) the universal negative proposition and the particular affirmative proposition (resp. the universal affirmative proposition and the particular negative proposition) contradict each other (if one of them is true, the other is false, and vice versa). Further, from (1) we can infer (2), from (3) we can infer (4)), and so on.

Leibniz' intention to construct the universal characteristic was extremely grandiose. In his days, people trusted in unlimited powers of reason. It seemed to them a bit later that the whole world will be completely investigated.

An automaton (abstract machine) that would simulate any physical and biological process is said to be a *Kabbalistic–Leibnizian automaton*. In this chapter, we will consider the possibilities of construction of such automata.

The main assumptions of Kabbalistic–Leibnizian automata are as follows:

- There is only information. Any chemical, physical, or biological process is an unconventional computer. Thus, any chemical, physical, or biological process can be simulated by an abstract machine.
- Any chemical, physical, or biological process can be considered as transition in massive-parallel proofs. Therefore, any chemical, physical, or biological phenomenon may be completely reduced to logical relations, for example, to logical inference rules.
- Any theoretical terms of chemistry, physics, or biology can be eliminated if we replace them by logical inference rules.

As a theoretical basis of Kabbalistic–Leibnizian automata, we will consider the theory of massive-parallel proofs, where inference rules play the role of cellular-automatic local transition functions. This novel approach is characterized as follows:

- Deduction is considered as a transition in cellular automata, where the states of cells are regarded as well-formed formulas of a logical language.
- We build up derivations without using axioms, and therefore there is no sense in distinguishing between logic and theory (i.e., logical and nonlogical axioms), derivable and provable formulas, and so on.
- In deduction, we do not obtain derivation trees and instead we find out derivation traces, that is, a linear evolution of each singular premise.
- Some derivation traces are circular, that is, some premises are derivable from themselves.
- Some derivation traces are infinite.

11.3
Proof-Theoretic Cellular Automata

For any logical language $\mathcal{L}$, we can construct a *proof-theoretic cellular automaton* (instead of conventional deductive systems) simulating massive-parallel proofs.

Definition 1. A proof-theoretic cellular automaton is a 4-tuple $\mathcal{A} = \langle \mathbb{Z}^d, S, N, \delta \rangle$, where

- $d \in \mathbb{N}$ is the number of dimensions and the members of $\mathbb{Z}^d$ are referred to as Cells;
- S is a finite or infinite set of elements called the states of an automaton $\mathcal{A}$, the members of $\mathbb{Z}^d$ take their values in S, and the set S is collected from well-formed formulas of a language $\mathcal{L}$;
- $N \subset \mathbb{Z}^d \setminus 0^d$ is a finite ordered set of n elements, and N is said to be a neighborhood;
- $\delta: S^{n+1} \to S$ that is δ is the inference rule of a language $\mathcal{L}$, and it plays the role of local transition function of an automaton $\mathcal{A}$.

As we see, an automaton is considered on the endless d-dimensional space of integers, that is, on $\mathbb{Z}^d$. Discrete time is introduced for $t = 0, 1, 2, \dots$ fixing each step of inferring.

For any given $z \in \mathbb{Z}^d$, its neighborhood is determined by $z + N = z + \alpha: \alpha \in N$. There are two often-used neighborhoods:

- Von Neumann neighborhood $N_{VN} = z \in \mathbb{Z}^d: \sum_{k=1}^d |z_k| = 1$
- Moore neighborhood $N_M = z \in \mathbb{Z}^d: \max_{k=\overline{1,d}} |z_k| = 1 = -1, 0, 1^d \setminus 0^d$.

For example, if $d = 2$, $N_{VN} = \{(-1, 0), (1, 0), (0, -1), (0, 1)\}$; $N_M = \{(-1, -1), (-1, 0), (-1, 1), (0, -1), (0, 1), (1, -1), (1, 0), (1, 1)\}$.

In the case $d = 1$, von Neumann and Moore neighborhoods coincide. It is easily seen that $|N_{VN}| = 2d$, $|N_M| = 3^d - 1$.

At the moment t, the *configuration* of the whole system (or the *global state*) is given by the mapping $x^t: \mathbb{Z}^d \to S$, and the *evolution* is the sequence $x^0, x^1, x^2, \ldots$ defined as follows: $x^{t+1}(z) = \delta(x^t(z), x^t(z + \alpha_1), \ldots, x^t(z + \alpha_n))$, where $\langle \alpha_1, \ldots, \alpha_n \rangle \in N$. Here, x^0 is the initial configuration, and it fully determines the future behavior of the automaton. It is the set of all premises (not axioms).

We assume that δ is an inference rule, that is, a mapping from the set of premises (their number cannot exceed $n = |N|$) to a conclusion. For any $z \in \mathbb{Z}^d$, the sequence $x^0(z), x^1(z), \ldots, x^t(z), \ldots$ is called a derivation trace from a state $x^0(z)$. If there exists t such that $x^t(z) = x^l(z)$ for all $l > t$, then a derivation trace is *finite*. It is *circular/cyclic* if there exists l such that $x^t(z) = x^{t+l}(z)$ for all t.

Definition 2. In case all derivation traces of a proof-theoretic cellular automaton $\mathcal{A}$ are circular, this automaton $\mathcal{A}$ is said to be *reversible*.

Notice that x^{t+1} depends only upon x^t, that is, the previous configuration. It enables us to build the function $G_\mathcal{A}: C_\mathcal{A} \to C_\mathcal{A}$, where $C_\mathcal{A}$ is the set of all possible configurations of the cellular automaton $\mathcal{A}$ (it is the set of all mappings $\mathbb{Z}^d \to S$, because we can take each element of this set as the initial configuration x^0, though not every element can arise in the evolution of some other configuration). $G_\mathcal{A}$ is called the global function of the automaton.

Example 1. [modus ponens] Consider a propositional language $\mathcal{L}$ that is built in the standard way with the only binary operation of implication $\supset$. Let us suppose that well-formed formulas of that language are used as the set of states for a proof-theoretic cellular automaton $\mathcal{A}$. Further, assume that *modus ponens* is a transition rule of this automaton $\mathcal{A}$ and it is formulated for any $\varphi, \psi \in \mathcal{L}$ as follows:

$$x^{t+1}(z) = \begin{cases} \psi, & \text{if } x^t(z) = \varphi \supset \psi \text{ and } \varphi \in (z + N); \\ x^t(z), & \text{otherwise.} \end{cases}$$

Further dynamics will depend on the neighborhood. If we assume the Moor neighborhood in the two-dimensional space, this dynamics will be exemplified by the evolution of cell states in Figures 11.1–11.3.

$(p \supset q) \supset r$	$p \supset (p \supset q)$	$p \supset q$	$(p \supset q) \supset (p \supset q)$	$(r \supset p) \supset r$
$(p \supset r) \supset (q \supset r)$	$p \supset q$	p	$p \supset (p \supset q)$	$r \supset p$
$p \supset r$	p	$p \supset (q \supset (p \supset q))$	p	r
$p \supset (q \supset r)$	$p \supset p$	$p \supset q$	$(p \supset r) \supset (q \supset p)$	$p \supset r$
$p \supset q$	$p \supset (q \supset p)$	q	$p \supset r$	p

Figure 11.1 Initial configuration of a proof-theoretic cellular automaton $\mathcal{A}$ with the Moor neighborhood in the two-dimensional space, its states run over formulas set up in a propositional language $\mathcal{L}$ with the only binary operation $\supset$, $t = 0$. Notice that p, q, r are propositional variables.

r	p ⊃ q	q	p ⊃ q	r
q ⊃ r	q	p	p ⊃ q	p
r	p	q ⊃ (p ⊃ q)	p	r
q ⊃ r	p	q	q ⊃ p	r
p ⊃ q	p ⊃ (q ⊃ p)	q		p

Figure 11.2 Evolution of $\mathcal{A}$ described in Figure 11.1 at the time step $t = 1$.

r	q	q	q	r
r	q	p	q	p
r	p	q	p	r
r	p	q	p	r
q	p	q	r	p

Figure 11.3 Evolution of $\mathcal{A}$ described in Figure 11.1 at the time step $t = 3$. Its configuration cannot vary further.

Example 1 shows that, first, we completely avoid axioms and, second, we take premises from the cell states of the neighborhood according to a transition function. As a result, we do not come across proof trees in our novel approach to deduction taking into account that a cell state has just a linear dynamics (the number of cells and their location do not change). This allows us evidently to simplify deductive systems.

Now we try to consider a cellular-automaton presentation of two basic deductive approaches: Hilbert's type and sequent ones.

Example 2. [Hilbert's inference rules] Suppose a propositional language $\mathcal{L}$ contains two basic propositional operations: negation and disjunction. As usual, the set of all formulas of $\mathcal{L}$ is regarded as the set of states of an appropriate proof-theoretic cellular automata. In that, we will use the exclusive disjunction of the following five inference rules converted from Joseph R. Shoenfield's deductive system:

$$x^{t+1}(z) = \begin{cases} \psi \vee \varphi, & \text{if } x^t(z) = \varphi; \\ \varphi, & \text{if } x^t(z) = \varphi \vee \varphi; \\ (\chi \vee \psi) \vee \varphi, & \text{if } x^t(z) = \chi \vee (\psi \vee \varphi); \\ \chi \vee \psi, & \text{if } x^t(z) = \varphi \vee \chi \text{ and } (\neg \varphi \vee \psi) \in (z + N); \\ \chi \vee \psi, & \text{if } x^t(z) = \neg \varphi \vee \psi \text{ and } (\varphi \vee \chi) \in (z + N). \end{cases}$$

Example 3. [sequent inference rules] Let us take a sequent propositional language $\mathcal{L}$ in which the classical propositional language with negation, conjunction, disjunction, and implication is extended by adding the sequent relation $\hookrightarrow$. Recall that a *sequent* is an expression of the form $\Gamma_1 \hookrightarrow \Gamma_2$, where $\Gamma_1 = \{\varphi_1, \ldots, \varphi_j\}$, and $\Gamma_2 = \{\psi_1, \ldots, \psi_i\}$ are finite sets of well-formed formulas of the standard propositional language, which has the following interpretation: $\Gamma_1 \hookrightarrow \Gamma_2$ is logically valid iff

$$\bigwedge_j \varphi_j \supset \bigvee_i \psi_i$$

is logically valid. Let S denote the set of all sequents of $\mathcal{L}$; furthermore, let us assume that this family S is regarded as the set of states for a proof-theoretic cellular automaton $\mathcal{A}$. The transition rule of $\mathcal{A}$ is an exclusive disjunction of the 14 singular rules (six structural rules and eight logical rules):

$$x^{t+1}(z) = \Gamma_1 \hookrightarrow \Gamma_2, \begin{cases} \text{if } \Gamma_1 \hookrightarrow \Gamma_2 \text{ is a result of applying to } x^t(z) \\ \text{eather one of structural rules} \\ \text{or the left (right) introduction of negation} \\ \text{or the left introduction of conjunction} \\ \text{or the right introduction of disjunction} \\ \text{or the right introduction of implication.} \end{cases}$$

$$x^{t+1}(z) = \begin{cases} \Gamma \hookrightarrow \Gamma', \psi \wedge \chi, & \text{if } x^t(z) = \Gamma \hookrightarrow \Gamma', \psi \text{ and} \\ & (\Gamma \hookrightarrow \Gamma', \chi) \in (z + N) \\ \Gamma, \psi \vee \chi \hookrightarrow \Gamma', & \text{if } x^t(z) = \Gamma, \psi \hookrightarrow \Gamma' \text{ and} \\ & (\Gamma, \chi \hookrightarrow \Gamma') \in (z + N); \\ \psi \supset \chi, \Gamma, \Delta \hookrightarrow \Gamma', \Delta', & \text{if } x^t(z) = \Gamma \hookrightarrow \Gamma', \psi \text{ and} \\ & (\chi, \Delta \hookrightarrow \Delta') \in (z + N). \end{cases}$$

Example 4. [Brotherston's cyclic proofs] We extend the sequent language used in the previous example by adding predicates N, E, O and appropriate inference rules of Figure 11.4 for them. Further, let us extend also the automaton of Example 3 in the same way by representing inference rules of Figure 11.4 in the cellular-automatic form.

Now we assume that a cell has an initial state $[\Gamma, N(z) \hookrightarrow \Delta, O(z), E(z)]$ and its neighbor cell an initial state $[\Gamma, z = 0 \hookrightarrow \Delta]$ that is equal to $[\Gamma, z = 0 \hookrightarrow \Delta, O(z), E(z)]$ for any $t = 4, 14, 24, \ldots$ and to $[\Gamma, z = 0 \hookrightarrow \Delta, E(z), O(z)]$ for any $t = 9, 19, 29, \ldots$. Then we will have the following infinite cycle:

$$[\Gamma, N(z) \hookrightarrow \Delta, O(z), E(z)] \to^{\text{(substitution)}} [\Gamma, N(y) \hookrightarrow \Delta, O(y), E(y)]$$
$$\to [\Gamma, N(y) \hookrightarrow \Delta, O(y), O(y+1)] \to [\Gamma, N(y) \hookrightarrow \Delta, E(y+1), O(y+1)]$$
$$\to [\Gamma, z = (y+1), N(y) \hookrightarrow \Delta, O(z), E(z)] \to^{\text{(case N)}} [\Gamma, N(z)$$
$$\hookrightarrow \Delta, E(z), O(z)] \to \cdots$$

$$\frac{\Gamma \hookrightarrow N(x)}{\Gamma \hookrightarrow N(x+1)}, \quad \frac{\Gamma \hookrightarrow N}{\Gamma \hookrightarrow \Delta, N(0)}, \quad \frac{\Gamma \hookrightarrow E(x)}{\Gamma \hookrightarrow O(x+1)}, \quad \frac{\Gamma \hookrightarrow O(x)}{\Gamma \hookrightarrow E(x+1)}, \quad \frac{\Gamma \hookrightarrow \Delta}{\Gamma \hookrightarrow \Delta, E(0)},$$

$$\frac{N(x) \hookrightarrow \Delta}{N(x+1) \hookrightarrow \Delta}, \quad \frac{\Gamma \hookrightarrow N}{\Gamma, N(0) \hookrightarrow \Delta}, \quad \frac{E(x) \hookrightarrow \Delta}{O(x+1) \hookrightarrow \Delta}, \quad \frac{O(x) \hookrightarrow \Delta}{E(x+1) \hookrightarrow \Delta}, \quad \frac{\Gamma \hookrightarrow \Delta}{\Gamma, E(0) \hookrightarrow \Delta},$$

$$\frac{\Gamma, t = 0 \hookrightarrow \Delta \quad \Gamma, t = x+1, N(x) \hookrightarrow \Delta}{\Gamma, N(t) \hookrightarrow \Delta} \quad \text{(Case N), where } x \notin FV(\Gamma \cup \Delta \cup \{N(t)\}),$$

$$\frac{\Gamma \hookrightarrow N}{\Gamma[x] \hookrightarrow \Delta[x]} \quad \text{(Substitution).}$$

Figure 11.4 Inference rules for predicates N ("being a natural number"), E ("being an even number"), O ("being an odd number"), see Ref. [7].

Another instance of cyclic proof (see [21–23, 25, 27, 28]) is given in Example 5. As we see, the possibility of cyclic derivation traces depends on configuration of cell states.

Traditional tasks concerning proof theory, such as completeness and independence of axioms, lose their sense in massive-parallel proof theory, although it can be readily shown that we can speak about consistency:

Proposition 1 *Proof theories given in Examples 2 and 3 are consistent, that is, we cannot deduce a contradiction within them.*

Now we can define Kabbalistic–Leibnizian automata:

Definition 3. Any simulation of a chemical, physical, and biological process by means of proof-theoretic cellular automata is called a Kabbalistic–Leibnizian automaton.

11.4
The Proof-Theoretic Cellular Automaton for Belousov–Zhabotinsky Reaction

Kabbalistic–Leibnizian automata assume that massive-parallel computing is observed everywhere in natural systems. There are different approaches to nature-inspired computing: reaction-diffusion computing [8–10]; chemical computing [11]; biological computing [12, 13]; and so on. In all these computational models, parallel inferring and concurrency are assumed as key notions. In [14], a hypothesis is put forward that the paradigm of parallel and concurrent computation caused by rejecting the set-theoretic axiom of foundation can be widely applied in modern physics. In this section, we analyze simulating the Belousov–Zhabotinsky reaction within the framework of our theory of massive-parallel proofs.

Let us consider a proof-theoretic cellular automaton with circular proofs for the Belousov–Zhabotinsky reaction containing feedback relations. The mechanism of this reaction (namely cerium(III) $\longleftrightarrow$ cerium(IV) catalyzed reaction) is very complicated: its recent model contains 80 elementary steps and 26 variable species concentrations. Let us consider a simplification of Belousov–Zhabotinsky reaction by assuming that the set of states consists just of the following reactants: Ce^{3+}, $HBrO_2$, BrO_3^-, H^+, Ce^{4+}, H_2O, $BrCH(COOH)_2$, Br^-, $HCOOH$, CO_2, $HOBr$, Br_2, and $CH_2(COOH)_2$, which interact according to inference rules (reactions) 11.1–11.7. In this reaction, we observe sudden oscillations in color from yellow to colorless, allowing the oscillations to be observed visually. In spatially nonhomogeneous systems (such as a simple Petri dish), the oscillations propagate as spiral wavefronts. The oscillations last for about 1 min and are repeated over a long period. The color changes are caused by alternating oxidation and reduction in which cerium changes its oxidation state from cerium(III) to cerium(IV) and vice versa: $Ce^{3+} \to Ce^{4+} \to Ce^{3+} \to \ldots$.

When Br^- has been significantly lowered, the reaction pictured by inference rule 11.1 causes an exponential increase in bromous acid ($HBrO_2$) and the oxidized form of the metal ion catalyst and indicator, cerium(IV). Bromous acid is subsequently converted to bromate (BrO_3^-) and HOBr (the step 11.3). Meanwhile, the step 11.2 reduces cerium(IV) to cerium(III) and simultaneously increases bromide (Br^-) concentration. Once the bromide concentration is high enough, it reacts with bromate (BrO_3^-) and HOBr in 11.4 and 11.6 to form Br_2. Further, Br_2 reacts with $CH_2(COOH)_2$ to form $BrCH(COOH)_2$ and the process begins again. Thus, parallel processes in 11.1–11.7 have several cycles which are performed synchronously (see example 4 for another case of cyclic proofs).

The proof-theoretic simulation of Belousov–Zhabotinsky reaction can be defined as follows:

Definition 4. Consider a propositional language $\mathcal{L}$ with the only the binary operation $\oplus$, and it is built in the standard way over the set of variables $S = \{Ce^{3+}, HBrO_2, BrO_3^-, H^+, Ce^{4+}, H_2O, BrCH(COOH)_2, Br^-, HCOOH, CO_2, HOBr, Br_2, CH_2(COOH)_2\}$. Let S be the set of states of proof-theoretic cellular automaton $\mathcal{A}$. The inference rule of the automaton is presented by the conjunction 4of singular inference rules 11.1–11.7:

$$11.1 \wedge 11.2 \wedge 11.3 \wedge 11.4 \wedge 11.5 \wedge 11.6 \wedge 11.7.$$

The operation $\oplus$ has the following meaning: $A \oplus B$ defines a probability distribution of events A and B in neighboring cells participating in the reaction causing the appearance of $A \oplus B$. Then $\mathcal{A}$ simulates the Belousov–Zhabotinsky reaction.

Definition 5. Let $p, s_i, s_{i+1} \in \{Ce^{3+}, HBrO_2, BrO_3^-, H^+, Ce^{4+}, H_2O, BrCH(COOH)_2, Br^-, HCOOH, CO_2, HOBr, Br_2,$ and $CH_2(COOH)_2\}$. A state p is called a premise for deducing s_{i+1} from s_i by the inference rule $11.1 \wedge 11.2 \wedge \cdots \wedge 11.7$ iff

- p is s_i or
- in a neighboring cell we find out an expression of the form $p \oplus A$, $B \oplus C$, where A, B, C are propositional metavariables, that is, they run over either the empty set or the set of states closed under the operation $\oplus$. Thus, we assume that each premise should occur in a separate cell. This means that, if we find out an expression $p_i \oplus p_j$, $B \oplus C$ or $p_i \oplus A$, $p_j \oplus C$ in a neighboring cell and both p_i and p_j are needed for deducing, whereas p_i, p_j do not occur in other neighboring cells, then p_i, p_j cannot be considered as premises.

$$x^{t+1}(z) = \begin{cases} (1)\ Ce^{4+} \oplus HBrO_2 \oplus H_2O, & \text{if } x^t(z) \in \{Ce^{3+}\} \text{ and} \\ \quad \text{premises } HBrO_2, BrO_3^-, H^+ \in (z+N); \\ (2)\ x^t(z), & \text{otherwise.} \end{cases} \quad (11.1)$$

$$x^{t+1}(z) = \begin{cases} (1)\ Br^- \oplus Ce^{3+} \oplus HCOOH \oplus CO_2 \oplus H^+, & \text{if } x^t(z) \in \{Ce^{4+}\} \\ \quad \text{and premises } BrCH(COOH)_2, H_2O \in (z+N); \\ (2)\ x^t(z), & \text{otherwise.} \end{cases} \quad (11.2)$$

$$x^{t+1}(z) = \begin{cases} (1)\ \text{HOBr} \oplus \text{BrO}_3^- \oplus \text{H}^+, & \text{if } x^t(z) \in \{\text{HBrO}_2\}; \\ (2)\ x^t(z), & \text{otherwise.} \end{cases} \quad (11.3)$$

$$x^{t+1}(z) = \begin{cases} (1)\ \text{HOBr} \oplus \text{HBrO}_2, & \text{if } x^t(z) \in \{\text{BrO}_3^-\} \text{ and} \\ & \text{premises } \text{Br}^-, \text{H}^+ \in (z+N); \\ (2)\ x^t(z), & \text{otherwise.} \end{cases} \quad (11.4)$$

$$x^{t+1}(z) = \begin{cases} (1)\ \text{HOBr}, & \text{if } x^t(z) \in \{\text{Br}^-\} \text{ and} \\ & \text{premises } \text{HBrO}_2, \text{H}^+ \in (z+N); \\ (2)\ x^t(z), & \text{otherwise.} \end{cases} \quad (11.5)$$

$$x^{t+1}(z) = \begin{cases} (1)\ \text{Br}_2 \oplus \text{H}_2\text{O}, & \text{if } x^t(z) \in \{\text{HOBr}\} \text{ and} \\ & \text{premises } \text{Br}^-, \text{H}^+ \in (z+N); \\ (2)\ x^t(z), & \text{otherwise.} \end{cases} \quad (11.6)$$

$$x^{t+1}(z) = \begin{cases} (1)\ \text{Br}^- \oplus \text{H}^+ \oplus \text{BrCH(COOH)}_2, & \text{if } x^t(z) \in \{\text{Br}_2\} \\ \text{and premises } \text{CH}_2(\text{COOH})_2 \in (z+N); \\ (2)\ x^t(z),\ \text{otherwise.} \end{cases} \quad (11.7)$$

Example 5. [Belousov–Zhabotinsky's cyclic proofs] We can simplify the automaton defined above by assuming that $\oplus$ is a metatheoretic operation with the following operational semantics:

$$\frac{A \oplus B}{A} \qquad \frac{A \oplus B}{B},$$

where A and B are metavariables defined on S. The informal meaning of that operation is that we can ignore one of the variables coupled by $\oplus$. In the cellular automaton $\mathcal{A}$, this metaoperation will be used as follows:

$$x^{t+1}(z) = \begin{cases} X, Y, & \text{if } x^t(z) = A \oplus B \text{ and according to rules 11.1–11.7,} \\ & X \text{ changes from } A \text{ and } Y \text{ changes from } B; \\ X, & \text{if } x^t(z) = A \oplus B \text{ and according to rules 11.1–11.7,} \\ & X \text{ changes from } A \text{ and } B \text{ does not change}; \\ Y, & \text{if } x^t(z) = A \oplus B \text{ and according to rules 11.1–11.7,} \\ & Y \text{ changes from } B \text{ and } A \text{ does not change}; \\ A \oplus B, & \text{if } x^t(z) = A \oplus B \text{ and rules 11.1–11.7} \\ & \text{cannot be applied to } A \text{ or } B. \end{cases}$$

$$(11.8)$$

Let us suppose now that X, Y run over the set of states closed under the operation $\oplus$.

$$x^{t+1}(z) = \begin{cases} X, Y, & \text{if (i) } x^t(z) = X, Y \text{ and (ii) both } X \text{ and } Y \\ & \text{are simultaneously usable (not usable)} \\ & \text{as premises in at least two different rules} \\ & \text{of 11.1–11.7 (see definition 4);} \\ X, & \text{if (i) } x^t(z) = X, Y \text{ and (ii) only } X \text{ is usable} \\ & \text{as a premise in at least one rule of 11.1–11.7;} \\ Y, & \text{if (i) } x^t(z) = X, Y \text{ and (ii) only } Y \text{ is usable} \\ & \text{as a premise in at least one rule of 11.1–11.7.} \end{cases} \quad (11.9)$$

idempotency: $A ::= A, A.$ (11.10)

commutativity: $A, B ::= B, A.$ (11.11)

Hence, we cannot ignore any of the two variables coupled by $\oplus$ and should accept both of them if in the neighborhood there are reactants that catenate both variables and change them. This rule is the simplest interpretation of $A \oplus B$ in definition 3. We have three cases: (i) both variables are catenated with reactants from the neighborhood, in which case we mean that the probability distribution of events A and B is the same and equal to 0.5 and, as a result, we cannot choose one of them and accept both; (ii) only A is catenated with reactants from the neighborhood, in which case the probability distribution of event A is equal to 1.0 and that of B to 0.0; (iii) only B is catenated with reactants from the neighborhood, in which case the probability distribution of event B is equal to 1.0 and that of A to 0.0. Thus, $A \oplus B$ is a function that associates either exactly one value with its arguments (i.e., either A or B) or simultaneously both values (i.e., A and B).

This simplified version of the automaton $\mathcal{A}$ is exemplified in Figure 11.5.

Evidently, reducing the complicated dynamics of Belousov–Zhabotinsky reaction to conventional logical proofs is a task that cannot be solved in an easy way differently from simulating within massive-parallel proofs.

11.5
The Proof-Theoretic Cellular Automaton for Dynamics of *Plasmodium* of *Physarum polycephalum*

The dynamics of plasmodium of *Physarum polycephalum* can be regarded as another simple example of the natural proof-theoretic automata (Kabbalistic–Leibnizian automata), see [18–20, 24, 26]. The point is that, when the plasmodium is cultivated on a nutrient-rich substrate (agar gel containing crushed oat flakes), it exhibits uniform circular growth similar to the excitation waves in the excitable Belousov–Zhabotinsky medium (Figure 11.6). If the growth substrate lacks nutrients, for example, the plasmodium is cultivated on a non-nutrient-rich and repellent-containing gel, a wet filter paper or even glass surface localizations emerge and branching patterns become clearly visible (Figures 11.7 and 11.8).

The plasmodium continues its spreading, reconfiguration, and development as long as there are enough nutrients. When the supply of nutrients is over, the

(I) Initial configuration, $t=0$	$HBrO_2$ $BrCH(COOH)_2$ Br^-	BrO_3^- Ce^{3+} $CH_2(COOH)_2$	H^+ H_2O BrO_3^-

⇓

(II) $t=1$	$HOBr \oplus BrO_3^- \oplus H^+$ $BrCH(COOH)_2$ Br^-	BrO_3^- $Ce^{4+} \oplus HBrO_2 \oplus H_2O$ $CH_2(COOH)_2$	H^+ H_2O BrO_3^-

⇓

(III) $t=2$	$HOBr \oplus BrO_3^- \oplus H^+$ $BrCH(COOH)_2$ Br^-	BrO_3^- $Br^- \oplus Ce^{3+} \oplus HCOOH \oplus CO_2 \oplus H^+$, $HOBr \oplus BrO_3^- \oplus H^+$ $CH_2(COOH)_2$	H^+ H_2O BrO_3^-

⇓ (IV) $t=3$

$HOBr \oplus BrO_3^- \oplus H^+$ $BrCH(COOH)_2$ Br^-	$HOBr \oplus HBrO_2$ $Br^- \oplus Ce^{3+} \oplus HCOOH \oplus CO_2 \oplus H^+, Br_2 \oplus H_2O$, $HOBr \oplus HBrO_2$ $CH_2(COOH)_2$	H^+ H_2O BrO_3^-

⇓ (V) $t=4$

$HOBr \oplus BrO_3^- \oplus H^+$ $BrCH(COOH)_2$ Br^-	$Br_2 \oplus H_2O, HOBr \oplus BrO_3^- \oplus H^+$ $Ce^{4+} \oplus HBrO_2 \oplus H_2O, Br_2 \oplus H_2O, HOBr \oplus BrO_3^- \oplus H^+$ $HOBr, Br \oplus H^+ \oplus BrCH(COOH)_2$ $CH_2(COOH)_2$	H^+ H_2O BrO_3^-

⇓ (VI) $t=5$

$Br_2 \oplus H_2O, HOBr \oplus HBrO_2$ $BrCH(COOH)_2$ Br^-	$Br_2 \oplus H_2O, HOBr \oplus HBrO_2$ $Br_2 \oplus H_2O, Br^- \oplus Ce^{3+} \oplus HCOOH \oplus CO_2 \oplus H^+$, $HOBr \oplus BrO_3^- \oplus H^+, HOBr \oplus HBrO_2$, $Br^- \oplus H^+ \oplus BrCH(COOH)_2$ $CH_2(COOH)_2$	H^+ H_2O BrO_3^-

⇓
...

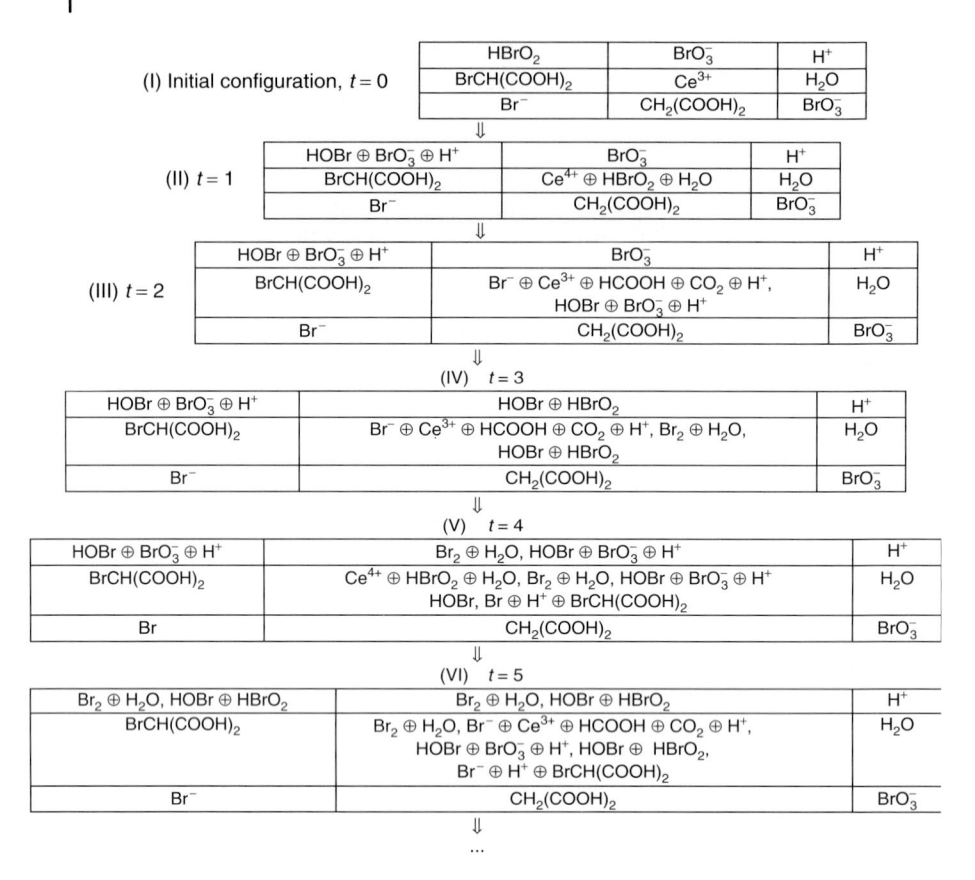

Figure 11.5 Evolution of a reversible proof-theoretic cellular automaton $\mathcal{A}$ with the Moor neighborhood in the two-dimensional space for the Belousov–habotinsky reaction. This automaton simulates the circular feedback $Ce^{3+} \to Ce^{4+} \to Ce^{3+} \to \cdots$ (more precisely temporal oscillations in a well-stirred solution): Ce^{3+} is colorless and Ce^{4+} is yellow. The initial configuration of $\mathcal{A}$-cells described in (I) occurs in the same form in further steps and the cycle repeats several times. For entailing (I) → (II), we have just used inference rule 11.1 (row 2, column 2) and inference rule 11.3 (row 1, column 1); for entailing (II) → (III) inference rules 11.2, 11.3, and 11.8 (row 2, column 2); for entailing (III) → (IV) inference rule 11.4 (row 1, column 2) and inference rules 11.4, 11.6, 11.8, and 11.9 (row 2, column 2); for entailing (IV) → (V) inference rules 11.3 and 11.6 (row 1, column 2) inference rules 11.1, 11.3, 11.5–11.7, 11.8, and 11.9 (row 2, column 2); for entailing (V) → (VI) inference rules 11.4, 11.6 (row 1, column 1), inference rules 11.4, 11.6, 11.10 (row 1, column 2), inference rules 11.2–11.4, 11.6, 11.7, 11.9–11.11 (row 2, column 2).

plasmodium either switches to the fructification state (if the level of illumination is high enough), when sporangia are produced, or forms sclerotium (encapsulates itself in a hard membrane) if in darkness.

The pseudopodium propagates in a manner analogous to the formation of wave fragments in subexcitable Belousov–Zhabotinsky systems. Starting from the initial

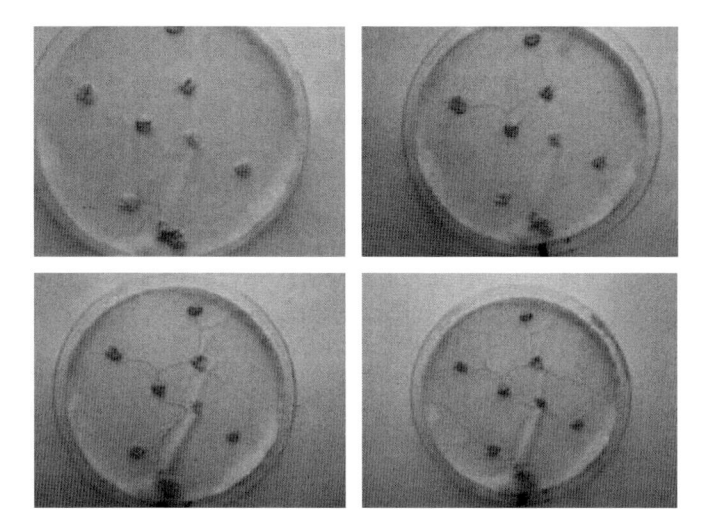

Figure 11.6 Example of computational process in a *Physarum* machine. Photographs are taken with a time lapse of circa 24 h. *Courtesy of Andy Adamatzky.*

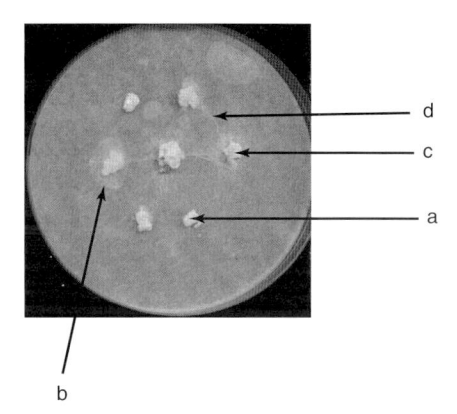

Figure 11.7 Basic components of *Physarum* and its environment: (a) oat flake, (b) propagating pseudopodium, plasmodium's Wave fragment, (c) oat flake colonized by plasmodium, (d) protoplasmic tube. *Courtesy of Andy Adamatzky.*

conditions, the plasmodium exhibits foraging behavior, searching for sources of nutrients (Figure 11.6). When such sources are located and taken over, the plasmodium forms characteristic veins of protoplasm, which contract periodically. The Belousov–Zhabotinsky reaction and plasmodium are light sensitive, which gives us the means to program them. *Physarum* exhibits articulated negative phototaxis, and the Belousov–Zhabotinsky reaction is inhibited by light. Therefore, by using masks of illumination one can control the dynamics of localizations in these media. Light sensitivity of plasmodium has been already explored in the design of robotic controllers [15, 16].

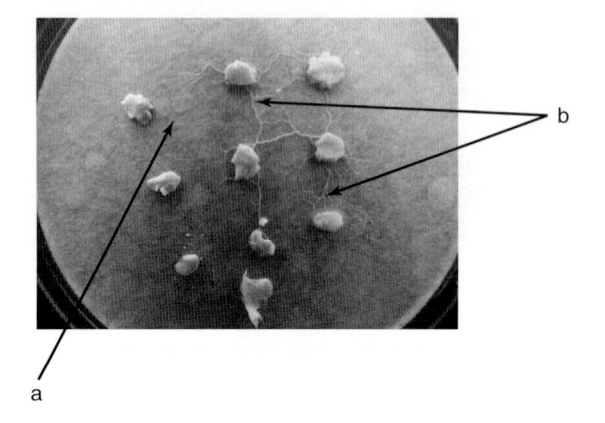

Figure 11.8 Snapshot of experimental dish with propagating Plasmodium. New activated propagating zone (a) and sites of branching pseudopodia, junctions of protoplasmic tubes (b) are shown by arrows. *Courtesy of Andy Adamatzky.*

(I) Initial configuration, $t = 0$

A_{11}	Z_{12}	A_{31}
R_{12}	Z_{22}	Z_{32}
R_{13}	Z_{23}	P_{33}

$\Downarrow$

(II) $t = 1$

A_{11}	Z_{12}	A_{31}
R_{12}	Z_{22}	$Z_{32} \oplus C_{32}$
R_{13}	Z_{23}	P_{33}

$\Downarrow$

(III) $t = 2$

A_{11}	Z_{12}	$A_{31} \oplus C_{31}$
R_{12}	Z_{22}	$Z_{32} \oplus C_{32}$
R_{13}	Z_{23}	P_{33}

$\Downarrow$

(IV) $t = 3$

A_{11}	Z_{12}	P_{31}
R_{12}	Z_{22}	$Z_{32} \oplus C_{32}$
R_{13}	Z_{23}	P_{33}

Figure 11.9 Evolution of a proof-theoretic cellular automaton $\mathcal{A}$ with the Moor neighborhood in the two-dimensional space for *Physarum polycephalum*.

Experiments with *Physarum polycephalum* were carried out by Prof. Adamatzky as follows. The plasmodia of *Physarum polycephalum* were cultured on wet paper towels, fed with oat flakes, and moistened regularly. The plasmodium was subcultured every 5–7 days.

Experiments were performed in standard Petri dishes, 9 cm in diameter. Depending on the particular experiment, we used 2% agar gel or moistened filter paper, nutrient-poor substrates, and 2% oatmeal agar, nutrient-rich substrates (Sigma–Aldrich). All experiments were conducted in a room with diffusive light of 3–5 cd/m, 22°C temperature. In each experiment, an oat flake colonized by the plasmodium was placed on a substrate in a Petri dish, and few intact oat flakes were distributed on the substrate. The intact oat flakes acted as source of nutrients, attractants for the plasmodium. Petri dishes with plasmodium were scanned on a standard HP scanner. The only editing done to the scanned images was color enhancement: increase of saturation and contrast.

Repellents were implemented with illumination domains using blue electroluminescent sheets, see details in [17]. Masks were prepared from black plastic: that is, a triangle was cut in the plastic, and when this mask was placed on top of the electroluminescent sheet; light passed only through the cuts [17].

Results of experiments may be described in terms of proof-theoretic cellular automata. Let us assume that its set of states consists of the entities from the following sets (Figure 11.7):

1) The set of *neutral zones*, $\{Z_1, Z_2, \ldots\}$, where nothing goes;
2) The set of *growing pseudopodia*, $\{P_1, P_2, \ldots\}$, localized in *active zones* (see an example on Figure 11.7a). On a nutrient-rich substrate, plasmodium propagates as a typical circular target wave, while on the nutrient-poor substrates localized wave fragments are formed;
3) The set of *attractants* $\{A_1, A_2, \ldots\}$, they are sources of nutrients, on which the plasmodium feeds. It is still a subject of discussion how exactly plasmodium feels the presence of attractants; indeed, diffusion of some kind is involved. Based on previous experiments by Prof. Adamatzky, we can assume that if the whole experimental area is about 8–10 cm in diameter, then the plasmodium can locate and colonize near the sources of nutrients;
4) The set of *repellents* $\{R_1, R_2, \ldots\}$. Plasmodium of *Physarum* avoids light. Thus, domains of high illumination are repellents such that each repellent R is characterized by its position and intensity of illumination, or force of repelling;
5) The set of *protoplasmic tubes* $\{C_1, C_2, \ldots\}$. Typically, plasmodium scans sources of nutrients with protoplasmic tubes/veins (Figure 11.7). The plasmodium builds a planar graph, where nodes are sources of nutrients, for example, oat flakes, and edges are protoplasmic tubes.

Hence, the set of states in the proof-theoretic cellular automaton for the dynamics of Plasmodium of *Physarum polycephalum* is equal to $\{Z_1, Z_2, \ldots\} \cup \{P_1, P_2, \ldots\} \cup \{A_1, A_2, \ldots\} \cup \{R_1, R_2, \ldots\} \cup \{C_1, C_2, \ldots\}$.

The proof-theoretic simulation of *Physarum polycephalum* (an example is provided in Figure 11.9) is defined as follows:

Definition 6. Consider a propositional language $\mathcal{L}$ with the only binary operation $\oplus$, and it is built in the standard way over the set of variables $S = \{Z_1, Z_2, \ldots, P_1, P_2, \ldots, A_1, A_2, \ldots, R_1, R_2, \ldots, C_1, C_2, \ldots\}$. Let S be the set of states of the proof-theoretic cellular automaton $\mathcal{A}$. The inference rule of the automaton is presented by the conjunction of singular inference rules 11.12–11.18:

$$11.12 \wedge 11.13 \wedge 11.14 \wedge 11.15 \wedge 11.16 \wedge 11.17 \wedge 11.18.$$

The operation $\oplus$ has the following meaning: $A \oplus B$ defines a probability distribution of events A and B in cells. The operation $\oplus$ is idempotent, commutative, and associative. Then $\mathcal{A}$ simulates the dynamics of Plasmodium of *Physarum polycephalum*.

Definition 7. Let $p, s_i, s_{i+1} \in \{Z_1, Z_2, \ldots, P_1, P_2, \ldots, A_1, A_2, \ldots, R_1, R_2, \ldots, C_1, C_2, \ldots\}$. A state p is called a premise for deducing s_{i+1} from s_i by inference rules 11.12–11.18 iff

1) p is s_i or
2) in a neighboring cell we find out an expression of the form $p \oplus X$, where X is a propositional metavariable, that is, it runs over either the empty set or the set of states closed under the operation $\oplus$. Thus, we assume that each premise should occur in a separate cell. This means that if we find out an expression $p_i \oplus p_j \oplus X$ in a neighboring cell and both p_i and p_j are needed for deducing, whereas p_i, p_j do not occur in other neighbor cells, then p_i, p_j could not be considered as premises. This restriction is just for rule 11.18.

$$x^{t+1}(z) = \begin{cases} (1) \; X & \text{if } x^t(z) = X \oplus A_i \text{ and } X = Y \oplus P_j \\ (2) \; x^t(z), & \text{otherwise.} \end{cases} \quad (11.12)$$

$$x^{t+1}(z) = \begin{cases} (1) \; X \oplus P_j & \text{if } x^t(z) = X \oplus C_n \oplus A_i \\ (2) \; x^t(z), & \text{otherwise.} \end{cases} \quad (11.13)$$

$$x^{t+1}(z) = \begin{cases} (1) \; X \oplus P_j & \text{if } x^t(z) = X \oplus P_j \oplus A_i \\ (2) \; x^t(z), & \text{otherwise.} \end{cases} \quad (11.14)$$

$$x^{t+1}(z) = \begin{cases} (1) \; X \oplus P_i \vee X \oplus P_j & \text{if } x^t(z) = X \oplus P_i \oplus P_j \\ (2) \; x^t(z), & \text{otherwise.} \end{cases} \quad (11.15)$$

$$x^{t+1}(z) = \begin{cases} (1) \; X \oplus C_i \vee X \oplus C_j & \text{if } x^t(z) = X \oplus C_i \oplus C_j \\ (2) \; x^t(z), & \text{otherwise.} \end{cases} \quad (11.16)$$

$$x^{t+1}(z) = \begin{cases} (1) \; X \oplus C_i, & \text{if } x^t(z) = X \text{ and} \\ & \text{premises } R_i \notin (z+N), \\ & \text{premises } C_j, A_i \in (z+N); \\ (2) \; x^t(z), & \text{otherwise.} \end{cases} \quad (11.17)$$

$$x^{t+1}(z) = \begin{cases} (1) \; X \oplus C_i, & \text{if } x^t(z) = X \text{ and} \\ & \text{premises } R_i \notin (z+N), \\ & \text{premises } P_j, A_i \in (z+N); \\ (2) \; x^t(z), & \text{otherwise.} \end{cases} \quad (11.18)$$

11.6
Unconventional Computing as a Novel Paradigm in Natural Sciences

In natural sciences, there are many theoretical terms that, as it seems to us, allow understanding reality better, but actually they replace true reality by the images made up. For example, in understanding the Belousov–Zhabotinsky reaction, the following theoretical terms of chemistry are used: matter, chemical element, chemical reaction, autowave, valence, and so on. In understanding the dynamics of Plasmodium of *Physarum polycephalum*, the following theoretical terms of biology are assumed: contiguous living system, cell, protoplasm, physiology, morphology, and so on.

If we share positions of unconventional computing and agree that any natural process is a computation and nature as a whole is a computer, theoretical terms of physics, chemistry, and biology will lose sense. We can eliminate them from the language of scientific theories.

Theoretical terms cannot be verified: that is, they are not reduced to observation terms. However, within the limits of unconventional computing, we can substitute logical terms simulating the process of computation for theoretical terms. Logical terms do not require ready verification. They form only the structure and algorithmization of the observation terms.

Let us suppose that we created the universal language $\mathcal{L}_\mathcal{U}$ of *Kabbalistic–Leibnizian automata*, that is, the universal logical language of unconventional computing that allows us to consider any natural process as a computation process. Then the language $\mathcal{L}_\mathcal{U}$, first, eliminates all theoretical terms of natural sciences, replacing them by logical terms of massively parallel proof theory; and, second, instead of the huge number of scientific theories of physics, chemistry, biology, we obtain an opportunity to have the universal logical theory that simulates the universe.

Let us show how we can eliminate theoretical terms in the language $\mathcal{L}_\mathcal{U}$. Take n theoretical terms: $T_1, T_2, \ldots, T_n$, linked to the observation terms by correspondence rules of an appropriate theory and m observation terms, that is, their references are found out immediately: $O_1, O_2, \ldots, O_m$. The complete theoretical statement is a union consisting of T and O terms:

$$\langle T_1, T_2, \ldots, T_n; O_1, O_2, \ldots, O_m \rangle.$$

This formula can be replaced by another, identical to the first in the way that all theoretical terms are replaced by logical terms $U_1, U_2, \ldots, U_k$ which describe the process of computation of an appropriate automata with features $\langle O_1, O_2, \ldots, O_m \rangle$:

$$\langle U_1, U_2, \ldots, U_k; O_1, O_2, \ldots, O_m \rangle.$$

In such a statement, empirical values of theoretical terms are positioned, that is, they obtain references. These references are expressed by observation terms. Such a representation of theoretical expressions has a number of advantages: for example, sophisticated theoretical terms of natural sciences could find the classes of references. For example, it is obvious that the statements describing Earth's rotation have the set {the Earth} as a class of references. However, it is not clear which set of physical bodies realizes the statements describing the concept "electron".

In a proposition of language $\mathcal{L}_\mathcal{U}$, the sophisticated expression setting the theoretical terms is transformed into an expression in which there are only observation terms as well as terms of logic and mathematics, and therefore the concept "electron" may be eliminated from the theory expressions.

In our simulation of the Belousov–Zhabotinsky reaction as observation terms, the following states of an appropriate cellular automaton were used: Ce^{3+}, $HBrO_2$, BrO_3^-, H^+, Ce^{4+}, H_2O, $BrCH(COOH)_2$, Br^-, $HCOOH$, CO_2, $HOBr$, Br_2, $CH_2(COOH)_2$, and in the simulation of the dynamics of Plasmodium of *Physarum polycephalum* the following states were considered: $\{Z_1, Z_2, \ldots, P_1, P_2, \ldots, A_1, A_2, \ldots, R_1, R_2, \ldots, C_1, C_2, \ldots\}$.

The theoretical terms needed for understanding the Belousov–Zhabotinsky reaction were replaced by inference rules 11.1–11.7, and theoretical terms needed for understanding the dynamics of Plasmodium of *Physarum polycephalum* by inference rules 11.12–11.18.

Both examples show that, for designing the language $\mathcal{L}_\mathfrak{U}$, it is necessary to solve Leibniz' problem to find out characters (atoms of any information): that is, to find out the minimum set of observation terms sufficient for an elimination of any theoretical terms. This problem is similar to the problem of looking for Higgs boson. Its solution will allow the creation of the universal language of natural sciences, to construct a supercomputer "Earth" to come up with the Answer to The Ultimate Question of Life, the Universe, and Everything.

11.7
Conclusion

In this paper, the problem of presenting the Universe as logical automaton has been considered. As theoretical basis of such a presentation, we have proposed the massively parallel proof theory.

Acknowledgments

I would like to express my warm gratitude to Prof. Andy Adamatzky for his kind permission to use his results of experiments with *Physarum polycephalum* and appropriate photographs. Many brainstorming sessions with him on the topic of reaction-diffusion computing allowed me to take a new outlook at logic and to construct the massively parallel proof theory.

References

1. Wescott, W.W. (1887) Sepher Yetzirah or the Book of Creation, tr.
2. Schumann, A. (ed) (2010) *Judaic Logic*, Gorgias Press.
3. Schumann, A. (ed.) (2011) *Modern review of judaic logic*. Special issue of History and Philosophy of Logic
4. Leibniz. (1677) Preface to the general science.
5. Leibniz. (1685) The Art of Discovery, Wiener 51.
6. Leibniz. (1966) Zur allgemeinen Charakteristik, *Hauptschriften zur Grundlegung der Philosophie. Philosophische Werke*. Band 1, Felix Meiner, Hamburg.
7. Brotherston, J. (2005) Cyclic proofs for first-order logic with inductive definitions, in *TABLEAUX 2005*, LNAI, vol. 3702, (ed. B. Beckert), Springer-Verlag, pp. 78–92.
8. Adamatzky, A. (2001) *Computing in Nonlinear Media and Automata*

Collectives, Institute of Physics Publishing.

9. Adamatzky, A., Wuensche, A., and De Lacy Costello, B. (2006) Glider-based computation in reaction-diffusion hexagonal cellular automata. *Chaos, Solitons Fractals*, **27**, 287–295.

10. Schumann, A. and Adamatzky, A. (2009) Towards semantical model of reaction-diffusion computing. *Kybernetes*, **38**, 1518–1531.

11. Berry, G. and Boudol, G. (1992) The chemical abstract machine. *Teor. Comput. Sci.*, **96**, 217–248.

12. Ivanitsky, G.R., Kunisky, A.S., and Tzyganov, M.A. (1984) Study of 'target patterns' in a phage-bacterium system, in *Self-organization: Autowaves and Structures Far From Equilibrium* (ed. V.I. Krinsky), Heidelberg-Springer, pp. 214–217.

13. Prajer, M., Fleury, A., and Laurent, M. (1997) Dynamics of calcium regulation in Paramecium and possible morphogenetic implication. *J. Cell Sci.*, **110**, 529–535.

14. Khrennikov, A. and Schumann, A. (2009) *Physics beyond the set-theoretic axiom of foundation*. AIP Conference Proceedings – March 10, Vaxjo, Sweden, vol. 1101, pp. 374–380.

15. Adamatzky, A. (2007) Physarum machines: encapsulating reaction-diffusion to compute spanning tree. *Naturwisseschaften*, **94**, 975–980.

16. Adamatzky, A. and Jones, J. (2010) Programmable reconfiguration of Physarum machines. *Nat. Comput.*, **9** (1), 219–237.

17. Adamatzky, A. (2010) Steering plasmodium with light: dynamical programming of physarum machine. *New Math. Nat. Comput.*, Accepted arXIv:0908:0850 (2009).

18. Nakagaki, T., Yamada, H., and Toth, A. (2001) Path finding by tube morphogenesis in an amoeboid organism. *Biophys. Chem.*, **92**, 47–52.

19. Adamatzky, A., De Lacy Costello, B., and Asai, T. (2005) *Reaction-Diffusion Computers*, Elsevier.

20. Adamatzky, A. (2007) Physarum machine: implementation of a Kolmogorov-Uspensky machine on a biological substrate. *Parallel Process. Lett.*, **17**, 455–467.

21. Khrennikov, A. and Schumann, A. (2009) p-Adic physics, non-well-founded reality and unconventional computing p-adic numbers. *Ultramet. Anal. Appl.*, **1** (4), 297–306.

22. Brotherston, J. (2006) Sequent Calculus Proof Systems for Inductive Definitions. PhD thesis. University of Edinburgh, November.

23. Brotherston, J. and Simpson, A. (2007) *Complete Sequent Calculi for Induction and Infinite Descent*, LICS-22, IEEE Computer Society, pp. 51–60.

24. Nakagakia, T., Yamada, H., and Ueda, T. (2000) Interaction between cell shape and contraction pattern in the Physarum plasmodium. *Biophys. Chem.*, **84**, 195–204.

25. Santocanale, L. (2002) A calculus of circular proofs and its categorical semantics, in *Proceedings of FoSSaCS 2002, Grenoble, Apr. 2002*, LNCS, 2303 (eds M. Nielsen and U. Engberg), Springer-Verlag, pp. 357–371.

26. Schumann, A. and Adamatzky, A. (2011) Physarum spatial logic. *New Math. Nat. Comput.*, **7** (3), 483–498.

27. Schumann, A. (2008) Non-well-founded probabilities on streams, in *Soft Methods for Handling Variability and Imprecision*, Advances in soft computing 48 (eds D. Dubois, et al.), Springer, pp. 59–65.

28. Schumann, A. (2010) Towards Theory of Massive-Parallel Proofs, Cellular Automata Approach, Bulletin of the Section of Logic.

12
Approaches to Control of Noise in Chemical and Biochemical Information and Signal Processing
Vladimir Privman

12.1
Introduction

We review recent experimental advances in the realizations of chemical [1–4] and biomolecular [5–12] systems that process information in a "digital" fashion, carrying out binary gate functions, for example, AND, OR, and so on, by utilizing (bio)chemical processes. The present chapter is an updated "reprint" of a review article [1] written for the same publisher. The information processing and storing "units" need not be limited to molecules. They can include [1–13] natural and synthetic supramolecular, biomolecular, and/or catalytic structures (enzymes, DNA, etc.), as well as whole cells. Present chemical or biochemical "information processing" or "signal processing," to be termed *computing* for brevity, systems have not been versatile and practical enough to compete with the conventional silicon (Si) computers. The short-term applications have rather been envisaged in offering additional functionalities for multisignal sensing [14, 15] and interfacing/actuation [15–17] when excessive wiring to Si computers and power sources is not desirable, for example, in biomedical implants, diagnostic patches, and so on.

Beyond the design of various binary gates, chemical and biochemical computing face the challenge of developing functionalities to connect the gates and other network components to enable fault-tolerant information processing of increasing complexity [18–20]. Reported results for networks [1–4, 18–20] of (bio)chemical information-processing units have included systems performing elements of basic arithmetic operations [21, 22], multifunctional molecules [23–25], DNA-based gates and circuits [26, 27], and enzyme-catalyzed reactions realizing concatenated gates [19, 20, 28, 29].

Here we introduce, by illustrative model examples, concepts in noise reduction and control for scalability in biochemical computing. The approach has been tested in experimental realizations for enzyme-reaction-based logic gates and networks [10, 12, 14, 17, 19, 30]. However, the reported theoretical ideas generally apply to a broad range of chemical and biomolecular information-processing systems, presently suggesting that typical networks of up to 10 binary gates can operate with the acceptable level of noise [10, 12, 17], similar to findings in networking of neurons

Molecular and Supramolecular Information Processing: From Molecular Switches to Logic Systems,
First Edition. Edited by Evgeny Katz.
© 2012 Wiley-VCH Verlag GmbH & Co. KGaA. Published 2012 by Wiley-VCH Verlag GmbH & Co. KGaA.

[31, 32]. For networks of more than order 10 binary steps, additional nonbinary network elements, as well as proper network design to utilize redundancy for digital error correction, will be needed for fault-tolerant operation [10, 12, 17, 30].

The level of noise in systems envisaged for applications of future chemical [1–4] and biomolecular [5–12, 14, 17, 19, 20, 31–34] computing systems is quite high as compared to the internal noise in electronic information-processing devices. Indeed, both the input/output signals and the "gate machinery" chemical concentrations can typically fluctuate several percent or more, on the scale normalized to the digital **0–1** range. Avoiding noise amplification by careful design and parameter selection for gates and networks is therefore quite important even for relatively small networks. Here we do not address the *origin/sources* of stochastic and environmental noise in (bio)chemical reactions. We also do not survey experimental findings, which are, however, extensively cited; see also Ref. [1]. Instead, we consider solvable rate equation models that serve to illustrate recently developed concepts in (bio)chemical computing gate design for noise suppression.

Theoretical considerations reviewed here apply to numerous reported chemical and biochemical information-processing systems. Indeed, chemical processes can be cast [35, 36] in the language of computing operations, with signals represented by changes [1–4, 36–48] in structural, chemical, or physical properties resulting from physical [49–76], chemical [77–84], or more than one type [53, 85–87] of input. The output signals can be detected spectroscopically [88–93] or electrically/electrochemically [94–96]. Chemical computing can be done in the bulk, for example, in solution [97, 98] or at surfaces/interfaces [14–17, 99–102], such as at electrodes or on Si chips. Supramolecular structures have also been considered as switchable devices for logic operations [103].

Chemical computing equivalents of standard binary gates, such as AND [104, 105], OR [106], XOR [103, 107], NOR [91, 93, 108–111], NAND [111–113], INHIB [114–117], and XNOR [118, 119] have been realized. Reversibility [120, 121], reconfigurability [122–125], and resettability [126, 127] of the resulting gates have been explored. Digital logic functions of several-gate device components [128–134] have been realized, such as keypad lock and memory units [85, 135–145]. Chemical computing systems can function as single-molecule [146] nanoscale devices [147], as well as perform parallel computations using numerous molecules [148].

Chemical computing shows great promise [149–151]; however, as most unconventional computing approaches [152], it is not being developed as an alternative to the speed and versatility of Si computers. The short-term focus has been on novel functionalities and applications: microrobotics, multi-input (bio)sensors/actuators, and implantable device components. The main challenge for chemical computing has been networking of basic gates for achieving scalable, fault-tolerant information-processing networks [153]. Small networks performing basic operations have been realized [21, 22, 24], for example, adder/subtractor and their subunits [154–159]. Multisignal response to chemical or physical inputs has been demonstrated [23–25], and attempts at scaling up the complexity according to biological principles have been reported [74].

Biomolecular or *biochemical* information processing, to be termed *biocomputing* for brevity, constitutes a branch of chemical computing. It has drawn a lot of recent interest, because biomolecules offer natural specificity when used in complex "chemical soup" environments, as well as biocompatibility, the latter important for biomedical/biotechnological applications. Biomolecules may also be suitable for scalability paradigms borrowing ideas from nature. Furthermore, the biocatalytic nature of many utilized biomolecular processes offers certain advantages for analog noise control in the binary gate circuit design paradigm [17]. Proteins/enzymes [5–12, 14–17, 19, 20, 160–164] and DNA/RNA/DNAzymes [6, 7, 165–181] have been extensively used for gates, for realizing small networks and computational units, and for systems motivated [182] by applications.

This chapter is organized as follows. In Section 12.2, we describe general concepts for considering (bio)chemical binary gates. Gate design for decreasing noise amplification is addressed in Section 12.3. Section 12.4 addresses optimization of AND gates. Section 12.5 describes gate design as part of a network. Section 12.6 is devoted to summary and discussion of future challenges.

12.2
From Chemical Information-Processing Gates to Networks

Processing of large quantities of information at high levels of complexity requires utilization of a paradigm of scalable networking of simple gates. Recent chemical and biochemical computing literature has usually implicitly assumed an approach similar to that used in Si-chip electronic devices [183, 184] of designing fault-tolerant systems that can avoid buildup of noise without prohibitive use of resources. However, with biomolecules, one could perhaps also use design concepts borrowed from processes in living organisms [185]. Hybrid solutions can be expected, with bio-inspired elements supplementing the electronics designs. Other approaches include massive parallelism [181], specifically with DNA [186].

Thus far, the vast majority of the recent enzyme-based biocomputing realizations and analyses reported [10, 12, 14–17, 19, 20, 30], and similarly the rest of the biomolecular computing literature, have at best realized only small networks of gates, even though the aim has been to follow the digital approach based on analog gates and other elements connected in increasing-complexity networks [183, 184]. (Bio)chemical computing is presently also far from the complexity of coupled biochemical reaction sets needed for mimicking processes in living organisms. Near-future applications of moderate-complexity (bio)chemical "computing" systems will likely be in novel sensor systems [15], processing several input signals, and yielding yes/no digital outputs, corresponding to "sense/respond" or "sense/diagnose/treat" actions. Therefore, both the chemical steps and the output transduction to electrodes/electronic computers for the "action" step suggest the use of the binary yes/no digitalization.

More importantly, the binary/digital information processing paradigm offers an approach for control of the level of noise buildup in complex networks.

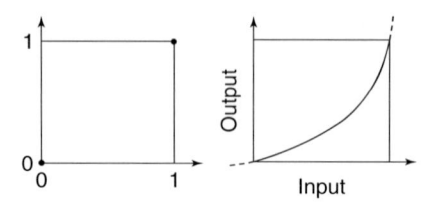

Figure 12.1 (a) The identity binary function: digital **0** and **1** are mapped to the same values. (b) A possible response curve.

Chemical and biochemical systems operate as larger levels of noise than electronic computers. As already mentioned, the input reactants and the "gate machinery" chemical concentrations, for example, catalysts, typically fluctuate within at least a couple of percent of the range of values identified as the binary **0** and **1**, and careful attention to the control of noise buildup is required for networks as small as two to three gates [10, 15, 19, 20].

While we address "digital" information processing, the network elements are actually always analog. Figure 12.1 offers an illustration for the simplest "gate" function: the identity, means signal transmission, conversion, or transduction. A possible analog response is also shown. The input and output signals are actually not limited to two values, or to the range bounded by them, of the digital/binary **0** and **1** selected as appropriate for a specific application. The analog signals can also be considered beyond the "digital" range, if physically allowed, as shown by the broken line extensions. Chemical concentrations can only be nonnegative, but the binary **0** does not have to always be at the physical zero.

A simple model is offered by an irreversible diatomic chemical reaction described within the rate equation approximation with the rate constant k, of the species A, of initial concentration $A(0) = A_0$, combining to yield the two-species product, C, of concentration $C(t)$ at time $t \geq 0$, with, initially, $C(0) = 0$:

$$A + A \xrightarrow{k} C \tag{12.1}$$

$$\frac{dA}{dt} = -2kA^2 \tag{12.2}$$

$$C(t) = \frac{A_0 - A(t)}{2} = \frac{kA_0^2 t}{1 + 2kA_0 t} \tag{12.3}$$

We assume that the information-processing application identifies a reference value, $A_{\max}$, of the input as the digital **1**, and, in the simplest case, the physical zero as the digital **0** input. The product of the reaction constitutes the output signal used/measured at the "gate time" t_g. The binary values for the output are then set by the gate itself: **0** and **1** will be, respectively, $C = 0$ and

$$C_{\max} = \frac{kA_{\max}^2 t_g}{1 + 2kA_{\max} t_g} \tag{12.4}$$

The logic range variables, x and z, represent the input, $A(0) = A_0$, and the output, $C(t_g)$, normalized to the "digital" range of values:

$$x = A_0/A_{\max} \qquad (12.5)$$

$$z = C(t_g)/C_{\max} \qquad (12.6)$$

With these definitions, we have the gate response function, shown in Figure 12.2

$$z(x) = \frac{(1+2p)x^2}{1+2px} \qquad (12.7)$$

which depends on the parameter combination

$$p = kA_{\max}t_g \qquad (12.8)$$

The digital **1** of the input, $A_{\max}$, is generally determined by the specific application or other network elements to which the present gate is connected. However, we can to some extent vary the reaction rate constant, k, by altering the physical and chemical conditions of the system, within the range allowed by the operational environment. We can also possibly adjust the reaction time, t_g. This allows a certain degree of control of the "response function shape," which can be used for gate design and optimization, as elaborated on in the following sections.

The considered chemical reaction generally yields concave shapes, shown in Figure 12.2. Furthermore, as illustrated in the figure, the actual shape of the gate response function cannot be varied significantly by just "tweaking" the parameters. Indeed, *order of magnitude variations* in the parameter values – which might not be

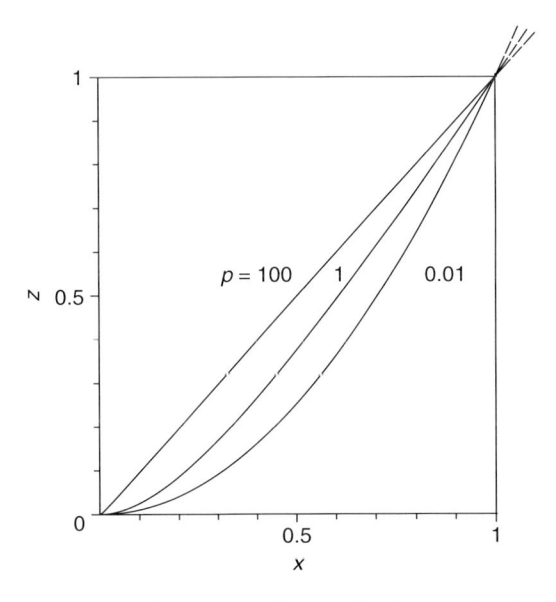

Figure 12.2 The response function for the reaction $A + A \rightarrow C$, see Eqs. (12.7) and (12.8), for three values of the parameter p, in increments of 2 orders of magnitude. All the curves are concave.

practical in many applications – are needed to achieve a qualitatively different shape. This difficulty [10, 12, 19] is shared by most, but especially catalytic, (bio)chemical computing gates and can be traced to that "activities" of reagents and catalysts are effectively cancelled out in the leading, linear order in defining the rescaled "logic range" variables (Eqs. (12.5) and (12.6)). Finally, we note that both variables in Figure 12.2 need not be limited to [0,1]; the curves are well defined and shown for x and z larger than 1 as well.

12.3
Noise Handling at the Gate Level and Beyond

Topics of noise amplification and filtering are considered here on the example of the "identity" gate just introduced. Two-input/one-output AND gates are addressed in the next section. The following reaction exemplifies a response more realistic of typical chemical kinetics than that considered in Section 12.2:

$$A + B \xrightarrow{K} C \tag{12.9}$$

with the rate constant K and initial conditions $A(0) = A_0$, $B(0) = B_0$, and $C(0) = 0$, and the output signal measured as $C(t_g)$. This reaction can be considered as a two-input process. However, here we regard A_0 as the input set by the environment in which the gate is used, whereas $B_0 (< A_0)$ will be for now assumed as a small (so that it limits rather than drives the output) "gate machinery" chemical, the supply of which can be controlled.

The rate equations and their solutions are summarized in

$$\frac{dA}{dt} = -KAB = -KA(A - A_0 + B_0) \tag{12.10}$$

$$C(t) = \frac{A_0 B_0 [1 - e^{-(A_0 - B_0)Kt}]}{A_0 - B_0 e^{-(A_0 - B_0)Kt}} \tag{12.11}$$

Equations (12.5) and (12.6) are then used to rescale the input and output in terms of the "logic" variables:

$$z(x) = \frac{x(1 - e^{-ax+b})(a - be^{-a+b})}{(1 - e^{-a+b})(ax - be^{-ax+b})} \tag{12.12}$$

This expression depends on two dimensionless combinations,

$$a = KA_{\max} t_g \text{ and } b = KB_0 t_g \tag{12.13}$$

These parameters can be controlled, yielding a response illustrated in Figure 12.3, by changing the physical and chemical conditions (vary the rate constant K); the "gate machinery" chemical supply, B_0; and the reaction time, t_g.

One can prove [187] by algebraic considerations that the function in Eq. (12.12) is always a monotonically increasing *convex* (Figure 12.3). Indeed, for catalytic (bio)chemical reactions and many other (bio)chemical processes, convex response curves – and surfaces, for more than one input – are generally expected. The

12.3 Noise Handling at the Gate Level and Beyond

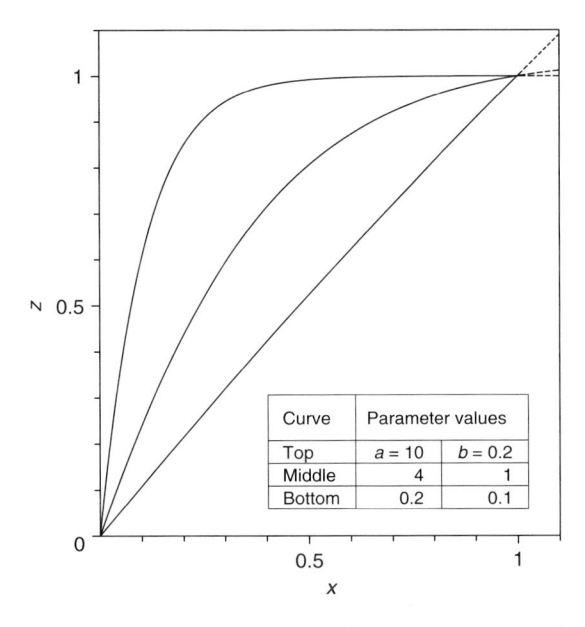

Figure 12.3 The convex response function (Eq. (12.12)) for the reaction $A + \cdots \to C$, where the omitted reactant is not considered a variable input, here shown for three different choices of the parameters a and b.

Curve	Parameter values	
Top	$a = 10$	$b = 0.2$
Middle	4	1
Bottom	0.2	0.1

product of the reaction – the output – is typically proportional to (linear in) the input-signal chemical concentration(s) for *small* inputs. For *large* inputs, the output is usually limited, for example, by the reactivity of the available (bio)catalyst, or, in our case, the availability of the second reactant, B. Therefore, the output signal reaches saturation in the large-input limit.

There are many possible sources of error in gate functioning. The most obvious noise is that in the inputs, which is actually quite large in chemical and biochemical environments. The gate function will transfer the resulting distribution of the input values into noise in the output. In addition, the binary **0** and **1** signal values need not be sharply defined. In applications, input/output signals in certain ranges of values may sometimes constitute **0** or **1**. For example, a range of normal physiological concentrations can be **0**, whereas an interval of pathophysiological values can be **1**, and these ranges need not even be bounded and can overlap, the latter necessitating additional statistical analyses or multi-input testing/sensing not addressed here. The gate function can also be noisy, and the distribution of its values can be displaced away from the desired digital values/ranges: in our notation, noise and fluctuations in concentrations and physical parameters of the system can lead to a distribution of the values of $z(x)$, for each x, rather than a sharply defined function such as in Eqs. (12.7–12.12). The mean value of this distribution need not pass precisely through the expected logic values at the "logic" inputs.

We will term *analog* the noise due to the spread of the output signal about the reference "digital" values (or ranges). In order to prevent buildup of noise as gates

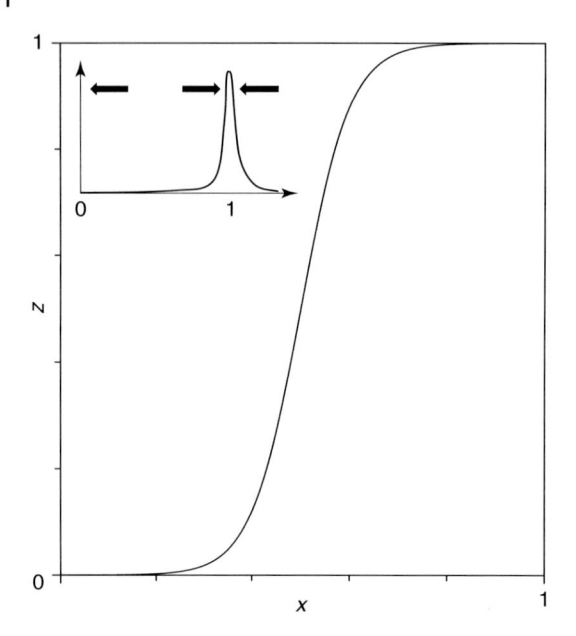

Figure 12.4 Sigmoidal response for the "identity" gate to act as a filter. The central inflection region should be narrow and positioned away from both logic **0** and **1**, and the slopes near both binary values should be zero or very small. Here, we assume that the signals can only be nonnegative; the sigmoidal curve continues smoothly (not shown) past the point $x, z = 1, 1$. The *inset* illustrates a possible input, x, distribution, here spread about **1** – assumed to be the intended input/output digital value. The filtering will drive the output, z, values originating from x close to the peak of the distribution, toward the correct digital answer, **1** (shown by the pair of facing arrows). The values in the tail of the distribution will be driven toward the wrong digital answer, **0** (shown by the unpaired arrow); this results in a small-probability "digital" error. Similarly (not shown), distributions peaked near **0**, when it is the expected digital input/output, will also be sharpened, but the tail values will be driven to the wrong digital answer.

and other network elements are connected, we have to pass our signals through "filters" with response close to that shown in Figure 12.4. Ideally we would like to have the sigmoidal property – small slopes/gradients at and around the digital values – in all or most of our gates. Filters can also be used as separate network elements. There is evidence that filtering for suppressing analog noise buildup is utilized by nature [188, 189]. Experimental attempts to realize a biochemical filter have only recently shown successes [190, 191].

The inset in Figure 12.4 points out that filtering can push those values that are far from the correct digital result even closer to the wrong answer. Thus, the process of digitalization itself also introduces the "digital" type of noise: small probability of a wrong binary value. Digital errors are not very probable and only become important to correct for large enough networks. Standard techniques based on redundancy are available [192] for digital error correction.

For enzyme-based gates studied by our group, for the presently realized network sizes, analog error correction is important and has recently received significant attention [10, 12, 14, 15, 17, 19, 20, 30]. It has been estimated [10, 12] that up to order 10 such gates can be connected in a network before digital error correction is warranted.

Experimental realizations of the sigmoidal behavior (Figure 12.4) have been an ongoing effort [190, 191], based on the ideas [1, 10, 187, 189] that an additional reactant, F, which depletes the product, but can only consume (react fast with) a small quantity of it, will suppress the response at small inputs without voiding the saturation property at large inputs, thus yielding a sigmoidal response. In connection with the system of the type defined in Eq. (12.9), we can consider the reaction $C + F \xrightarrow{\rho} \cdots$, with a fast rate, ρ, and with $\cdots$ denoting inert chemicals. This added reaction, however, delays the saturation at large inputs. Another option is $C + F \xrightarrow{\rho} A + \cdots$, which, however, introduces a feedback loop – the effects of which have not been studied – by regenerating some of the input. Variants realized experimentally [190, 191] were for systems more complicated than the single reaction in Eq. (12.9). In the present nomenclature, the ideas utilized in one approach [190], for instance, can be described by that the output of the added process, $C + F \xrightarrow{\rho} S + \cdots$, included a "restoration" of a secondary input chemical, S, involved in the original biocatalytic reaction and leading to the output signal detection. The closest equivalent for the system described by Eq. (12.9) would be to have $C + F \xrightarrow{\rho} B + \cdots$. Interestingly, the system of rate equations obtained by adding this reaction to the one in Eq. (12.9) is still solvable in a closed form in quadratures, because the solution steps lead to a single differential equation, which, while nonlinear, is of the Bernoulli form [193]. However, the expressions obtained are sufficiently complicated so that the closed-form results are unilluminating, and numerical evaluation is needed to make them illustrative/tractable, which is outside the scope of this presentation.

Ultimately, the output signal of several (bio)chemical information-processing steps in near-term sensor applications of the "decision-making" type [14, 15] will likely be fed into conventional electronic circuitry. Well-developed research, not reviewed here, addresses the interfacing of biomolecular logic with "smart" signal-responsive [16, 59, 74, 101, 194–204] materials and with electrodes and bioelectronic devices [16, 205–211]. The interfacing/transduction of (bio)chemical signals to electronic ones can also incorporate a filtering "sigmoidal" property, as has been recently experimentally demonstrated [16].

The above discussion reveals that while (bio)chemical filtering is a desirable property for gates and network elements, its direct experimental realization has thus far been only preliminary. Therefore, efforts have also been devoted to minimize noise amplification for gates with *convex* response curves/surfaces of the "standard" (bio)chemical reaction type. For single-input/single-output gates, such as the illustrated "identity" function, Figure 12.3, noise amplification (increase in the spread of the noise distribution due to the gate function) is simply related to the maximum of the two slopes at the binary points, and it can be minimized by having both slopes close to 1, that is, a nearly straight line response curve (Figure 12.3).

However, a danger – also identified in designing filtering systems [190, 191] – with this approach has been that the near-linear behavior is realized straightforwardly when the reaction is far from saturation. The latter regime corresponds to weak output signal, and therefore while there is no noise *amplification*, another source of relative sensitivity to noise is introduced, that of the random "environmental" external noise being comparable to the spread of the binary **0** and **1** signal reference values.

One solution has been to "drive" the reaction by flooding the system with reagent(s) that will increase the process rates. For example, for the reaction scheme considered in this section (Eqs. (12.9–12.13)), we could effectively increase the overall time-dependent rate, $KB(t)$, of the consumption of the input, A, by selecting $B_0 \gg A_0$ (instead of the originally assumed regime $B_0 < A_0$) for all the relevant inputs: with $B_0 \gg A_{\max}$, we have $b \gg a$, and the response curve (Eq. (12.12)), then reduced to the linear one, $z(x) \approx x$ (for all $x \in [0,1]$). This is perhaps not an interesting "curve" to consider, but it does the job of avoiding/minimizing noise amplification. The situation with the two-input/one-output gate functions is more complicated and is discussed in the next section.

12.4
Optimization of AND Gates

AND is the most studied gate in the (bio)chemical computing literature, and practically the only one explored in detail for the optimization of its response. Since the truth table for the AND gate is that the output **1** is obtained *only* when *both* inputs are **1**, AND is a natural outcome as a product of a two-input chemical reaction. The AND gates themselves are not universal, but they become such if supplemented by NOT: NAND (not-AND) gates can be networked to realize an arbitrary binary function. Indeed, the NOT version of filtering [16] – the vertically flipped sigmoidal curve as compared to Figure 12.4 – would be particularly interesting to realize and widely incorporate in networked biochemical processes.

We consider a simple model for the AND gate in chemical computing: we now regard the reaction in Eq. (12.9): $A + B \to C$, as a two-input, one-output process. We introduce the "logic range" variable for the input B, rescaled to the binary interval [0,1],

$$y = B_0/B_{\max} \tag{12.14}$$

paralleling the definitions in Eqs. (12.5) and (12.6). Here $B_{\max}$ is the reference value for logic **1** of the B-input. The quantity z defined in Eq. (12.6) is now a two-variable function, $z(x, y)$, describing the AND gate response *surface* shape. The solution of the rate equations, given by Eq. (12.11), is now recast in terms of the new set of the "logic range" variables to yield

$$z(x,y) = \frac{xy(\alpha e^\alpha - \beta e^\beta)(e^{\alpha x} - e^{\beta y})}{(e^\alpha - e^\beta)(\alpha x e^{\alpha x} - \beta y e^{\beta y})} \tag{12.15}$$

where we defined

$$\alpha = KA_{\max}t_g, \quad \beta = KB_{\max}t_g \qquad (12.16)$$

These are similar to the (dimensionless) parameters in Eq. (12.13), but we now have less control over their values, because their ratio is set by the application (the environment) of the gate, which in most cases predetermines both $A_{\max}$ and $B_{\max}$. Only the product Kt_g can be externally adjusted.

The response surfaces are illustrated in Figures 12.5 and 12.6. Recall that the noise in the input signals is passed to the output, with the added noise effects due to the gate function: imprecise (on average) and fluctuating values of $z(x, y)$. In addition to designing gates as precise as possible, we can also minimize the propagation of analog noise, and hopefully avoid noise amplification, by finding

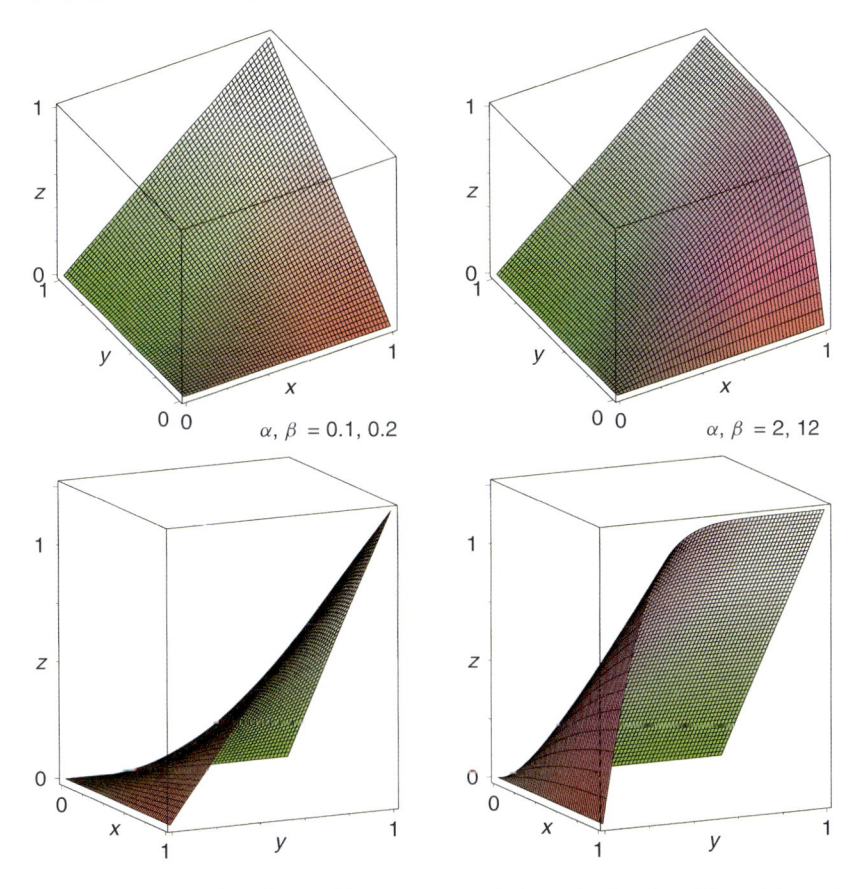

Figure 12.5 Examples of smoothly varying response surfaces, $z(x,y)$, for the AND gate function representing the reaction $A + B \to C$ (Eqs. (12.15) and (12.16)) for two choices of the parameters α, β (assumed as independently adjustable, which need not be the case in practice). The top images give the frontal view, whereas the bottom images show the back view for the two surfaces.

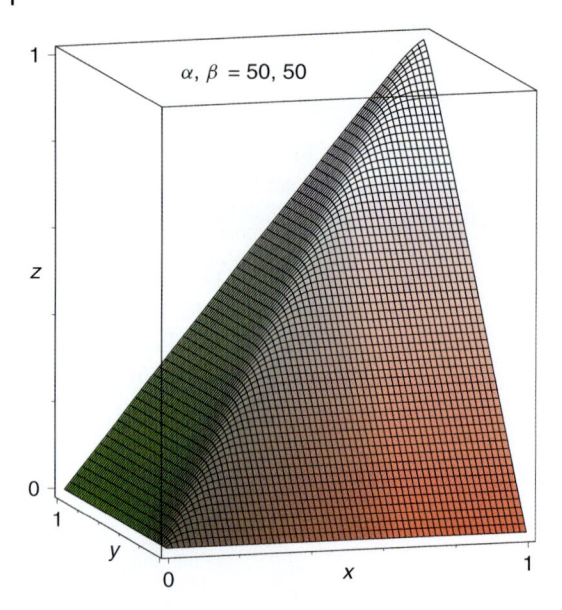

Figure 12.6 Same as in Figure 12.5, but for the fast-reaction case (large, here equal, α, β). The emergence of a response surface with nonsmooth features is seen: formation of a ridge (here, along the diagonal), and also the shrinking of the region for which the value of the gradient near the point (0,0) remains small. Note that the gradient *at* the origin, $\left|\vec{\nabla}z(x,y)\right|_{x=0,y=0}$, is zero for all three surfaces shown in both figures. Nonuniformities also set in along the ridge region, including near the point (1,1). A forming, off-diagonal ridge can already be seen in the right images in Figure 12.5, because of a relatively large value of β there.

parameters (such as Kt_g) that yield gates that suppress, or at least diminish spread in the input signals, by having small slopes near the logic points. Indeed, the absolute value of the gradient vector, $\left|\vec{\nabla}z(x,y)\right|$, calculated at the logic points, measures the noise spread amplification or suppression. This is relevant only if the input noise distribution is narrow and also provided the gate function $z(x,y)$ is smooth in those regions near the logic points, which are approximately the size of the spread of the noise distributions. Relatively smooth $z(x,y)$ shapes are illustrated in Figure 12.5. We can try to identify parameter values for which the largest of the four gradients, $\left|\vec{\nabla}z(x,y)\right|_{x=0,y=0}$, $\left|\vec{\nabla}z(x,y)\right|_{x=1,y=0}$, $\left|\vec{\nabla}z(x,y)\right|_{x=0,y=1}$, $\left|\vec{\nabla}z(x,y)\right|_{x=1,y=1}$ is as small as possible (note that $\left|\vec{\nabla}z\right|_{00}$ is always zero for this particular model). For this calculation, let us for now assume that both α and β can be adjusted independently. By numerical calculation we then find that for moderate values of α and β, the minimum is obtained for $\alpha = \beta \approx 0.4966$ and is given by $\left|\vec{\nabla}z\right|_{10} = \left|\vec{\nabla}z\right|_{01} = \left|\vec{\nabla}z\right|_{11} \approx 1.1796$.

It turns out that gate functions of this type amplify analog noise even under optimal conditions. Noise amplification in the best case scenario is about 18%.

Studies [10, 12, 19] of enzyme-based AND gates, which have utilized more realistic (and thus more complicated and not exactly solvable) rate equation models appropriate for biocatalytic reactions, found similar estimates. Experimental data were fitted, and results were numerically analyzed by using both the rate equation approach and more phenomenological shape-fitting forms, the latter exemplified in the next section. If not optimized, then for experimental parameter values selected randomly for convenience, smooth, convex gates corresponding to typical (bio)chemical reactions can have very large noise amplification, 300–500%. Reaching optimal conditions is not always straightforward, primarily because the gate function shape depends only weakly on parameter values. Even under optimal conditions, at least about 20% noise amplification is to be expected.

For fast enough reactions, the maximal gradient can be smaller than ~ 1.2. However, as illustrated in Figure 12.6, the gate function surface then develops sharp features, and the gradients can no longer be used as measures of noise amplification, because they remain close to the logic point values only in tiny regions near these points, as compared to the noise spread of at least several percentage typical for (bio)chemical signals. Generally, when the spread of the noise is larger than the x and y scales over which the gate function or its derivatives vary significantly, one can assume a certain shape of the *input* noise distribution, such as a product of approximately Gaussian distributions in x and y for inputs at each of the logic points, or half-Gaussian, if the logic zero is exactly at the physical zero. Given a model for the gate response function, one can then numerically calculate [10] the *output* signal distribution for each of the inputs and thus estimate the noise amplification factor [10, 12, 14].

The "ridged" gate response function (e.g., Figure 12.6) was first encountered [12] in a study of an enzymatic system, which also realized a smooth-response counterpart when a different chemical was used as one of the inputs [12]. The reaction kinetics was more complicated than in the present model, but the finding has confirmed the general expectations: The optimal conditions are obtained with a symmetrically (diagonally) positioned ridge, as in Figure 12.6, and the noise amplification factor is then only slightly larger than 1, estimated by considering distributions. Thus, noise amplification is practically avoided. However, such gates do not have the noise suppression (filter) property.

Figure 12.7 presents a schematic of an AND gate response sigmoidal in only one of the two inputs, which was recently explored and experimentally realized [14]. Many allosteric enzymes have such a "self-promoter" property with respect to one of their substrates (input chemical species). A key finding [14] has been that the single-sided sigmoidal shape can be tuned by parameter adjustment to have the noise amplification factor only slightly above 1, so that there is practically no noise amplification. However, a desirable two-sided sigmoid response, also shown schematically in Figure 12.7, has not to our knowledge been realized at the level of a single AND gate, in the chemical or biomolecular computing literature. Certain biochemical processes in nature, which are much more complex than our synthetic AND gate systems, do realize [188] a two-sided sigmoidal response.

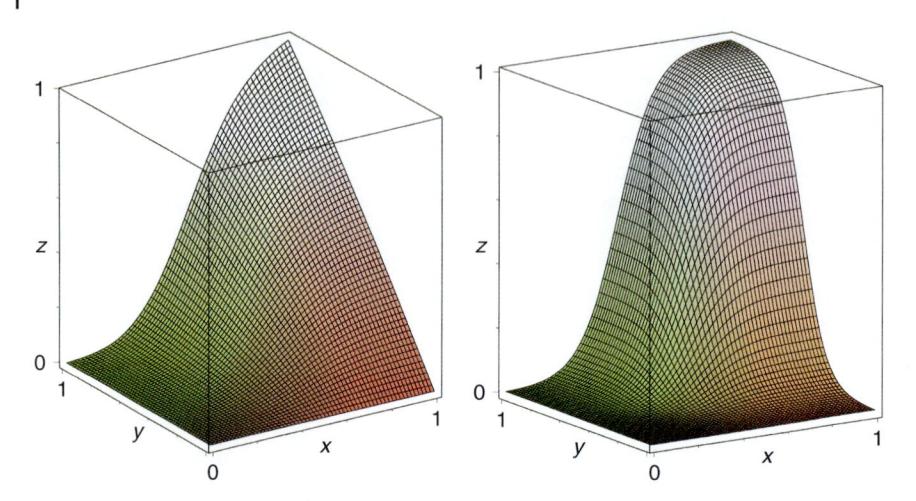

Figure 12.7 (a) Schematic of a one-sided sigmoidal behavior. (b) Schematic of a desirable two-sided sigmoidal response for AND gates.

12.5
Networking of Gates

We have seen that optimization of (bio)chemical gates one at a time is not straightforward. In most cases, a rather large variation of the controllable parameters is needed: physical and chemical conditions, reactant concentrations, and in some cases choice of (bio)chemical species, which may not be experimentally feasible. The actual detailed kinetic modeling of the reactions involved, especially for biomolecular systems, is in itself a challenging task [10, 12, 14, 15, 17]: the kinetics of most biomolecular processes, specifically those used for AND gates, is complex and not well studied. The quality of the experimental data for the gate response function is limited because of the noise in the gate function itself, limited lifetime for constant activity of the biocatalytic species, and so on. Thus, multiparameter complex reaction schemes are difficult to substantiate by data fitting in the gate design context, which requires models to work for a large range of adjustable parameters.

An alternative approach involves optimization of the *relative* gate functioning in a network, whereby each gate is modeled within a very approximate, phenomenological curve/surface-fitting approach. These ideas have recently been tested [19] for coupled enzymatic reactions that include steps common in sensor development [212] for maltose and its sources. A modular network representation of the biocatalytic processes involved is possible in terms of three AND gates (Figure 12.8). This "cartoon" representation is actually approximate, because it obscures some of the complexity of the constituent processes [19].

The approach involves first proposing a phenomenological fitting function for the gate response surface in terms of as few parameters as possible, enough to

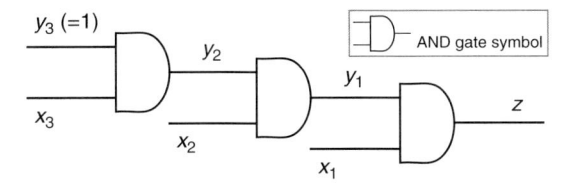

Figure 12.8 The three-gate network [19], with varied inputs $x_{1,2,3}$ and constant y_3.

capture the expected *qualitative* features of the shape. For a typical convex "identity" gate, the fitting function is conveniently written as

$$z(x) \approx \frac{sx}{(s-1)x+1} \qquad (12.17)$$

This is a single parameter, s, rational form that "looks" qualitatively correct, provided we assume

$$s > 1 \qquad (12.18)$$

Indeed, the curve is then convex and has slope s at $x, z(x) = 0, 0$, and $1/s$ at $x, z(x) = 1, 1$.

For each AND gate, we then use the two-parameter, $s > 1$ and $u > 1$, product

$$z(x, y) \approx \frac{(sx)(uy)}{[(s-1)x+1][(u-1)y+1]} \qquad (12.19)$$

The gradient values are $|\vec{\nabla}z|_{00} = 0$, $|\vec{\nabla}z|_{10} = u$, $|\vec{\nabla}z|_{01} = s$, and $|\vec{\nabla}z|_{11} = \sqrt{s^{-2}+u^{-2}}$. The minimum of the largest of the last three values is obtained for $s = u = \sqrt[4]{2} \approx 1.189$, which is also the *value* of the gradient, consistent with the earlier reported empirical expectation that smooth convex AND gates can typically be optimized at best to yield noise amplification somewhat under 20%.

Having introduced our approximate fitting functions, we now experimentally vary selective inputs in the network (Figure 12.8). In the experiment [19], each of the three inputs $x_{1,2,3}$ was separately varied between 0 (corresponding to the binary 0) and the reference value predefined as 1, while all the other inputs (including y_3) were at their reference 1 values. In fact, when the parameterization of Eq. (12.19) is applied to all three gates in our network of Figure 12.8, we get a rational expression for z as a function of all the four inputs ($x_{1,2,3}$ and y_3). Setting all but a single x input to 1, we get the parameterization for the measurement with that input varied; we only keep that varying argument of $z(\cdots)$ for simplicity:

$$z(x_1) = \frac{s_1 x_1}{(s_1 - 1)x_1 + 1} \qquad (12.20)$$

$$z(x_2) = \frac{s_2 u_1 x_2}{(s_2 u_1 - 1)x_2 + 1} \qquad (12.21)$$

$$z(x_3) = \frac{s_3 u_1 u_2 x_3}{(s_3 u_1 u_2 - 1)x_3 + 1} \qquad (12.22)$$

Interestingly, each data set depends only on a single parameter (s_1, $s_2 u_1$, or $s_3 u_1 u_2$).

While we only get partial information on gate functioning, we can attempt to "tweak" the relative gate activities in the network to improve the stability. If the proposed approximate description is semiquantitatively accurate for a given gate, then the parameters s and u for that gate will be functions of adjustable quantities, such as the gate time, input concentrations of some of the chemicals, and reaction rates (which can in turn be controlled by the physical and chemical conditions). In addition, s and u can depend on other quantities that are not controllable.

Without detailed rate equation kinetic modeling, this parameter dependence is not known. However, examination of the fitted quantities (s_1, $s_2 u_1$, $s_3 u_1 u_2$) still provides useful information on the *relative* effect that the gates have, by their contribution to the gradients at various logic points, when compared to the optimal values ($s_1 = 2^{1/4}$, $s_2 u_1 = 2^{1/2}$, $s_3 u_1 u_2 == 2^{3/4}$). The initial sets of data [19] were collected with the experimentally convenient, randomly selected values for the "gate machinery" and other parameters. Examination of the results [19] has lead to a semiquantitative conclusion that the deviations from the optimal values could largely be attributed to the gate that is the closest to the output in Figure 12.8 ($z = x_1$ AND y_1): it was too "active" as compared to the other two gates (i.e., its biocatalytic reaction was too fast). A new experiment was then devised [19] with the concentration of the enzyme catalyzing this gate's function reduced by an order of magnitude (approximately 11-fold). The new data collected for the modified network, when fitted, yielded s_1, $s_2 u_1$, $s_3 u_1 u_2$ values significantly closer to the optimal [19].

12.6
Conclusions and Challenges

We reviewed aspects of and approaches to gate optimization for control of analog noise amplification, which is important for connecting gates in small networks. For larger networks, digital error correction by redundancy will also have to be implemented, and various network elements will have to be devised for filtering; signal splitting; signal balancing; gate-to-gate connectivity; memory; interfacing with external input, output, and control mechanisms; and so on.

We used simple rate equation models that allow exact solvability, to illustrate and motivate the discussion. We avoided presenting experimental data and their numerical analysis, which can be found in the cited articles. Our presentation has been limited to AND gates and related systems. Indeed, all the recent studies, with one exception, an XOR gate [213, 214], of noise control in (bio)chemical computing have thus far been for AND gates and, furthermore, again with just few exceptions presented, that is, Ref. [215], only for those with the binary **0** set at the physical zeros of chemical concentrations. While these limitations are natural for chemical kinetics, they are definitely not typical for applications envisaged, notably, multi-input sensing [15].

As new experiments on mapping out (bio)chemical gate functioning and network designs are reported, new features of noise and error control will be explored.

Specifically, noise in the gate function itself, including spread of its values and imprecise mean values – not exactly at the expected reference output **0** or **1**, with deviations possibly also different for various inputs that should ideally yield the same logic output – will have to be considered and corrected, most likely by filtering. Indeed, we conclude by emphasizing that, while longer term, network design and scaling up will be crucial, the shorter term challenges in (bio)chemical information processing have been to design and experimentally realize versatile and effective *(bio)chemical filter processes* and other nonbinary network elements that can be concatenated with various binary logic gates.

Acknowledgments

The author acknowledges research funding by the United States National Science Foundation (grants CBET-1066397 and CCF-1015983).

References

1. Privman, V. (2010) *Isr. J. Chem.*, **51**, 118–131.
2. Credi, A. (2007) *Angew. Chem. Int. Ed.*, **46**, 5472–5475.
3. de Silva, A.P. and Uchiyama, S. (2007) *Nat. Nanotechnol.*, **2**, 399–410.
4. Szaciłowski, K. (2008) *Chem. Rev.*, **108**, 3481–3548.
5. Shao, X.G., Jiang, H.Y., and Cai, W.S. (2002) *Prog. Chem.*, **14**, 37–46.
6. Saghatelian, A., Volcker, N.H., Guckian, K.M., Lin, V.S.Y., and Ghadiri, M.R. (2003) *J. Am. Chem. Soc.*, **125**, 346–347.
7. Ashkenasy, G. and Ghadiri, M.R. (2004) *J. Am. Chem. Soc.*, **126**, 11140–11141.
8. Baron, R., Lioubashevski, O., Katz, E., Niazov, T., and Willner, I. (2006) *J. Phys. Chem. A*, **110**, 8548–8553.
9. Baron, R., Lioubashevski, O., Katz, E., Niazov, T., and Willner, I. (2006) *Angew. Chem. Int. Ed.*, **45**, 1572–1576.
10. Privman, V., Strack, G., Solenov, D., Pita, M., and Katz, E. (2008) *J. Phys. Chem. B*, **112**, 11777–11784.
11. Strack, G., Pita, M., Ornatska, M., and Katz, E. (2008) *ChemBioChem*, **9**, 1260–1266.
12. Melnikov, D., Strack, G., Pita, M., Privman, V., and Katz, E. (2009) *J. Phys. Chem. B*, **113**, 10472–10479.
13. Flood, A.H., Ramirez, R.J.A., Deng, W.Q., Muller, R.P., Goddard, W.A., and Stoddart, J.F. (2004) *Aust. J. Chem.*, **57**, 301–322.
14. Privman, V., Pedrosa, V., Melnikov, D., Pita, M., Simonian, A., and Katz, E. (2009) *Biosens. Bioelectron.*, **25**, 695–701.
15. Katz, E., Privman, V., and Wang, J. (2010) in *Proceeding Conference IC-QNM 2010* (eds V. Ovchinnikov and V. Privman), IEEE Computer Society Conference Publishing Services, Los Alamitos, CA, pp. 1–9.
16. Privman, M., Tam, T.K., Pita, M., and Katz, E. (2009) *J. Am. Chem. Soc.*, **131**, 1314–1321.
17. Katz, E. and Privman, V. (2010) *Chem. Soc. Rev.*, **39**, 1835–1857.
18. Wagner, N. and Ashkenasy, G. (2009) *Chem. Eur. J.*, **15**, 1765–1775.
19. Privman, V., Arugula, M.A., Halamek, J., Pita, M., and Katz, E. (2009) *J. Phys. Chem. B*, **113**, 5301–5310.
20. Arugula, M.A., Halamek, J., Katz, E., Melnikov, D., Pita, M., Privman, V., and Strack, G. (2009) in *Proceeding Conference CENICS 2009* (eds K.B. Kent, P. Dini, O. Franza, T. Palacios, C. Reig, J.E. Maranon, A. Rostami, D. Zammit-Mangion, W.C.

Hasenplaugh, F. Toran, and Y. Zafar), IEEE Computer Society Conference Publishing Services, Los Alamitos, CA, pp. 1–7.
21. Brown, G.J., de Silva, A.P., and Pagliari, S. (2002) *Chem. Commun.*, **2002**, 2461–2463.
22. Pischel, U. (2007) *Angew. Chem. Int. Ed.*, **46**, 4026–4040.
23. Raymo, F.M. and Giordani, S. (2001) *J. Am. Chem. Soc.*, **123**, 4651–4652.
24. Guo, X., Zhang, D., Zhang, G., and Zhu, D. (2004) *J. Phys. Chem. B*, **108**, 11942–11945.
25. Liu, Y., Jiang, W., Zhang, H.Y., and Li, C.J. (2006) *J. Phys. Chem. B*, **110**, 14231–14235.
26. Stojanovic, M.N. and Stefanovic, D. (2003) *Nat. Biotechnol.*, **21**, 1069–1074.
27. Macdonald, J., Li, Y., Sutovic, M., Lederman, H., Pendri, K., Lu, W.H., Andrews, B.L., Stefanovic, D., and Stojanovic, M.N. (2006) *Nano Lett.*, **6**, 2598–2603.
28. Niazov, T., Baron, R., Katz, E., Lioubashevski, O., and Willner, I. (2006) *Proc. Natl. Acad. U.S.A.*, **103**, 17160–17163.
29. Strack, G., Ornatska, M., Pita, M., and Katz, E. (2008) *J. Am. Chem. Soc.*, **130**, 4234–4235.
30. Fedichkin, L., Katz, E., and Privman, V. (2008) *J. Comput. Theor. Nanosci.*, **5**, 36–43.
31. Feinerman, O. and Moses, E. (2005) *J. Neurosci.*, **26**, 4526–4534.
32. Feinerman, O., Rotem, A., and Moses, E. (2008) *Nat. Phys.*, **4**, 967–973.
33. Benenson, Y., Paz-Elizur, T., Adar, R., Keinan, E., Livneh, Z., and Shapiro, E. (2001) *Nature*, **414**, 430–434.
34. Fu, P. (2007) *Biotechnol. J.*, **2**, 91–101.
35. Dittrich, P. (2005) *Lect. Notes Comput. Sci.*, **3566**, 19–32.
36. Shipway, A.N., Katz, E., and Willner, I. (2001) in *Molecular Machines and Motors*, Structure and Bonding, Vol. **99** (ed. J.P. Sauvage), Springer, Berlin, pp. 237–281.
37. Weinberger, D.A., Higgins, T.B., Mirkin, C.A., Stern, C.L., Liable-Sands, L.M., and Rheingold, A.L. (2001) *J. Am. Chem. Soc.*, **123**, 2503–2516.
38. Tseng, H.R., Vignon, S.A., Celestre, P.C., Perkins, J., Jeppesen, J.O., Di Fabio, A., Ballardini, R., Gandolfi, M.T., Venturi, M., Balzani, V., and Stoddart, J.F. (2004) *Chem. Eur. J.*, **10**, 155–172.
39. Zhao, Y.L., Dichtel, W.R., Trabolsi, A., Saha, S., Aprahamian, I., and Stoddart, J.F. (2008) *J. Am. Chem. Soc.*, **130**, 11294–11295.
40. Suresh, M., Ghosh, A., and Das, A. (2007) *Tetrahedron Lett.*, **48**, 8205–8208.
41. Crowley, J.D., Leigh, D.A., Lusby, P.J., McBurney, R.T., Perret-Aebi, L.E., Petzold, C., Slawin, A.M.Z., and Symes, M.D. (2007) *J. Am. Chem. Soc.*, **129**, 15085–15090.
42. Iwata, S. and Tanaka, K. (1995) *J. Chem. Soc. Chem. Commun.*, **1995**, 1491–1492.
43. Lee, S.H., Kim, J.Y., Kim, S.K., Leed, J.H., and Kim, J.S. (2004) *Tetrahedron*, **60**, 5171–5176.
44. Magri, D.C., Brown, G.J., McClean, G.D., and de Silva, A.P. (2006) *J. Am. Chem. Soc.*, **128**, 4950–4951.
45. Sadhu, K.K., Bag, B., and Bharadwaj, P.K. (2007) *J. Photochem. Photobiol. A*, **185**, 231–238.
46. Katz, E., Shipway, A.N., and Willner, I. (2002) in *Photoreactive Organic Thin Films*, Chapter II-7 (eds S. Sekkat and W. Knoll), Elsevier, San Diego, CA, pp. 219–268.
47. Balzani, V. (2003) *Photochem. Photobiol. Sci.*, **2**, 459–476.
48. Giordani, S., Cejas, M.A., and Raymo, F.M. (2004) *Tetrahedron*, **60**, 10973–10981.
49. Lion-Dagan, M., Katz, E., and Willner, I. (1994) *J. Am. Chem. Soc.*, **116**, 7913–7914.
50. Willner, I., Lion-Dagan, M., Marx-Tibbon, S., and Katz, E. (1995) *J. Am. Chem. Soc.*, **117**, 6581–6592.
51. Raymo, F.M. and Giordani, S. (2002) *Proc. Natl. Acad. U.S.A.*, **99**, 4941–4944.
52. Liu, Z.F., Hashimoto, K., and Fujishima, A. (1990) *Nature*, **347**, 658–660.

53. Doron, A., Portnoy, M., Lion-Dagan, M., Katz, E., and Willner, I. (1996) *J. Am. Chem. Soc.*, **118**, 8937–8944.
54. Liu, N.G., Dunphy, D.R., Atanassov, P., Bunge, S.D., Chen, Z., Lopez, G.P., Boyle, T.J., and Brinker, C.J. (2004) *Nano Lett.*, **4**, 551–554.
55. Ashton, P.R., Ballardini, R., Balzani, V., Credi, A., Dress, K.R., Ishow, E., Kleverlaan, C.J., Kocian, O., Preece, J.A., Spencer, N., Stoddart, J.F., Venturi, M., and Wenger, S. (2000) *Chem. Eur. J.*, **6**, 3558–3574.
56. Katz, E. and Shipway, A.N. (2005) in *Bioelectronics: From Theory to Applications*, Chapter 11 (eds I. Willner and E. Katz), Wiley-VCH Verlag GmbH, Weinheim, pp. 309–338.
57. Bonnet, S. and Collin, J.P. (2008) *Chem. Soc. Rev.*, **37**, 1207–1217.
58. Thanopulos, I., Kral, P., Shapiro, M., and Paspalakis, E. (2009) *J. Mod. Opt.*, **56**, 686–703.
59. Katz, E., Sheeney-Haj-Ichia, L., Basnar, B., Felner, I., and Willner, I. (2004) *Langmuir*, **20**, 9714–9719.
60. Katz, E., Baron, R., and Willner, I. (2005) *J. Am. Chem. Soc.*, **127**, 4060–4070.
61. Hsing, I.M., Xu, Y., and Zhao, W.T. (2007) *Electroanalysis*, **19**, 755–768.
62. Hirsch, R., Katz, E., and Willner, I. (2000) *J. Am. Chem. Soc.*, **122**, 12053–12054.
63. Katz, E., Sheeney-Haj-Ichia, L., and Willner, I. (2002) *Chem. Eur. J.*, **8**, 4138–4148.
64. Willner, I. and Katz, E. (2003) *Angew. Chem. Int. Ed.*, **42**, 4576–4588.
65. Wang, J. and Kawde, A.N. (2002) *Electrochem. Commun.*, **4**, 349–352.
66. Laocharoensuk, R., Bulbarello, A., Mannino, S., and Wang, J. (2007) *Chem. Commun.*, **2007**, 3362–3364.
67. Wang, J. (2008) *Electroanalysis*, **20**, 611–615.
68. Lee, J., Lee, D., Oh, E., Kim, J., Kim, Y.P., Jin, S., Kim, H.S., Hwang, Y., Kwak, J.H., Park, J.G., Shin, C.H., Kim, J., and Hyeon, T. (2005) *Angew. Chem. Int. Ed.*, **44**, 7427–7432.
69. Wang, J., Scampicchio, M., Laocharoensuk, R., Valentini, F., Gonzalez-Garcia, O., and Burdick, J. (2006) *J. Am. Chem. Soc.*, **128**, 4562–4563.
70. Loaiza, O.A., Laocharoensuk, R., Burdick, J., Rodriguez, M.C., Pingarron, J.M., Pedrero, M., and Wang, J. (2007) *Angew. Chem. Int. Ed.*, **46**, 1508–1511.
71. Zheng, L. and Xiong, L. (2006) *Colloids Surf., A*, **289**, 179–184.
72. Riskin, M., Basnar, B., Katz, E., and Willner, I. (2006) *Chem. Eur. J.*, **12**, 8549–8557.
73. Riskin, M., Basnar, B., Chegel, V.I., Katz, E., Willner, I., Shi, F., and Zhang, X. (2006) *J. Am. Chem. Soc.*, **128**, 1253–1260.
74. Chegel, V.I., Raitman, O.A., Lioubashevski, O., Shirshov, Y., Katz, E., and Willner, I. (2002) *Adv. Mater.*, **14**, 1549–1553.
75. Poleschak, I., Kern, J.M., and Sauvage, J.P. (2004) *Chem. Commun.*, **2004**, 474–476.
76. Le, X.T., Jégou, P., Viel, P., and Palacin, S. (2008) *Electrochem. Commun.*, **10**, 699–703.
77. Ashton, P.R., Ballardini, R., Balzani, V., Baxter, I., Credi, A., Fyfe, M.C.T., Gandolfi, M.T., Gomez-Lopez, M., Martinez-Diaz, M.V., Piersanti, A., Spencer, N., Stoddart, J.F., Venturi, M., White, A.J.P., and Williams, D.J. (1998) *J. Am. Chem. Soc.*, **120**, 11932–11942.
78. Richmond, C.J., Parenty, A.D.C., Song, Y.F., Cooke, G., and Cronin, L. (2008) *J. Am. Chem. Soc.*, **130**, 13059–13065.
79. Shiraishi, Y., Tokitoh, Y., Nishimura, G., and Hirai, T. (2005) *Org. Lett.*, **7**, 2611–2614.
80. Coutrot, F., Romuald, C., and Busseron, E. (2008) *Org. Lett.*, **10**, 3741–3744.
81. Nishimura, G., Ishizumi, K., Shiraishi, Y., and Hirai, T. (2006) *J. Phys. Chem. B*, **110**, 21596–21602.
82. Zhou, W.D., Li, J.B., He, X.R., Li, C.H., Lv, J., Li, Y.L., Wang, S., Liu, H.B., and Zhu, D.B. (2008) *Chem. Eur. J.*, **14**, 754–763.
83. Fahlman, R.P., Hsing, M., Sporer-Tuhten, C.S., and Sen, D. (2003) *Nano Lett.*, **3**, 1073–1078.

84. Shiraishi, Y., Tokitoh, Y., and Hirai, T. (2005) *Chem. Commun.*, **2005**, 5316–5318.
85. Baron, R., Onopriyenko, A., Katz, E., Lioubashevski, O., Willner, I., Wang, S., and Tian, H. (2006) *Chem. Commun.*, **2006**, 2147–2149.
86. Katz, I., Willner, B., and Willner, I. (1997) *Biosens. Bioelectron.*, **12**, 703–719.
87. Biancardo, M., Bignozzi, C., Doyle, H., and Redmond, G. (2005) *Chem. Commun.*, **2005**, 3918–3920.
88. Giordani, S. and Raymo, F.M. (2003) *Org. Lett.*, **5**, 3559–3562.
89. Raychaudhuri, B. and Bhattacharyya, S. (2008) *Appl. Phys. B*, **91**, 545–550.
90. Straight, S.D., Andrasson, J., Kodis, G., Bandyopadhyay, S., Mitchell, R.H., Moore, T.A., Moore, A.L., and Gust, D. (2005) *J. Am. Chem. Soc.*, **127**, 9403–9409.
91. Wang, Z., Zheng, G., and Lu, P. (2005) *Org. Lett.*, **7**, 3669–3672.
92. Fang, C.J., Zhu, Z., Sun, W., Xu, C.H., and Yan, C.H. (2007) *New J. Chem.*, **31**, 580–586.
93. Straight, S.D., Liddell, P.A., Terazono, Y., Moore, T.A., Moore, A.L., and Gust, D. (2007) *Adv. Funct. Mater.*, **17**, 777–785.
94. Li, F., Shi, M., Huang, C., and Jin, L. (2005) *J. Mater. Chem.*, **15**, 3015–3020.
95. Wen, G., Yan, J., Zhou, Y., Zhang, D., Mao, L., and Zhu, D. (2006) *Chem. Commun.*, **2006**, 3016–3018.
96. Szaciłowski, K., Macyk, W., and Stochel, G. (2006) *J. Am. Chem. Soc.*, **128**, 4550–4551.
97. Qian, J.H., Xu, Y.F., Qian, X.H., and Zhang, S.Y. (2008) *ChemPhysChem*, **9**, 1891–1898.
98. Fioravanti, G., Haraszkiewicz, N., Kay, E.R., Mendoza, S.M., Bruno, C., Marcaccio, M., Wiering, P.G., Paolucci, F., Rudolf, P., Brouwer, A.M., and Leigh, D.A. (2008) *J. Am. Chem. Soc.*, **130**, 2593–2601.
99. Nitahara, S., Terasaki, N., Akiyama, T., and Yamada, S. (2006) *Thin Solid Films*, **499**, 354–358.
100. de Silva, A.P. (2008) *Nature*, **454**, 417–418.
101. Mendes, P.M. (2008) *Chem. Soc. Rev.*, **37**, 2512–2529.
102. Gupta, T. and van der Boom, M.E. (2008) *Angew. Chem. Int. Ed.*, **47**, 5322–5326.
103. Credi, A., Balzani, V., Langford, S.J., and Stoddart, J.F. (1997) *J. Am. Chem. Soc.*, **119**, 2679–2681.
104. de Silva, A.P., Gunaratne, H.Q.N., and McCoy, C.P. (1993) *Nature*, **364**, 42–44.
105. de Silva, A.P., Gunaratne, H.Q.N., and McCoy, C.P. (1997) *J. Am. Chem. Soc.*, **119**, 7891–7892.
106. de Silva, A.P., Gunaratne, H.Q.N., and Maguire, G.E.M. (1994) *J. Chem. Soc. Chem. Commun.*, **1994**, 1213–1214.
107. de Silva, A.P. and McClenaghan, N.D. (2002) *Chem. Eur. J.*, **8**, 4935–4945.
108. de Silva, A.P., Dixon, I.M., Gunaratne, H.Q.N., Gunnlaugsson, T., Maxwell, P.R.S., and Rice, T.E. (1999) *J. Am. Chem. Soc.*, **121**, 1393–1394.
109. Turfan, B. and Akkaya, E.U. (2002) *Org. Lett.*, **4**, 2857–2859.
110. Cheng, P.N., Chiang, P.T., and Chiu, S.H. (2005) *Chem. Commun.*, **2005**, 1285–1287.
111. Zhou, J., Arugula, M.A., Halámek, J., Pita, M., and Katz, E. (2009) *J. Phys. Chem. B*, **113**, 16065–16070.
112. Baytekin, H.T. and Akkaya, E.U. (2000) *Org. Lett.*, **2**, 1725–1727.
113. Zong, G., Xiana, L., and Lua, G. (2007) *Tetrahedron Lett.*, **48**, 3891–3894.
114. Gunnlaugsson, T., Mac Dónaill, D.A., and Parker, D. (2000) *Chem. Commun.*, **2000**, 93–94.
115. Gunnlaugsson, T., Mac Dónaill, D.A., and Parker, D. (2001) *J. Am. Chem. Soc.*, **123**, 12866–12876.
116. de Sousa, M., de Castro, B., Abad, S., Miranda, M.A., and Pischel, U. (2006) *Chem. Commun.*, **2006**, 2051–2053.
117. Li, L., Yu, M.X., Li, F.Y., Yi, T., and Huang, C.H. (2007) *Colloids Surf., A*, **304**, 49–53.
118. Luxami, V. and Kumar, S. (2008) *New J. Chem.*, **32**, 2074–2079.
119. Qian, J.H., Qian, X.H., Xu, Y.F., and Zhang, S.Y. (2008) *Chem. Commun.*, **2008**, 4141–4143.
120. Pérez-Inestrosa, E., Montenegro, J.M., Collado, D., Suau, R., and

120. Casado, J. (2007) *J. Phys. Chem. C*, **111**, 6904–6909.
121. Remón, P., Ferreira, R., Montenegro, J.M., Suau, R., Pérez-Inestrosa, E., and Pischel, U. (2009) *ChemPhysChem*, **10**, 2004–2007.
122. Coskun, A., Deniz, E., and Akkaya, E.U. (2005) *Org. Lett.*, **7**, 5187–5189.
123. Jiménez, D., Martínez-Máñez, R., Sancenón, F., Ros-Lis, J.V., Soto, J., Benito, A., and García-Breijo, E. (2005) *Eur. J. Inorg. Chem.*, **2005**, 2393–2403.
124. Sun, W., Xu, C.H., Zhu, Z., Fang, C.J., and Yan, C.H. (2008) *J. Phys. Chem. C*, **112**, 16973–16983.
125. Li, Z.X., Liao, L.Y., Sun, W., Xu, C.H., Zhang, C., Fang, C.J., and Yan, C.H. (2008) *J. Phys. Chem. C*, **112**, 5190–5196.
126. Zhou, Y., Wu, H., Qu, L., Zhang, D., and Zhu, D. (2006) *J. Phys. Chem. B*, **110**, 15676–15679.
127. Sun, W., Zheng, Y.R., Xu, C.H., Fang, C.J., and Yan, C.H. (2007) *J. Phys. Chem. C*, **111**, 11706–11711.
128. Pischel, U. and Heller, B. (2008) *New J. Chem.*, **32**, 395–400.
129. Andreasson, J., Straight, S.D., Bandyopadhyay, S., Mitchell, R.H., Moore, T.A., Moore, A.L., and Gust, D. (2007) *J. Phys. Chem. C*, **111**, 14274–14278.
130. Amelia, M., Baroncini, M., and Credi, A. (2008) *Angew. Chem. Int. Ed.*, **47**, 6240–6243.
131. Pérez-Inestrosa, E., Montenegro, J.M., Collado, D., and Suau, R. (2008) *Chem. Commun.*, **2008**, 1085–1087.
132. Andreasson, J., Straight, S.D., Moore, T.A., Moore, A.L., and Gust, D. (2008) *J. Am. Chem. Soc.*, **130**, 11122–11128.
133. Margulies, D., Felder, C.E., Melman, G., and Shanzer, A. (2007) *J. Am. Chem. Soc.*, **129**, 347–354.
134. Suresh, M., Ghosh, A., and Das, A. (2008) *Chem. Commun.*, **2008**, 3906–3908.
135. Katz, E. and Willner, I. (2005) *Chem. Commun.*, **2005**, 5641–5643.
136. Chatterjee, M.N., Kay, E.R., and Leigh, D.A. (2006) *J. Am. Chem. Soc.*, **128**, 4058–4073.
137. Pita, M., Strack, G., MacVittie, K., Zhou, J., and Katz, E. (2009) *J. Phys. Chem. B*, **113**, 16071–16076.
138. Katz, E. and Willner, I. (2006) *Electrochem. Commun.*, **8**, 879–882.
139. Galindo, F., Lima, J.C., Luis, S.V., Parola, A.J., and Pina, F. (2005) *Adv. Funct. Mater*, **15**, 541–545.
140. Bandyopadhyay, A. and Pal, A.J. (2005) *J. Phys. Chem. B*, **109**, 6084–6088.
141. Fernandez, D., Parola, A.J., Branco, L.C., Afonso, C.A.M., and Pina, F. (2004) *J. Photochem. Photobiol. A*, **168**, 185–189.
142. Pina, F., Roque, A., Melo, M.J., Maestri, I., Belladelli, L., and Balzani, V. (1998) *Chem. Eur. J.*, **4**, 1184–1191.
143. Will, G., Rao, J.S.S.N., and Fitzmaurice, D. (1999) *J. Mater. Chem.*, **9**, 2297–2299.
144. Hiller, J. and Rubner, M.F. (2003) *Macromolecules*, **36**, 4078–4083.
145. Pina, F., Lima, J.C., Parola, A.J., and Afonso, C.A.M. (2004) *Angew. Chem. Int. Ed.*, **43**, 1525–1527.
146. Stadler, R., Ami, S., Joachim, C., and Forshaw, M. (2004) *Nanotechnology*, **15**, S115–S121.
147. de Silva, A.P., Leydet, Y., Lincheneau, C., and McClenaghan, N.D. (2006) *J. Phys. Condens. Matter*, **18**, S1847–S1872.
148. Adamatzky, A. (2004) *IEICE Trans. Electron.*, **E87C**, 1748–1756.
149. Bell, G. and Gray, J.N. (1997) in *Beyond Calculation: The Next Fifty Years of Computing*, Chapter 1 (eds P.J. Denning and R.M. Metcalfe), Copernicus/Springer, New York, pp. 5–32.
150. Fu, L., Cao, L.C., Liu, Y.Q., and Zhu, D.B. (2004) *Adv. Colloid Interface Sci.*, **111**, 133–157.
151. Zauner, K.P. (2005) *Crit. Rev. Solid State Mater. Sci.*, **30**, 33–69.
152. Calude, C.S., Costa, J.F., Dershowitz, N., Freire, E., and Rozenberg, G. (eds) (2009) *Unconventional Computation*, Lecture Notes in Computer Science, Vol. **5715**, Springer, Berlin.

153. Shiva, S.G. (1998) *Introduction to Logic Design*, 2nd edn, Marcel Dekker, New York.
154. Qu, D.H., Wang, Q.C., and Tian, H. (2005) *Angew. Chem. Int. Ed.*, **44**, 5296–5299.
155. Andréasson, J., Straight, S.D., Kodis, G., Park, C.D., Hambourger, M., Gervaldo, M., Albinsson, B., Moore, T.A., Moore, A.L., and Gust, D. (2006) *J. Am. Chem. Soc.*, **128**, 16259–16265.
156. Andréasson, J., Kodis, G., Terazono, Y., Liddell, P.A., Bandyopadhyay, S., Mitchell, R.H., Moore, T.A., Moore, A.L., and Gust, D. (2004) *J. Am. Chem. Soc.*, **126**, 15926–15927.
157. Lopez, M.V., Vazquez, M.E., Gomez-Reino, C., Pedrido, R., and Bermejo, M.R. (2008) *New J. Chem.*, **32**, 1473–1477.
158. Margulies, D., Melman, G., and Shanzer, A. (2006) *J. Am. Chem. Soc.*, **128**, 4865–4871.
159. Kuznetz, O., Salman, H., Shakkour, N., Eichen, Y., and Speiser, S. (2008) *Chem. Phys. Lett.*, **451**, 63–67.
160. Ashkenazi, G., Ripoll, D.R., Lotan, N., and Scheraga, H.A. (1997) *Biosens. Bioelectron.*, **12**, 85–95.
161. Sivan, S. and Lotan, N. (1999) *Biotechnol. Progr.*, **15**, 964–970.
162. Sivan, S., Tuchman, S., and Lotan, N. (2003) *Biosystems*, **70**, 21–33.
163. Deonarine, A.S., Clark, S.M., and Konermann, L. (2003) *Future Generat. Comput. Syst.*, **19**, 87–97.
164. Unger, R. and Moult, J. (2006) *Proteins*, **63**, 53–64.
165. Stojanovic, M.N., Stefanovic, D., LaBean, T., and Yan, H. (2005) in *Bioelectronics: From Theory to Applications*, Chapter 14 (eds I. Willner and E. Katz), Wiley-VCH Verlag GmbH, Weinheim, pp. 427–455.
166. Lewin, D.I. (2002) *Comput. Sci. Eng.*, **4**, 5–8.
167. Benenson, Y., Gil, B., Ben-Dor, U., Adar, R., and Shapiro, E. (2004) *Nature*, **429**, 423–429.
168. Ezziane, Z. (2006) *Nanotechnology*, **17**, R27–R39.
169. Rinaudo, K., Bleris, L., Maddamsetti, R., Subramanian, S., Weiss, R., and Benenson, Y. (2007) *Nat. Biotechnol.*, **25**, 795–801.
170. Win, M.N. and Smolke, C.D. (2008) *Science*, **322**, 456–460.
171. Ogawa, A. and Maeda, M. (2009) *Chem. Commun.*, **2009**, 4666–4668.
172. Stojanovic, M.N., Mitchell, T.E., and Stefanovic, D. (2002) *J. Am. Chem. Soc.*, **124**, 3555–3561.
173. Benenson, Y. (2009) *Curr. Opin. Biotechnol.*, **20**, 471–478.
174. Lederman, H., Macdonald, J., Stefanovic, D., and Stojanovic, M.N. (2006) *Biochemistry*, **45**, 1194–1199.
175. Schlosser, K. and Li, Y. (2009) *Chem. Biol.*, **16**, 311–322.
176. Elbaz, J., Shlyahovsky, B., Li, D., and Willner, I. (2008) *ChemBioChem*, **9**, 232–239.
177. Li, T., Wang, E., and Dong, S. (2009) *J. Am. Chem. Soc.*, **131**, 15082–15083.
178. Moshe, M., Elbaz, J., and Willner, I. (2009) *Nano Lett.*, **9**, 1196–1200.
179. Willner, I., Shlyahovsky, B., Zayats, M., and Willner, B. (2008) *Chem. Soc. Rev.*, **37**, 1153–1165.
180. Soreni, M., Yogev, S., Kossoy, E., Shoham, Y., and Keinan, E. (2005) *J. Am. Chem. Soc.*, **127**, 3935–3943.
181. Xu, J. and Tan, G.J. (2007) *J. Comput. Theor. Nanosci.*, **4**, 1219–1230.
182. Kahan, M., Gil, B., Adar, R., and Shapiro, E. (2008) *Physica D*, **237**, 1165–1172.
183. Weste, N.H.E. and Harris, D. (2004) *CMOS VLSI Design: A Circuits and Systems Perspective*, Pearson/Addison-Wesley, Boston.
184. Wakerly, J. (2005) *Digital Design: Principles and Practices*, Pearson/Prentice Hall, Upper Saddle River.
185. Alon, U. (2007) *An Introduction to Systems Biology. Design Principles of Biological Circuits*, Chapman & Hall/CRC Press, Boca Raton, FL.
186. Braich, R.S., Chelyapov, N., Johnson, C., Rothemund, P.W.K., and Adleman, L. (2002) *Science*, **296**, 499–502.
187. Privman, V. (2011) *J. Comput. Theor. Nanosci.*, **8**, 490–502.
188. Setty, Y., Mayo, A.E., Surette, M.G., and Alon, U. (2003) *Proc. Natl. Acad. U.S.A.*, **100**, 7702–7707.

189. Buchler, N.E., Gerland, U., and Hwa, T. (2005) *Proc. Natl. Acad. U.S.A.*, **102**, 9559–9564.
190. Privman, V., Halamek, J., Arugula, M.A., Melnikov, D., Bocharova, V., and Katz, E. (2010) *J. Phys. Chem. B*, **114**, 14103–14109.
191. Pita, M., Privman, V., Arugula, M.A., Melnikov, D., Bocharova, V., and Katz, E. (2011) *Phys. Chem. Chem. Phys.*, **13**, 4507–4513.
192. Pretzel, O. (1992) *Error-Correcting Codes and Finite Fields*, Oxford University Press, Oxford.
193. Bernoulli, J. (1695) *Acta Eruditorum (Lipsiae)*, **MDCXCV**, 537–553.
194. Glinel, K., Dejugnat, C., Prevot, M., Scholer, B., Schonhoff, M., and Klitzing, R.V. (2007) *Colloids Surf., A*, **303**, 3–13.
195. Ahn, S.K., Kasi, R.M., Kim, S.C., Sharma, N., and Zhou, Y.X. (2008) *Soft Matter*, **4**, 1151–1157.
196. Tokarev, I. and Minko, S. (2009) *Soft Matter*, **5**, 511–524.
197. Pita, M., Minko, S., and Katz, E. (2009) *J. Mater. Sci.- Mater. Med.*, **20**, 457–462.
198. Willner, I., Doron, A., and Katz, E. (1998) *J. Phys. Org. Chem.*, **11**, 546–560.
199. Wang, X., Gershman, Z., Kharitonov, A.B., Katz, E., and Willner, I. (2003) *Langmuir*, **19**, 5413–5420.
200. Luzinov, I., Minko, S., and Tsukruk, V.V. (2004) *Prog. Polym. Sci.*, **29**, 635–698.
201. Minko, S. (2006) *Polymer Rev.*, **46**, 397–420.
202. Tokarev, I., Gopishetty, V., Zhou, J., Pita, M., Motornov, M., Katz, E., and Minko, S. (2009) *ACS Appl. Mater. Interfaces*, **1**, 532–536.
203. Motornov, M., Zhou, J., Pita, M., Gopishetty, V., Tokarev, I., Katz, E., and Minko, S. (2008) *Nano Lett.*, **8**, 2993–2997.
204. Motornov, M., Zhou, J., Pita, M., Tokarev, I., Gopishetty, V., Katz, E., and Minko, S. (2009) *Small*, **5**, 817–820.
205. Krämer, M., Pita, M., Zhou, J., Ornatska, M., Poghossian, A., Schöning, M.J., and Katz, E. (2009) *J. Phys. Chem. B*, **113**, 2573–2579.
206. Zhou, J., Tam, T.K., Pita, M., Ornatska, M., Minko, S., and Katz, E. (2009) *ACS Appl. Mater. Interfaces*, **1**, 144–149.
207. Wang, X., Zhou, J., Tam, T.K., Katz, E., and Pita, M. (2009) *Bioelectrochemistry*, **77**, 69–73.
208. Tam, T.K., Zhou, J., Pita, M., Ornatska, M., Minko, S., and Katz, E. (2008) *J. Am. Chem. Soc.*, **130**, 10888–10889.
209. Tam, T.K., Ornatska, M., Pita, M., Minko, S., and Katz, E. (2008) *J. Phys. Chem. C*, **112**, 8438–8445.
210. Amir, L., Tam, T.K., Pita, M., Meijler, M.M., Alfonta, L., and Katz, E. (2009) *J. Am. Chem. Soc.*, **131**, 826–832.
211. Tam, T.K., Pita, M., Ornatska, M., and Katz, E. (2006) *Bioelectrochemistry*, **76**, 4–9.
212. Shirokane, Y., Ichikawa, K., and Suzuki, M. (2000) *Carbohydr. Res.*, **329**, 699–702.
213. Privman, V., Zhou, J., Halamek, J., and Katz, E. (2010) *J. Phys. Chem. B*, **114**, 13601–13608.
214. Halamek, J., Bocharova, V., Arugula, M.A., Strack, G., Privman, V., and Katz, E. (2011) *J. Phys. Chem. B*, **115**, 9838–9845.
215. Melnikov, D., Strack, G., Zhou, J., Windmiller, J.R., Halamek, J., Bocharova, V., Chuang, M.C., Santhosh, P., Privman, V., Wang, J., and Katz, E. (2010) *J. Phys. Chem. B*, **114**, 12166–12174.

13
Electrochemistry, Emergent Patterns, and Inorganic Intelligent Response

Saman Sadeghi and Michael Thompson

13.1
Introduction

Biological information processing at the organism or macroscopic level involving a population of interacting and communicating cells is elegantly evident in sensory response, whether signals are created internally or externally. Pain, for example, often has its genesis at the internal level, whereas visual, olfactory, and gustatory processes, for example, are instigated in response to the external environment. Research on sensory response over many years has elicited the fact that the brain produces neural patterns in terms of both spatial and temporal representation. Much of this intelligent response is associated with emergent phenomena, including patterns, which form in systems that function under far-from-equilibrium conditions. Such patterns involve a large number of elements that follow nonlinear laws of stimulation and communication. Pattern formation in these systems is often the underlying mechanism of an active response to environmental changes and can be interpreted as a result of the distributed parallel information processing that takes place within the material. Interestingly, these patterns are not limited to biological entities; indeed there is a wide range of complex nonlinear dissipative systems that exhibit interesting emergent patterns within a range of parameters.

To best understand the cause of the emergence and evolution of these patterns, the system must be examined and tested in a holistic sense. If such a system is studied through conventional reductionist scientific reasoning, the same dynamics and capabilities would not be produced. Accordingly, a clearer understanding of the elements responsible for composing the system is an essential part of generating a glimpse at the inner workings of intelligent matter, be it far-removed topics such as interacting molecules in an inorganic system compared to the behavior of neurons in the brain. However, the real clue to appreciation of the secrets of pattern formation as a cognitive and intelligent response mechanism lies in the examination of the actual process that depends on the forces that operate within the entire system. Unfortunately, computer simulations run on digital transistor-based machines that respond to input information in a very different and deterministic way. Because of their serial architecture, computers can for the most part only process a single or possibly a few variables at a time. This imposes a huge limiting factor and difficulty when dealing with complex multivariable input information, such as three-dimensional recognition of objects, which the brain performs effortlessly and sometimes subconsciously. A system closer to an "organic" representation of the brain function can be performed via experimentation on a neuronal culture obtained from natural sources. However, the results of this type of study may well lead to misleading conclusions as small subsets of organic media may have vastly different characteristics than the entire system. This aspect often appears to lead those in the medical community to question the value of physical science as it relates to medical research.

As we emphasize, emergent complex and transient patterns are mechanisms by which the higher organic systems such as the brain process multivariable information. Emulating this process on a physical system would then present a means to study, or even mimic, biological information processing. As mentioned above, emergent pattern phenomena are also observed in inorganic systems. One clear example that provides a viable experimental platform is the detection of emergent phenomena associated with surface electrochemical processes. This allows the system to respond to input information through evolving patterns in space and time. Associative mapping of this sort offers the opportunity to devise an electrochemical cognitive system, where pattern formation can be looked at as a macroscopic phenomenon resulting from the extensive distributive computing that occurs at the microscopic level.

13.2
Patten Formation in Complex Systems

Even relatively simple outside-equilibrium systems can be driven into forming complex spatial or temporal patterns when the states of the system deform by large amounts in response to small perturbations. Reaction diffusion equations, for example, exhibit nonlinearity through the intrinsic forces on each element and local communication between them. In such a system, spontaneous emergence of coherence or structure, without specific control from outside, may arise in the absence of spatial degrees of freedom, and complex behavior may persist with only a few variables.

Pattern formation in a complex system is the result of emergent phenomena that leads to appearance of structures that cannot be reduced and predicted by looking at individual components or the instantaneous forces that act on the system. This collective behavior, which appears at both temporal and spatial scales that are not defined dynamics of individual elements, emerges over time from interaction of subsystems with no directing hand or prior design.

Theoretical work on nonlinear dynamical systems has made much progress in the last few decades in describing and predicting pattern formation mechanisms in a complex far-from-equilibrium medium. By reducing a system with an infinite degree of freedom to a relatively small finite number of variables it is possible to successfully describe a large variety of spatiotemporal patterns and even bifurcations in the phase space. Theoretical models have shown that when a bifurcation occurs, the associated unstable trajectories move away from the originally stable state to a vicinity of a low-dimensional subspace of the full phase space [1]. This subspace can effectively describe the dynamic behavior of the entire system since trajectories starting elsewhere converge to it, making the degrees of freedom outside the attracting subspace effectively irrelevant to the final pattern that forms within the material. This low dimensionality of the attracting subspace in complex systems near their thresholds of instability is a useful result that has enabled theoreticians to describe many pattern-formation problems in systems that may be composed of very large number of dynamic variables.

Theoretical consideration of dynamical systems can also describe more complex patterns than simple fixed point states and bifurcations, such as oscillatory or turbulent flows. Oscillatory behavior is generally described as a flow on a limit cycle (or closed loop) in phase space, and chaotic states may be represented by more complex sets of strange attractors. Strange attractors, which are discussed again later in this chapter, hold a great potential for information processing and can be understood through an expression of Lyapunov exponents. These exponents describe the local exponential divergence or convergence rates between two nearby trajectories, and when averaged over time, if the exponent is positive, nearby orbits will separate from each other exponentially in time and converge for negative exponents. The exponent is a measure of the rate of separation of infinitesimally close trajectories

$$|d(t)| = e^{\lambda t} |d(0)|$$

For λ representing the Lyapunov exponent, d(0) is the infinitesimally small initial separation and d(t) is the separation of trajectories after time (t). The largest Lyapunov exponent can be calculated as

$$\lambda = \lim_{t \to \infty} \frac{1}{t} \ln \frac{|d(t)|}{|d(0)|}$$

In a system where the phase space of an attracting set is bounded, if the trajectory of the attractor possesses a positive Lyapunov exponent, the only possibility is for the set to be fractal. Fractal patterns and strange attractors that possess both positive and negative Lyapunov exponents are useful information-processing tools since the infinitely long and nonrepeating trajectory makes them ideal dynamical information storage units, while on the other hand they can, at the same time, serve as information compressors by being confined to a recognizable subset of the entire phase space of the system.

The formation of patterns that emerge intrinsically within the material can support information-processing capacity of the system and make additional functionalities available to it. Information processing in a natural system is the underlying mechanism of its behavior. To better understand such information processing, a systematic approach to analyzing multidimensional information embedded in emergent patterns is needed. Although techniques for pattern detection exist, such as those employed to detect emergent patterns in the brain [2] and the electrochemical platform reported in this chapter [3], there does not yet appear to be tools available to thoroughly describe and define the structure of emergent phenomena in a such way that allows the prediction, detection, measurement, and quantification of their complexity.

13.3
Intelligent Response and Pattern Formation

Given a set of initial conditions, there often exists a finite set of decisions or actions that are plausible for a particular situation. Accordingly, our cognitive ability may be reduced to problem classification and selection of appropriate action. This is very much analogous to pattern recognition, whereby we are confronted with either a very complex, distorted, or incomplete set of information, where a classification decision is needed. In all cases, because of either the sheer complexity of the data or simple incomplete presentation, we are faced, or must resort to, making a decision from a subset of information. We can see here that the problem of feature extraction, where the essential bits of information necessary for classification and recognition are recognized, is inexorably intertwined with information processing and cognitive ability.

An abstract model of how a pattern-forming system might perform computation in a manner similar to and inspired by the brain, is illustrated in Figure 13.1. There are a number features in the scheme proposed here, which form the basis of a biologically inspired information-processing system. Such a configuration has the capability to compress a set of similar input information to a discrete number of

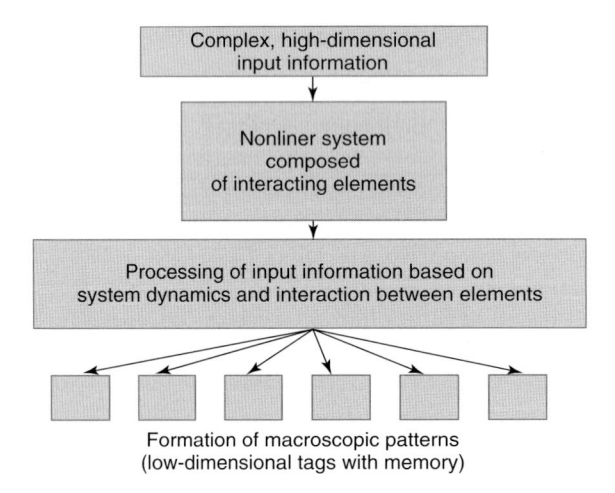

Figure 13.1 Basis for information processing based on emergent patterns. (Reprinted with kind permission from Elsevier B.V., Amsterdam.)

recognition states. The mapping of data in this fashion allows both for recognition from incomplete sets of information and a recall memory associated with the data.

The power of a system that can perform the proposed scheme will be related to its ability to process a large amount of information in a fast and reliable manner, in the same way as the brain deals with data from sensory transducers, in a fast and reliable manner. For this purpose, we have to look to a physical system, where defining properties or unique spatiotemporally distributed states can serve as the recognition flags and memory for particular sets of information. The idea of data storage using localized structures of physical systems [4] and emergent computation through spatially extended systems such as cellular automata has been reported in the literature [5–7]. Such a physical system may be regarded as an adaptive filter whose properties are changed by the input signals that are provided to it [8]. In this way, a particular set of input signals would act as keys that would evoke recognition responses and stored information associated with a stimulus. Nonlinear dynamical systems possess the capability to exhibit these properties and will be investigated as the medium to create such a system.

13.3.1
Self-Organization in Systems Removed from the Equilibrium State

Pattern formation, which we are proposing as a foundation for the process of cognition, possesses a profound connection with nonlinear phenomena found in complex physical systems. A good body of theoretical research has been conducted in this field during the past few decades, and several models have been proposed. These are based either on outside-equilibrium thermodynamics, which point to the onset of dissipative structures [9, 10], or kinetic models that make use of simplifications and compromises in order to describe and predict the behavior of the nonlinear dynamic system [11–15].

Systems that can be described by differential equations can exhibit changes from simple to complex behavior in response to changes in their control parameters. For example, chaotic and periodic motion can be seen even in the case of a periodically forced and damped pendulum, which has only a few degrees of freedom [16]. Furthermore, sequences of increasingly complex spatiotemporal patterns can be observed if we look to dynamical systems with many degrees of freedom described through partial differential equations. Such equations can model the interrelated evolution of variables that describe the system. Mathematical models of these systems often have multiple steady-state solutions for a set of initial conditions and have the capability to jump from one solution to the next or choose a preferred steady state based on changes in the parameters that have an influence on the system. The steady states are often described as attractors in phase space, which is a multidimensional theoretical space that is spanned by the dynamical variables of the system. In this way, every unique point in the phase space describes a particular set of values or a state, which portrays the system in an instant of time. Thus, we can visualize the dynamics of the system in terms of trajectories. These trajectories can help us track the systems evolution from one state to the next. An attractor is then a steady state that can range from a homogenous arrangement of the system to temporal or spatial patterns ranging from simple one-frequency oscillations to more complex arrangements. It is the utilization and application of these preferred attractor patterns in response to specific parameters that is part of the focus of this discussion.

13.3.2
Patterns in Nature

Living organisms are examples of open, complex systems that through evolution were displaced from an equilibrium state to build the diversity of organic compounds that are the basis of life. Nonequilibrium animate systems found in nature are open systems that exchange matter and energy and process information to carry out activities essential to maintaining life and responding to changes in environmental conditions. The basis of an active response to environment and information processing is the emergence of what Prigogine termed *dissipative* patterns [17] and structure forms in polymer chemistry that are the inevitable consequence of critically complex, far-from-equilibrium chemical systems. It has further been suggested that with a sufficient diversity and concentration of enzymes and organic substrate, a supercritical explosion of diversity of organic compounds would be achieved and that this self-extending supercritical system is the very basis of emergence of life on earth [18].

Many examples of emergent patterns exist in nature, which leads specifically to concepts associated with work in the field of phase transitions and critical phenomena. Emergent patterns in nonlinear dissipative dynamical systems can account for a large variety of phenomena, including examples as diverse as the markings on a butterfly's wings, a leopard's spots, the shape of a spleenwort fern, palpitations of a living heart, the ring of nerve cells within the brain, and the Belousov Zhabotinsky (BZ) reaction (Figure 13.2) [19]. Even systems with short-range interaction of simple elements can exhibit long-range spatiotemporal correlated patterns, which can eventually emerge as stationary or transient expressions. Organic systems have the flexibility and adaptability to form complex spatiotemporal patterns in response to environmental conditions. Such changes demand critical dynamics of a system, which will allow it to respond with ease and smoothness in fractions of a second. One can say this exemplifies the collective state of the system at the edge of chaos [20]. Such examples have been documented in species evolution, ant colony behavior and foraging, swarm models, bacterial populations, and flock formations [21, 22].

13.3.3
Functional Self-Organizing Systems

An ideal cognitive device that can potentially outperform digital computers will exhibit the ability to classify complex input information in a content-addressable fashion. The recognition of content and indexing can be performed in a

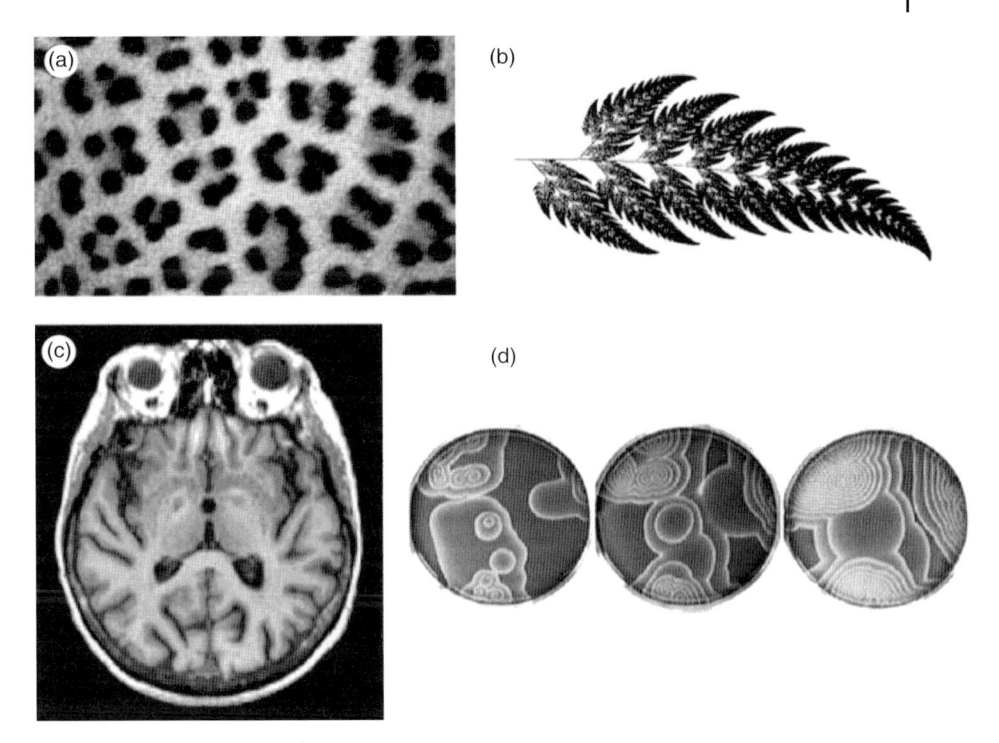

Figure 13.2 A large variety of phenomena can be understood by means of nonlinear-dissipative-dynamical-systems theory. Examples are (a) a leopard's spots, (b) the shape of a spleenwort fern, (c) the firing of nerve cells within the brain, and (d) the Belousov Zhabotinsky reaction. (Reprinted by kind permission from the Royal Society of London, UK.)

fault-tolerant manner based on the dynamics of the system, which are sensitive to input parameters. These dynamics would be performing the role of massively parallel computers operating in order to recognize the input data with retrieval of associated memory.

Emergent pattern formation in nonlinear systems can perform this task of recognition from imperfect data through evolution of their dynamics toward a particular attractor. The attractor is chosen from a set of coexisting attractors that form the phase space of the system. This is an illustration of phase space dynamics, where, depending on the input perturbation, the system finds itself in one of the many basins of attraction and thus converges to evolve on the specific evolution path of that attractor. The stable solutions of the differential equations proposed in the previous section can represent just such attractors. Attempts have been made to characterize the maximum information that can be stored in a medium by relating the formation of such patterns to generalizations of Shannon's channel capacity [23].

As the attractor occupies only a portion of the phase space, there is an inherent compression of the input information. This compression is the result of the reduced degrees of freedom as the initial stimulus is forced to evolve from a set of N degrees of freedom, forming the information phase space to that of D_i degrees of freedom, which defines the information dimension of the attractor (i) toward which the stimulus relaxes. In the case of strange attractors that exhibit a noninteger dimension, such as the reported 2.05 dimension of a Lorenz attractor [24], a compression factor of N-D_i/N of 32% is afforded. This compression factor ensures a measure of fault tolerance during information processing by a system. A progressive increase in compression accommodates a larger basin of attraction for a particular attractor, allowing stimuli with larger Hamming distances to relax to the same recognition state.

The information dimension of a particular attractor can be thought of as the level of abstraction that allows segmentation of large variety of external world information into a few significant categories. However, there is an obvious advantage of compressibility and ease of recognition with simpler attractors. Ideal compression systems are, for example, stable steady-state solutions or singular nodes in the phase space, which would have a 0 information dimension, while oscillatory limit cycles would have a dimension of 1. It is clear, however, that homogeneous states and attractor patterns have no capacity to store information and that a periodic pattern only contains information about a wave number and the position of one vertex. The information dimension of the attractor can play the role of dynamically retrievable memory, which shows up in the form of patterns produced by the attractor. The experimental results reported subsequently in the present chapter point to the existence of multiple coexisting strange attractors in, for example, an electrochemical system. These attractors possess positive and negative Lyapunov exponents λ, with $|\lambda_-| > |\lambda_+|$ [25]. These strange attractors, while still squeezing the dimension of input information due to the larger $|\lambda_-|$, have the capacity to create a specific and nontrivial response based on the emergent pattern of the attractor. Such attractors, in comparison to steady-state patterns that perform poorly as memory storage devices, can exhibit associative memory recall functionality.

13.3.4
Emergent Patterns and Associative Memory

Associative memory is closely linked with pattern formation and recognition [26], and can be defined as a system that stores a set of patterns, $x^{(p)}$, $p = 1, 2, \ldots, k$ in its internal state, and produces the copy of a particular set $x^{(r)} = \left(\xi_1^{(r)}, \xi_2^{(r)}, \ldots, \xi_n^{(r)}\right)$, $r \in \{1, 2, \ldots, k\}$ to the output when presented with any signal from the set of $x^{(i)} = (\xi_1, \xi_2, \ldots, \xi_n)$ in which a specified subset of values ξ_i matches with the corresponding subset of the $\xi_i^{(r)}$.

However, for a system such as the one considered here, the reply is stored in the form of attractors $x_t^{(r)} = \left(\omega_1^{(r)}, \omega_2^{(r)}, \ldots, \omega_m^{(r)}\right)$ for m being the number of tags associated with $x^{(r)}$, in which every element of the reduced dimensional, time-dependant signal $x_t^{(r)}$ is a tag of $x^{(r)}$ associated with a particular aspect of $x^{(r)}$.

In this way, we can ensure invariant representation of equivalent stimuli such as geometrical transformation of input information I giving rise to the same tags $x_t^{(r)}$.

This also ensures the optimized and efficient utilization of the available memory states, whereby only the most significant aspects of $x^{(r)}$ are stored and recalled through associative memory. This kind of memory and recall association is also a commonly employed method in biological systems. For example, we can recognize the face of a friend or a general emotion such as anger through a few distinct features. These features are tags that are constant in pictures that may have been taken under different conditions or, in case of emotions, from different individuals.

The tags are, in effect, the patterns generated by the electrochemical system that are simpler and more invariant than the original occurrence of information provided to the system. They can be looked at as discrete eigen features of the input information, and may be understood as the result of discrimination by a system that automatically categorizes all input data with similar prominent features, thus distinguishing aspects of input information. Classification of new occurrences with a set composed of existing tags from other $x_t^{(r)}$ can be looked at as making an educated guess. This is analogous to producing a link, or being unable to distinguish between stimuli, which share a common set of features and are provided under constraints of limited or imperfect cues. Furthermore, linking relations between input information can be established by looking at the number of shared tags between various data. The input information $x^{(i)}$ can be in the form of a multidimensional array I, such as a two-dimensional array σ^2 for an image or a three-dimensional array σ^3 with an added dimension of time.

We can understand the mapping dynamics through a changing internal state S of the system, which when presented with a set of input information I, produces a time varying output signal $x_t^{(r)}$, which is a function of the internal state and I at time t. To formalize this, let us consider two mapping functions: f for the internal state, where

$$f: S^{(t+1)} = f(S^{(t)}, I^{(t)})$$

and g for the output, where

$$g: x_t^{(r)} = g(S^{(t)}, I^{(t)})$$

Here, the internal state of the system S is complex and depends on all variables that are required to fully describe the electrochemical system. The state $x_t^{(r)}$ on the other hand is the set of observable tags that have associations attached to them and can be significantly simpler than I and S. The processing power of the system is essentially encompassed in the state S and the mapping functions g and f.

The input information may or may not have time dependency. If the input information, I, is time independent, then the tags associated with I may be spread in time and emerging with some time delay, all describing a particular aspect of I. In the case of a time-dependant $I^{(t)}$, given $I(t=0)$ the full output time series $\left\{x_t^{(r)}\right\}_{t=0 \to t}$ may be generated and in this way the future behavior of $I(t)$ predicted from the system response with the system presented with a single element of $I(0)$. Furthermore, time delay is analogous and continuous as opposed to finite time delay in digital circuits.

The input information may have a particular context C associated with it, whereby the information presented to the system is composed of $(C, I(t))$ where C is presented by the DC offset of the input signal in the electrochemical system, and I by the AC and spatial component of the perturbation. Different C factors are equivalent to the experiments performed under different control parameters, where the system may evolve differently, giving rise to distinguishable patterns $\left\{x_t^{(r)}\right\}_{t=0 \to t'}$, and thus discriminating between input information based on the context in which the information was presented to the system.

When a sufficient superset of $\{x^{(r)}\}$ tags are associated with input information superset $\{I\}$, and the system is presented with new information I_1, which is inherently different in elements but sufficiently similar to sets I_2 and I_3 such that it can invoke an output $x^{(r)}$ that is a subset of combined sets $x^{(2)} + x^{(3)}$. The system can be thought of as synthesizing a new kind of information based on innovative association. Such innovative recognition has been suggested in the form of unintended attractors that have been observed in simulations of recurrent attractor neural networks [27].

13.4
Artificial Cognitive Materials

We have reviewed the theoretical and experimental link between emergent patterns in physical nonlinear dynamical systems and a capacity for biologically inspired information processing. A cognitive device designed on this principle would be capable of compressing and simulating a physical phenomenon through spontaneous emergence of spatial and temporal coherence among its elements. Understanding and controlling self-organization within complex nonlinear dynamical media can potentially open up possibilities of creating technologies that are a departure from traditional approaches to information processing, toward creating a type of hardware that functions based entirely on its internal dynamics. This offers a new platform for creating intelligent systems, one that has the potential to be orders of magnitude more powerful for classification and recognition.

There are several examples of laboratory-based *controlled* experimental pattern formation systems such as Belousov-Zhabotinsky chemical reactions, pattern formation in a neuronal petri dish, and chemical logic gates. However, these platforms either lack the diversity and complexity of patterns necessary for a practical information-processing unit or are difficult to maintain, control, and interact with. The process of emergent patterns and category formation in the system must be based on a set of coexisting attractors, which, through the basin of their attraction, are able to compress large subsets of initial conditions presented to the device.

To achieve, recognize, and categorize external stimuli, an information processor must spontaneously be able to produce a wide variety of distinct spatiotemporal patterns and associate them with the input stimuli provided to the system. There must further be simple recognition alerts that are triggered by the input stimuli and further in-depth memories associated with the data presented to the system. Nonlinear dynamical systems, through the use of strange attractors, can give rise to easily recognizable patterns through identification of attractors that occupy a certain subset of the phase space and effectively reduce the initial number of the degrees of freedom. Meanwhile, they can carry out complex functional tasks through recalling memories and producing information and increasing resolution by following the trajectory of the attractor.

An appraisal of some of the underlying mechanisms in electrochemical processes reveals features that are entirely appropriate and necessary for a complex configuration to behave as a pattern-forming system:

1) Far-from-equilibrium conditions and criteria for nonlinearity are evident in the kinetic equations of electrochemistry.
2) Instant, long-range communication, and parallel dynamics are made possible by spatial transport of ions through the classic equations of reaction, diffusion, and convection, as well as the migration of charges inside the electrolyte. Temporal patterns that affect the changing of the double layer potential are dependent on a coupled differential equation, which can be derived from both mass transfer and charge balance [28, 29].
3) Feedback mechanisms exist, which, together with the dynamics of electrochemical pattern-forming system, lead to phase transitions, multiple stability points, sustained oscillations, strange attractors, and chaotic behavior.

Accordingly, tandem measurements and control of electrochemical pattern-forming systems offer the harnessing of the computational capabilities that arise from the underlying nonlinear reaction, diffusion, and drift dynamics with long-range interconnection that take place between these processes three dimensionally inside the electrolyte and on the surface. This highly distributed activity leads to the formation of patterns in a reasonable time, which is the analog of the computational mapping of the input information that had affected the processes. In summary, we can formalize the idea of pattern formation as a means to facilitate information recognition by describing recognition as formation of localized phase structure with identifiable features that can effectively categorize and store tags on input information. This is tantamount to a processing power that arises from within the dynamics of the material.

13.5
An Intelligent Electrochemical Platform

One possible approach to effect the achievement of computation via electrochemistry is through the emergent patterns that can be instigated during the electrodissolution of metals. Experimental evidence for the existence of a wide variety of temporal patterns in such systems was observed as early as 1978 [30, 31], and there have been more recent studies on spatial patterns [32, 33]. This has been complimented with more recent theoretical work that has established a basis for formation of temporal and spatial patterns during electrodissolution of metals, based on coupled ordinary differential equations and partial differential equations describing elements of the system such as concentration and double layer potential [34].

The occurrence of spatiotemporal patterns during electrochemical reactions is related to the interplay between reaction rates given by the Butler-Volmer equation and the mass action diffusion given by Fick's laws, which bring a time dependency into the picture. This is further complicated by the structure of the double layer and its time dependence, which gives rise to changing migration currents. Furthermore, surface reactions, geometry of the employed electrochemical cell and electrode, as well as external controls must be taken into consideration.

Pattern-forming electrochemical systems display a region of negative differential resistance in their current–potential characteristic [35], which in addition to the natural kinetic driving force of any increase in potential, manifests an activator–inhibitor system. In this type of system, the potential takes on the role of an activator, while the inhibitor is a chemical quantity, such as the concentration of the electroactive species or the coverage of the electrode. Spatial inhomogeneities in activator and inhibitor can induce migration and diffusion fluxes, respectively, which in turn lead to the occurrence of electrochemical pattern formation. The main difference from a reaction-diffusion system, such as that of Turing's, is that in contrast to diffusion, migration represents a long-range spatial coupling [36]. Furthermore, the transport mechanism in electrochemical systems is mostly influenced by migration rather than by diffusion [37]. In this way, local perturbations in the double layer potential are mediated through the electric field in the liquid phase [38]. Thus, spatial inhomogeneities in the double layer potential are felt not only by the nearest neighbors but also by a whole range of neighboring sites.

The representative equations for the electrochemical system examined here are related to reaction-diffusion type dynamics, since they can be described by a local function and a spatial coupling term. Furthermore, global coupling is routinely present in electrochemical experiments and could play a stabilizing or destabilizing role in the dynamics, acting as an activator as well as an inhibitor, depending on the electrochemical reaction under consideration [39]. The strength of this global coupling may be readily varied, since it is introduced by an external control circuit [40].

A theoretical model for the electrochemical kinetics of metal electrode dissolution can be considered through a model based on coupled differential equations that take time variance of defining parameters into account. There are a number of conditions in electrochemical systems that depend on coupled parameters that are governed by nonlinear dynamical laws. When a metal electrode is immersed in liquid electrolyte, ions from the metal surface will dissolve until a junction potential is created because fo the charge imbalance. This potential, E_0^0, under standard conditions is specific to each metal dissolution process and is related to the free energy of cell reaction by considering that during the completion of the cell reaction, the transfer of charge, nF, is affected. Under equilibrium conditions, this potential is influenced by metal ions activity, a_M, in the electrolyte, as can be derived from basic thermodynamic principles, and is given by the Nernst equation:

$$E_0 = E_0^0 + \frac{RT}{nF} Ln\left(\frac{a_{M^{n+}}}{a_M}\right) \tag{2.1}$$

We can remove this system away from equilibrium to promote the formation of dissipative patterns through the application of an overpotential $\Delta\phi$. The resulting kinetics describing the nonlinear effect of polarization on the current density due to charge transfer polarization are described by the Butler–Volmer equation. However, when the concentration of the reacting species at the electrode surface is lower than that of the electrolyte, the flux of ions becomes the rate controlling factor. Mass transfer may be due to either diffusion, which depends on concentration gradients, or migration, which is the movement of charged species under an electric field. Convection, which is due to fluid motion, is not discussed here since the electrolyte solution in our experiments was not stirred. As discussed, mass transfer plays an important role in providing local and long-range coupling and communication in pattern formation during electrochemical dissolution.

When the electrode processes take place according to diffusion kinetics, the steady state current passing through the working electrode is given by Fick's first law:

$$J = -D \left.\frac{\partial c}{\partial x}\right|_{x \to 0}$$

where J is the diffusion flux (moles per square meter per second), c is the concentration (moles per cubic meter), and D is the diffusion coefficient (square meters per second) with temperature dependence of

$$D = D_0 e^{-\frac{E_A}{RT}}$$

where E_A represents the activation energy for diffusion (Joules per mole). $\left.\frac{\partial c}{\partial x}\right|_{x \to 0}$ is the concentration gradient in the immediate vicinity of the electrode.

The time variance of concentration plays a role in pattern formation (Figure 13.3), and we can relate the time differential to the concentration gradient through Fick's second law, which can be derived from the first law of diffusion and mass balance:

$$\frac{\partial c}{\partial t} = -\frac{\partial}{\partial x}J = D\frac{\partial^2}{\partial x^2}$$

or $D\nabla^2 c$ in three dimensions, considering that the diffusion coefficient does not change with coordinates. In the case of steady state, Fick's second law gives a linear concentration profile near the working electrode. It is therefore necessary, for complex temporal behavior, that the concentration should exhibit a minimum second-order variation with respect to spatial coordinates.

Similarly, migration current flux is the result of potential gradient described as

$$J_m = -\frac{zFDc}{RT}\frac{d\phi}{dx}$$

Further to these coupled differential equations describing the relation of concentration, potential, and current, a phase boundary is formed when a metal electrode is brought into contact with the liquid electrolyte. The subsequent rearrangement of electric charges to achieve equilibrium in each phase due to charge transfer and adsorption of ions among other factors gives rise to a structured and distributed electric potential difference in the boundary layer known as the *double layer*. At a given potential, hydrated ions will accumulate in front of the working electrode and will balance an equal excess charge in the metal. The thickness of the charge layer in the metal is less than 1 Å and can safely be assumed to be zero thickness. A diffuse layer connects the outer plane to the bulk solution. The thickness of the diffuse layer depends on the concentration of the solution.

We can solve for the structure of the double layer by considering

$$\frac{d^2\phi}{dx^2} = -\frac{\rho(x)}{\varepsilon\varepsilon_0}$$

for $\rho(x)$ representing the charge density and $\phi(x)$ potential obeying Poisson's equation. From this relation, a differential equation for the inner potential can be derived, which can be linearized to give

$$\frac{d^2\phi}{dx^2} = \kappa^2 \phi(x)$$

where κ represents the Debye inverse length.

The dynamics and coupling of the aforementioned controlling parameters can illustrate the theoretical basis of pattern formation ranging from oscillatory to spatial and temporal patterns, spanning from chaos to orderly regimes and transitions in phase dynamics [41, 42].

The stability of the system defined in this way, through a set of linear differential equations, requires an evaluation of the change in small perturbations of both the double layer potential and concentration at the stationary points through the Jacobian matrix.

$$\begin{pmatrix} \frac{d\delta\phi_{dl}}{dt} \\ \frac{d\delta c}{dt} \end{pmatrix} = J_{\phi_{dl}^s, c^s} \begin{pmatrix} \delta\phi_{dl} \\ \delta c \end{pmatrix}$$

If the eigenvalues of J are both negative, the fixed point is stable; the steady state is otherwise unstable. The stability of the system can also be studied with an evaluation of the largest Lyapunov exponents, which is calculated for the results presented here. Through control and perturbation of parameters that influence the electrochemical system, it is possible to cause a sign shift in the eigenvalues of J and direct the evolution of the dynamics and bifurcations in the system.

The evolution of the system can be pictured in a phase space whose coordinates are concentration and double layer potential. Figure 13.3a shows the null clines curves ($d/dt = 0$) of ϕ_{dl} and concentration. It is simple to see that the steady state of the system is where the two curves intersect, since at such a point both dc/dt and $d\phi_{dl}/dt$ are zero. Furthermore, the time derivate of each variable, which determines the direction of movement in phase space, changes sign as the corresponding null cline curve is crossed. Since the double layer potential changes much more rapidly than concentration, being off the steady state, the system will move much

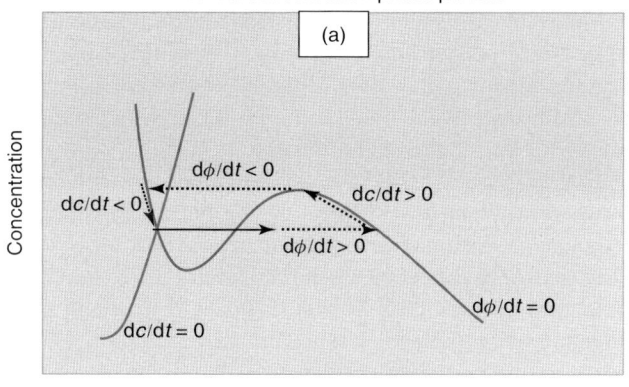

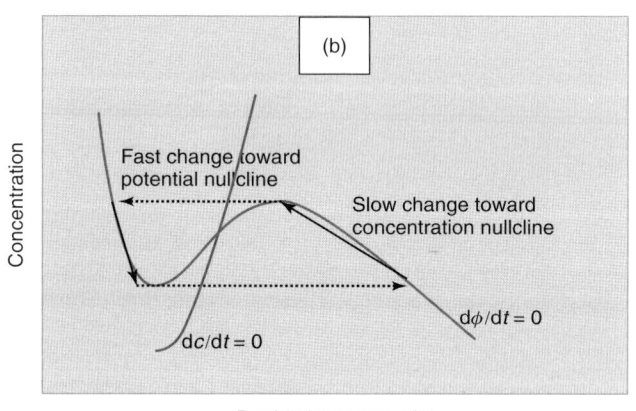

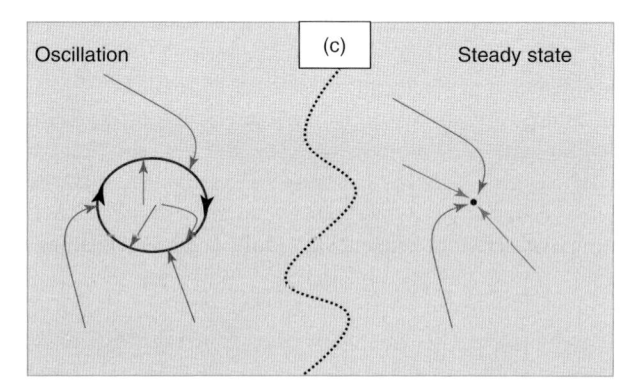

Figure 13.3 (a) Representation of a perturbation leading to steady state. Solid arrow indicates a perturbation, and the dashed arrows show the path of flow and evolution of the system following a potential perturbation. (b) Representation of possible oscillations mechanism in the phase portrait. (c) Representation of basins of attraction for the two variable temporal model of the electrochemical system. (Reprinted by kind permission from Elsevier B.V., Amsterdam.)

faster in the horizontal direction to reach the double layer steady state than in the vertical direction. Furthermore, the direction will be dictated depending on which side of the null cline the system resides at a given time. Once on the potential null cline curve, the system will slowly move toward the concentration null cline. This perturbation off the steady state is illustrated in Figure 13.3a, where the final relaxation point of the system is the single steady-state value.

If, however, a situation as in Figure 13.3b is considered, then it can be seen that the intersection occurs on the negative slope of the ϕ_{dl} nullcline, where the direction in which ϕ_{dl} changes in this neighborhood points away from the steady state. On the contrary, the change in concentration points toward the steady state. Since the ratio of potential to concentration time scale is small, the dynamics are such that ϕ_{dl} changes much more rapidly than concentration. This promotes an oscillatory behavior, where the system evolves on the potential null cline and jumps from one branch to another without ever reaching the steady state.

Considering the dynamics of the system in the phase plane as illustrated in Figure 13.3c, one can see a division of two different basins of attraction. Any system parameters starting or being perturbed to one side of the separatrix, which divides the phase space into different basins, will arrive at a particular attractor. That is to say it will either oscillate or arrive at a steady state.

The spatially homogenous dynamics that we have considered so far have ignored the evolution of other variables that affect the dynamics at the electrode/electrolyte interface. Such variables can include the coverage of the adsorbates on the surface of the electrode, adsorption processes, migration, surface reactions, and structural phase transition processes at the electrode surface. These and other extensions and refinements of this described model can be found in the literature [43, 44].

The number of coupled time-dependent elements in electrochemical pattern-forming systems makes a full analytical model rather impossible to solve without the aid of simplifying assumptions. However, from an experimental perspective, these relations provide the needed ingredients to form a sufficiently complex system capable of producing a range of spatiotemporal patterns. Furthermore, the coupling can be fine-tuned, which permits control over the appearance of patterns and transitions between them.

The patterns exhibited by pattern-forming electrochemical systems range from oscillatory to chaotic and require signal-processing techniques for quantitative feature extraction. The time delay plot in Figure 13.4 shows an example of unsteady attractors that the temporal patterns can exhibit. As the only acquired variable from the electrochemical experiment is current and there are other variables such as concentration and double layer potential that also influence the dynamics of the system, time delay plots are employed to extract more data from the single acquired variable [45].

These patterns are complimented by inherent short-term history dependence in dynamics, because of the location of the system in the phase space at any moment [46]. Furthermore, longer term history dependence is possible as a result of surface changes on the working electrode. Feedback training mechanism can be created whereby, upon recognition of the input of information, perturbations

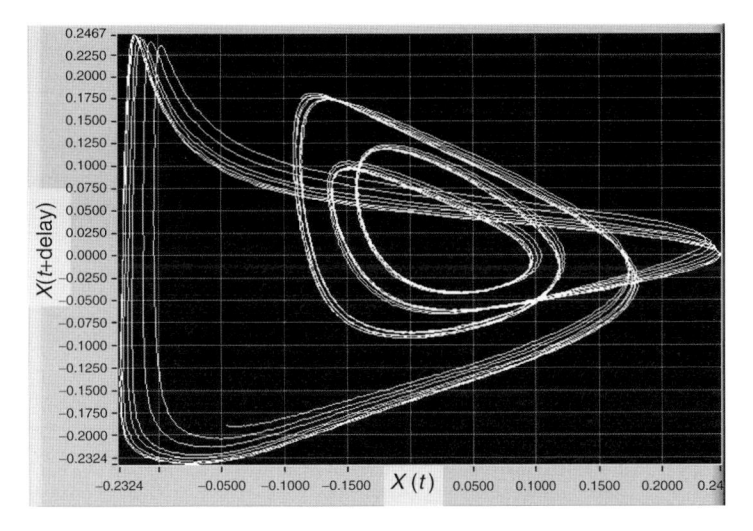

Figure 13.4 N-periodic phase space dynamics of instabilities in electrochemical pattern forming systems showing multiple loops on a two-dimensional time delay plot. (Reprinted by kind permission from the Royal Society of Chemistry, Cambridge, UK.)

are repeated to strengthen a particular pattern by making use of slow variations in the internal resistance of a ferroelectric layer that can be embedded beneath the working electrode or the use of software weighing factors.

13.6
From Chemistry to Brain Dynamics

13.6.1
Understanding the Brain

The older and controversial notions of the brain, such as encapsulated in mentalistic approaches to the analysis of brain function are still very much prevalent in the modern science of dealing with the mind [47]. However, to make real progress we must focus our efforts on an attempt to discover the neural basis of behavior, the activities that animals, including humans, actually perform in their daily lives. What we know for certain is that our brain is capable of modeling the world around us and recognizing and responding to slight changes to any one of the many models it has stored [48]. At any given moment the brain is dealing with a large number of variables provided to it by the body's sensory systems and yet is capable of simultaneously processing this time varying multivariable information and forming an appropriate command in response to this changing environment.

Figure 13.5 represents a simplified and useful portrait of the role of the brain as an information-processing system working in conjunction with the body to recognize and respond to the environment. Such a model serves as the inspiration

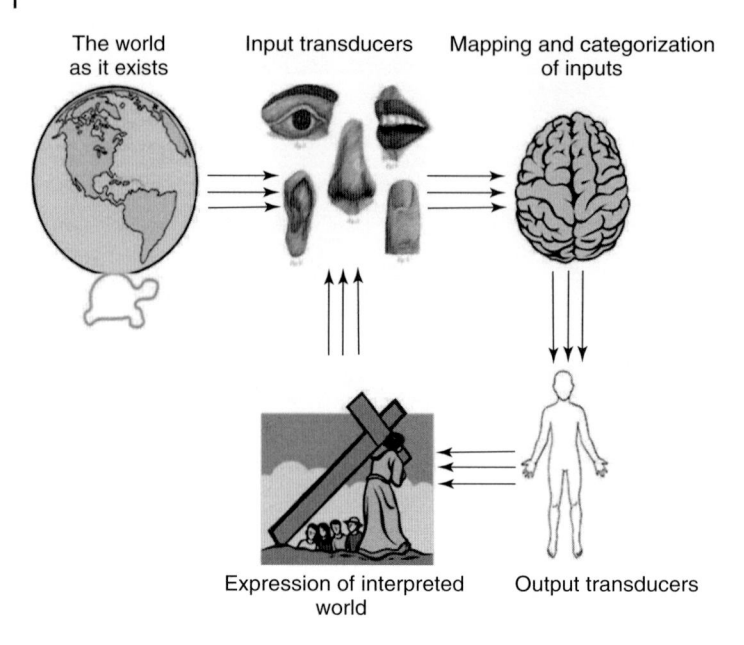

Figure 13.5 Brain and cognition.

of the endeavor in this discussion to produce a material entity that can process information based on its internal dynamics. In Figure 13.5, the world represents information to which the system or the body is exposed. Although the exact nature of this information, independent of our observations, is perhaps in part a mystery, the representation of this information is provided to the brain in the form of neural activity received from the body's sensory arsenal. The senses play the role of transducers, which translate the information as best as possible, to a language with which the information-processing system can work. This information is then recognized as belonging to certain categories with memories and links associated with each. There may perhaps be extended layers of such categories and complex links forming between them. The information and memories called up from certain recognized states may then evoke specific neuronal signals that demand a physiological response regarding the information provided to the brain. This expression of the brain's interpretation is carried out by the body, which can be thought of as output transducers enabling action in the physical world. Understanding this process helps draw several parallels with the biologically inspired electrochemical information-processing system proposed here and discussed in this chapter.

The postulate in this report has not been that the current understanding in the basic laws of chemistry and physics constitute a firm foundation on which brain function can be explained. The alternative view, that the scientific power of reduction and prediction of our physical world, places too far a constraint on an immensely complex system such as the brain to explain consciousness can perhaps be adhered to with equal or greater validity [49]. Although we may argue that the function of the brain is a consequence of a sequence and hierarchy of emergent patterns, it does not suffice to say self-organization or investigation by physiology on the neuronal level can reduce the dynamic patterns in the brain to a low-dimensional equation [50, 51].

13.6.2
Brain Dynamics

Spatiotemporal dynamics within the brain are not clearly characterized or understood. Nonlinear dynamics have been used to model and study the interaction of neurons that creates higher brain functions. These studies have motivated intense theoretical activity within the scientific community regarding the role of biological attractors in information processing, memory, and cognition. The transient nature of complex spatiotemporal patterns in the brain and the striking variability of the signals obtained by recording neuronal activity [52–54] are good indicators that the dynamics of the nervous system are nonlinear. It is certainly difficult to correlate a clear implication of the underlying mechanisms for the diverse chaotic, multifractal, and long-range scaling that is observed in the experimental data obtained from the brain. This is, however, a welcomed challenge for physicists to tackle, and the realization of the problem has benefited the field of computational biology and neuroscience by bringing people from different fields together and speaking the same language.

Although a number of formal models with dynamic and chaotic properties have enjoyed success as analogs for biological networks, their effectiveness in pattern recognition, storing, and retrieval have been drastically short of what nature can accomplish. This is a clear sign that none of the current descriptions fully capture the features of complex biological systems. The plausible yet incomplete theory is the inevitable involvement of mechanisms of emergent complex phenomena evident in dynamical systems. Furthermore, as the largest number of metastable states exists at a point near the transition, the brain can be accessing the largest repertoire of behaviors in a flexible way, near a critical point of a second-order phase transition [55].

The discussion in this section gives rise to an intriguing question: whether the higher cognitive abilities of the brain and other biological systems can be reduced and explained by the physical laws that govern the elements of the system, or if the collective dynamics is the result of something more than the sum of the elements? We do not strictly subscribe one way or the other because of lack of evidence for either case. One is based on the reduction of cognitive capacity to the neuronal level, while the other argues that there may need to exist physics as of yet undiscovered to explain the extraordinary range of abilities of the brain and functioning of the mind. However, we can say that self-organization plays a role, and these complex media may within their inherent dynamics possess abilities beyond the sum of their elemental capacity.

Owing to difficulty of direct control experimentation with large biological systems composed of the large network of interacting elements, simulations have prevailed over experimental studies of nonlinear properties of large cortical networks and higher brain functions. Yet, evidence of chaos and strange attractors has been obtained at the level of small populations of coupled cells [56]; it can be extrapolated that signals in the brain are distributed according to chaotic patterns at all levels of hierarchical structure. It is evident that better controlled experimental paradigms, ones that can capture the large number of elements and their nonlinear interactions found in complex biological organisms, are needed to advance our understanding of the information processing in these systems.

13.6.3
Electrochemical Dynamics

Electrochemical experiments, because of their inherent nonlinearity and ease of implementation and control, make an ideal platform to physically study pattern formation without abstraction or simplification, which may not reveal the full capacity of the system. Many oscillating electrochemical systems have been reported in the literature [57–60], and several, including anodic reactions and oxidation of Fe [61], Co [62], and Ni [63], have been studied under sinusoidal potential modulation. Anodic electrodissolution of Ni and particularly the Ni|H_2SO_4 system is of interest, as it has been extensively studied [64, 65] with oscillatory patterns observed under various parameters [66, 67].

The frequency power spectrum analysis of the data obtained from an electrochemical pattern-forming system is shown in Figure 13.6a. The broader peaks in the power spectrum point to the unsteady nature of each oscillation, occupying a band of activity as opposed to a single sharp oscillation. The other interesting aspect of this spectrum is in the changing nature of the power spectrum with respect to increasing frequency. Plotted on the logarithmic amplitude scale and linear frequency graph, the decrease in amplitude with increasing frequency is reminiscent of $1/f$ amplitude frequency relation encountered in biological systems. This energy drop in the power spectrum is another indication of the complexity of temporal patterns as is often observed in chaotic systems [68]. For comparative measures, the frequency spectrum of electroencephalogram (EEG) obtained from a human brain is included (Figure 13.6b). The power drop with respect to frequency on the semilog plot suggests the possibility of the presence of instabilities ranging from low-dimensional steady states, limit cycles, and simple transient structures in phase space to higher dimensional attractors with more complex frequency structures in the system.

The electrochemical experiments reported here are examples of patterns in nonlinear systems that can evolve from initially stable homogenous structures, given small perturbations. Following the introduction of instability, the nonlinear dynamics by which the components of a system interact will define its pattern-forming evolution. Furthermore, the new patterns are entirely distinct from the initial instabilities and their linear functions introduced to the system. It is this very ability of far-from-equilibrium nonlinear systems to map a set of potentially complex inputs to a collection of categorized patterns that is proposed to be exploited.

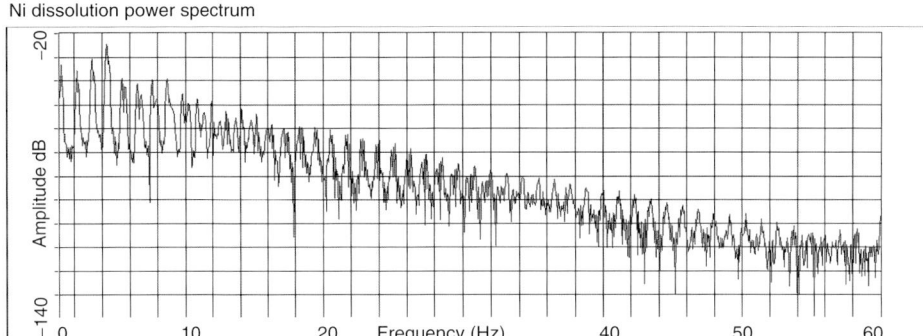

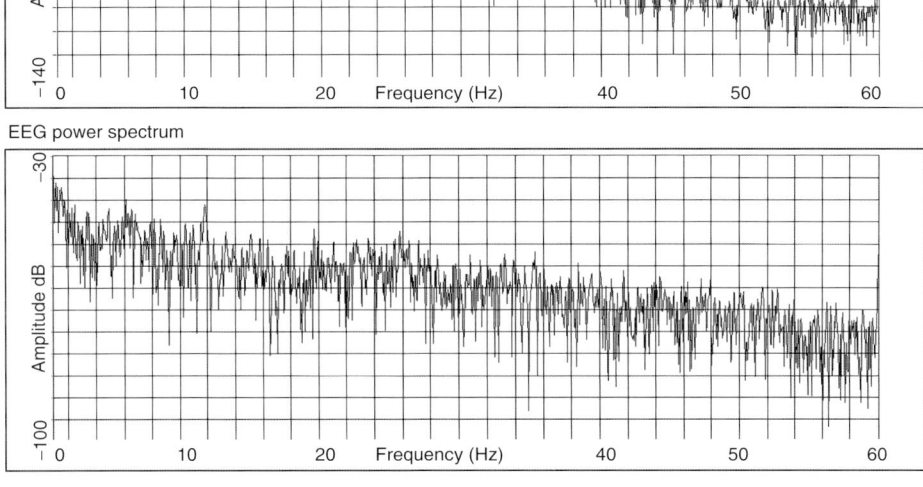

Figure 13.6 (a) A power spectrum analysis of Ni dissolution patterns in the oxygen evolution region, and (b) the same analysis applied to EEG brain signal obtained from a human subject.

Figure 13.7 is an example of strange attractor dynamics in the system response where it can be clearly seen that there is a positive Lyapunov exponent. Such attractor dynamics allow for a level of abstraction that can segment a large variety of external world information into a few significant categories, while at the same time they can store information and exhibit a memory function through their information dimension.

There are a number of features in the scheme proposed here, which form the basis of a biologically inspired information-processing system. Such a configuration has the capability to compress a set of similar input information to a discrete number of recognition states. The mapping of data in this manner allows both for recognition from incomplete sets of information and a recall memory associated with the data.

13.6.4
Experimental Paradigm for Information Processing in Complex Systems

Self-organization dynamics are the underlying principal behind much of the interesting processes found in complex animate materials. However, biological

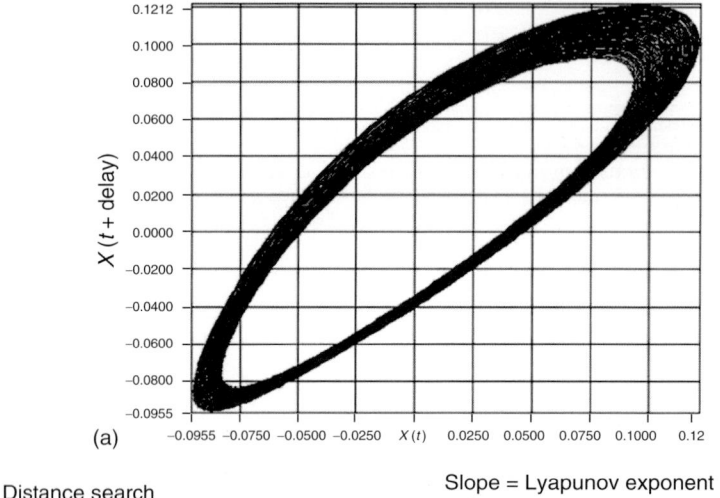

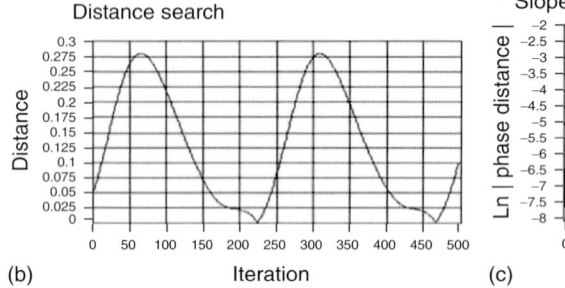

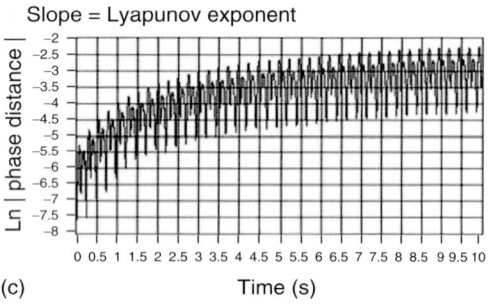

Figure 13.7 Lyapunov exponent and phase space distance graphs of invoked dynamics within the oxygen evolution region. (a) Two-dimensional phase space of the attractor. (b) Trace used to find the iteration where the trajectory comes to an infinitesimally close distance from its starting position, representing the period of oscillation. This information is used in (c), which tracks the distance between two infinitesimally close points from two separate trajectories plotted on a logarithmic scale as a function of time. The Lyapunov exponent is the slope of this graph. (Reprinted by kind permission from Elsevier B.V., Amsterdam.)

dynamics do not always give rise to simple patterns. The discovery of chaotic itinerancy in high-dimensional dynamical systems such neural activity in the cortex are evidence of transitory dynamics and complex attractors [69]. Thus, any realistic biologically inspired information-processing system must be able to incorporate and recreate these higher dimensional dynamics.

In the context of self-organization and the ability to produce dynamic *functional* patterns, complex systems, whether inorganic or biological, may give rise to emergent patterns. This is why the study of chemical systems, whose elements, dynamics, and nonlinear interactions are perhaps better understood and easier to replicate and control than the biological neuronal system, is so beneficial. Emergent patterns in chemical and electrochemical systems in particular offer a robust and engineered experimental paradigm to emulate what might be happening in other biological systems, functioning on the basis of self-organization and pattern formation.

The behavior of complex biological information-processing systems is best understood through an experimental platform that truly functions as an ensemble whose response is the totality of its overall states while all the components of the system are modified at the same time to form attractors, trajectories, and bifurcations in the phase space. In such a system, inputs do not dictate the complete internal state of the medium, rather they merely perturb its intrinsic dynamics to represent the problem or the change in environment that demands an appropriate response.

To the dismay of some neurophysiologists and engineers who have tried to simulate brain function [70], the brain is not a static machine with a functional map that can be invariably correlated to physiological activities of our body. Although general outlines of brain maps exist, there is, in fact, a large variability in these schemes that have been identified in various human and animal subjects. Furthermore, the wiring of our brain is more dynamic and fluid than it was once thought. The plasticity of the cortex to change its representation after an injury has been documented, although not much is understood about the dynamics of these changes. Here is another area where an open inorganic system, such as the electrochemical one described here, whose dynamics are readily accessible for investigation, can find application in shedding light on the changes in the landscape of the brain's cognitive phase space and the path and projection of attractors toward basins of emergent patterns.

In contrast to man-made systems designed to perform particular tasks, the brain would rely on self-organized (and chaotic) processes to generate relevant activity patterns, and perception would be a dynamic and creative course of actions rather than following predetermined set of logical rules [71]. Similarly, an appropriate dynamical systems theory and experimental paradigm should strive to understand the unfolding of cognitive processes over time and identify what external and internal factors influence this unfolding.

13.7
Final Remarks

Biological evolution has obviously resulted in incredibly refined systems and structures that have consistently inspired scientists and engineers to mimic biological "technology" for many years. This specific field of endeavor has often been termed *biomimetics*. The reason for this activity, outside of simple curiosity, lies in the superior performance of biological systems over man-engineered technologies. Specific characteristics of interest, among many, are multifunctional capability, mechanisms of self-repair and regeneration, and adaptability to widely varying environments. There have been many examples of attempts at biological mimicry, too numerous to outline here, such as polymeric materials that display the flexibility and strength of biological matrices, visual devices based on artificial receptors for the detection of electromagnetic (EM) radiation, and transistor-based devices that can detect chemical compounds, for example, the so-called electronic nose. One very evident characteristic of these research and development activities has been the involvement of individuals from many disciplines such as engineering, biology, physics and chemistry, nanotechnology, and materials science. An excellent example, which has received significant attention, is the lipid-membrane-based biosensor. Here, chemists prepare lipid monolayers or bilayers in collaboration with biologists who incorporate molecular receptors in the layers. These are deposited on electronic devices produced by engineers, and a detailed understanding of the mechanisms of signal development is produced by physicists. However, the key point is that the sensing configuration has at its heart the lipid membrane much as found in olfactory and gustatory sensing in biology.

The technology most relevant to this article, of course, is the artificial neural network (ANN). In this application, the concept is the mimicking of biological neuron arrangements using artificial systems via a serial computer. The technology has a long history, dating back to the end of the nineteenth century. Many innovations have been introduced over the years, such as Hebbian learning, backpropagation, Hopfield networks, and parallel distributed processing. The structures that often employ artificial neural inner, hidden, and outer layers have been used in both attempts at examination of genuine biological networks and in information processing. With respect to the latter, there have been a number of technical successes such as speech recognition, image analysis, and adaptive control. Despite these advances, it is fair to say that ANN does not come close to the "real thing." Even a relatively modest network can occupy enormous amounts of CPU. Accordingly, an approach that does not necessarily mimic biology, but rather emulates the crucial ingredients associated with the massive interconnectivity of neurons. The present work attempts to catch at least the essence of a biological neural network in terms of its nonlinear dynamical properties. This is achieved in a typical inorganic configuration that is widely known to display such a property, electrochemistry at electrode surfaces. One would hope that in the future, a better understanding of this physics coupled to advances such as electrode array technology could lead to a wholly new concept for computer-based information processing.

References

1. Guckenheimer, J. and Holmes, P. (1983) *Nonlinear Oscillations, Dynamical Systems, and Bifurcations of Vector Fields*, Springer-Verlag.
2. Sadeghi, S., MacKay, W.A., van Dam, R.M., and Thompson, M. (2011) Algorithm for real-time detection of signal patterns using phase synchrony: an application to an electrode array. *Meas. Sci. Technol.*, **22** (2), 025802.
3. Sadeghi, S. and Thompson, M. (2010) Temporal patterns and oscillatory

voltage perturbation during an electrochemical process. *Phys. Chem. Chem. Phys.*, **12**, 6795–6809.
4. Coullet, P., Riera, C., and Tresser, C. (2004) A new approach to data storage using localized structures. *Chaos*, **14** (1), 193–198.
5. Crutchfield, J.P. (1994) The calculi of emergence – computation, dynamics and induction. *Phys. D*, **75** (1–3), 11–54.
6. Crutchfield, J.P. and Mitchell, M. (1995) The evolution of emergent computation. *Proc. Natl. Acad. Sci. U.S.A.*, **92** (23), 10742–10746.
7. Delgado, J. and Sole, R.V. (1997) Collective-induced computation. *Phys. Rev. E*, **55** (3), 2338–2344.
8. Kohonen, T. (1989) *Self-Organization and Associative Memory*, Springer-Verlag, Berlin, New York.
9. Nicolis, G. and Prigogine, I. (1977) *Self-Organization in Nonequilibrium System: From Dissipative Structures to Order Through Fluctuations*, John Wiley & Sons, Inc., New York.
10. Maselko, J. (2003) Pattern formations in chemical systems. *Adv. Complex Syst.*, **6** (1), 3–14.
11. Kiss, L. (1988) *Kinetics of Electrochemical Metal Dissolution*, Elsevier, Amsterdam, New York.
12. Cross, M.C. and Hohenberg, P.C. (1993) Pattern-formation outside of equilibrium. *Rev. Mod. Phys.*, **65** (3), 851–1112.
13. De Wit, A. (1999) Spatial patterns and spatiotemporal dynamics in chemical systems. *Adv. Chem. Phy.*, **109**, 435–513.
14. Ertl, G. (2002) Dynamics of surface reactions. *Faraday Discuss.*, **121**, 1–15.
15. Burke, J. and Knobloch, E. (2006) Localized states in the generalized Swift-Hohenberg equation *Phys. Rev. E*, **73** (5), 056211-1–056211-15.
16. Gollub, J.P. and Langer, J.S. (1999) Pattern formation in nonequilibrium physics. *Rev. Mod. Phys.*, **71** (2), S396.
17. Nicolis, G. and Prigogine, I. (1989) *Exploring Complexity: An Introduction*, W.H. Freeman, New York.

18. Farmer, J.D., Kauffman, S.A., and Packard, N.H. (1986) Autocatalytic replication of polymers. *Phys. D: Nonlin. Phenom.*, **22** (1–3), 50–67.
19. Coveney, P.V. (2003) Self-organization and complexity: a new age for theory, computation and experiment. *Philos. Trans. R. Soc. Lond. Ser. A: Math. Phys. Eng. Sci.*, **361** (1807), 1057–1079.
20. Kauffman, S.A. (1993) *The Origins of Order: Self-Organization and Selection in Evolution*, 1st edn, Oxford University Press, New York.
21. Holland, J.H. (1999) *Emergence: From Chaos To Order*, Basic Books.
22. Morowitz, H.J. (2004) *The Emergence of Everything: How the World Became Complex*, Oxford University Press, New York.
23. Coullet, P., Toniolo, C., and Tresser, C. (2004) How much information can one store in a nonequilibrium medium? *Chaos*, **14** (3), 839–844.
24. Grassberger, P. and Procaccia, I. (1983) Characterization of strange attractors. *Phys. Rev. Lett.*, **50** (5), 346–349.
25. Eckmann, J.P. and Ruelle, D. (1985) Ergodic-theory of chaos and strange attractors. *Rev. Mod. Phys.*, **57** (3), 617–656.
26. Vanderwolf, C.H. (1998) Brain, behavior, and mind: what do we know and what can we know? *Neurosci. Biobehav. Rev.*, **22** (2), 125–142.
27. Yoon, R.S., Borrett, D.S., and Kwan, H.C. (1995) Three-dimensional object recognition using a recurrent attractor neural network. Engineering in Medicine and Biology Society, 1995, IEEE 17th Annual Conference, Vol. - 1, pp. 379–380.
28. Krischer, K., Mazouz, N., and Grauel, P. (2001) Fronts, waves, and stationary patterns in electrochemical systems. *Angew. Chem. Int. Ed.*, **40** (5), 851–869.
29. Christoph, J. and Eiswirth, M. (2002) Theory of electrochemical pattern formation. *Chaos*, **12** (1), 215–230.
30. Frank, U. (1978) Chemical oscillations. *Angew. Chem. Int. Ed. Engl.*, **17** (1), 1–15.
31. Lev, O., Wolffberg, A., Sheintuch, M., and Pismen, L.M. (1988) Bifurcations to periodic and chaotic motions in anodic

nickel dissolution. *Chem. Eng. Sci.*, **43** (6), 1339–1353.
32. Ertl, G. (1991) Oscillatory kinetics and spatiotemporal self-organization in reactions at solid-surfaces. *Science*, **254** (5039), 1750–1755.
33. Hudson, J.L. and Bassett, M.R. (1991) Oscillatory electrodissolution of metals. *Rev. Chem. Eng.*, **7** (2), 109–170.
34. Flatgen, G. and Krischer, K. (1995) A general-model for pattern-formation in electrode-reactions. *J. Chem. Phys.*, **103** (13), 5428–5436.
35. Koper, M.T.M. (1998) Non-linear phenomena in electrochemical systems. *J. Chem. Soc., Faraday Trans.*, **94** (10), 1369–1378.
36. Mazouz, N. and Krischer, K. (2000) A theoretical study on turing patterns in electrochemical systems. *J. Phys. Chem. B*, **104** (25), 6081–6090.
37. Mazouz, N., Flatgen, G., and Krischer, K. (1997) Tuning the range of spatial coupling in electrochemical systems: from local via nonlocal to global coupling. *Phys. Rev. E*, **55** (3), 2260–2266.
38. Levart, E. and Schuhman, D. (1970) Migration-diffusion coupling and concept of electrochemical impedance. *J. Electroanal. Chem.*, **24** (1), 41–44–.
39. Krischer, K., Mazouz, N., and Flatgen, G. (2000) Pattern formation in globally coupled electrochemical systems with an S-Shaped current-potential curve. *J. Phys. Chem. B*, **104** (31), 7545–7553.
40. Kiss, I.Z., Wang, W., and Hudson, J.L. (1999) Experiments on arrays of globally coupled periodic electrochemical oscillators. *J. Phys. Chem. B*, **103** (51), 11433–11444.
41. Krischer, K., Lubke, M., Wolf, W., Eiswirth, M., and Ertl, G. (1991) Chaos and interior crisis in an electrochemical reaction. *Ber. Bunsen-Ges. Phys. Chem.*, **95** (7), 820–823.
42. Jaeger, N.I., Otterstedt, R.D., Birzu, A., Green, B.J., and Hudson, J.L. (2002) Evolution of spatiotemporal patterns during the electrodissolution of metals: experiments and simulations. *Chaos*, **12** (1), 231–239.
43. Bertram, M. and Mikhailov, A.S. (2001) Pattern formation in a surface chemical reaction with global delayed feedback. *Phys. Rev. E*, **63** (6), 066102-1–066102-13.
44. Bertram, M. and Mikhailov, A.S. (2003) Pattern formation on the edge of chaos: mathematical modeling of CO oxidation on a Pt(110) surface under global delayed feedback. *Phys. Rev. E*, **67** (3), 036208-1–036298-9.
45. Fraser, A.M. and Swinney, H.L. (1986) Independent coordinates for strange attractors from mutual information. *Phys. Rev. A*, **33** (2), 1134–1140.
46. Krischer, K., Lubke, M., Eiswirth, M., Wolf, W., Hudson, J.L., and Ertl, G. (1993) A hierarchy of transitions to mixed-mode oscillations in an electrochemical system. *Phys. D*, **62** (1–4), 123–133.
47. Dulany, D. (2011) What should be the roles of conscious states and brain states in theories of mental activity?FNx08. *Mens Sana Monogr.*, **9** (1), 93.
48. Barlow, H. (1990) The mechanical mind. *Annu. Rev. Neurosci.*, **13**, 15–24.
49. Eccles, S.J.C. (1977) *The Understanding of the Brain*, McGraw-Hill.
50. Rossler, O. and Rossler, R. (1994) Chaos in physiology. *Integr. Psychol. Behav. Sci.*, **29** (3), 328–333.
51. Rossler, O.E. (1987) Chaos in coupled optimizers. *Ann. N. Y. Acad. Sci.*, **504**, 229–240.
52. Haken, H. (2002) *Brain Dynamics: Synchronization and Activity Patterns in Pulse-Coupled Neural Nets with Delays and Noise*, Springer, Berlin, New York.
53. Kelso, J.A.S. (1995) *Dynamic Patterns: The Self-Organization of Brain and Behavior*, The MIT Press.
54. Nolte, G., Bai, O., Wheaton, L., Mari, Z., Vorbach, S., and Hallett, M. (2004) Identifying true brain interaction from EEG data using the imaginary part of coherency. *Clin. Neurophysiol.*, **115** (10), 2292–2307.
55. Chialvo, D.R. (2010) Emergent complex neural dynamics. *Nat. Phys.*, **6** (10), 744–750.
56. Korn, H. and Faure, P. (2003) Is there chaos in the brain? II. Experimental

evidence and related models. *C. R. Biol.*, **326** (9), 787–840.
57. Wojtowicz, J. (1972) Oscillatory behavior in electrochemical systems, in *Modern Aspects of Electrochemistry*, (eds. Conway, B.E and Bockris, J.O'M.) vol. 8, Plenum Press, Neywork, p. 47–120.
58. Koper, M.T.M. and Sluyters, J.H. (1994) Instabilities and oscillations in simple-models of electrocatalytic surface-reactions. *J. Electroanal. Chem.*, **371** (1–2), 149–159.
59. Lou, W. and Ogura, K. (1995) Current oscillations observed on a stainless-steel electrode in sulfuric-acid-solutions with and without chromic-acid. *Electrochim. Acta*, **40** (6), 667–672.
60. Russell, P. and Newman, J. (1986) Anodic-dissolution of iron in acidic sulfate electrolytes. 1. Formation and growth of a porous salt film. *J. Electrochem. Soc.*, **133** (1), 59–69.
61. Pagitsas, M., Karantonis, A., and Sazou, D. (1992) Application of periodic forcing on the simplified Franck-Fitzhugh model for the electrochemical oscillations observed during the electrodissolution of iron in sulfuric-acid-solutions. *Electrochim. Acta*, **37** (6), 1047–1059.
62. Sazou, D. and Pagitsas, M. (1995) Experimental bifurcation-analysis of the cobalt phosphoric-acid electrochemical oscillator. *Electrochim. Acta*, **40** (6), 755–766.
63. Berthier, F., Diard, J.P., Le Gorrec, B., and Montella, C. (2004) Study of the forced Ni|1 M H2SO4 oscillator. *J. Electroanal. Chem.*, **572** (2), 267–281.
64. Epelboin, I. and Keddam, M. (1972) Kinetics of formation of primary and secondary passivity in sulfuric aqueous-media. *Electrochim. Acta*, **17** (2), 177–186.
65. Jouanneau, A., Keddam, M., and Petit, M.C. (1976) General model of anodic behavior of nickel in acidic media. *Electrochim. Acta*, **21** (4), 287–292.
66. Haim, D., Lev, O., Pismen, L.M., and Sheintuch, M. (1992) Modeling spatiotemporal patterns in anodic nickel dissolution. *Chem. Eng. Sci.*, **47** (15–16), 3907–3913.
67. Lev, O., Wolffberg, A., Pismen, L.M., and Sheintuch, M. (1989) The structure of complex behavior in anodic nickel dissolution. *J. Phys. Chem.*, **93** (4), 1661–1666.
68. Strogatz, S.H. (1994) *Nonlinear Dynamics and Chaos: with Applications to Physics, Biology, Chemistry, and Engineering*, Addison-Wesley Pub. Co, Reading, PA.
69. Érdi, P. (2001) How to construct a brain theory? *Behav. Brain Sci.*, **24** (5), 815.
70. Hawkins, J. and Blakeslee, S. (2004) *On Intelligence*, Times Books (Adapted).
71. Beer, R.D. (2000) Dynamical approaches to cognitive science. *Trends Cogn. Sci.*, **4** (3), 91–99.

14
Electrode Interfaces Switchable by Physical and Chemical Signals Operating as a Platform for Information Processing
Evgeny Katz

14.1
Introduction

Modified electrodes functionalized with various organic monolayers and thin films attached to their conducting surfaces have found numerous applications in electrocatalysis, sensors, and fuel cells [1–7]. Particularly, active research was directed to the applications of modified electrodes in different bioelectrochemical systems [8, 9], including biosensors [10–13] and biofuel cells [14–17]. In the past decade, different modified electrodes functionalized with signal-responsive molecules [18], polymers [19], or supramolecular complexes [20] were developed to "switch on demand" electrochemical properties of electrode surfaces. Their applications in switchable biosensors [21], fuel cells [22], and electrochemical systems processing information [23] have been suggested. Various physical and/or chemical signals as well as their combinations were used to switch electrochemical properties of the modified interfaces between active and inactive states for specific electrochemical, electrocatalytic, and bioelectrocatalytic reactions. Light signals (irradiation of electrodes with visible or ultraviolet light) [24–31], magnetic field applied at electrode surfaces loaded with magnetic particles or magnetic nanowires [32–43], and electrical potentials producing chemical changes at the electrode surfaces [44–48] were used to reversibly alternate electrochemical properties of the modified electrodes. Chemical [29, 49–51] or biochemical [52] signals resulting in reversible changes of the interfacial properties were also used to switch the electrode activity ON/OFF for specific electrochemical transformations. The present chapter gives an overview of different signal-responsive electrochemical interfaces, emphasizing the importance of scaling up the complexity of the signal-processing systems and their applications in unconventional information-processing systems.

Molecular and Supramolecular Information Processing: From Molecular Switches to Logic Systems,
First Edition. Edited by Evgeny Katz.
© 2012 Wiley-VCH Verlag GmbH & Co. KGaA. Published 2012 by Wiley-VCH Verlag GmbH & Co. KGaA.

14.2
Light-Switchable Modified Electrodes Based on Photoisomerizable Materials

Electrodes with switchable interfacial properties controlled by light signals are usually based on photochemically induced isomerization processes in monolayers or thin films immobilized on their surfaces [53]. Photoisomerizable molecular or supramolecular systems are considered to be important components of various photochemical information recording/processing systems and optical switches [54, 55]. Monolayers or thin films containing photoisomerizable entities of spiropyran (SP) [24–27], azobenzene [30, 31], diarylethene [56–59], or phenoxynaphthacenequinone [28] were immobilized on electrode surfaces to control interfacial electrochemical properties by light-induced reactions at the modified surfaces. Highly sophisticated photoswitchable supramolecular complexes were immobilized on electrode surfaces to operate as the light-driven molecular "machines" affecting interfacial properties and facilitating electron transport at their specific conformations [20, 60–62]. Electrodes modified with photoisomerizable species were suggested as bases of optoelectronic devices for information processing (e.g., flip-flop memory units [63]).

Different mechanisms could be involved in the regulation of photocontrolled electrochemical processes. In some cases, the immobilized photoisomerizable species (diarylethene and phenoxynaphthacenequinone derivatives) change their redox activity upon photoinduced intramolecular transformations directly affecting the electrochemical properties of the modified interfaces. Other photoisomerizable derivatives (SP and azobenzene derivatives) change their charge/conformation, affecting the interfacial electrochemical processes for diffusional redox probes, resulting in so-called command interfaces [53]. Reversible photoisomerization of the surface-confined diarylethene and phenoxynaphthacenequinone derivatives between redox inactive and active states allowed light-induced switching of the modified electrodes between electrochemically mute and active states; the latter were used to activate electrocatalytic cascades amplifying the interfacial redox changes [28]. Reversible light-induced conformational changes of azobenzene moieties at electrode surfaces resulted in light-modulated accessibility of the electrodes for diffusional redox species [30] or activated electron-transporting molecular "machinery" associated with the photoisomerizable entities [60–62]. Protonation/deprotonation processes following the primary photoinduced isomerization of the surface-confined SP moieties resulted in the different charges of the isomeric states: neutral SP produced upon irradiation with visible light and positively charged protonated merocyanine (MRH^+) generated by ultraviolet irradiation [26]. The different charge states generated at the modified interfaces allowed the charge-controlled discrimination of electrochemical processes [29]: the neutral SP monolayer did not discriminate diffusional redox species by their charges, while the positively charged MRH^+ monolayer inhibited the electrochemical processes for positively charged redox species because of their electrostatic repulsion from the electrode surface (Frumkin effect [64]). This photoswitchable charge-controlled discrimination of electrochemical processes was exemplified with cytochrome c (Cyt)

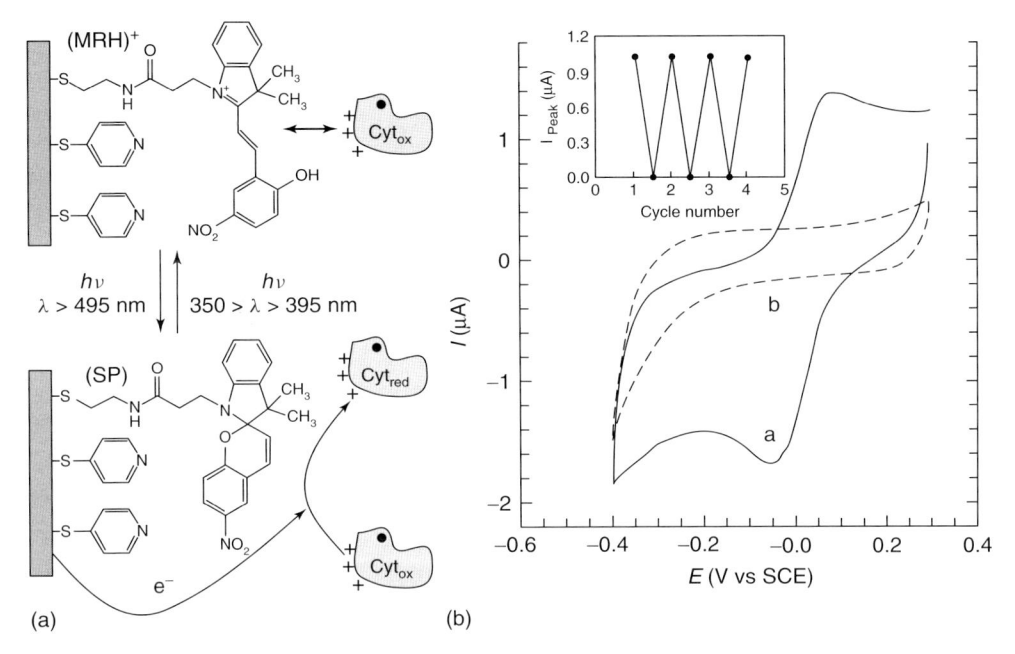

Figure 14.1 (a) The electrochemical process of Cyt discriminated by the charge at the photoswitchable electrode surface functionalized with pyridine units and spiropyran (SP)/merocyanine (MRH$^+$) photoisomerizable entities. (b) Cyclic voltammograms obtained for the Cyt solution at different isomeric states of the electrode interface: (a) SP produced by visible light and (b) MRH$^+$ generated by ultraviolet irradiation. Inset: reversible cycle between the activated and inhibited electrochemical process of Cyt modulated by light signals. (Adapted from Ref. [26], with permission; copyright American Chemical Society, 1995.)

electrochemical reactions [26]. A mixed self-assembled monolayer composed of p-mercaptopyridine units (promoting Cyt electrochemistry) and photoisomerizable nitrospiropyran moieties (switchable between the neutral SP and positively charged MRH$^+$ states) was used to modulate the Cyt electrochemical responses by light signals (Figure 14.1a). The neutral SP state of the photoisomerizable monolayer did not restrict the association of the positively charged Cyt molecules with the pyridine promoter units at the electrode surface, thus allowing the reversible electrochemical process of the protein heme center (Figure 14.1b, curve a). When the electrode surface was irradiated with ultraviolet light (350 nm $< \lambda <$ 395 nm), the positively charged MRH$^+$ state was produced in the monolayer, resulting in the repulsion of the positively charged Cyt and inhibiting its electrochemical process (Figure 14.1b, curve b). The electrode illumination with visible light ($\lambda >$ 495 nm) regenerated the neutral SP state at the surface, thus reactivating the Cyt electrochemical process. The reversible cycling between the active and inactive states of the interface for the Cyt electrochemical reaction was achieved (Figure 14.1b, inset). The photoswitchable electrochemical activity of Cyt was amplified by secondary enzymatic reactions driven by Cyt redox transformations: the electrochemical reduction of

Cyt was coupled with O_2 reduction biocatalyzed by cytochrome oxidase, while the electrochemical oxidation of Cyt was coupled with lactate oxidation biocatalyzed by lactate dehydrogenase. Both bioelectrocatalytic cascades were activated only when the primary electrochemical activity of Cyt was allowed at the neutral SP state of the monolayer [26].

Many other examples of the photoswitchable electrochemical reactions controlled by isomeric states of electrode interfaces were demonstrated – all of them were activated by light signals applied on the electrode surfaces [24–31, 56–59]. The designed photoswitchable electrode interfaces could be used as a platform for performing logic operations and other digital functions controlled by external light signals.

14.3
Magnetoswitchable Electrodes Utilizing Functionalized Magnetic Nanoparticles or Nanowires

Magnetic particles [33–38] or magnetic nanowires [39–42] functionalized with organic shells were associated with electrode surfaces to change their interfacial properties and control electrochemical and bioelectrochemical processes by applying an external magnetic field. Magnetic microbeads chemically modified with different redox species were reversibly translocated by an external magnetic field between a solution-suspended state, where the redox species did not communicate with the electrode surface, and a surface-confined state allowing electron transfer between the conducting surface and the redox species [35–37]. The magneto-controlled electron transport between the redox species linked to the magnetic beads and the electrode surface was used to reversibly activate/inhibit bioelectrocatalytic cascades mediated by the redox species [35–37]. Lateral translocation of DNA-functionalized magnetic beads between conducting and nonconducting domains of an electrode array allowed magneto-controlled activation and inhibition of the DNA redox process [38]. Adaptive multifunctional magnetic nanowires were designed for controlling on demand the operation of electrochemical sensors and microchip devices and for triggering and modulating electrocatalytic and bioelectrocatalytic processes by the application of an external magnetic field [39–42].

An impressive effect on the interfacial properties of electrodes was demonstrated for magnetic nanoparticles functionalized with hydrophobic organic shells upon their magnetoinduced translocation between a nonaqueous liquid phase (e.g., toluene) being above the aqueous electrolyte solution and the electrode surface [33, 34, 65]. The magnetoinduced attraction of the hydrophobic magnetic nanoparticles to the electrode surface resulted in the formation of a thin film isolating the conducting interface from the aqueous solution, thus inhibiting all diffusional electrochemical processes for soluble redox probes [34] (Figure 14.2a). Magnetoinduced lifting of the nanoparticles from the electrode surface and their redispersion in the toluene solution resulted in the cleaning of the conducting interface and reactivation of diffusional redox reactions (Figure 14.2b). The impedance spectra

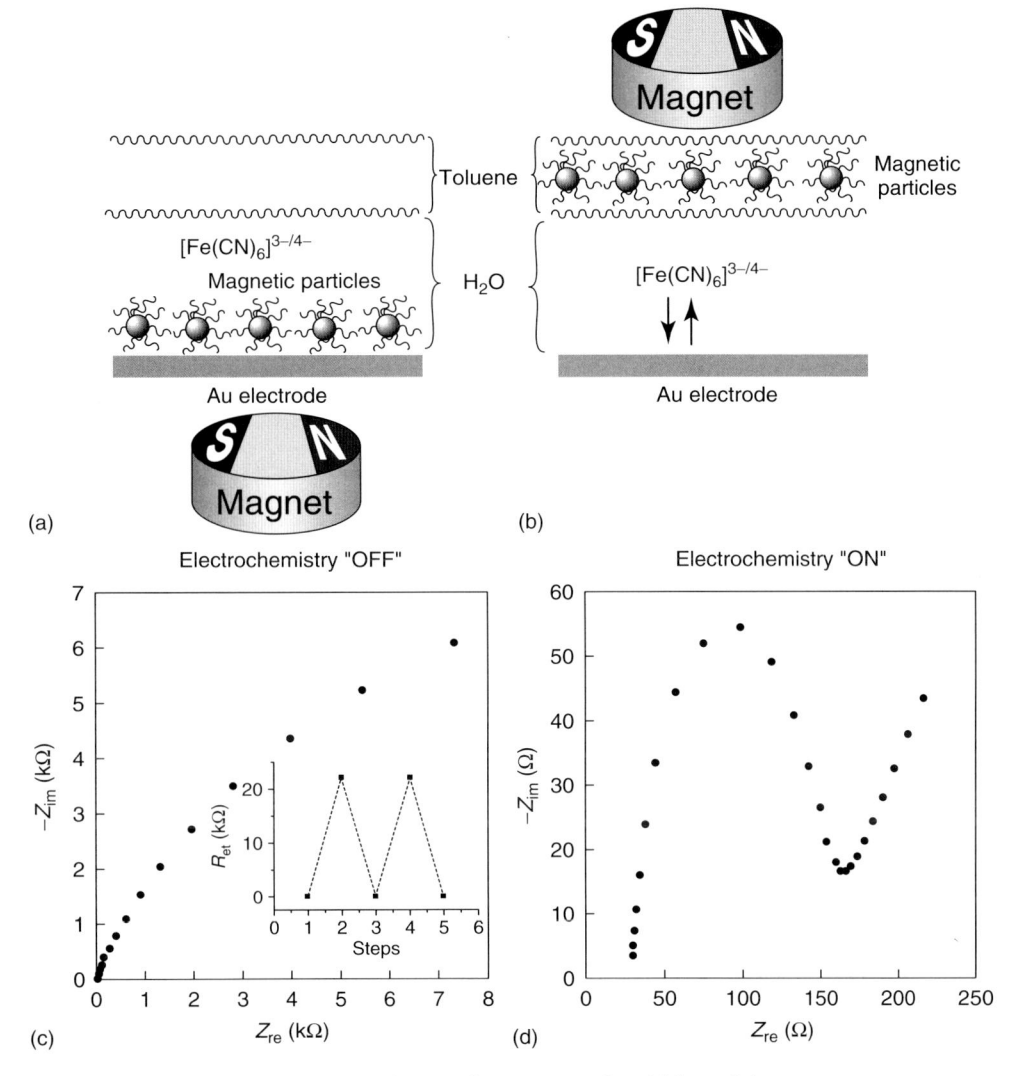

Figure 14.2 Diffusional electrochemical processes switched OFF (a) and ON (b) by the hydrophobic magnetic nanoparticles attracted to and removed from the electrode surface upon application of external magnetic signals. The impedance spectra obtained for the system with the electrode surface blocked with the hydrophobic magnetic nanoparticles (c) and for the electrode surface after the magnetoinduced lifting of the nanoparticles from the surface (d). Inset (c): Changes of the electron transfer resistance derived from the impedance spectra on reversible cycle between the ON and OFF electrochemical states modulated by the magnetic signals. (Adapted from Ref. [34], with permission; copyright American Chemical Society, 2004.)

obtained for the blocked and free electrode surface states produced by the magnetic signals show very high and low electron transfer resistances, respectively (Figure 14.2c,d). The reversible cycle between the OFF and ON electrochemical states was achieved by repeated application of the magnetic field in different directions moving the nanoparticles up and down (Figure 14.2c, inset). At the same time, redox species directly linked to the electrode surface preserved their redox activity being in intimate contact with the conductive support even in the presence of the hydrophobic magnetic nanoparticles on the surface [33]. The surface-confined redox species demonstrated different mechanisms of their electrochemical reactions upon attraction–retraction of the magnetic nanoparticles: an aqueous environment was produced at the electrode surface in the absence of the magnetic nanoparticles, allowing protonation of the redox species in the course of the reduction process, while a nonaqueous environment was generated in the presence of the hydrophobic magnetic nanoparticles, thus restricting the proton transfer process [33]. The discrimination of the diffusional and surface-confined electrochemical processes as well as the alteration of the electrochemical mechanisms induced by the magnetic signals applied on the electrode allowed switching between different bioelectrocatalytic [33] and photoelectrocatalytic [66] reactions. The magneto-controlled interfaces associated with the hydrophobic magnetic nanoparticles were also used to develop biosensing systems switchable between different analytes [67], to achieve interfacial charging processes switchable between single/multielectron transfer mechanisms [68] and to design "write–read–erase" systems for information processing [69, 70].

Large-scale aligned arrays of conductive nanowires formed on the electrode surface in the presence of an external magnetic field [71] are beneficial for improvement of charge-carrier collection and overall efficiency of diffusional electrochemical and bioelectrocatalytic reactions. Au-shell/$CoFe_2O_4$-magnetic-core nanoparticles were self-assembled in the presence of a magnetic field to yield "forest" of standing nanowires, increasing the electrode surface area (Figure 14.3a) and thus amplifying electrochemical responses of a diffusional redox probe by about sixfold, followed by cyclic voltammetry (Figure 14.3b). The process was reversed when the magnetic field was switched off, resulting in disaggregation of the conducting nanowires. The primary electrochemical reaction of the electron relay was coupled with the bioelectrocatalytic oxidation of glucose in the presence of soluble glucose oxidase (GOx), resulting in the amplification of the biocatalytic cascade controlled by the nanostructured assembly on the electrode surface (Figure 14.3a). The studied nanoelectrode array was suggested as a general platform for electrochemical biosensors with the enhanced current outputs controlled by the structure of the self-assembled nanowires. The most important advantage of the present system was an easy control over the generated nanoelectrode assembly on the electrode surface by varying the time intervals for the nanowires growth and by switching ON and OFF the magnetic field, thus resulting in the nanowires self-assembling and disassembling, respectively.

The rapid development of magneto-controlled electrochemical and bioelectrochemical systems resulted in the design of very sophisticated adaptive bioelectronic

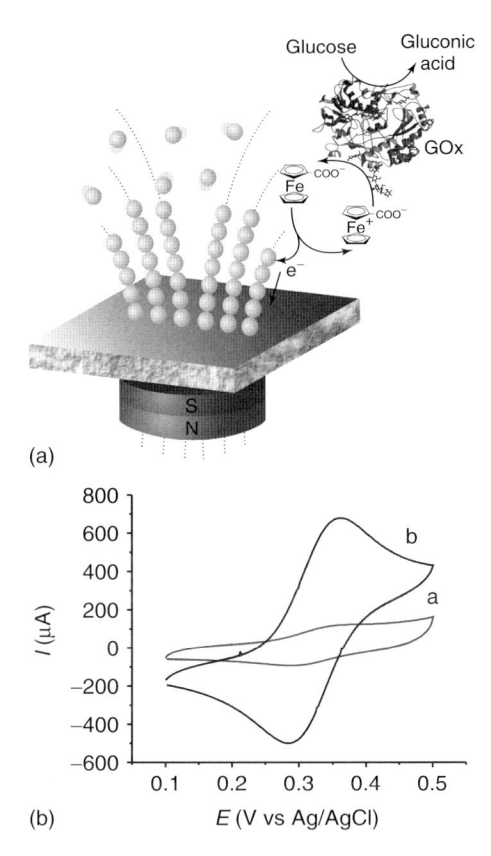

Figure 14.3 (a) Magnetic-field-controlled reversible assembling of Au-coated magnetic nanoparticles and their use as a nanostructured electrode with the enhanced ability of electrochemical oxidation of ferrocenemonocarboxylic acid coupled with glucose oxidation biocatalyzed by GOx. (b) Cyclic voltammograms (CVs) of ferrocenemonocarboxylic acid (0.1 mM) obtained in the absence (a) and presence (b) of the magnetic field. (Adapted from Ref. [71], with permission; copyright American Chemical Society, 2008.)

systems operating when magnetic signals were applied on them [40]. Application of magneto-controlled electrochemical interfaces in various signal-processing systems for unconventional computing is highly feasible.

14.4
Potential-Switchable Modified Electrodes Based on Electrochemical Transformations of Functional Interfaces

Electrical potentials applied on an electrode can result in chemical changes of the electrode surface, significantly affecting electrochemical processes proceeding at the interface [72]. For example, bare metal surfaces (including noble metals, e.g., Au, Pt) can change their electrochemical properties upon formation of thin films of oxides at positive potentials [73]. Redox materials (e.g., redox polymers) immobilized on electrode surfaces can significantly enhance interfacial changes upon application of various potentials transforming the immobilized materials into different oxidation states [74]. Electrochemical, optical, and wettability properties of polymer-modified electrodes were modulated by the application of different potentials on the conducting supports [75, 76]. Modified electrodes where the interfacial resistance can be switched between low and high values upon electrochemical formation of metallic clusters in a polymeric matrix and their dissolution, respectively, represent particular interest [44, 47]. For example, a polymer system with EDTA units in the polymeric chains was deposited layer by layer onto an electrode surface and loaded with Cu^{2+} ions [44]. The Cu^{2+} ions associated with the polymer system because of their complexation with EDTA ligands were electrochemically reduced upon application of −0.6 V (vs Ag/AgCl sat.), resulting in the formation of metallic clusters with a high electronic conductivity across the polymer film (Figure 14.4a), while application of 0.3 V resulted in electrochemical dissolution of the clusters and regeneration of the ionic Cu^{2+} state with a poor electronic conductivity (Figure 14.4b). The interfacial conductivity of the thin film was probed by Faradaic impedance spectroscopy demonstrating low and high values of the electron transfer resistance, for the metallic (Cu^0) and ionic (Cu^{2+}) states of copper in the polymer film, respectively (Figure 14.4c,d). Complexation of Cu^{2+} ions with EDTA ligands prevented their leakage from the polymer film, preserving the cationic species for the reversible modulation of the interfacial properties when the potential changes. The interfacial conductivity at the modified electrode was reversibly cycled between the low and high electron transfer resistance values upon application of different potentials as the input signals (Figure 14.4c, inset). Similar results were obtained for an electrode modified with polyacrylic acid (PAA) containing Cu^{2+} ions entrapped in the film [47]. This approach was used to design a biofuel cell with a switchable and tunable power output controlled by the external electrical signals alternating the interfacial electron transfer resistance [77].

Functionalization of electrode surfaces with signal-responsive materials aided establishing an entirely new electrode switching behavior [19, 78, 79]. The most frequently used approach is based on surface-confined pH-responsive polyelectrolytes reversibly switchable between a charged hydrophilic form permeable for redox species of the opposite charge and a neutral hydrophobic state that is not permeable for ionic species [19, 80, 81]. These polyelectrolytes can be switched between permeable and nonpermeable forms by local interfacial pH changes generated by electrochemical reactions, being thus switchable by potentials applied on the modified electrode surface [82, 83]. For example, a poly(4-vinyl pyridine) (P4VP)-brush-modified indium tin oxide (ITO) electrode was used to switch reversibly the interfacial activity upon electrochemical signals [82]. Application of an external potential electrochemically reducing O_2 resulted in the concomitant consumption of hydrogen ions at the electrode interface, thus yielding a higher pH value and triggering deprotonation and restructuring of the P4VP brush on the electrode surface (Figure 14.5a). The initial swollen state of the protonated P4VP brush (pH 4.4) was permeable for anionic $[Fe(CN)_6]^{4-}$ redox species (Figure 14.5b, curve a),

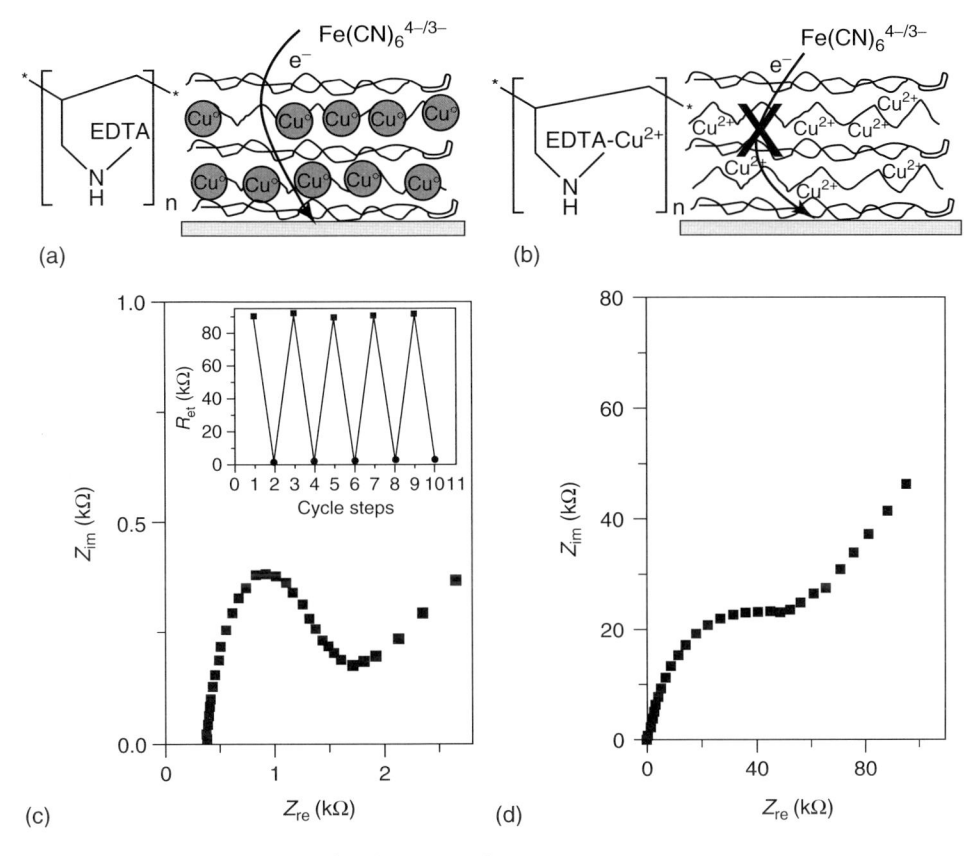

Figure 14.4 The switchable interfacial conductivity of the copper-loaded polymer thin film upon electrochemical formation of the metallic Cu^0 clusters (a) and their electrochemical dissolution (b). The Faradaic impedance spectra corresponding to the electronically conducting interface containing Cu^0 clusters (c) and the nonconducting polymer with Cu^{2+} ionic state (d). Inset (c): the reversible modulation of the interfacial electron transfer resistance by the application of the reductive and oxidative potentials on the modified electrode. (Parts (c) and (d) are adapted from Ref. [44], with permission.)

while the electrochemically produced local pH of 9.1 resulted in the deprotonation of the polymer brush. The produced hydrophobic shrunken state of the polymer brush was impermeable for the anionic redox species, thus fully inhibiting its redox process at the electrode surface (Figure 14.5b, curve b). It should be noted that the pH change was generated locally at the modified interface, while the bulk pH value had very little changes. The interface return to the electrochemically active state was achieved by disconnecting the applied potential followed by stirring the electrolyte solution or by slow diffusional exchange of the electrode-adjacent thin layer with the bulk solution. The developed approach allowed the electrochemically triggered reversible inhibition ("closing") of the electrode interface (Figure 14.5b, inset). The opposite "opening" process was also electrochemically triggered when

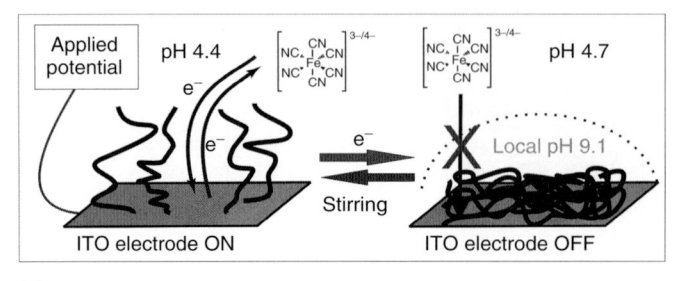

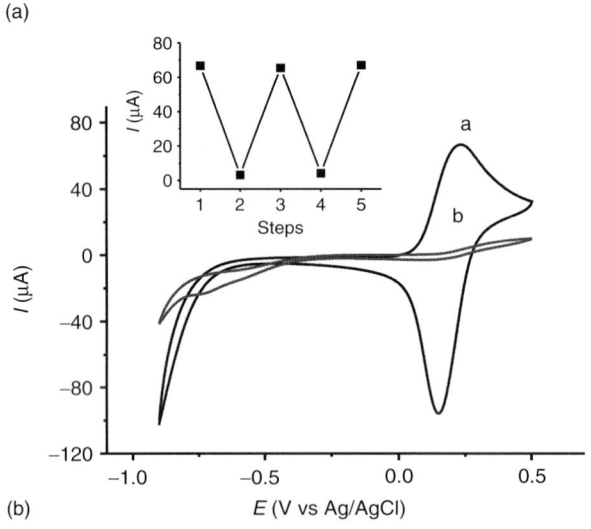

Figure 14.5 (a) pH-controlled reversible switching of the P4VP brush between ON (left) and OFF (right) states, allowing and restricting the anionic species penetration to the electrode surface, thus activating and inhibiting their redox process. (b) CVs obtained on the P4VP-brush-modified ITO electrode in the presence of 0.5 mM $K_4[Fe(CN)_6]$: (a) before the application of the potential on the electrode and (b) after application of -0.85 V to the electrode for 20 min. The background electrolyte was composed of 1 mM lactic buffer (pH 4.4) and 100 mM sodium sulfate saturated with air. Inset: reversible switching of the peak current value upon "closing" the interface by the electrochemical signal and restoring electrode activity by solution stirring. (Adapted from Ref. [82], with permission; copyright American Chemical Society, 2010.)

a mixed polymeric brush composed of poly(2-vinyl pyridine) (P2VP) and PAA was associated with an electrode surface [83]. Similar to the previous example, the local interfacial pH value was increased upon electrochemical reduction of O_2 resulting in the transition of the polyelectrolyte thin film from a neutral state to a negatively charged form due to dissociation of the PAA component. This resulted in the switch of the modified electrode surface from a hydrophobic inactive state to a hydrophilic negatively charged state permeable for cationic redox species (e.g., $[Ru(NH_3)_6]^{3+}$). Application of this approach to different interfacial systems will allow a vast range of switchable electrodes with the externally controlled activity useful for applications in biosensors and biofuel cells.

Many other systems with the interfacial conductivity (as well as other interfacial properties) modulated by the potential signals applied on the electrodes were designed, including nanostructured assemblies self-assembled on modified surfaces [45, 46, 84]. Their application in electrochemically controlled interfacial systems for unconventional computing is straightforward, being particularly beneficial for designing memory systems with "write–read–erase" properties.

14.5
Chemically/Biochemically Switchable Electrodes and Their Coupling with Biomolecular Computing Systems

Various approaches were used to design modified electrodes selectively sensitive to chemical and biochemical signals. One of the approaches is based on the use of receptor units immobilized on an electrode surface to bind selectively analyte species, resulting in changes of interfacial properties (electrical, optical, piezoelectric, etc.) [85–87]. Further advances in the design of signal-responsive materials and their immobilization on solid supports [88, 89] resulted in the new switchable electrodes reversibly activated and deactivated by chemical or biochemical signals [19, 50]. For example, a nanostructured membrane prepared from a polyelectrolyte, quarternized poly(2-vinylpyridine (qP2VP) cross-linked with 1,4-diiodobutane (DIB), was deposited on an ITO electrode surface, and the membrane pores were reversibly closed and opened upon reacting the membrane with cholesterol and washing it out, respectively [50] (Figure 14.6a,b). Cholesterol binding to the polymer chains through the formation of hydrogen bonds resulted in the swelling of the membrane and closing of the pores (Figure 14.6b), while washing of cholesterol from the membrane restored its initial state with the open pores (Figure 14.6a). The reversible opening/closing of the membrane pores was followed by measuring the electrode electron transfer resistance for a diffusional redox probe using Faradaic impedance spectroscopy (Figure 14.6c). The cyclic addition and removal of cholesterol to and from the membrane resulted in the modulated interfacial resistance controlled by chemical signals (Figure 14.6c, inset).

pH-sensitive polymer brushes [88] tethered to electrode surfaces demonstrated pH-switchable interfacial behavior when a neutral state of the polyelectrolyte was impermeable for ionic redox species, keeping the electrode mute, while the ionized state of the polymer brush was permeable for ionic redox species of the opposite charge, allowing their access to the conducting support and activating the electrochemical process [19, 90, 91]. Protonation–deprotonation of the polymer brush was controlled by the pH value, which could be changed in the bulk solution [19, 90, 91] or varied locally at the interface by means of electrochemistry (Section 14.4) [82, 83]. Further sophistication of the pH-switchable interfaces was achieved by electrode modification with mixed-polymer systems [19, 90, 91]. For example, P2VP and PAA, tethered to an ITO electrode as a mixed brush, demonstrated three differently charged states controlled by an external pH value: positively charged because of protonation of the P2VP component (pH 3), neutral when the charges of P2VP and

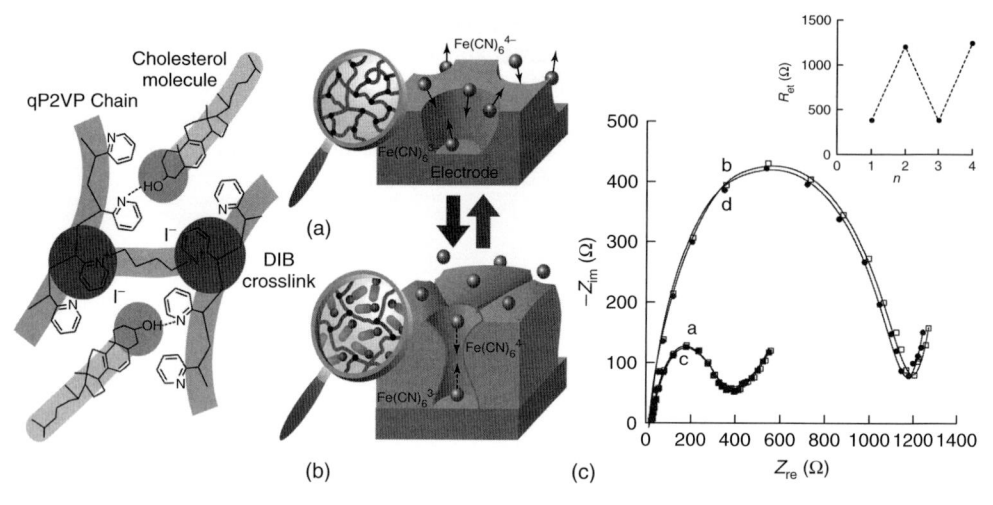

Figure 14.6 Structure of the cross-linked polyelectrolyte gel membrane and the reversible change of the chemical stimuli-responsive membrane in the absence (a) and presence (b) of cholesterol. (c) Faradaic impedance spectra obtained for the membrane-modified electrode in the absence (curves a,c) and presence (curves b,d) of cholesterol, corresponding to the open and closed pores, respectively. Inset (c): reversible cyclic changes of the electron transfer resistance derived from the impedance spectra upon addition and removal of cholesterol. (Adapted from Ref. [50], with permission; copyright American Chemical Society, 2007.)

PAA are compensated (pH 4.5), and negatively charged when the PAA component is dissociated (pH 6) (Figure 14.7a) [90, 91]. The pH-controlled switching between the different charges allowed discrimination between electrochemical reactions of oppositely charged redox species: $[Fe(CN)_6]^{4-}$ and $[Ru(NH_3)_6]^{3+}$. The negatively charged species were allowed to access the conducting support and demonstrate their electrochemical activity only when the mixed-polymer brush was positively charged; on the contrary, the positively charged species accessed the electrode being electrochemically active only when the modified interface was negatively charged, and the neutral state of the modified interface was not permeable for all ionic species keeping the electrode mute for all electrochemical reactions. Gradual pH changes in the supporting electrolyte solution demonstrated reversible transition from the electrochemical reaction of $[Fe(CN)_6]^{4-}$ to the redox process of $[Ru(NH_3)_6]^{3+}$ and back (Figure 14.7b,c) [90, 91]. This switchable/tunable behavior of the modified interface was considered as a prototype of future on-demand drug releasing systems [90] and as a 2-to-1 multiplexer for unconventional chemical information-processing systems [91].

Even more interesting electrochemical properties were discovered for the polymer brush tethered to an electrode surface and functionalized with bound redox species [51]. $Os(dmo-bpy)_2$ redox groups (dmo-bpy=4,4'-dimethoxy-2,2'-bipyridine) were covalently bound to the P4VP brush chains grafted on an ITO electrode. The polymer-bound redox species were found to be electrochemically active at

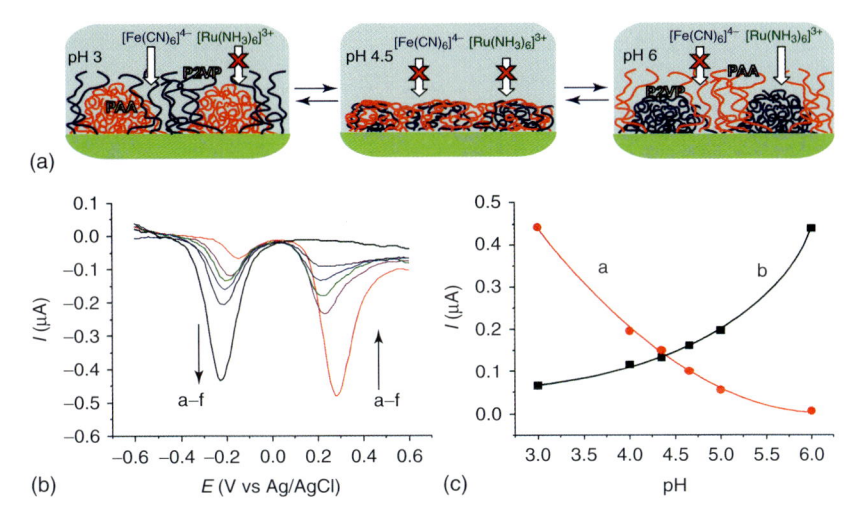

Figure 14.7 (a) The polymer brush permeability for differently charged redox probes controlled by a solution pH value: (left) the positively charged protonated P2VP domains allow electrode access for the negatively charged redox species, (middle) the neutral hydrophobic polymer thin film inhibits electrode access for all ionic species, and (right) the negatively charged dissociated PAA domains allow electrode access for the positively charged redox species. (b) The differential pulse voltammograms (DPVs) obtained for the mixed-polymer brush in the presence of $[Fe(CN)_6]^{4-}$, 0.5 mM, and $[Ru(NH_3)_6]^{3+}$, 0.1 mM, at variable pH of the solution: (a) 3.0, (b) 4.0, (c) 4.35, (d) 4.65, (e) 5.0, and (f) 6.0. The background solution was composed of 0.1 M phosphate buffer titrated to the specified pH values. (c) The peak current dependence on the pH value for the anionic, $[Fe(CN)_6]^{4-}$, (curve a) and cationic, $[Ru(NH_3)_6]^{3+}$, (curve b) species, as derived from the DPVs measured at the variable pH values. (Adapted from Ref. [91], with permission.)

pH < 5 when the polymer is protonated, swollen, and the chains are flexible. Upon changing the pH to values higher than pH 6, the polymer chains lose their charge and the produced shrunken state of the polymer does not show electrochemical activity. This was explained by the poor mobility of the polymer chains in the shrunken state restricting the direct contact between the Os-complex units and the conducting support. A low density of the Os-complex in the polymer film does not allow electron hopping between the redox species, and their electrochemical activity can be achieved only upon quasi-diffusional translocation of the polymer chains in the swollen state, bringing the redox species to a short distance from the conducting support (Figure 14.8a). The reversible activation and deactivation of the modified electrode was followed by cyclic voltammetry measured at pH 4 (ON state) and pH 6 (OFF state) (Figure 14.8b). The reversible transition of the Os-complex-functionalized polymer brush between the swollen and shrunken states upon varying pH values allowed the modulation of the electrode activity between the ON and OFF states, respectively (Figure 14.8b, inset). Bioelectrocatalytic oxidation of glucose in the presence of soluble GOx was mediated by the

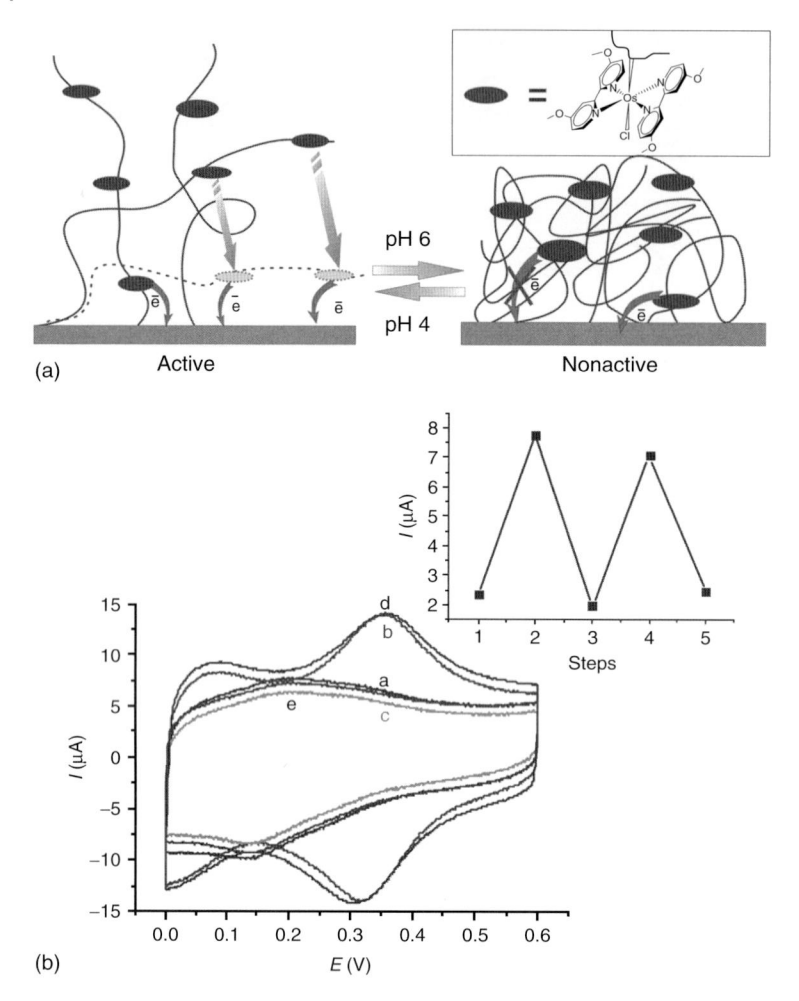

Figure 14.8 (a) Reversible pH-controlled transformation of the Os-P4VP-polymer brush on the electrode surface between electrochemically active and inactive states. (b) Cyclic voltammograms of the Os-P4VP-modified electrode obtained on stepwise measurements performed in neutral and acidic solutions: (a) pH 7.0, (b) pH 3.0, (c) again pH 7.0, (d) again pH 3.0, and (e) again pH 7.0. Inset: reversible switching of the Os-P4VP-modified electrode activity. (Adapted from Ref. [51], with permission; copyright American Chemical Society, 2008.)

Os-complex-functionalized electrode being in the ON state, while muting the OFF state of the electrode for the bioelectrochemical reaction [51]. The bioelectrocatalytic cascade activated by the interfacial changes allowed amplification of the primary process being potentially useful in unconventional computing systems as a signal-amplifying tool.

The pH-controlled reversible transition of the Os-complex-functionalized electrode between the electrochemically active and inactive states [51] was used to activate/deactivate the bioelectrocatalytic glucose oxidation upon performing enzymatic reactions changing pH values [52]. Two soluble hydrolytic enzymes, esterase and urease, were used to change the pH value *in situ*, thus affecting the bioelectrocatalytic activity of the modified electrode for glucose oxidation in the presence of soluble GOx. Addition of ethyl butyrate to the solution resulted in the formation of butyric acid biocatalyzed by esterase. The generated butyric acid yielded a pH of about 3.8 when the Os-complex-functionalized electrode is electrochemically active, thus enabling the bioelectrocatalytic oxidation of glucose (Figure 14.9a). Then the addition of urea to the solution resulted in the formation of ammonia biocatalyzed by urease, restoring the initial pH of about 6.5 and disabling glucose oxidation at the modified electrode. Electrode switching between ON/OFF states upon the addition of two chemical input signals could be presented as a system mimicking electronic enable-reset circuitry operating in the digital form, where the absence/presence of the input signals are digitally presented as **0/1** inputs and the resulting electrode OFF/ON states are considered as **0/1** states [52]. The bioelectrocatalytically active and inactive states of the modified electrode switched *in situ* by the enzymatic reactions were followed by cyclic voltammetry (Figure 14.9b). The reversible transition between the active and inactive states demonstrated the possibility to modulate the bioelectrocatalytic activity by biochemical signals (Figure 14.9b, inset).

It should be noted that the system described above has a disadvantage of using soluble enzymes for processing chemical signals to switch the electrode interface between ON and OFF states [52]. Obviously, more advanced systems should include these enzymes integrated with the switchable interface. This was achieved by enzyme immobilization on the pH-switchable polymer brush using Au nanoparticles as a platform for the seamless enzyme integration with the switchable interface [92]. Two immobilized enzymes, esterase and urease, were activated by the corresponding substrates, ethyl butyrate and urea. The biocatalyzed reactions produced *in situ* butyric acid or ammonia, decreasing or increasing the pH value, thus resulting in electrode activation or inactivation, respectively (Figure 14.10a). Importantly, the electrode switching ON and OFF was performed upon local interfacial pH changes without alteration of the bulk pH value and was followed by cyclic voltammetry in the presence of a diffusional redox probe (Figure 14.10b,c).

One more step forward in the increasing sophistication of the switchable electrode systems was attained when different activating signals described above where integrated in one multifunctional system. The novel system included pH-switchable polymer brush associated with an ITO electrode surface and magnetic nanoparticles functionalized with GOx enzyme being activated with the glucose substrate [93]. When the GOx-modified magnetic nanoparticles were suspended in the solution, glucose addition resulted in the formation of gluconic acid in the bulk volume in a small quantity, which was not enough to change the pH value for switching the electrode properties. However, when the nanoparticles were concentrated on the electrode surface, the local interfacial pH decrease generated upon biocatalytic oxidation of glucose and formation of gluconic acid was already enough for P4VP

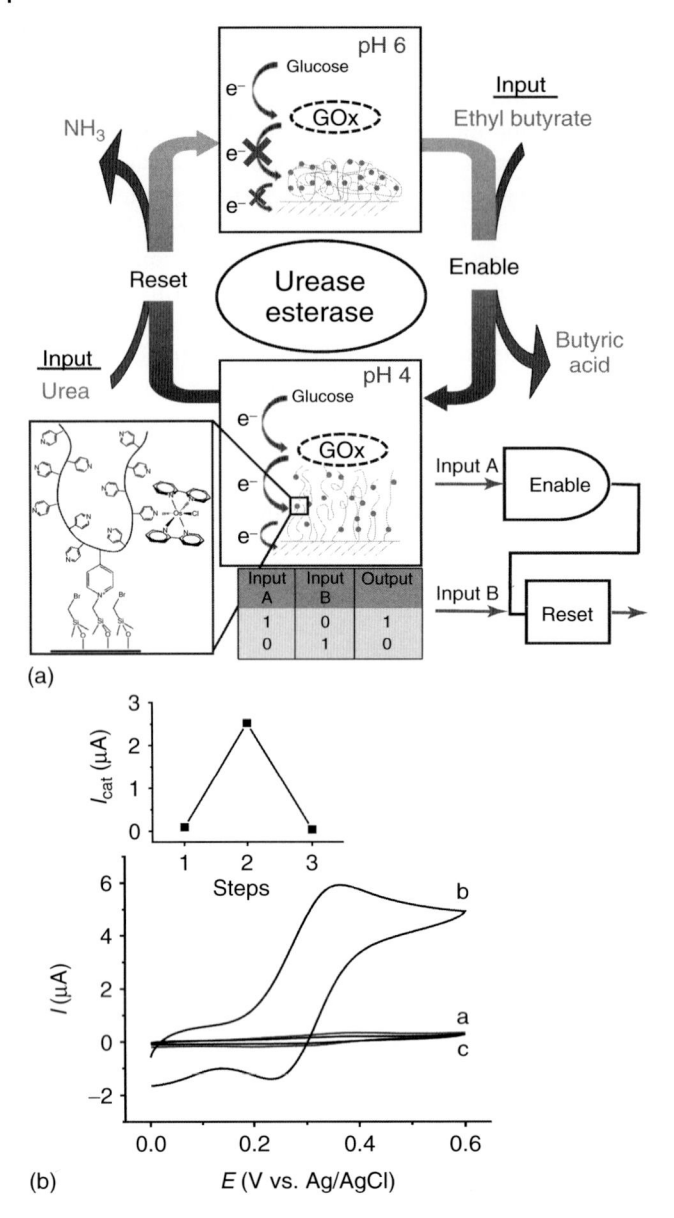

Figure 14.9 (a) Switchable bioelectrocatalytic oxidation of glucose controlled by external enzymatic reactions. (b) Cyclic voltammograms obtained for the switchable bioelectrocatalytic glucose oxidation when the system is (a) in the initial OFF state, pH about 6.5; (b) enabled by the ethyl butyrate input signal, pH about 3.8; and (c) inhibited by the urea reset signal, pH about 7.5. Inset: switchable bioelectrocatalytic current: (step 1) initial OFF state, (step 2) enabled ON state, and (step 3) reset to the OFF state. (Adapted from Ref. [52], with permission; copyright American Chemical Society, 2008.)

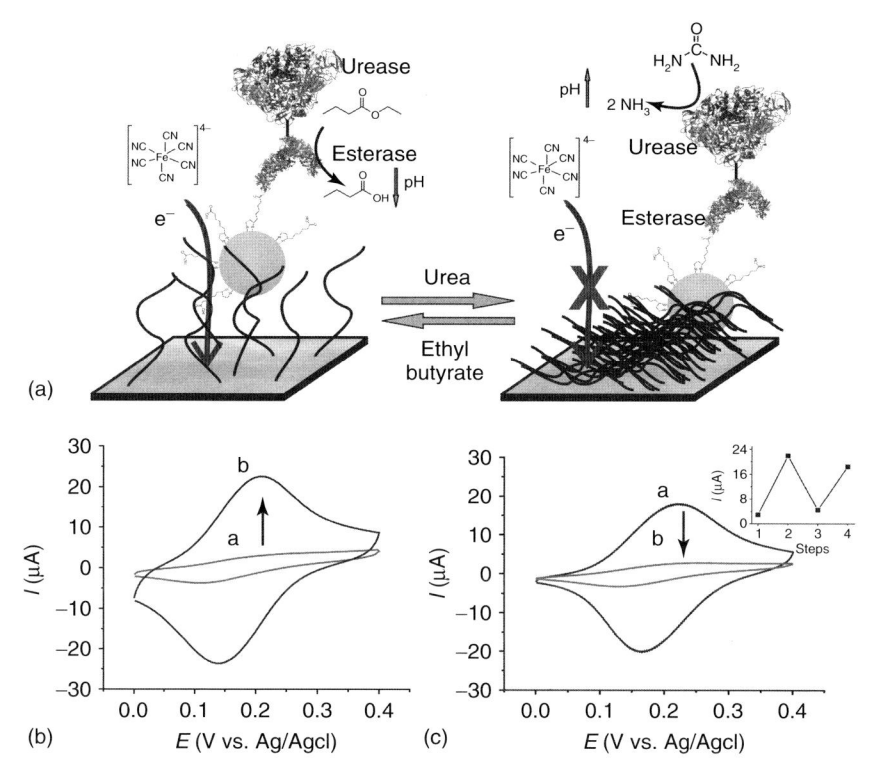

Figure 14.10 (a) Switching the electrode ON and OFF by biocatalyzed reactions of the immobilized enzymes. (b,c) CVs obtained on the enzymes-P4VP-modified ITO electrode in the presence of 0.2 mM $K_4[Fe(CN)_6]$ upon application of chemical signals. (b) Na_2SO_4 solution, 0.1 M, with the bulk pH 6: before (a) and 30 min after (b) the application of ethyl butyrate, 10 mM. (c) Na_2SO_4 solution, 0.1 M, with the bulk pH 4.5: before (a) and 30 min after (b) the application of urea, 2 mM. Arrows show the direction of the biocatalytically induced changes. Inset (c): reversible changes of the peak current upon cyclic ON–OFF transformations induced by the biochemical signals. (Adapted from Ref. [92], with permission; reproduced with permission of the Royal Society of Chemistry.)

brush protonation and for opening of the electrode interface for an electrochemical reaction of a diffusional redox probe (Figure 14.11a). The opening of the electrode surface occurred only upon cooperative application of two signals: the magnetic field to collect the nanoparticles on the interface and addition of glucose to activate the biocatalytic reaction yielding the pH change. The interfacial changes were followed by Faradaic impedance spectroscopy, which demonstrated the electron transfer resistance, R_{et}, decrease corresponding to the open state of the interface only after both activating signals (magnetic field and glucose) were applied (Figure 14.11b). Thus, the system was mimicking Boolean AND logic operation with two signals of different nature: magnetic and chemical.

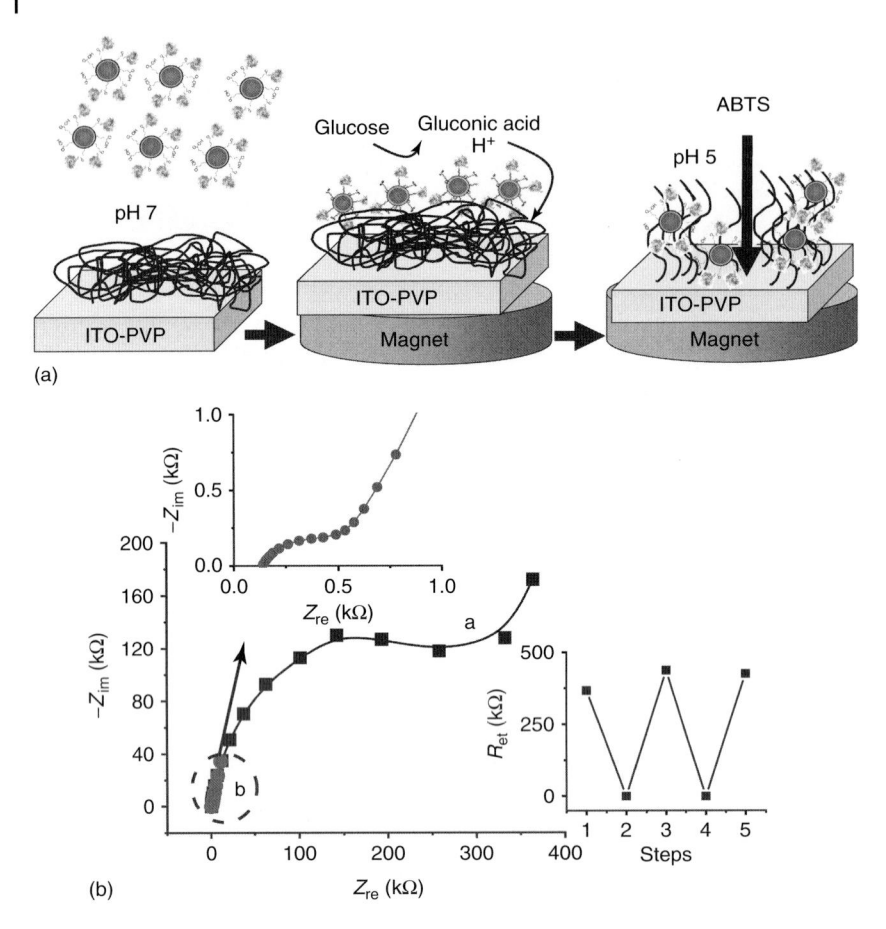

Figure 14.11 (a) Magnetoinduced concentration of the GOx-functionalized Au-shell/$CoFe_2O_4$-magnetic-core nanoparticles on the electrode surface modified with P4VP brush to open the interface for a diffusional redox probe upon local pH changes produced in course of a biocatalytic reaction. (b) Impedance spectra (Nyquist plots) obtained on the P4VP-modified electrode with the GOx nanoparticles magnetoconfined at the electrode: (a) in the absence of glucose and (b) in the presence of 10 mM glucose (also shown at a smaller scale). Inset: reversible switching of the R_{et} by adding and removing glucose. The impedance spectra were recorded in the presence of 2,2′-azino-*bis*(3-ethylbenzothiazoline-6-sulfonic acid) (ABTS), 0.1 mM, used as a soluble redox probe. (Adapted from Ref. [93], with permission; copyright American Chemical Society, 2009.)

14.6
Summary and Outlook

Extensive work performed in the area of switchable electrode interfaces has led to numerous systems controlled by a large variety of physical and chemical signals. Particularly important results are expected when research advances in signal-responsive materials, modified electrodes, and chemical computing are integrated into a new research area. Further scaling up the complexity of the chemical information-processing systems will allow the next level of sophistication when switchable electrodes are controlled by various chemical processes. Particularly interesting would be a combination of biomolecular computing systems and switchable modified electrodes. Coupling of "smart" switchable electrodes with sophisticated multistep biochemical pathways could be envisaged in the continuing research. Integration of the signal-responsive electrodes with the information-processing systems might be used to develop "smart" multisignal responsive biosensors and actuators controlled by complex biochemical environment. The biochemically/physiologically controlled switchable electrodes will operate as an interface between biological and electronic systems in future micro/nanorobotic devices.

Switchable electrodes controlled by signal-processing enzyme-based logic systems were already integrated in "smart" biofuel cells producing electrical power dependent on complex variations of biochemical signals [94, 95]. Enzyme-biocatalytic [96, 97] and immune-biorecognition [98] systems have been developed to control performance of switchable biofuel cells. Future implantable biofuel cells producing electrical power on demand depending on physiological conditions are feasible as the result of the present research. Further development of sophisticated enzyme-based biocomputing networks will be an important phase in the development of "smart" bioelectronic devices. Scaling up the complexity of biocomputing system controlling biofuel cell activity will be achieved by networking immune- and enzyme-based logic gates responding to a large variety of biochemical signals. Biofuel cells switchable by enzyme-based [99] or immunosystem-based [100] keypad lock systems have been designed to operate as self-powered biomolecular information security systems. The correct biomolecular "password" introduced into the keypad lock as a sequence of biomolecular input signals resulted in the activation of the biofuel cell, while all other "wrong" permutations of the molecular inputs preserved the "OFF" state of the biofuel cell. Further research directed to the increasing stability and robustness of the information-processing biocatalytic electrodes could result in many practical applications, including, for example, bioelectrocatalytic barcode generation [101].

The present developments and future expectations are based on the application of a multidisciplinary approach, which will require further collaborative contribution from electrochemists, specialists in materials science, and unconventional molecular and biomolecular computing.

Acknowledgments

This research was supported by the National Science Foundation (Awards CBET-1066397, CCF-1015983), by ONR (Grant N00014-08-1-1202), and by the Semiconductor Research Corporation (Award 2008-RJ-1839G).

References

1. Albery, W.J. and Hillman, A.J. (1981) *Annu. Rep. Prog. Chem. Sect. C*, **78**, 377–437.
2. Murray, R.W. (1984) in *Electroanalytical Chemistry*, vol. **13** (ed. A.J. Bard), Marcel Dekker, New York, pp. 191–368.
3. Murray, R.W. (1980) *Acc. Chem. Res.*, **13**, 135–141.
4. Wrighton, M.S. (1986) *Science*, **231**, 32–37.
5. Abruña, H.D. (1988) *Coord. Chem. Rev.*, **86**, 135–189.
6. Chen, D. and Li, J.H. (2006) *Surf. Sci. Rep.*, **61**, 445–463.
7. Zen, J.M. and Senthil Kumar, S., and Tsai, D.M. (2003) *Electroanalysis*, **15**, 1073–1087.
8. Rusling, J.F. and Forster, R.J. (2003) *J. Colloid Interface Sci.*, **262**, 1–15.
9. Willner, I. and Katz, E. (2000) *Angew. Chem. Int. Ed.*, **39**, 1180–1218.
10. Wang, J. (1999) *J. Pharm. Biomed. Anal.*, **19**, 47–53.
11. Wang, J. (2008) *Talanta*, **75**, 636–641.
12. Gooding, J.J. (2008) *Electroanalysis*, **20**, 573–582.
13. Wollenberger, U., Spricigo, R., Leimkuhler, S., and Schronder, K. (2008) in *Biosensing for the 21st Century, Advances in Biochemical Engineering/Biotechnology*, Vol. **109** (eds R. Renneberg and F. Lisdat), Springer, New York, pp. 19–64.
14. Moehlenbrock, M.J. and Minteer, S.D. (2008) *Chem. Soc. Rev.*, **37**, 1188–1196.
15. Davis, F. and Higson, S.P.J. (2007) *Biosens. Bioelectron.*, **22**, 1224–1235.
16. Bullen, R.A., Arnot, T.C., Lakeman, J.B., and Walsh, F.C. (2006) *Biosens. Bioelectron.*, **21**, 2015–2045.
17. Barton, S.C., Gallaway, J., and Atanassov, P. (2004) *Chem. Rev.*, **104**, 4867–4886.
18. Katz, E., Willner, B., and Willner, I. (1997) *Biosens. Bioelectron.*, **12**, 703–719.
19. Motornov, M., Sheparovych, R., Katz, E., and Minko, S. (2008) *ACS Nano*, **2**, 41–52.
20. Flood, A.H., Ramirez, R.J.A., Deng, W.Q., Muller, R.P., Goddard, W.A., and Stoddart, J.F. (2004) *Aust. J. Chem.*, **57**, 301–322.
21. Laocharoensuk, R., Bulbarello, A., Hocevar, S.B., Mannino, S., Ogorevc, B., and Wang, J. (2007) *J. Am. Chem. Soc.*, **129**, 7774–7775.
22. Wang, J., Musameh, M., Laocharoensuk, R., Gonzalez-Garcia, O., Oni, J., and Gervasio, D. (2006) *Electrochem. Commun.*, **8**, 1106–1110.
23. Shipway, A.N., Katz, E., and Willner, I. (2001) in *Structure and Bonding, Molecular Machines and Motors*, Vol. **99** (ed. J.-P. Sauvage), Springer-Verlag, Berlin, pp. 237–281.
24. Lion-Dagan, M., Katz, E., and Willner, I. (1994) *J. Am. Chem. Soc.*, **116**, 7913–7914.
25. Katz, E., Lion-Dagan, M., and Willner, I. (1995) *J. Electroanal. Chem.*, **382**, 25–31.
26. Willner, I., Lion-Dagan, M., Marx-Tibbon, S., and Katz, E. (1995) *J. Am. Chem. Soc.*, **117**, 6581–6592.
27. Willner, I., Lion-Dagan, M., and Katz, E. (1996) *Chem. Commun.*, 623–624.
28. Doron, A., Portnoy, M., Lion-Dagan, M., Katz, E., and Willner, I. (1996) *J. Am. Chem. Soc.*, **118**, 8937–8944.
29. Doron, A., Katz, E., Tao, G.L., and Willner, I. (1997) *Langmuir*, **13**, 1783–1790.
30. Liu, N.G., Dunphy, D.R., Atanassov, P., Bunge, S.D., Chen, Z., Lopez, G.P., Boyle, T.J., and Brinker, C.J. (2004) *Nano Lett.*, **4**, 551–554.
31. Liu, Z.F., Hashimoto, K., and Fujishima, A. (1990) *Nature*, **347**, 658–660.
32. Hsing, I.M., Xu, Y., and Zhao, W.T. (2007) *Electroanalysis*, **19**, 755–768.
33. Katz, E., Baron, R., and Willner, I. (2005) *J. Am. Chem. Soc.*, **127**, 4060–4070.
34. Katz, E., Sheeney-Haj-Ichia, L., Basnar, B., Felner, I., and Willner, I. (2004) *Langmuir*, **20**, 9714–9719.
35. Willner, I. and Katz, E. (2003) *Angew. Chem. Int. Ed.*, **42**, 4576–4588.

36. Katz, E., Sheeney-Haj-Ichia, L., and Willner, I. (2002) *Chem. Eur. J.*, **8**, 4138–4148.
37. Hirsch, R., Katz, E., and Willner, I. (2000) *J. Am. Chem. Soc.*, **122**, 12053–12054.
38. Wang, J. and Kawde, A.N. (2002) *Electrochem. Commun.*, **4**, 349–352.
39. Laocharoensuk, R., Bulbarello, A., Mannino, S., and Wang, J. (2007) *Chem. Commun.*, 3362–3364.
40. Wang, J. (2008) *Electroanalysis*, **20**, 611–615.
41. Loaiza, O.A., Laocharoensuk, R., Burdick, J., Rodriguez, M.C., Pingarron, J.M., Pedrero, M., and Wang, J. (2007) *Angew. Chem. Int. Ed.*, **46**, 1508–1511.
42. Wang, J., Scampicchio, M., Laocharoensuk, R., Valentini, F., Gonzalez-Garcia, O., and Burdick, J. (2006) *J. Am. Chem. Soc.*, **128**, 4562–4563.
43. Lee, J., Lee, D., Oh, E., Kim, J., Kim, Y.P., Jin, S., Kim, H.S., Hwang, Y., Kwak, J.H., Park, J.G., Shin, C.H., Kim, J., and Hyeon, T. (2005) *Angew. Chem. Int. Ed.*, **44**, 7427–7432.
44. Zheng, L. and Xiong, L. (2006) *Colloids Surf. A*, **289**, 179–184.
45. Riskin, M., Basnar, B., Katz, E., and Willner, I. (2006) *Chem. Eur. J.*, **12**, 8549–8557.
46. Riskin, M., Basnar, B., Chegel, V.I., Katz, E., Willner, I., Shi, F., and Zhang, X. (2006) *J. Am. Chem. Soc.*, **128**, 1253–1260.
47. Chegel, V.I., Raitman, O.A., Lioubashevski, O., Shirshov, Y., Katz, E., and Willner, I. (2002) *Adv. Mater.*, **14**, 1549–1553.
48. Le, X.T., Jégou, P., Viel, P., and Palacin, S. (2008) *Electrochem. Commun.*, **10**, 699–703.
49. Hou, K.Y., Yu, L., Severson, M.W., and Zeng, X.Q. (2005) *J. Phys. Chem. B*, **109**, 9527–9531.
50. Tokarev, I., Orlov, M., Katz, E., and Minko, S. (2007) *J. Phys. Chem. B*, **111**, 12141–12145.
51. Tam, T.K., Ornatska, M., Pita, M., Minko, S., and Katz, E. (2008) *J. Phys. Chem. C*, **112**, 8438–8445.
52. Tam, T.K., Zhou, J., Pita, M., Ornatska, M., Minko, S., and Katz, E. (2008) *J. Am. Chem. Soc.*, **130**, 10888–10889.
53. Katz, E., Shipway, A.N., and Willner, I. (2002) in *Photoreactive Organic Thin Films*, Chapter II-7 (eds S. Sekkat and W. Knoll), Academic Press, pp. 219–268.
54. Katz, E. and Shipway, A.N. (2005) in *Bioelectronics: From Theory to Applications*, Chapter 11 (eds I. Willner and E. Katz), Wiley-VCH Verlag GmbH, Weinheim, pp. 309–338.
55. Katsonis, N., Lubomska, M., Pollard, M.M., Feringa, B.L., and Rudolf, P. (2007) *Prog. Surf. Sci.*, **82**, 407–434.
56. Wesenhagen, P., Areephong, J., Landaluce, T.F., Heureux, N., Katsonis, N., Hjelm, J., Rudolf, P., Browne, W.R., and Feringa, B.L. (2008) *Langmuir*, **24**, 6334–6342.
57. Browne, W.R., Kudernac, T., Katsonis, N., Areephong, J., Hielm, J., and Feringa, B.L. (2008) *J. Phys. Chem. C*, **112**, 1183–1190.
58. Areephong, J., Browne, W.R., Katsonis, N., and Feringa, B.L. (2006) *Chem. Commun.*, 3930–3932.
59. Nakashima, N., Deguchi, Y., Nakanishi, T., Uchida, K., and Irie, M. (1996) *Chem. Lett.*, 817–818.
60. Willner, I., Pardo-Yissar, V., Katz, E., and Ranjit, K.T. (2001) *J. Electroanal. Chem.*, **497**, 172–177.
61. Katz, E., Lioubashevsky, O., and Willner, I. (2004) *J. Am. Chem. Soc.*, **126**, 15520–15532.
62. Katz, E., Sheeney-Ichia, L., and Willner, I. (2004) *Angew. Chem. Int. Ed.*, **43**, 3292–3300.
63. Baron, R., Onopriyenko, A., Katz, E., Lioubashevski, O., Willner, I., Wang, S., and Tian, H. (2006) *Chem. Commun.*, 2147–2149.
64. Delahay, P. (1965) in *Double Layer and Electrode Kinetics: Advances in Electrochemistry and Engineering*, Chapter 3 (eds P. Delahay and C.W. Tobias), Wiley Interscience, New York.
65. Willner, I. and Katz, E. (2006) *Langmuir*, **22**, 1409–1419.
66. Katz, E. and Willner, I. (2005) *Angew. Chem. Int. Ed.*, **44**, 4791–4794.

67. Katz, E. and Willner, I. (2005) *Chem. Commun.*, 4089–4091.
68. Katz, E., Lioubashevski, O., and Willner, I. (2006) *Chem. Commun.*, 1109–1111.
69. Katz, E. and Willner, I. (2005) *Chem. Commun.*, 5641–5643.
70. Katz, E. and Willner, I. (2006) *Electrochem. Commun.*, **8**, 879–882.
71. Jimenez, J., Sheparovych, R., Pita, M., Narvaez Garcia, A., Dominguez, E., Minko, S., and Katz, E. (2008) *J. Phys. Chem. C*, **112**, 7337–7344.
72. Vetter, K.J. (1967) *Electrochemical Kinetics: Theoretical Aspects*, Academic Press, New York.
73. Woods, R. (1976) in *Electroanalytical Chemistry*, vol. **9**, Chapter 1 (ed. A.J. Bard), Marcel Dekker, New York, pp. 1–162.
74. Bard, A.J., Stratmann, M., Rubinstein, I., Fujihira, M., and Rusling, J.F. (eds) (2007) *Encyclopedia of Electrochemistry: Modified Electrodes*, vol. **10**, Wiley-VCH Verlag GmbH, Weinheim.
75. Raitman, O.A., Katz, E., Willner, I., Chegel, V.I., and Popova, G.V. (2001) *Angew. Chem. Int. Ed.*, **40**, 3649–3652.
76. Chegel, V., Raitman, O., Katz, E., Gabai, R., and Willner, I. (2001) *Chem. Commun.*, 883–884.
77. Katz, E. and Willner, I. (2003) *J. Am. Chem. Soc.*, **125**, 6803–6813.
78. Combellas, C., Kanoufi, F., Sanjuan, S., Slim, C., and Tran, Y. (2009) *Langmuir*, **25**, 5360–5370.
79. Choi, E.Y., Azzaroni, O., Cheng, N., Zhou, F., Kelby, T., and Huck, W.T.S. (2007) *Langmuir*, **23**, 10389–10394.
80. Harris, J.J. and Bruening, M.L. (2000) *Langmuir*, **16**, 2006–2013.
81. Park, M.K., Deng, S.X., and Advincula, R.C. (2004) *J. Am. Chem. Soc.*, **126**, 13723–13731.
82. Tam, T.K., Pita, M., Trotsenko, O., Motornov, M., Tokarev, I., Halámek, J., Minko, S., and Katz, E. (2010) *Langmuir*, **26**, 4506–4513.
83. Tam, T.K., Pita, M., Motornov, M., Tokarev, I., Minko, S., and Katz, E. (2010) *Adv. Mater.*, **22**, 1863–1866.
84. Riskin, M., Katz, E., and Willner, I. (2006) *Langmuir*, **22**, 10483–10489.
85. Diamond, D. and McKervey, M.A. (1996) *Chem. Soc. Rev.*, **25**, 15–24.
86. Yang, D.H., Ju, M.-J., Maeda, A., Hayashi, K., Toko, K., Lee, S.-W., and Kunitake, T. (2006) *Biosens. Bioelectron.*, **22**, 388–392.
87. Gabai, R., Sallacan, N., Chegel, V., Bourenko, T., Katz, E., and Willner, I. (2001) *J. Phys. Chem. B*, **105**, 8196–8202.
88. Minko, S. (2006) *Polym. Rev.*, **46**, 397–420.
89. Luzinov, I., Minko, S., and Tsukruk, V.V. (2004) *Prog. Polym. Sci.*, **29**, 635–698.
90. Motornov, M., Tam, T.K., Pita, M., Tokarev, I., Katz, E., and Minko, S. (2009) *Nanotechnology*, **20**, 434006.
91. Tam, T.K., Pita, M., Motornov, M., Tokarev, I., Minko, S., and Katz, E. (2010) *Electroanalysis*, **22**, 35–40.
92. Bocharova, V., Tam, T.K., Halámek, J., Pita, M., and Katz, E. (2010) *Chem. Commun.*, **46**, 2088–2090.
93. Pita, M., Tam, T.K., Minko, S., and Katz, E. (2009) *ACS Appl. Mater. Interfaces*, **1**, 1166–1168.
94. Katz, E. (2010) *Electroanalysis*, **22**, 744–756.
95. Katz, E. and Pita, M. (2009) *Chem. Eur. J.*, **15**, 12554–12564.
96. Tam, T.K., Pita, M., Ornatska, M., and Katz, E. (2009) *Bioelectrochemistry*, **76**, 4–9.
97. Amir, L., Tam, T.K., Pita, M., Meijler, M.M., Alfonta, L., and Katz, E. (2009) *J. Am. Chem. Soc.*, **131**, 826–832.
98. Tam, T.K., Strack, G., Pita, M., and Katz, E. (2009) *J. Am. Chem. Soc.*, **131**, 11670–11671.
99. Halámek, J., Tam, T.K., Strack, G., Bocharova, V., Pita, M., and Katz, E. (2010) *Chem. Commun.*, **46**, 2405–2407.
100. Halámek, J., Tam, T.K., Chinnapareddy, S., Bocharova, V., and Katz, E. (2010) *J. Phys. Chem. Lett.*, **1**, 973–977.
101. Strack, G., Luckarift, H.R., Nichols, R., Cozart, K., Katz, E., and Johnson, G.R. (2011) *Chem. Commun.*, **47**, 7662–7664.

15
Conclusions and Perspectives
Evgeny Katz

The different topics reviewed in the book highlight the tremendous progress in the area of unconventional chemical computing. The various research subdirections outlined in the book, which include (i) chemical design of complex multifunctional molecules, supramolecular systems and nanospecies for mimicking operation of logic gates and other information-processing units (Chapters 2–7); (ii) excitation of diffusional medium for computing processes (Chapters 8 and 11); (iii) theoretical work aiming at the development of novel unconventional computing paradigms and algorithms (Chapters 9–12); and finally (iv) integration of chemical computing systems with switchable interfaces and devices (Chapters 13 and 14), demonstrate diversity of the research area. While primary efforts have been directed to the formulation of switchable molecules with complex structure and multifunctional behavior controlled by various input signals (optical, electrical, magnetic, and chemical) [1], the extension of this work resulted in networking of the switchable components, allowing some computational operations performed at the molecular level [2]. As the result, molecular full-adder and full-subtractor have been designed [3]. Further progress was directed to the molecular engineering of auxiliary elements required for information processing, such as molecular comparators [4], digital multiplexers/demultiplexers [5], encoders/decoders [6], as well as flip-flop and write/read/erase memory units [7]. Some functional devices, for example, molecular keypad lock systems [8], have been designed for illustrating the power and applicability of the unconventional information-processing approach.

All these advances promise a bright and prosperous future for the field. Did the field reach the level of maturation where only technological applications may be envisaged? The answer to this question is definitely "NO." New concepts and methods are constantly introduced to the field, suggesting the development of new routes in the ground of unconventional chemical computing. Despite rapid progress in the design of chemical systems for signal processing and logic decision-making, practical results are still far away from the potentially possible achievements. Massively parallel information-processing reactions [9], presently illustrated with some diffusional processes (e.g., Belousov–Zhabotinsky reactions) [10], potentially allow solutions of hard-to-solve computational problems [11], being future competitors for silicon-based electronic computers. Optimistic predictions

Molecular and Supramolecular Information Processing: From Molecular Switches to Logic Systems,
First Edition. Edited by Evgeny Katz.
© 2012 Wiley-VCH Verlag GmbH & Co. KGaA. Published 2012 by Wiley-VCH Verlag GmbH & Co. KGaA.

look forward to the possibility of mimicking and computing any physical, chemical, or biological phenomena with the use of molecular systems upon involving 10^{23} molecules performing computations in parallel. Still these predictions are mostly kept at the theoretical level; however, they inspire chemical researchers to design real systems capable of fulfilling theoretical expectations. Theoretical work [12], having roots in Leibniz's ideas and even having some connections with the philosophy of Kabbalah [13], is an important part of unconventional chemical computing, which directs chemists to the design of functional molecules and their ensembles, demonstrating properties that are necessary for computations. Some of the aspects in this work could benefit from direct analogy with electronic computing systems. For example, noise analysis and management might be performed similar to how it is done in electronic systems, obviously using the conceptual approach used in electronics, but realizing it with an unconventional molecular "hardware" [14].

New fresh ideas resulting in further progress in the area of unconventional chemical computing are coming from biology. For example, artificial abiotic systems mimicking elementary properties of neuron networks were developed demonstrating that scaling up the complexity of chemical computing systems can be inspired by biological principles [15]. Particularly interesting results are expected in the area of chemical computing with incorporation of biomolecules into information-processing systems. This complementary area of biomolecular computing is covered in another new book of Wiley-VCH: *Biomolecular Information Processing: From Logic Systems to Smart Sensors and Actuators* – E. Katz, Editor. Integration of molecular information-processing systems, particularly with the use of biomolecules, and electronic transducing elements (electrodes, field-effect transistors, microchips, etc.), recently developed in the novel area of bioelectronics [16], will result in technological advances moving unconventional chemical/biochemical computing from the presently studied wet chemistry systems to real operational devices [17]. The complexity of these "smart" devices controlled by unconventional chemical computing systems might be scaled up to the level of machines [18] and even autonomously operating robots [19].

Another push for technological advances in chemical computing will certainly come from nanotechnology providing miniaturization of chemical information-processing systems [20]. While the obvious advantage of chemical systems is computing at the molecular level, the presently studied systems are usually represented by bulk chemical reactions. Nanotechnology will allow scaling down the size of the computing elements providing excitation, operation, and result read-out with the use of a single molecule [21].

Advances in unconventional chemical computing and practical applications that originated from research efforts in the field promise the continuous prosperity of this scientific topic. Incorporation of biotechnology and nanotechnology concepts and novel "smart" nanostructured materials into the domain of unconventional information-processing systems paves the way to new challenges and highlights the long-term and continuous interest in the field. It is expected that interdisciplinary efforts of computer scientists, chemists, biologists, physicists, material scientists, and electronic engineers will result in exciting scientific accomplishments in the coming years.

References

1. (a) de Silva, A.P. and Uchiyama, S. (2007) *Nat. Nanotechnol.*, **2**, 399–410; (b) Szacilowski, K. (2008) *Chem. Rev.*, **108**, 3481–3548; (c) Credi, A. (2007) *Angew. Chem. Int. Ed.*, **46**, 5472–5475; (d) Pischel, U. (2010) *Aust. J. Chem.*, **63**, 148–164; (e) Andreasson, J. and Pischel, U. (2010) *Chem. Soc. Rev.*, **39**, 174–188.
2. (a) de Silva, A.P., Uchiyama, S., Vance, T.P., and Wannalerse, B. (2007) *Coord. Chem. Rev.*, **251**, 1623–1632; (b) Pischel, U. (2007) *Angew. Chem. Int. Ed.*, **46**, 4026–4040.
3. (a) Margulies, D., Melman, G., and Shanzer, A. (2006) *J. Am. Chem. Soc.*, **128**, 4865–4871; (b) Kuznetz, O., Salman, H., Shakkour, N., Eichen, Y., and Speiser, S. (2008) *Chem. Phys. Lett.*, **451**, 63–67.
4. Pischel, U. and Heller, B. (2008) *New J. Chem.*, **32**, 395–400.
5. (a) Andreasson, J., Straight, S.D., Bandyopadhyay, S., Mitchell, R.H., Moore, T.A., Moore, A.L., and Gust, D. (2007) *J. Phys. Chem. C*, **111**, 14274–14278; (b) Amelia, M., Baroncini, M., and Credi, A. (2008) *Angew. Chem. Int. Ed.*, **47**, 6240–6243.
6. Andreasson, J., Straight, S.D., Moore, T.A., Moore, A.L., and Gust, D. (2008) *J. Am. Chem. Soc.*, **130**, 11122–11128.
7. (a) Chatterjee, M.N., Kay, E.R., and Leigh, D.A. (2006) *J. Am. Chem. Soc.*, **128**, 4058–4073; (b) Baron, R., Onopriyenko, A., Katz, E., Lioubashevski, O., Willner, I., Wang, S., and Tian, H. (2006) *Chem. Commun.*, 2147–2149.
8. (a) Margulies, D., Felder, C.E., Melman, G., and Shanzer, A. (2007) *J. Am. Chem. Soc.*, **129**, 347–354; (b) Suresh, M., Ghosh, A., and Das, A. (2008) *Chem. Commun.*, 3906–3908.
9. Adamatzky, A. (2011) *J. Comput. Theor. Nanosci.*, **8**, 295–303.
10. Adamatzky, A., De Lacy Costello, B., Bull, L., and Holley, J. (2011) *Isr. J. Chem.*, **51**, 56–66.
11. Matiyasevich, Y. (1987) *Problems of Cybernetics*, vol. **131**, Nauka, Moscow; (English translation in: Kreinovich, V. and Mints, G. (eds) (1996) *Problems of Reducing the Exhaustive Search*, American Mathematical Society, Providence, pp. 75–77).
12. (a) Matsumaru, N., Centler, F., di Fenizio, P.S., and Dittrich, P. (2007) *Int. J. Unconventional Comput.*, **3**, 285–309; (b) Schumann, A. and Adamatzky, A. (2009) *Kybernetes*, **38**, 1518–1531.
13. Schumann, A. (2011) *Hist. Philos. Logic*, **32**, 1–8.
14. (a) Privman, V. (2010) *Isr. J. Chem.*, **51**, 118–131; (b) Privman, V. (2011) *J. Comput. Theor. Nanosci.*, **8**, 490–502.
15. Pina, F., Melo, M.J., Maestri, M., Passaniti, P., and Balzani, V. (2000) *J. Am. Chem. Soc.*, **122**, 4496–4498.
16. Willner, I. and Katz, E. (eds) (2005) *Bioelectronics: from Theory to Applications*, Wiley-VCH Verlag GmbH, Weinheim.
17. Chang, B.-Y., Crooks, J.A., Chow, K.-F., Mavré, F., and Crooks, R.M. (2010) *J. Am. Chem. Soc.*, **132**, 15404–15409.
18. Adamatzky, A. and Jones, J. (2010) *Nat. Comput.*, **9**, 219–237.
19. (a) Adamatzky, A. and Jones, J. (2008) *J. Bionic Eng.*, **5**, 348–357; (b) Adamatzky, A. (2008) *Kybernetes*, **37**, 258–264.
20. Seminario, J.M., Ma, Y.F., and Tarigopula, V. (2006) *IEEE Sens. J.*, **6**, 1614–1626.
21. (a) Stadler, R., Ami, S., Joachim, C., and Forshaw, M. (2004) *Nanotechnology*, **15**, S115–S121; (b) Pototschnig, M., Chassagneux, Y., Hwang, J., Zumofen, G., Renn, A., and Sandoghdar, V. (2011) *Phys. Rev. Lett.*, **107**, 063001.

Index

a

acenes 136–137
– and acene-like structures and self-organization motifs 137–141
– application in organic electronic devices 141–142
algorithms 2, 4–5, 6, 210–211
– feasible 211–212, 213
– unfeasible 212–213
all-photonic multifunctional molecular logic device 75
AND logic 14, 15, 18, *63*, *64*, 66, 67, 83, 90, 100, 101, 102, 108, 113, 115, *116*, 194
– consolidating 84–87
– optimization 290–294
arithmetic circuits, in subexcitable chemical media
– awakening gates 175–176
– Belousov–Zhabotinsky (BZ) medium
– – localizations in 176–180
– – memory cells with discs 201–203
– – regular and irregular disc networks 193–201
– – vesicles 180–181
– – binary ladder 186–188
– – carry out 191–193
– – sum 188–190
– collision-based computing 176
– interaction between wave fragments 181–183
– universality and polymorphism 183–186
artificial molecular machines 1
artificial neural network (ANN) 328

b

Belousov–Zhabotinsky (BZ) medium 1, 2
– and chemical kinetics 240
– localizations in 176–180
– memory cells with discs 201–203
– regular and irregular disc networks 193–201
– proof-theoretic cellular automata for 268–271
– vesicles 180–181
bidirectional half subtractor and reversible logic device 28–32
binary ladder 186–188
– carry out 191–193
– sum 188–190
binary logic, with synthetic molecular and supramolecular species
– chemical computers and 26–27
– combinational logic gates and circuits
– – all-optical integrated logic operations based on communicating molecular switches 38–41
– – bidirectional half subtractor and reversible logic device 28–32
– – concepts 27–28
– – encoder/decoder based on ruthenium tris(bipyridine) 36–38
– – unimolecular multiplexer–demultiplexer 32–36
– information processing 25–26
– sequential logic circuits
– – concepts 41–42
– – memory effect in communicating molecular switches 42–43
– – molecular keypad lock 43–45
– – set–reset memory device based on a copper rotaxane 46–48
biocomputing 281–283, 293, 294
biomedical applications 281, 283
biomimetics 327
biomolecular systems. *See* biocomputing

Boolean operations 3, 4, 61–64
Brotherston's cyclic proofs 267–268

c

carminic acid (CA) 157
catechol 154
cellular automaton 176, 180, 186–187, 188, 189, 190, 191, 204
chemical computers 26–27
chemical computing 1, 2, 5–7
chemical kinetics
– Belousov–Zhabotinsky reaction 240
– and chemical computing 237–238
– – as theoretical challenge 238
– discussion 244–245
– dynamical systems 241
– equations 239–240
– external noise effect 245
– limited accuracy 242
– limited x_i values 242
– natural hypothesis 240
– need to consider auxiliary chemical substances 242–244
– proof 246–256
– time factor 241
– until late 1950s 240
– solving equations of 211–212
chemically/biochemically switchable electrodes, and coupling with biomolecular computing systems 343–350
collision-based computing 176
combinational logic gates and circuits
– all-optical integrated logic operations, based on communicating molecular switches 38–41
– bidirectional half subtractor and reversible logic device 28–32
– concepts 27–28
– encoder/decoder, based on ruthenium tris(bipyridine) 36–38
– unimolecular multiplexer–demultiplexer 32–36
communicating molecular switches, all-optical integrated logic operations based on 38–41
cucurbiturils (CBs) 110–116
cyclic voltammetry 91

d

Davydov splitting 155
diagonal interaction 137
digital information processing 281–285, 287–289
double-throw switches 60

e

electrochemical atomic layer deposition (ECALD) 131–132
electrochemical atomic layer epitaxy (ECALE) 126
electrochemical deposition 125–133
– nanoheterostructure preparation 133–135
– nanoparticles directed self-assembly 135–136
electrochemistry 305
– artificial cognitive materials 314–315
– brain
– – dynamics 323–324
– – understanding of 321–323
– electrochemical dynamics 324–325
– experimental paradigms for information processing in complex systems 325–327
– intelligent electrochemical platform 315–321
– intelligent response and pattern formation 308–309
– – emergent patterns and associative memory 312–314
– – functional self-organizing systems 310–312
– – patterns in nature 310
– – self-organization in systems removed from equilibrium state and 309–310
– pattern formation in complex systems 306–308
electrode interfaces switchable by physical and chemical signals 333
– chemically/biochemically switchable electrodes and coupling with biomolecular computing systems 343–350
– light-switchable modified electrodes, based on photoisomerizable materials 334–336
– magnetoswitchable electrodes 336–339
– potential-switchable modified electrodes 339–343
encoder/decoder, based on ruthenium tris(bipyridine) 36–38
encoders and decoders 69–71
engineering luminescent molecules 79–83
– AND logic consolidation 84–87
– lab-on-a-molecule systems 87–90
– logic gates with the same modules in different arrangements 83–84
– redox-fluorescent logic gates 90–95

f

fluorescence 100, 101, 102–108, 110, 112–115

fluorophore-spacer-receptor modular configuration 81
Fukui function 150
functional integration 65

h
half adder 198–201
– and half subtractors 65–68
high-concentration chemical computing techniques, for hard-to-solve problems 209–210
– algorithms 210–211
– – feasible 211–212, 213
– – unfeasible 212–213
– problem significance 213–214
– – class NP 215
– – class P and P=NP problem 215–216
– – description 214
– – engineering 215
– – exhaustive search 216
– – NP-complete problems 216–217
– – theoretical physics 215
– propositional satisfiability 217–228
– relation to optimization
– – importance 228–229
– – main idea 229–231
– – relation to freedom of choice 233–234
– – relation to neural computing 231–232
– – relation to reasoning under uncertainty 232–233
– – relational to numerical optimization 231
Hilbert's inference rules 266
hybrid semiconducting materials 121–122
– digital devices based on PEPS effect 161–167
– electrochemical deposition 125–133
– – nanoheterostructure preparation 133–135
– – nanoparticles directed self-assembly 135–136
– organic semiconductors 136–137
– – acenes application in organic electronic devices 141–142
– – self-organization motifs exhibited by acenes and acene-like structures 137–141
– photocurrent switching phenomena mechanisms 142–143
– – composite semiconductor materials 144–148
– – neat semiconductor 143–144
– – optoelectronic devices based on organic molecules/semiconductors 160–161
– – semiconductor–absorbate interactions 148–152
– – surface-modified semiconductor 152–159
– semiconducting thin layers and nanoparticles 122–123
– – chemical bath deposition 124–125
– – nanoparticle microwave synthesis 123
hydrogen-bonded supramolecular assemblies, as logic devices 102–103

i
information processing 1–7, 25–26. *See also individual entries*
INHIBIT (INH) gate 62, 90, 106, 107, 108, 111, 113
internal charge transfer (ICT) 81

k
Kabbalistic–Leibnizian automata, for universe simulation 259
– historical background 259–264
– proof-theoretic cellular automata 264–268
– – for Belousov–Zhabotinsky reaction 268–271
– – for *plasmodium of Physarum polycephalum* 271–276
– unconventional computing, as novel paradigm in natural sciences 276–278
Kröger's theory 127

l
lab-on-a-molecule systems 87–90, 95
ligand, commonly used 124–125
light-switchable modified electrodes, based on photoisomerizable materials 334–336
logic. *See also individual entries*
– expanding 16–17
– generalizing 15–16
– processing 4, 5
– utilizing 17–19
logic gates 15, 16, 19, 351
– combinational, and circuits
– – all-optical integrated logic operations based on communicating molecular switches 38–41
– – bidirectional half subtractor and reversible logic device 28–32
– – concepts 27–28
– – encoder/decoder based on ruthenium tris(bipyridine) 36–38
– – unimolecular multiplexer–demultiplexer 32–36
– elementary 194–197

logic gates (contd.)
– with the same modules in different arrangements 83–84
luminescence 29, 31, 32, 36–7, 39, 41, 42, 45. See also engineering luminescent molecules
luminescent switching systems, designing 11–13

m

magnetoswitchable electrodes 336–339
Matiyasevich's chemical computer description 219–222
medical diagnostics 80, 95
memory effect, in communicating molecular switches 42–43
metal ion inputs recognition, by crown ethers 100–102
molecular computation 16, 19
molecular computational identification (MCID) method 19
molecular computer 26–27
molecular keypad lock 43–45
molecular logic gates 19
– with [2]pseudorotaxane- and [2]rotaxane-based switches 103–110
molecular sensors 12–13
moleculators 68
molecule-to-band charge transfer 155
Moore neighborhood 264, 265
multiplexer–demultiplexer 68–69
– unimolecular 32–36
multiplicity, of logic types 28

n

NAND logic 103, 195
nanoparticles 336–339, 337, 339, 347, 349, 350
– directed self-assembly 135–136
– semiconducting thin layers and 122–123
– – chemical bath deposition 124–125
– – nanoparticle microwave synthesis 123
nanotechnology 11
nanowires 336, 338
neural computing 231–232
noise control approaches, in chemical and biochemical information and signal processing 281–282
– AND gates optimization 290–294
– gate level and beyond 286–290
– gates and networks 283–286, 294–296
NOR gate 90
NOT logic gates 82

o

optoelectronic devices, based on organic molecules/semiconductors 160–161
organic semiconductors 136–137
– acenes application in organic electronic devices 141–142
– self-organization motifs exhibited by acenes and acene-like structures 137–141
OR gate 195

p

pentacene 136, 142
9-phenyl-benzol[1,2]quinolizino[3,4,5,6-fed]phenanthidinylium (PQP) 140
photochemistry 11
photochromic molecules 54–55
photocurrent switching phenomena mechanisms 142–143
– composite semiconductor materials 144–148
– neat semiconductor 143–144
– optoelectronic devices based on organic molecules/semiconductors 160–161
– semiconductor–absorbate interactions 148–152
– surface-modified semiconductor 152–159
photoelectrochemical photocurrent switching (PEPS) 122, 148
– digital devices based on 161–167
photoinduced electron transfer (PET) 11, 12, 80, 81
photonically switched molecular logic devices 53
– advanced logic functions 64–65
– – all-photonic multifunctional molecular logic device 75
– – encoders and decoders 69–71
– – half-adders and half subtractors 65–68
– – multiplexers and demultiplexers 68–69
– – sequential logic devices 71–75
– Boolean logic gates 61–64
– photochromic molecules 54–55
– photonic control
– – electron transfer 59–61
– – energy transfer 55–58
physical inputs 20
physical integration 65
polycyclic aromatic hydrocarbons (PAHs) 139, 140
porphyrins 56–62, 68, 72
potential-switchable modified electrodes 339–343

propositional satisfiability 217
– chemical computing solving hard-to-solve problem of 218, 225–226
– – application 219
– – auxiliary result 226–228
– – discrete-time version of equations 225
– – high-concentration chemical reaction usage 223–224
– – history 218–219
– – Matiyasevich's chemical computer description 219–222
– – resulting equations 224
– – simplified equations 229
– – simplified version 222
protein folding 213
pseudorotaxane supramolecular complex 3

r

reconfiguration 62, 71
redox-fluorescent logic gates 90–95
rubrene 141

s

Schottky theory 151
Schrödinger's equation 212
sensing/switching conversion, into logic 13–14
sensors 12, 283, 289, 294
sequential logic circuits
– concepts 41–42
– memory effect in communicating molecular switches 42–43
– molecular keypad lock 43–45
– set–reset memory device based on a copper rotaxane 46–48
sequential logic devices 71–75
sequential logic functions 28
sequent inference rules 266–267
set–reset memory device, based on a copper rotaxane 46–48
signal passing problem 198
signal-responsive electrodes 340, 343
stimuli-responsive membrane 344

sulfide ion precursors 124
superposition, of logic types 28
supramolecular assemblies, for information processing 99–100
– hydrogen-bonded supramolecular assemblies, as logic devices 102–103
– metal ion inputs recognition, by crown ethers 100–102
– molecular logic gates with [2]pseudorotaxane- and [2]rotaxane-based switches 103–110
– supramolecular host-guest complexes with cyclodextrins and cucurbiturils 110–116
supramolecular host-guest complexes, with cyclodextrins and cucurbiturils 110–116
supramolecular systems 1
surface potential 148
switching systems 11–13
– conversion to logic 13–14

t

titanium dioxide (TiO_2) 122, 133, 146, 147, 152, 153, 154, 157, 158–159, 162

u

underpotential deposition 127
universal characteristic 261–262, 263

v

Von Neumann neighborhood 264, 265

w

wave fragments, interaction between 181–183
Weller equation 82–83
wide-band-gap semiconductors 122, 133, 135, 148, 151, 153, 161, 162, 164, 167
work function 149

x

XNOR logic 105
XOR gate 28, 29, 66, 67, 102, 104, 105, 108, 164, 196